AF736360

Progress in Physical Chemistry Volume 3

Modern and Universal First-principles Methods for Many-electron Systems in Chemistry and Physics

by
Franz Michael Dolg (Ed.)

Series Editor: Helmut Baumgärtel

Oldenbourg Verlag München

Like all natural sciences Physical Chemistry also is strongly affected by a development which leads to an inflation of information details which cannot be thoroughly covered in its depth by the regular journal world. Therefore the editors and the publisher of Zeitschrift für Physikalische Chemie have decided to provide a platform for scientists to present their research and results on a broader basis. Thus the book series "Progress in Physical Chemsitry" has been created.

The first volume of the series is devoted to "Different Aspects of Intermolecular Interaction". It collects most important reviews on the topic published in Zeitschrift für Physikalische Chemie between 2004 and 2006.

Volume 2 covers the results of the Collaborative Research Center (SFB) 277 of the German Research Foundation (DFG).

Volume 3 covers the results fo the Priority Programm (SPP) 1145 of the German Research Foundation (DFG).

Helmut Baumgärtel
(Series Editor)

Bibliografische Information der Deutschen Nationalbibliothek

Die Deutsche Nationalbibliothek verzeichnet diese Publikation in der Deutschen Nationalbibliografie; detaillierte bibliografische Daten sind im Internet über <http://dnb.d-nb.de> abrufbar.

© 2010 Oldenbourg Wissenschaftsverlag GmbH
Rosenheimer Straße 145, D-81671 München
Telefon: (089) 45051-0
oldenbourg.de

Das Werk einschließlich aller Abbildungen ist urheberrechtlich geschützt. Jede Verwertung außerhalb der Grenzen des Urheberrechtsgesetzes ist ohne Zustimmung des Verlages unzulässig und strafbar. Das gilt insbesondere für Vervielfältigungen, Übersetzungen, Mikroverfilmungen und die Einspeicherung und Bearbeitung in elektronischen Systemen.

Lektorat: Birgit Zoglmeier
Herstellung: Anna Grosser
Coverentwurf: Kochan & Partner, München
Gedruckt auf säure- und chlorfreiem Papier
Druck: Grafik + Druck GmbH, München
Bindung: Thomas Buchbinderei GmbH, Augsburg

ISBN 978-3-486-59827-8

Die Bunsen-Gesellschaft

Ihr NETZWERK der PHYSIKALISCHEN CHEMIE

Die Bunsen-Gesellschaft verbindet seit mehr als 110 Jahren Hochschul- und Industrieforscher aus der physikalischen Chemie in ihrem Netzwerk. Mit mehr als 1600 Mitgliedern fördert die Bunsen-Gesellschaft den Nachwuchs, organisiert Veranstaltungen und engagiert sich in der Forschung.

DBG

Deutsche Bunsen-Gesellschaft
für Physikalische Chemie e. V.

www.bunsen.de

Preface

Volume **3** of the series 'Progress in Physical Chemistry' comprises 27 articles reporting method developments in the field of first-principles electronic structure theory in chemistry and physics. These papers, most of them with review character, have all been written by members of the german-wide Scientific Priority Program (SPP) 1145 'Modern and Universal First-principles Methods for Many-electron Systems in Chemistry and Physics' of the German Research Foundation (DFG). The SPP 1145 was funded by the DFG during three project periods from 2003 to 2010 and brought together scientists from theoretical chemistry, theoretical physics and applied mathematics. All contributions have also been published in a special topical issue of 'Zeitschrift für Physikalische Chemie' **(2010/03–04)**.

During the last symposium of the former DFG SPP 'Relativistic Effects in Heavy-Element Chemistry and Physics' (1994–2000, coordinator Bernd A. Hess†) in 2000 at Kloster Banz a small group of scientists, among them e.g. Werner Kutzelnigg, decided that it should be attempted to initiate a new purely theoretical priority program in the field of electronic structure theory. After dealing intensively and very successfully for six years within the ending priority program with the development of efficient and accurate approaches for the evaluation of relativistic contributions in first-principles calculations in chemistry and physics, it was decided that the emphasis of the new priority program should be on method development for the many-electron problem, i.e. mainly on electron correlation. The honor of writing the actual proposal to the DFG was left to a task force of, at that time, younger scientists (Eberhard Engel, Heinz-Jürgen Flad as well as the author), who assembled the contributions from numerous colleagues from theoretical chemistry, theoretical physics and applied mathematics in an application. Later, in 2002 the DFG granted funding of the SPP 1145 and after the initial symposium 2003 in Bonn 27 research projects were launched. For the second and third term 26 and 24 projects were granted in 2005 and 2007, respectively. The DFG funding consisted mainly in PhD student positions and a few postdoctoral positions for the participating research groups, funds for travelling and for holding yearly symposia for the entire SPP 1145 community as well as workshops on special topics.

The main idea of the SPP 1145 was to serve as a common platform for scientists from the various usually disjunct communities focussing on first-principles electronic structure calculations. It was especially intended to lead to an intense exchange of ideas between the method developers. The followers of density functional theory should learn from those of *ab initio* wavefunction theory and *vice versa*, and both communities should get in closer contact to mathematicians interested in the field. The range of applicability of existing approaches should be extended, their accuracy should be increased. Besides this less frequently applied as well as new methods should be further developed within the priority program. The contributions to this volume highlight various exciting theoretical developments in line with these intentions which were, at least in part, inspired and made possible by the SPP 1145.

The contributors of this volume thank the reviewers for their valuable comments and the editorial staff of *Zeitschrift für Physikalische Chemie* for their helpful support. The present and former members of the SPP 1145 are grateful to the DFG for the generous funding of a theory-only program. They acknowledge the continous interest, administrative advice and professional guidance by the DFG representatives Drs. Johanna Kowol-Santen and Frank-Dieter Kuchta. Special thanks are due to the members of the international review panels who did spend a lot of their precious time to review the applications, read the reports and select the projects to be funded. Their presence at the SPP 1145 symposia was highly appreciated and had without doubt positive effects on the program as a whole as well as on many of the projects refereed by them.

Finally, as a coordinator and spokesman I would like to thank my colleagues in the SPP 1145 steering group, i.e. Profs. Hubert Erbert, Jürgen Gauß, Andreas Görling, Eberhard Groß, Bernd Hess†, Willem Klopper, Hans-Joachim Werner and Priv.-Doz. Eberhard Engel, as well as Dr. Heinz-Jürgen Flad for their work initiating the SPP 1145 and making it successful. Last but not least I thank all authors of the present volume for their contributions and their efforts to provide overviews summarizing for their part of the field six years of exciting research efforts in modern electronic structure theory.

Cologne, March 2010
Michael Dolg

A Critical Evaluation of the Dynamical Thresholding Algorithm in Coupled Cluster Calculations

By Alexander A. Auer*

Institut für Chemie, Straße der Nationen 62, 09111 Chemnitz Germany

(Received June 30, 2009; accepted September 24, 2009)

Dynamical Thresholding Algorithm / Coupled Cluster Doubles

A new implementation of the Dynamical Thresholding (DT) coupled cluster doubles (CCD) approximation is presented. Using prescreening techniques for the intermediate quantities that occur in the CCD amplitude equations, a computationally efficient algorithm is obtained, in which one single threshold allows for control of accuracy. A critical evaluation of the accuracy and efficiency is carried out for two testcases. Furthermore, the issue of potential energy surface discontinuities and the usage of enveloping localized orbitals is discussed in detail.

1. Introduction

One of the major trends in modern developments of electronic structure theories is the evolution (one can also say the renaissance) of low-scaling correlation methods, that overcome the high computational cost in post-HF ab-initio methods. However, these approaches are only currently being applied to a larger variety of problems, and most of them are not yet used on a routine basis as black-box tools in quantum chemistry. Nevertheless, there is already a wide range of different approaches, of which the most practicable are probably yet to evolve from a longer process of trial and improvement.

One type of approximation follows from applying the early ideas of integral screening to the the wavefunction parameters that are expressed in a local basis. A screening scheme is then constructed for the amplitudes or coefficients which indirectly leads to a reduction of computational effort due to the sparsity in the parameters and integrals. [1–9] As a related ansatz one could also also count

* Current address: Max-Planck Institut für Eisenforschung GmbH, Max-Planck-Str. 1, 40237 Düsseldorf, Germany, E-mail: auer@mpie.de

© by Oldenbourg Wissenschaftsverlag, München

newer schemes in the framework of the coupled electron pair approximation based on pair natural orbitals (PNOs), in which a cutoff for the PNOs is used to screen wavefunction parameters [10] or certain implementations of coupled cluster (CC) methods with periodic boundary conditions. [46]

In the most successful type of approximations up to date, often simply referred to as "local methods" [11–17], the occupied and virtual orbitals are localized for example by the Pipek-Mezey (PM) scheme and by constructing projected atomic orbitals (PAOs).[17,19] Prior to the calculation, a limited set of wavefunction parameters is selected either based on distance, overlap or other criteria. Then, the working equations are solved for a subset of index pairs and within limited orbital domains, so that the number of parameters only grows linear with system size by construction. This scheme has been implemented for a wide variety of methods [20–35] and also for the calculation of molecular properties. [36–45]

A further approach to reduce the high computational complexity of post-Hartree-Fock methods, which is conceptually simple yet elegant and somewhat more coarse-grained are schemes in which the system is split into fragments of some kind and energies of fragments and interactions between fragments are calculated until the total energy converges. This includes incremental incremental methods [48–52] such as the "local ansatz" by Stollhoff, Stoll and Fulde [53–58] as well as divide-and-conquer strategies or related schemes for correlated methods. [47,59,60]

While screening based approaches are often associated with higher computational cost and might also suffer from convergence problems, they are expected to be robust and applicable as black-box methods. "Local methods" can be set up extremely efficiently, are fairly robust but suffer from problems due to localization or truncation schemes. [61–64] Incremental or divide-and-conquer approaches can easily be used for periodic systems and in QM/MM or QM/QM hybrid approaches, have the appeal of naturally low scaling and are easy to implement. However, the fact that no closed expression for a wavefunction is available in this case might render the calculation of certain properties difficult.

The Dynamical Thresholding (DT) algorithm is an alternative route to a local approximation for CC methods that falls under the screening-based methods. Here, local energy contributions are calculated at the MP2 level of theory and the size of each contribution is used as a screening criterion for the evaluation of the CC amplitudes. [5] This way, one single threshold can be used to truncate the wavefunction parameters and to go from the full CC treatment to an approximation that in the limit reduces to the MP2 result. The great advantage of this approach is, that no choice of special pairs, excitation domains or a division of the system into subunits are made prior to the calculation. As the truncation of the wavefunction parameters is guided by the MP2 guess it can be regarded as adapted to the system in question and qualifies this approximation as a black-box scheme in which only one parameter, the amplitude threshold, has to be set by the user. First benchmarks from a CCSD pilot implementation of the DT algorithm that has been devised using the TCE program generator [65–67] within the NWChem program package [68] were very promising. Especially the fact that the resulting error correlates well with the chosen threshold makes this approximation attractive. This scheme allows the user to adjust the threshold in such a way, that sufficient accuracy for a certain application is obtained.

In this contribution we describe the first implementation in which the reduction of the wavefunction parameters within the DT-screening scheme leads to a considerable reduction of computational effort. Furthermore, some fundamental issues for the implementation of the DT-coupled cluster doubles (CCD) approach are discussed like the intermediate problem in local CC implementations, the use of enveloping localized orbitals (ELOs) and potential energy surface discontinuities.

2. The DT-CCD implementation

Before focusing on the motivation for choosing the CCD method for an efficient implementation of the DT algorithm, we would first like to summarize the basic features of the scheme:

· After a HF-SCF calculation has been carried out, the MP2 energy of the system is calculated in the local basis. The current algorithm in principle supports any choice of local basis, ranging from pure atomic orbitals for representing the occupied/virtual space up to the common PM/PAO basis. However, in the following we will be focusing on the ELO/AO combination (for comparison, also the PM/AO combination will be considered). It should be noted, that for optimal performance in matrix multiply operations the local amplitudes are not treated element by element but rather in blocks of amplitudes.

· In the screening step it is decided whether a block of amplitudes is treated at the CC level of theory or approximated by perturbation theory, based on the MP2 energy contributions in the local basis. This way, two sets of amplitudes are obtained - the (optimally very small) set of CC amplitudes (denoted at CC-subset in the following) and the screened amplitudes, that will be termed as PT-subset (perturbative amplitudes) for simplicity in the following.

· For solving this set of hybrid equations, two schemes have been proposed, termed weakly and strongly coupled scheme. For both schemes all contractions in the CC equations are only carried out if the resulting residual corresponds to a block of amplitudes from the CC-subset. The two schemes differ in which set of amplitudes is used in the contractions: In the weakly coupled scheme, only amplitudes from the CC-subset are contracted into residual elements of the CC-subset. In the strongly coupled scheme all amplitudes, from the CC- and PT-subset, are used for the contractions into the CC-subset of the residual. A crucial point is the treatment of the amplitudes in the PT-subset. While one would assume that this part of the amplitudes is fixed during the CC iterations, experience shows that this inevitably leads to convergence problems in solving the hybrid equations. Preliminary studies have shown that for the DT scheme as described here, this is even the case if strong orthogonality is preserved and is due to the use of nonorthogonal or redundant orbitals for representing the virtual or occupied space in the combined equations. This problem, however, is elevated if an additional set of zeroth order equations is solved for the PT-subset of the amplitudes. For the strongly coupled algorithm, relatively smooth convergence to the exact result is observed, such that a series of calculations might even make extrapolation schemes applicable.

· After evaluation of the contractions in the amplitude equations the residual is transformed to the canonical basis, weighted by the denominator and back-transformed to the local basis to proceed to the next iteration (including DIIS convergence acceleration [69]). Detailed equations in the local basis for the CCD method can be found in Appendix A.

This way, an algorithm is obtained in which the screening of the wavefunction parameters is applied to the most expensive terms in the CC equations, with remaining steps for all amplitudes that have low prefactors and scale as N^5 at worse. The back and forth transformation of the local amplitudes might be regarded as inconvenient but compared to other N^5 steps it is comparably cheap and also serves another purpose. By transforming to the MO basis in every iteration the local amplitudes in the ELO/AO basis are projected onto the proper occupied and virtual spaces, which is related to the procedure Werner et al. use for the amplitude update in the LCC scheme. [20]

The aim for a computationally efficient local approximation in the spirit of the DT algorithm is a scheme that reduces the computational effort in the same way as the approximation reduces the number of amplitudes in the CC-subset. But while the number of amplitudes drops rapidly as the threshold is increased, especially for larger systems, the computational effort does not only depend on the number of non-zero t_2 amplitudes and integrals. In order to achieve optimal scaling of the algorithm, the CC equations have to be factorized. [70] In the factorized equations, intermediates occur, for which in principle no sparsity can be assumed. In order to illustrate this, let us consider a term from the t_2 equations which is quadratic in the t_2 amplitudes (for example the sixth term in equation 1 in Appendix A). If this term is factorized, two contractions are obtained as symbolically depicted in Fig. 1. In the first contraction the two-electron integrals are multiplied with the t_2 amplitudes to form an intermediate quantity. In the second contraction the intermediate is multiplied with the t_2 amplitudes to give the final contribution to the residual. For all quantities, the localized basis leads to sparsity in the elements, and for the residual the same sparsity can be assumed as for the amplitudes (it should be noted that in this example, the weakly coupled algorithm is assumed). In the case of a term like this, the first contraction leads to an intermediate that has a sparsity that is not as pronounced as in the original tensors - in the example of Fig. 1 this is illustrated by a simple "multiplication" of the two blocked matrices. If the intermediate is then contracted with the amplitudes into the residual, for which the non-zero blocks are known, several blocks

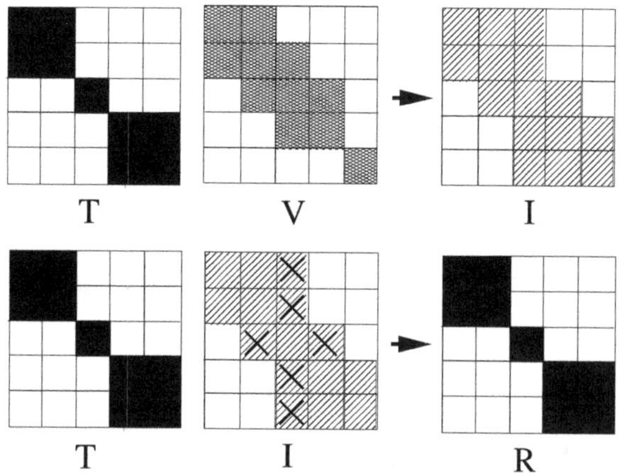

Fig. 1. Symbolic representation for a typical contribution of a CCD t_2 term quadratic in the amplitudes. Filled blocks depict large entries, empty blocks depict entries close to zero. Even if sparsity in the amplitudes (denoted as "T") and integrals (denoted as "V") is assumed, sparsity in the intermediate (denoted as "I") quantity does not automatically follow. If the intermediate is contracted into the residual (denoted as "R", same sparsity as for t_2 is assumed) certain blocks will not be needed that have been formed before (marked with an "x").

that are contained in the intermediate are not needed. Thus, even if the amplitudes and integrals are sparse, this property is lost in the intermediates.

For terms quadratic in the amplitudes a remedy for this is the following prescreening strategy: Before the actual contraction a "dry-run" is carried out at the block level, so that all blocks of the intermediates can be identified that will actually be used in the final contraction. This is easily possible, as in the second contraction the only other quantities are the residual and the amplitudes, for which the block structure is known. However, if a term is more than quadratic in the amplitudes, three or more steps are needed and several contractions will occur in which an intermediate is contracted with amplitudes to form further intermediates. This renders any kind of prescreening difficult if not impossible. While amplitudes and integrals might be sparse, efficiency is then lost in equations that include a hierarchy of intermediates for which there is no pronounced sparsity. A good example are CCSD t_2 terms that are cubic or quartic in t_1. This problem has been identified relatively early by Scuseria and Ayala who devised an AO-based CC algorithm.[71] Here, the intermediate problem is avoided by choosing Brueckner-CCD [72–74] as a starting point rather than conventional CCSD. This way, no terms have to be treated in which more than one intermediate occurs without having to neglect the influence of single excitations. In order to devise a computationally efficient DT-CC implementation we follow the same path, for simplicity focusing on CCD at first. In a previous publication we have presented a simple scheme for the conversion of the canonical CC equations into the ELO/AO representation.[5] Following this scheme, the CCD equations can easily be written in the local basis as shown in Appendix A.

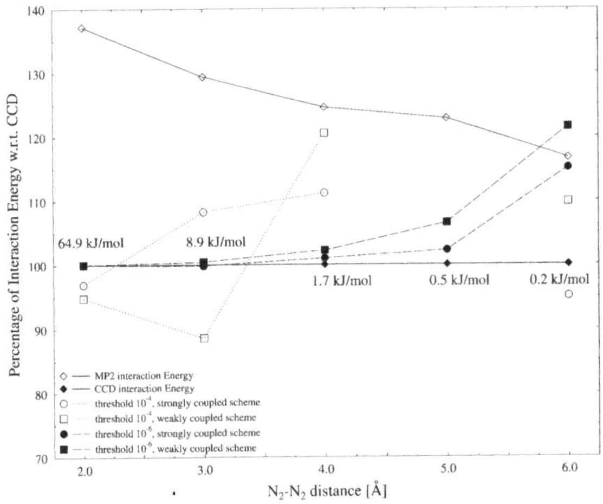

Fig. 2. Truncation error of the correlation contribution to the interaction energy of two N_2 molecules with respect to full CCD. Note that for a threshold of 10^{-4} the points for 5 Å have been omitted due to the very large deviations from the CCD and MP2 results.

3. Efficiency benchmark for the DT-CCD algorithm

While previous benchmarks for the DT scheme focused on errors in the absolute energy, the description of chemical reactions or stability of molecular structures requires reliable results of relative energies. Furthermore, a scheme for actual applications has to be computationally efficient, so that a central, though not necessarily the most important point is the reduction of computational effort. As testcase that includes the calculation of energy differences, the N_2-dimer in the side-on configuration, using the aug-cc-pVTZ [75,76] basis set has been chosen.[77] Furthermore, this system can also be viewed as a very simple model for distant, weakly interacting molecular fragments. For two N_2 molecules the correlation contribution to the interaction energy rapidly decreases with distance. At the CCD/aug-cc-pVTZ level of theory it is reduced from about 65 kJ/mol at 2 Å to about 0.2 kJ/mol at 6 Å (see Fig. 2).

As the DT algorithm is designed for a good threshold-error correlation, the error in the interaction energy should also correlate well with the threshold chosen. In the previous work it was found, that a given accuracy in the energy is reached at a threshold of approximately one order of magnitude higher.[5] If mH accuracy is desired, the threshold should be adjusted to around 10^{-5}. For the N_2 dimer the interaction energy should be reproduced well up to a distance of 2–3 Å at thresholds of 10^{-4}, thresholds of 10^{-6} should yield a good approximation over almost the full range of distances under consideration. If the strength of the interaction drops below the accuracy at the given threshold, the parameters that describe the weak interaction should be screened away and as a consequence the number of amplitudes in the CC-subset should reduce to approximately double the number of one N_2 molecule.

Fig. 2 shows the percentage of the full correlation contribution to the interaction energy for the DT-CCD algorithm with thresholds of 10^{-6} and 10^{-4} as a function of the intermolecular distance.[78] It should be noted that the scale ranges from 70 to 140 %, the zero line being the CCD correlation contribution to the interaction energy. The MP2 result is also given, which systematically overestimates the correlation contribution for the interaction energy. In this calculation, the scheme described in Section 2 has been applied, so that for the weakly coupled algorithm the sparsity information for amplitudes, integrals and the residual is used, while for the strongly coupled algorithm no sparsity information is used for the amplitudes. Before discussing the truncation error with threshold and intermolecular distance it should be mentioned that at a threshold of 10^{-6} the number of amplitudes in the CC-subset is 99 % of the full set of amplitudes at 2 Å, 74 % at 4 Å and drops to 30 % at 6 Å. For a threshold of 10^{-4} it is 86 % at 2 Å, 28 % at 4 Å and drops to about 15 % at 6 Å, which is approximately double the number of amplitudes of one N_2 molecule. Thus, the algorithm performs the desired adaption of the number of wavefunction parameters to the physics of the systems within the limits of the given threshold.

For a threshold of 10^{-6} the strongly and weakly coupled DT schemes reproduce the full CCD result within 10 % up to a distance of 5 Å (note that at this distance 10 % of the correlation contribution is one hundredth of a mH). At 6 Å the results practically reduce to the MP2 value. For a threshold of 10^{-4} it becomes obvious that the error is stable up to about 4 Å, the strongly coupled scheme being somewhat more systematic. However, as the absolute value of the correlation contribution of the interaction energy drops below 1 kJ/mol most of the wavefunction parameters that describe this subtle effect are screened away and the approximation breaks down. Here, very unsystematic results are obtained that are even far from the MP2 values the algorithm ultimatively converges to. Naturally, very large thresholds should be avoided, as in this case only very few but large contributions are refined in the CC-subset. The influence and the sign of these cannot be expected to be systematic in any respect.

The crucial point for an accurate yet efficient approximation is the reduction of the computational effort as more and more wavefunction parameters are discarded. Fig. 3 displays the timings for the CCD t_2 contributions in one iteration partitioned into the contributions from the different terms (see Appendix A) at a distance of 5 Å. The bar on the left is the DT algorithm without any truncation in the amplitude equations, the bar on the right is the same scheme for one N_2 molecule multiplied by a factor of two as a reference for the ideal linear scaling limit of two noninteracting systems. The full treatment is actually about a factor of 60 more expensive than the corresponding monomer calculation, which is in good agreement with the theoretical scaling of the CCD method that should yield a factor of 64. It should be noted that this example is still comparably small, so that two terms that scale as N^6 dominate the overall time, rather

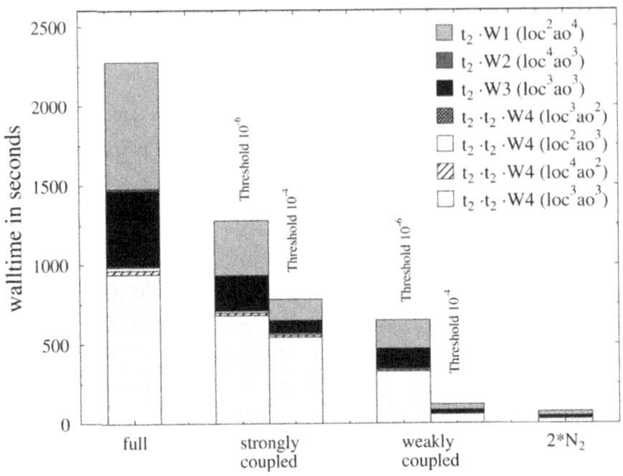

Fig. 3. Walltime in seconds for one CCD t_2 iteration for the N_2 dimer in an aug-cc-pVTZ basis at 5 Å distance. See Appendix A for an explicit formulation of the different CCD terms encoded in the barplot.

than just the particle-particle ladder term (that scales as occ^2 virt^4) as it is often the case in actual applications. [79]

At a distance of 5 Å and a threshold of 10^{-4}, the weakly coupled algorithm is approximately by a factor of three more expensive than the monomer calculation, which is reasonably close to the optimal behavior. The strongly coupled algorithm exhibits a factor of more than 20 for this threshold, which is due to the lack of sparsity information used for the t_2 amplitudes in the algorithm and applies to practically all terms of the t_2 equations. Thus, while being more robust in terms of truncation error this scheme is by far not as efficient computationally wise, even if weakly interacting systems are regarded.

While the N_2 dimer has been used as a proof of principle for the reduction of the computational effort, the actual scaling of the method with systems size has to be investigated in a more realistic setup. For this purpose we have chosen to investigate a simple alkane chain with increasing length (C_nH_{2n+2} with n = 2 to 12) at a threshold of 10^{-4}.[77] While this is a very common type of example - often glycine chains or chains of separate molecules are used - one should keep in mind that this is also a fairly simple testcase, as linear molecules have a more pronounced sparsity of wavefunction parameters than three dimensional structures. For the smallest molecule in this series, the DT algorithm does not screen any amplitudes at all, for C_6H_{14} 19 % are included in the CC-subset and for $C_{12}H_{24}$ only 6 % of the amplitudes are refined at the DT-CCD level of theory. Already starting from C_6H_{14} the additional number of amplitudes in the CC-subset for every further C_2H_4 unit is roughly constant (see second plot in Fig. 4). At the top and at the bottom of Fig. 4 the timings per iteration and the truncation error for the alkane chain using the weakly coupled DT-CCD algorithm (in a 6–31G* basis [80,81]) and a threshold of 10^{-4} are shown. For the largest example,

$C_{12}H_{24}$ the time required for one iteration is reduced by a factor of almost 40. [82] While not scaling linear, the computational scaling is far lower than N^6 - roughly doubling the system size from C_6H_{14} to $C_{12}H_{24}$ yields a factor of about 8 in computational time. The error in terms of absolute energy introduced by choosing a threshold of 10^{-4} is less than a mH for all cases. Nevertheless, it is growing with system size. While the basic construction of the DT algorithm should in principle yield a stable error for a broad variety of systems, this behavior is yet not surprising. On the one hand, as the system grows larger, also the number of screened parameter increases drastically, so that the neglected contributions might sum up to somewhat larger values. On the other hand, as it can be argued for any local method, as the energy is an extensive quantity, so should be the error in the energy made by an approximate scheme. [83] Second from the bottom of Fig. 4 the relative energy for the extension of the molecular chain by a C_2H_4 fragment is shown in oder to give an estimate for the accuracy of relative energies. It should be noted that the same scale on the y-axis has been used as for the absolute error plot below. In contrast to the error in the absolute energy the error of the relative energy is stable as the chain is extended. Overall, although this is a fairly simple testcase, the algorithm already shows a remarkable saving of the computational time with truncation errors that are below one mH for absolute and relative energies.

4. Enveloping localized orbitals (ELOs) and the issue of potential energy surface discontinuities

As mentioned before, a new local representation of the occupied orbitals - the ELOs - has been constructed for the usage in the DT scheme. These are a nonorthogonal set of linearly independent orbitals, that are atom centered and thus better localized than for example PM orbitals. Localization beyond occupied space localizability is achieved by mixing in virtual components yielding a set of AO-like orbitals, that are more numerous than the occupied orbitals but far less numerous than the AOs themself.

The procedure for constructing the ELOs is based on projecting the occupied orbitals on the AOs of a given center and extracting a minimal basis set for each atom. This way, for hydrogen one ELO is obtained, for first row atoms five ELOs are obtained etc. In further steps it is ensured that the set of new orbitals is strictly atom centered and exactly spans the occupied space.

In Fig. 5 four of the five ELOs for carbon are displayed (the fifth one not plotted being the 1s component) for the *sp*-carbon in carbon monoxide, the sp_2-carbon in ethene and the sp_3-carbon in fluoromethane for a cc-pVTZ basis set. [84] For all of these compounds the ELOs are very similar. Each of the ELOs is atom centered and in character mostly AO-like, although for carbon monoxide with a triple bond and a lone electron pair the ELOs do somewhat resemble hybrid orbitals.

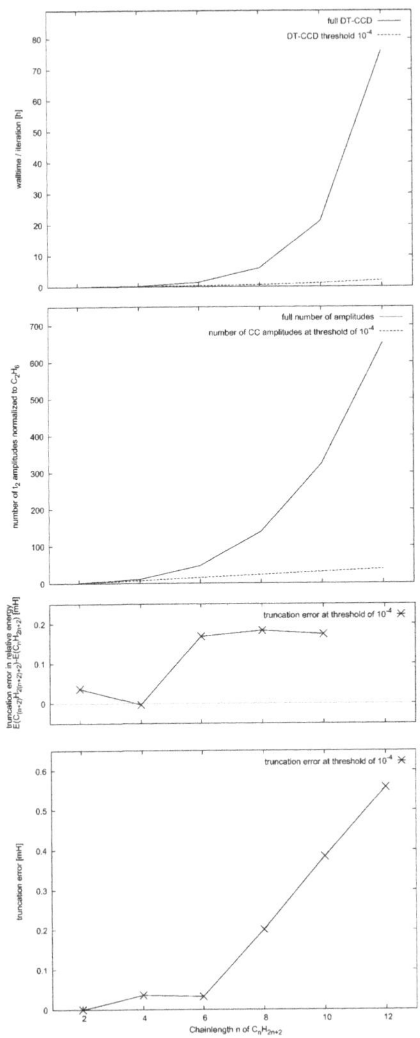

Fig. 4. Investigation of alkane chains C_nH_{2n+2} with n = 2 to 12 (given on the x-axis) for a threshold of 10^{-4} in a 6–31G* basis. The graph a) gives the walltime for one iteration (solid and dashed lines), the graph b) gives the number of the amplitudes in the full and truncated scheme (solid and dashed lines) relative to C_2H_6, the graph c) gives the truncation error in the relative energy between two alkanes that differ in chainlength by 2 in mH and the graph at the bottom gives the truncation error in the absolute energies in mH.

One important issue that has been raised by Russ and Crawford is the consequence of a local truncation scheme on the smoothness of potential energy surfaces (PES). In a detailed investigation the authors have shown that in the framework of LMP2 and LCCSD, potential energy surface discontinuities occur. For

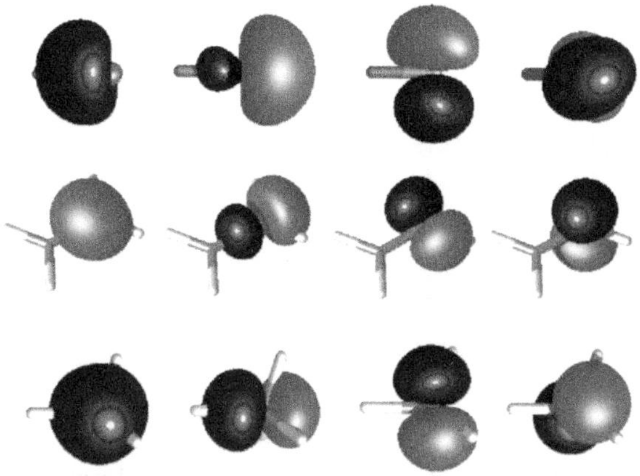

Fig. 5. ELOs for carbon in carbon monoxide (top), ethene (middle) and fluoromethane (bottom) in a cc-pVTZ basis. Note that four out of five carbon ELOs are depicted, the fifth one being the 1s-ELO which is the 1s core orbital.

the homolytic bond dissociation of singlet ketene for example, four discontinuities are found in a range of 1.0 to 2.2 Å of the CO dissociation coordinate (using the cc-pVDZ basis set). For this example, truncation errors are in the order of 3 mH and step heights of the discontinuities can amount to 1 mH. Pronounced discontinuities are found in the dissociation regime around 2 Å which are due to the changes in orbital domains in the course of bond breaking.[61] While this problem can be avoided by freezing the pair and domain structure,[63] this solution is somewhat unsatisfactory, as a bias towards one or the other structure is introduced. A more involved workaround was presented by Subotnik and Head-Gordon in which bump-functions are used to avoid sudden changes in the wavefunction by blending contributions in and out smoothly.[62,64]

In the DT scheme the choice of ELOs has been made as these are strictly atom centered and AO-like and are thus mostly independent of the system under investigation as shown above. Furthermore, as no a priori classification into pairs or domains is carried out, changes in the orbitals should have little effect as long as energy contributions from blocks of amplitudes change smoothly. While one should expect that this scheme should lead to a smooth PES even for PM orbitals, there is another issue in DT scheme that can lead to discontinuities, although for quite different reasons: In order to ensure matrix multiply operation efficiency in all contractions, it is necessary to block all local amplitudes. In the current scheme this is done by joining all amplitudes $t_{\mu\nu}^{\alpha\beta}$ into a block for which all orbitals α are located on a (super-)center A, all orbitals β are located on a (super-)center B, all orbitals μ are located on a (super-)center C and all orbitals ν are located on a (super-)center D, where a supercenter is a heavy atom with all neighboring hydrogen atoms. The screening procedure is then applied to energy

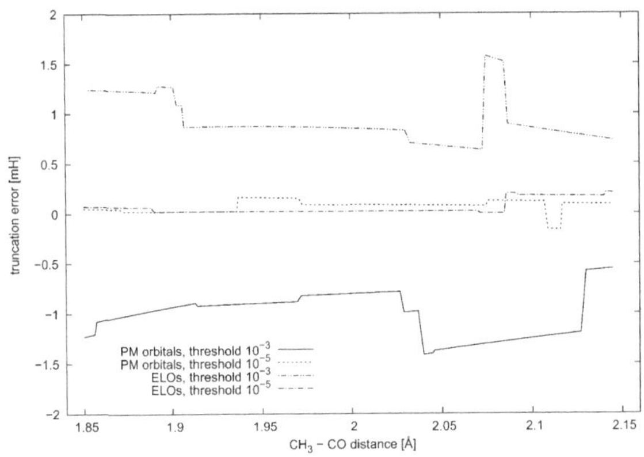

Fig. 6. Truncation error (in mH) in the potential energy surface of ketene as a function of the CO-dissociation distance (in Å) computed using the DT-CCD algorithm with thresholds of 10^{-3} and 10^{-5} based on ELO/AO and PM/AO localized occupied/virtual orbitals.

contributions of the blocks, rather than to contributions from single amplitudes. This way, the threshold also determines the incremental sizes of energy contributions which are screened.

In order to investigate the behavior of the DT algorithm, we have chosen to revisit the singlet ketene example discussed above.[85] In addition to the ELO/AO representation we have also monitored the PES using the PM/AO representation for the occupied and virtual space. The results for the C_s symmetric species and distances of the CO fragment between 1.85 and 2.15 Å using a cc-pVDZ basis [84] are shown in Fig. 6. From this plot it becomes obvious, that screening at the block level does indeed lead to discontinuities of the PES in the DT algorithm. In fact, these occur whenever the number of amplitudes refined in the CC-subset changes along the dissociation coordinate irrespective of the localized orbitals and the threshold applied. However, for smaller thresholds the step height of the discontinuities reduces significantly. As step heights observed at thresholds of 10^{-3} can amount to 1 mH, the largest steps observed for thresholds of 10^{-5} are 0.2 mH. [86] Here, a remedy would certainly be to first carry out the screening procedure and then block the remaining amplitudes, so that the energy increment fluctuation along the PES is much smaller due to the finer granularity of the quantities the screening is applied to.

5. Conclusions and outlook - the path to a production level implementation of the DT algorithm

We have outlined a new implementation of the DT-CCD algorithm based on the ELO/AO orbital representation that represents a milestone on the path to an efficient local CC approximation with controllable accuracy. Benchmark calculations for the N_2 dimer and a series of alkane chains show that in the algorithm the threshold determines the accuracy of the approximation and at the same time reduces computational cost according to the reduction of wavefunction parameters. An investigation of potential energy surface discontinuities shows that a favorable strategy is to carry out the screening of amplitudes at the amplitude level, rather than screening blocks of amplitudes.

For molecular chains the DT algorithm shows an impressive reduction of computational time as wavefunction parameters are screened and already for chains with 10 to 12 atoms the time per iteration reduces to less than 3 % of the full calculation. Due to the favorable scaling of the algorithm even better performance can be assumed for larger systems. While the performance benchmarks for the weakly coupled algorithm make this scheme favorable for actual applications, the convergence towards the full results is not very systematic and at times larger truncation errors are obtained. Thus, a compromise between the efficient weakly coupled and the robust strongly coupled scheme should be found. In the future a more efficient implementation for a production level code will further enhance the performance of the algorithm, the aim being the calculation of CCD and BCCD energies for molecular systems with about 50 atoms. Then, schemes like the DT-BCCD and DT-BCCD(T) are the next steps in the development of this new type of approximations, and also more approximate methods like the orbital invariant CEPA approaches [87] can easily be formulated in the framework of the DT algorithm.

6. Appendix A: CCD equations in the local basis

In the following the convention has been used that α and β denote external AO indices γ, δ, λ and κ denote internal indices in the AO basis, μ and ν denote external indices in the ELO basis and σ and ρ denote internal indices in the ELO basis.

For practical reasons, the CCD doubles equations have not been factorized in the conventional fashion used in MO based algorithms that ensures minimal computational effort. [70] The various terms are rather calculated sequentially in order to separately time and if necessary modify single contributions to the t_2 amplitudes. The residual is calculated as (note that the Einstein summation convention has been used):

$$\begin{aligned} R^{\alpha\beta}_{\mu\nu} &= \frac{1}{2}\, t^{\gamma\delta}_{\mu\nu}\, W^{1\,\alpha\beta}_{\ \ \gamma\delta} + \frac{1}{2}\, t^{\alpha\beta}_{\sigma\rho}\, W^{2\,\sigma\rho}_{\ \ \mu\nu} - t^{\alpha\gamma}_{\mu\rho}\, W^{3\,\rho\beta}_{\ \ \gamma\nu} - t^{\alpha\beta}_{\mu\rho}\, I^{1\,\rho}_{\ \ \nu} \\ &\quad + t^{\alpha\gamma}_{\mu\nu}\, I^{2\,\beta}_{\ \ \gamma} - \frac{1}{2}\, t^{\alpha\beta}_{\sigma\rho}\, I^{3\,\sigma\rho}_{\ \ \mu\nu} + \frac{1}{2}\, t^{\alpha\gamma}_{\mu\rho}\, I^{4\,\rho\beta}_{\ \ \gamma\nu} + \left[\hat{H}_0, T_2\right] \end{aligned} \tag{1}$$

the zeroth-order terms that are carried out fully for all amplitudes of the screening applied are:

$$\left[\hat{H}_0, T_2\right] \quad : \quad W^{5}\,{}^{\alpha\beta}_{\mu\nu} - t^{\alpha\beta}_{\sigma\nu}\, f^{\sigma}_{\mu} + t^{\alpha\gamma}_{\mu\nu}\, f^{\beta}_{\gamma} \tag{2}$$

where the intermediate quantities I are defined as:

$$I^{1}\,{}^{\rho}_{\nu} = \frac{1}{2}\, t^{\gamma\delta}_{\nu\sigma}\, W^{4}\,{}^{\rho\sigma}_{\gamma\delta} \qquad I^{2}\,{}^{\beta}_{\gamma} = -\frac{1}{2}\, t^{\beta\delta}_{\sigma\rho}\, W^{4}\,{}^{\sigma\rho}_{\gamma\delta}$$
$$I^{3}\,{}^{\sigma\rho}_{\mu\nu} = -\frac{1}{2}\, t^{\gamma\delta}_{\mu\nu}\, W^{4}\,{}^{\sigma\rho}_{\gamma\delta} \qquad I^{4}\,{}^{\rho\beta}_{\gamma\nu} = t^{\beta\delta}_{\nu\sigma}\, W^{4}\,{}^{\rho\sigma}_{\gamma\delta} \tag{3}$$

in the equations above, the two-electron integrals W, that are obtained from the AO-Integrals V, and the Fock matrix elements f in the local basis have been obtained by the following transformations:

$$W^{1}\,{}^{\alpha\beta}_{\gamma\delta} = d^{\alpha}_{\kappa}\, d^{\beta}_{\lambda}\, V^{\kappa\lambda}_{\gamma\delta} \qquad W^{2}\,{}^{\rho\sigma}_{\mu\nu} = c^{\rho}_{\alpha}\, c^{\sigma}_{\beta}\, e^{\gamma}_{\mu}\, e^{\delta}_{\nu}\, V^{\alpha\beta}_{\gamma\delta}$$
$$W^{3}\,{}^{\rho\beta}_{\gamma\nu} = c^{\rho}_{\alpha}\, d^{\beta}_{\delta}\, e^{\kappa}_{\nu} V^{\alpha\delta}_{\gamma\kappa} \qquad W^{4}\,{}^{\rho\sigma}_{\gamma\delta} = c^{\rho}_{\kappa}\, c^{\sigma}_{\lambda}\, V^{\kappa\lambda}_{\gamma\delta}$$
$$W^{5}\,{}^{\alpha\beta}_{\mu\nu} = d^{\alpha}_{\kappa} d^{\beta}_{\lambda} e^{\gamma}_{\mu} c^{\delta}_{\nu} V^{\kappa\lambda}_{\gamma\delta} \qquad f^{\beta}_{\gamma} = d^{\beta}_{\delta}\, f^{\delta}_{\gamma} \tag{4}$$
$$f^{\sigma}_{\mu} = c^{\sigma}_{\gamma}\, e^{\delta}_{\mu}\, f^{\gamma}_{\delta}$$

with d denoting density matrices and corresponding virtual quantities, c denoting the AO-MO transformation coefficients and e denoting products of density matrices and transformation coefficients. [5]

References

1. P. Y. Ayala, G. E. Scuseria, J. Chem. Phys. **110** (1999) 3660.
2. G. E. Scuseria, P. Y. Ayala, J. Chem. Phys. **111** (1999) 8330.
3. P. Y. Ayala, G. E. Scuseria, J. Comput. Chem. **21** (2000) 1524.
4. P. Constans, P. Y. Ayala, G. E. Scuseria, J. Chem. Phys. **113** (2000) 10451.
5. A. A. Auer, M. Nooijen, J. Chem. Phys. **125** (2006) 024104.
6. D. S. Lambrecht, B. Doser, C. Ochsenfeld, J. Chem. Phys. **123** (2005) 184102.
7. S. Schweizer, B. Doser, C. Ochsenfeld, J. Chem. Phys. **128** (2008) 209902.
8. B. Doser, D. S. Lambrecht, C. Ochsenfeld, Phys. Chem. Chem. Phys. **10** (2008) 3335.
9. B. Doser, D. S. Lambrecht, J. Kussmann, C. Ochsenfeld, J. Chem. Phys. **130** (2009) 064107.
10. F. Neese, F. Wennmohs, A. Hansen, J. Chem. Phys. **130** (2009) 114108 .
11. S. Saebø, P. Pulay, Annu. Rev. Phys. Chem. **44** (1993) 213.
12. P. Pulay, Chem. Phys. Lett. **100** (1983) 151.
13. S. Saebø, P. Pulay, Chem. Phys. Lett. **113** (1985) 13.
14. P. Pulay, S. Saebø, Theor. Chim. Acta **69** (1986) 357.
15. S. Saebø, P. Pulay, J. Chem. Phys. **86** (1987) 914.
16. S. Saebø, P. Pulay, J. Chem. Phys. **88** (1988) 1884.
17. S. Saebø, P. Pulay, J. Chem. Phys. **115** (2001) 3975.
18. J. Pipek, G. Mezey, J. Chem. Phys. **90** (1989) 4916.
19. J. W. Boughton, P. Pulay, J. Comput. Chem. **14** (1993) 736.
20. C. Hampel, H.-J. Werner, J. Chem. Phys. **104** (1996) 6286.
21. G. Reynolds, T. J. Martinez, E. A. Carter, J. Chem. Phys. **105** (1996) 6455.

22. G. Hetzer, P. Pulay, H.-J. Werner, Chem. Phys. Lett. **290** (1998) 143 .
23. M. Schütz, G. Hetzer, H.-J. Werner, J. Chem. Phys. **111** (1999) 5691 .
24. P. E. Maslen, M. Head-Gordon, J. Chem. Phys. **109** (1998) 7093.
25. M. S. Lee, P. E. Maslen, M. Head-Gordon, J. Chem. Phys. **112** (2000) 3592.
26. S. Li, Ma. Jing, Y. Jiang, J. Comput. Chem. **23** (2000) 237.
27. G. Hetzer, M. Schütz, H. Stoll, H.-J. Werner, J. Chem. Phys. **113** (2000) 9443.
28. M. Schütz, J. Chem. Phys. **113** (2000) 9986.
29. M. Schütz, H.-J. Werner, J. Chem. Phys. **114** (2001) 661.
30. M. Schütz, Phys. Chem. Chem. Phys. **4** (2002) 3941.
31. H.-J. Werner, F. R. Manby, P. J. Knowles, J. Chem. Phys. **118** (2003) 8149.
32. A. Venkatnathan, A. B. Szilva, D. Walter, R. Gdanitz, E. A. Carter, J. Chem. Phys. **120** (2004) 1693.
33. T. S. Chwee, A. B. Szilva, R. Lindh, E. A. Carter, J. Chem. Phys. **128** (2008) 224106.
34. H.-J. Werner, J. Chem. Phys. **129** (2008) 101103.
35. T. B. Adler, F. R. Manby, H.-J. Werner, J. Chem. Phys. **130** (2009) 054106.
36. A. El Azhary, G. Rauhut, P. Pulay, H.-J. Werner, J. Chem. Phys. **108** (1998) 5185.
37. J. Gauss, J.F. Stanton, Phys. Chem. Chem. Phys. **2** (2000) 2047 .
38. G.Rauhut, H.-J. Werner, Phys. Chem. Chem. Phys. **3** (2001) 4853 .
39. T. Korona, H.-J. Werner, J. Chem. Phys. **118** (2003) 3006.
40. N. J. Russ, T. D. Crawford, Chem. Phys. Lett. **400** (2004) 104.
41. T. Korona, K. Pflüger, H.-J. Werner, Phys. Chem. Chem. Phys. **6** (2004) 2059.
42. T. D. Crawford, Theor. Chem. Acc. **115** (2006) 227.
43. T. D. Crawford, M. C. Tam, M. L. Abrams, J. Phys. Chem. A **111** (2007) 12057.
44. N. J. Russ, T. D. Crawford, Phys. Chem. Chem. Phys. **10** (2008) 3345.
45. J. E. Subotnik, M. Head-Gordon, J. Phys.:Condens. Matter **20** (2008) 294211.
46. S. Hirata, R. Podeszwa, M. Tobita, R. J. Bartlett, J. Chem. Phys. **120** (2004) 2581.
47. N. Flocke, R. J. Bartlett, J. Chem. Phys. **121** (2004) 10935.
48. B. Paulus, Phys. Rep. **428** (2006) 1.
49. J. Friedrich, M. Hanrath, M. Dolg, J. Chem. Phys. **126** (2007) 154110 .
50. J. Friedrich, M. Hanrath, M. Dolg, Chem. Phys. **346** (2008) 266 .
51. J. Friedrich, M. Dolg, J. Chem. Phys. **129** (2008) 244105.
52. J. Friedrich, M. Dolg, J. Chem. Theory Comput. **5** (2009) 287.
53. G. Stollhoff, P. Fulde, Z. Phys. B **26** (1977) 251.
54. G. Stollhoff, P. Fulde, J. Chem. Phys. **73** (1980) 4548.
55. H. Stoll, Phys. Rev. B **46** (1992) 6700.
56. H. Stoll, Chem. Phys. Lett. **191** (1992) 548.
57. H. Stoll, J. Chem. Phys. **97** (1992) 8449.
58. G. Stollhoff, J. Chem. Phys. **105** (1996) 227.
59. M. Kobayashi, H. Nakai, J. Chem. Phys. **129** (2008) 044103.
60. M. Kobayashi, Y. Imamura, H. Nakai, J. Chem. Phys. **127** (2007) 074103 .
61. N. J. Russ, T. D. Crawford, J. Chem. Phys. **121** (2004) 691.
62. J. E. Subotnik, M. Head-Gordon, J. Chem. Phys. **123** (2005) 1.
63. R. A. Mata, H.-J. Werner, J. Chem. Phys. **125** (2006) 184110.
64. J. E. Subotnik, A. Sodt, M. Head-Gordon, J. Chem. Phys. **128** (2008) 034103.
65. S. Hirata, J. Phys. Chem. A **107** (2003) 9887.
66. G. Baumgartner, A. A. Auer, D. E. Bernholdt, A. Bibireata, V. Choppella, D. Cociorva, X. Gao, R. Harrison, S. Hirata, S. Krishanmoorthy, S. Krishnan, C.-C. Lam, M. Nooijen, R. Pitzer, J. Ramanujam, P. Sadayappan, , A. Sibiryakov, Proceedings of the IEEE, **93** (2005) 276.
67. A. A. Auer, G. Baumgartner, D. E. Bernholdt, A. Bibireata, V. Choppella, D. Cociorva, X. Gao, R. Harrison, S. Krishanmoorthy, S. Krishnan, Q. Lu, C.-C. Lam, M. Nooijen, R. Pitzer, J. Ramanujam, P. Sadayappan, A. Sibiryakov, J. Mol. Physics **104** (2006) 211.

68. E. Aprá, T. L. Windus, T. P. Straatsma, E. J. Bylaska, W. de Jong, S. Hirata, M. Valiev, M. Hackler, L. Pollack, K. Kowalski, R. Harrison, M. Dupuis, D. M. A. Smith, J. Nieplocha, V. Tipparaju, M. Krishnan, A. A. Auer, E. Brown, G. Cisneros, G. Fann, H. Fruchtl, J. Garza, K. Hirao, R. Kendall, J. Nichols, K. Tsemekhman, K. Wolinski, J. Anchell, D. Bernholdt, P. Borowski, T. Clark, D. Clerc, H. Dachsel, M. Deegan, K. Dyall, D. Elwood, E. Glendening, M. Gutowski, A. Hess, J. Jaffe, B. Johnson, J. Ju, R. Kobayashi, R. Kutteh, Z. Lin, R. Littlefield, X. Long, B. Meng, T. Nakajima, S. Niu, M. Rosing, G. Sandrone, M. Stave, H. Taylor, G. Thomas, J. van Lenthe, A. Wong, Z. Zhang, "NWChem, A Computational Chemistry Package for Parallel Computers, Version 4.7", , Pacific Northwest National Laboratory, Richland, Washington 99352–0999, USA. (2005).
69. P. Pulay, Chem. Phys. Lett. **73** (1980) 393.
70. J. F. Stanton, J. Gauss, J. D. Watts, R. J. Bartlett, J. Chem. Phys. **94** (1991) 4334.
71. G. E. Scuseria, P. Y. Ayala, J. Chem. Phys. **111** (1999) 8330.
72. K. A. Brueckner, A. M. Lockett, B. Rotenberg, Phys. Rev. **121** (1961) 255.
73. R. K. Nesbet, Phys. Rev. **109** (1958) 1632.
74. R. A. Chiles, C. E. Dykstra, J. Chem. Phys. **74** (1981) 4544.
75. D. E. Woon, T. H. Dunning, J. Chem. Phys. **100** (1994) 2975.
76. Comparisons of basis sets of double-zeta quality and of different triple-zeta basis sets including polarization functions have shown, that this behavior is almost independent of the basis set used.
77. All calculations have been carried out on dual Opteron 248 workstations with 4 GB RAM, all geometries used here were preoptimized at the B3-LYP/SVP [88–91] level of theory using the Turbomole program package. [92]
78. It should be noted that as this example rather serves for a proof of principle than the attempt to predict a molecular property, no BSSE correction has been calculated. Nevertheless, it would certainly be interesting to test the influence of the current truncation scheme on BSSE corrections in further work.
79. M. E. Harding, T. Metzroth, A. A. Auer, J. Gauss, J. Chem. Theory Comput. **4** (2008) 64.
80. W. J. Hehre, R. Ditchfield, J. A. Pople, J. Chem. Phys. **56** (1972) 2257.
81. P. C. Hariharan, J. A. Pople, Theor. Chim. Acta **28** (1973) 213 .
82. It should be noted, that even though the code used for the DT CCD computations presented here is still rather an experimental pilot implementation rather than a streamlined production level code, a molecular systems of this size roughly marks the crossover point with efficient state-of-the art canonical CC implementations.
83. It should be noted that this is also the reason why in the literature often only the percentage of the correlation energy captured by a local approximation is given.
84. T. H. Dunning, J. Chem. Phys. **90** (1989) 1007.
85. All geometries have been relaxed at the HF-SCF/cc-pVDZ level of theory. We have chosen this level of theory for simplicity, although Crawford et al. used the CCSD method to relax the geometries along the dissociation path. However, the HF-SCF method captures the main features of the structure and is used as a cost effective compromise.
86. It should be noted that for the PM orbitals another blocking scheme is required, as these are not atom centered and cannot naturally be devided in groups. Here, we have chosen to include all orbitals in one block that have the largest weight on the same center. Orbitals that are more delocalized and have comparable weights on more centers are put into separate tiles.
87. R. Bartlett, M. Musial, J. Chem. Phys. **125** (2006) 204105.
88. C. Lee, W. Yang, R. P. Parr, Phys. Rev. B **37** (1988) 785.
89. A. D. Becke, Phys. Rev. A **38** (1988) 3098.

90. A. D. Becke, J. Chem. Phys. **98** (1993) 5648.
91. A. Schäfer, H. Horn, R. Ahlrichs, J. Chem. Phys. **97** (1992) 2571 .
92. Turbomole : A. Ahlrichs, M. Bär, M. Häser, H. Horn, C. Kölmel, Chem. Phys. Lett. **162** (1989) 165.

An Additive Long-range Potential to Correct for the Charge-transfer Failure of Time-dependent Density Functional Theory

By Andreas Dreuw[1,*], Jürgen Plötner[1], Michael Wormit[1], Martin Head-Gordon[2], and Anthony Dean Dutoi[3]

[1] Institute of Physical and Theoretical Chemistry, Johann Wolfgang-Goethe University Frankfurt, Max-von-Laue-Str. 7, 60438 Frankfurt, Germany
[2] Department of Chemistry, University of California, Berkeley, and Chemical Science and Physical Bioscience Division, Lawrence Berkeley National Laboratory, Berkeley, California 94720, USA
[3] Institute of Physical and Theoretical Chemistry, Ruprecht Karls-University Heidelberg, Im Neuenheimer Feld 229, 69120 Heidelberg, Germany

(Received July 9, 2009; accepted September 24, 2009)

Excited States / Time Dependent Density Functional Theory / Charge Transfer / Long-range Corrected Potential / Hartree-Fock Exchange

Time-dependent density functional theory is one of the most widely used methods for the calculation of excited states of large molecules. However, it exhibits substantial problems with charge-transfer excited states when conventional exchange-correlation functionals are employed. Here, we introduce an additive long-range potential that can, in principle, be directly employed with any existing local, GGA or hybrid-functional, without re-fitting of the original functionals. The additive potential shifts the excitation energies to higher values and corrects for the wrong asymptotic behavior of their potential energy surfaces with respect to charge-separating coordinates. First examples of its successful application are shown.

1. Introduction

Linear-response time-dependent density functional theory, usually referred to as TDDFT [1–3], has become one of the most widely used quantum chemical methods to calculate electronic excited states of medium-sized and large molecules in all areas of chemistry and physics (for example, see [4–9]). In fact, excitation energies computed with TDDFT very often reach the accuracy of results obtained

* Corresponding author. E-mail: andreas.dreuw@theochem.uni-frankfurt.de

Z. Phys. Chem. **224** (2010) 311–324
© by Oldenbourg Wissenschaftsverlag, München

with sophisticated wavefunction-based ab initio methods. In particular, typical valence-excited $\pi\pi^*$ states lying energetically well below the first ionization potential can usually be described accurately and very reliably with TDDFT, using almost any available exchange-correlation functional [3,10]. Since such $\pi\pi^*$ states usually dominate optical spectra, i.e. absorption and fluorescence spectra, TDDFT very often reproduces experimental spectra quite nicely at very low computational cost compared with higher-level ab initio methods. This certainly is one of the major reasons for the success story of TDDFT.

In present-day TDDFT applications one generally relies on the so-called adiabatic local density approximation (ALDA) [1,2] to the time-dependent exchange-correlation potential functional

$$v_{xc}^{ALDA}(r,t) = v_{xc}[\rho(r,t)](r,r',t,t')\,\delta(t-t')\delta(r-r') \tag{1}$$

and, as a consequence, time-independent approximate ground state exchange-correlation (xc) functionals like SVWN [11,12], BLYP [13,14], or B3LYP [15] are applied in excited state calculations. Using B3LYP, the spatial locality of the xc-potential is lost due to the inclusion of non-local Hartree-Fock exchange. One major drawback of these approximate xc-functionals is that the corresponding potentials do not exhibit the correct $1/r$ asymptotic behavior, where r is the electron-electron distance [16]. This results in substantial errors for the excitation energies of Rydberg excited states in which one electron travels far away from the molecule. A possible correction to this problem has been provided by asymptotically corrected local functionals such as LB94, for example [17]. Due to the linear-response formalism of TDDFT, as will be described below, it is not possible to describe doubly-excited states within the ALDA approximation.

One may argue that Rydberg states and doubly-excited states are not important for the electronic spectra of many molecular systems, however, it has been noticed recently that TDDFT also gives substantial errors for the excitation energies of $\pi\pi^*$ states of molecules with extended π-systems [18,19], as well as for charge-transfer (CT) states [7,20–23]. The latter failure occurs for all kinds of charge-transfer states, intra-molecular or inter-molecular, but it is particularly drastic for long-range CT excited states in weakly interacting molecular complexes, whose excitation energies are underestimated by up to several eV [7,24]. Even if one is not interested in computing CT states, they might also interfere with locally excited states and affect the absorption spectrum by mixing of the wavefunctions. The potential energy curves of these states do not exhibit the correct $1/R$ asymptotic behavior, where R is a coordinate related to the distance between the separated charges of the CT state. Today, it is well understood that the wrong asymptotic behavior of the potential energy surfaces of CT excited states can be traced back to the full or partial neglect of exact non-local Hartree-Fock (HF) exchange [21,23,25], while the problematically low excitation energies at infinite separation are related to a missing discontinuity in the approximate xc-potentials occurring in the exact potential upon electron transfer [22].

In this work, first, the origin of the charge-transfer failure, resulting in excitation energies that are too low and incorrect asymptotic potentials will be discussed in detail in Section 2, where possible correction schemes are also described. Our own approach for an additive long-range correction potential for CT states is introduced and discussed in detail in Section 3. First numerical tests of the new potential are shown for the ethylene-tetrafluoroethylene complex as pedagogical example.

2. Origin of the CT failure in TDDFT

The starting point for the derivation of linear-response TDDFT is the time-dependent Kohn-Sham equation

$$i\frac{\partial}{\partial t}\phi_i(r,t) = \left\{-\frac{1}{2}\nabla_i^2 + v(r,t) + J[\rho(r,t)] + v_{xc}[\rho(r,t)](t,t',r)\right\}\phi_i(r,t), \qquad (2)$$

which can be derived via the Runge-Gross mapping theorem alone [26]. However, the xc-potential in eq.(2) is non-local in time and depends on all times, the past and future, and thus, for its exact application, information from a previously known solution of the time-dependent Schrödinger equation is always needed [27,28]. Since one generally does not know the future, one is forced to do the ALDA approximation described above(eq.(1)). Analysis of the linear-response of a Kohn-Sham system to an external oscillating electric field yields linear response equations that can be conveniently written in matrix notation in an RPA-type form

$$\begin{pmatrix} \mathbf{A} & \mathbf{B} \\ \mathbf{B} & \mathbf{A} \end{pmatrix}\begin{pmatrix} \mathbf{X} \\ \mathbf{Y} \end{pmatrix} = \omega \begin{pmatrix} \mathbf{1} & \mathbf{0} \\ \mathbf{0} & \mathbf{-1} \end{pmatrix}\begin{pmatrix} \mathbf{X} \\ \mathbf{Y} \end{pmatrix}, \qquad (3)$$

which is also known as Casida's equation [1]. If a general hybrid xc-functional is applied, i.e. one that also contains parts of HF exchange, the elements of the matrices **A** and **B** can be formally written (in the canonical basis) as

$$\begin{aligned} A_{ia\sigma,jb\tau} &= \delta_{\sigma\tau}\delta_{ij}\delta_{ab}(\epsilon_{a\sigma} - \epsilon_{i\tau}) \\ &+ (i_\sigma a_\sigma|j_\tau b_\tau) - \delta_{\sigma\tau}c_{HF}(i_\sigma j_\sigma|a_\tau b_\tau) \\ &+ (1 - c_{HF})(i_\sigma a_\sigma|f_{xc}|j_\tau b_\tau) \end{aligned} \qquad (4)$$

$$\begin{aligned} B_{ia\sigma,jb\tau} &= (i_\sigma a_\sigma|b_\tau j_\tau) - \delta_{\sigma\tau}c_{HF}(i_\sigma b_\sigma|a_\tau j_\tau) \\ &+ (1 - c_{HF})(i_\sigma a_\sigma|f_{xc}|b_\tau j_\tau), \end{aligned} \qquad (5)$$

where i,j are used for ground-state occupied orbitals, a,b for virtual orbitals, and σ and τ are spin variables and the 4-index integrals are given in Mulliken notation. As usual, ε_i refers to the ground-state orbital energy of orbital ϕ_i, and c_{HF} is the coefficient of the HF exchange part in the hybrid functionals, i.e. 0.2 for

B3LYP [15], for instance. The last terms of equations (3) and (4) contain the xc-kernel f_{xc}, which is defined as the functional derivative of the xc-potential with respect to the density

$$f_{xc}(r) = \frac{\delta v_{xc}[\rho](r)}{\delta \rho(r)}. \tag{6}$$

The explicit form is given, for example, in [29]. Diagonalization of the matrix given in eq. (2) delivers the excitation energies and excited state Kohn-Sham wavefunctions as eigenvalues and eigenvectors. For more details, the interested reader is referred to Ref. [3].

For simplicity let us consider a charge-transfer excited state between two practically non-interacting monomers, for which the CT failure is particularly pronounced; the underlying ideas, however, apply to any approximately CT state as well. In such a case, the occupied orbitals i, j are located on the electron donating molecule, while the virtual orbitals a, b reside on the electron-accepting monomer, i.e. the occupied and virtual orbitals do not overlap. As a direct consequence, all matrix elements in eqs.(3) and (4) containing products ia and jb vanish, and the matrix elements for a CT state reduce to

$$\begin{aligned} A_{ia\sigma,jb\tau} &= \delta_{\sigma\tau}\delta_{ij}\delta_{ab}(\epsilon_{a\sigma} - \epsilon_{i\tau}) - \delta_{\sigma\tau}c_{HF}(i_\sigma j_\sigma|a_\tau b_\tau) \\ B_{ia\sigma,jb\tau} &= 0. \end{aligned} \tag{7}$$

When a non-hybrid xc-functional is employed with $c_{HF} = 0$, then a TDDFT computation gives just the orbital energy difference between the occupied (electron-donating) and virtual (electron-accepting) orbitals of the ground-state Kohn-Sham calculation.

A good electrostatic estimate for excitation energy of CT excited states is given by

$$\omega_{CT} = IP - EA - \frac{1}{R}, \tag{8}$$

where IP is the ionization potential of the electron-donating monomer, EA is the electron affinity of the electron-accepting one and R is the distance between them. This equation is certainly a good estimate at large intermolecular separation but underestimates the excitation energies at shorter distances, due to the employed point charge model. To understand the failure of TDDFT for excited CT states, let us first briefly consider linear-response time-dependent Hartree Fock (TDHF), for which $c_{HF} = 1$ in eq. (6). In this case, the excitation energy of a CT excited state is given by

$$\omega_{CT} = \epsilon_a - \epsilon_i - (ii|aa). \tag{9}$$

Since the orbital energies are approximations to the ionization potential and electron affinity, owing to Koopmans' theorem, and the last term represents the electrostatic attraction between the transferred electron and the created hole, TDHF yields reasonable values for the excitation energies of CT excited states for physically correct reasons. It is worthwhile to note that the integral $(ii|aa)$ in eq. (8)

exactly cancels the integral $(aa|ii)$ which is contained in the orbital energy of the virtual orbital, stemming from the summation over all occupied orbitals in the Coulomb operator.

In TDDFT, on the contrary, only the orbital energy differences are obtained according to eq. (6) when c_{HF} is equal to zero. Since the orbital energies of the individual monomers of a non-interacting complex are essentially independent of the intermolecular distance coordinate R (which also corresponds to the charge-separation coordinate, in this case) the excitation energies of CT excited states are practically constant with respect to R. This is physically wrong, since the separation of a positive and a negative charge requires the potential energy surface to exhibit a $1/R$ dependence. This failure is due to the neglect of non-local Hartree-Fock exchange, whose linear response is the two-electron integral given in eq. (8), describing the electrostatic attraction between the transferred electron and the created hole. In other words, the transferred electron does not leave a positively charged monomer behind, as it should, but instead still interacts with itself. For this reason, this failure is also often referred to as electron-transfer self-interaction. Besides the wrong asymptotic of the potential energy surface, TDDFT also yields excitation energies for CT excited states that are way too low. This owes to the fact that Koopman's theorem applies only to the highest occupied molecular orbital in KS-DFT, while the orbital energies of the virtual MOs correspond to single-particle energies of excited electrons rather than additional electrons as in HF. This is intimately related to a missing discontinuity in the approximate xc-potential upon change of particle number [30]. It has been shown that the xc-potential jumps discontinuously by the value of *IP–EA* when an electron is transferred [22].

From this analysis, it is obvious that a large portion of the CT failure can be fixed when substantial amounts of non-local Hartree-Fock exchange are included in the xc-potentials, which has been realized in a couple of schemes so far [31–33]. In these schemes, the Coulomb operator is separated into short-range and long-range parts, for example, in [31], the error-function is employed

$$\frac{1}{r_{12}} = \frac{1 - erf(\mu r_{12})}{r_{12}} + \frac{erf(\mu r_{12})}{r_{12}}, \tag{10}$$

where $r_{12} = |r_1 - r_2|$. In these long-range separated functionals, the short-range electron interaction is evaluated using standard xc-potentials from DFT, e.g. BLYP [13,14] or B3LYP [15], while the long-range part is calculated with exact Hartree-Fock exchange. This idea has originally been suggested by Stoll and Savin already in 1985 [34–37] and has recently been successfully implemented, yielding long-range corrected LC-BLYP, for instance, which indeed improves the description of CT states by linear-response TDDFT dramatically [31,38–41]. A major drawback of this approach, however, is that the standard xc-functionals require a re-fitting of the parameters of the original xc-potentials. Also B3LYP [15] at short-range has been combined with increasing amounts of exact HF exchange at long-range resulting in the so-called CAM-B3LYP [32], which gives excellent CT excitation energies in comparison with benchmark calculations. However, since at most 60% HF exchange is included, the correct asymptotic behavior of the potential energy surface (PES) of CT states is not recovered [32]. A slightly different route is taken by Gritsenko and Baerends, who suggest a new long-range corrected xc-kernel that shifts the orbital energy of the acceptor orbital to a value related to the electron affinity. The wrong asymptotic behavior of the CT states is corrected by a distance-dependent Coulomb term that corrects for electron-transfer self-interaction [42]. Similar in spirit is the approach of Baer and Neuhauser, who also include long-range Hartree-Fock exchange, but who employ a different separation function for the Coulomb operator [33,43].

A second approach to include exact exchange is to use exact local Kohn-Sham exchange instead of non-local Hartree-Fock exchange, which is known as local Hartree-Fock (LHF), or also optimized effective potentials (OEPs), which have recently been introduced by Görling [44] and Ivanov et al. [45]. First tests have recently shown that they may well be capable of yielding the correct asymptotic $1/R$ dependence of excited CT states, since they possess singularities as soon as the overlap between the electron-donating and electron-accepting orbitals vanishes [46]. It is, however, numerically challenging to compensate for the vanishing overlap with the divergence of the xc-kernel f_{xc} such that the fourth term of the **A** matrix (equation (3)) cancels the electron-transfer self-interaction error correctly and recovers the correct $1/R$ asymptote.

Recently, Ziegler suggested an approach based on a revised electronic Hessian to improve the excitation energies of CT excited states within density functional theory [47]. Employing standard static perturbation theory, it was shown that those Coulomb integrals appear in second order that describe the electron-hole Coulomb interaction, i.e. the terms representing the linear response of Hartree-Fock exchange that are missing in linear-response TDDFT as described above. This is clearly due to the fact that the linear response of the Coulomb operator yields exchange like terms and vice versa. As a consequence, second order response of the Coulomb operator gives Coulombic terms again and the electron-hole interaction is recovered. In summary, it has been shown that one needs at least second order terms in a perturbative treatment to correct for the wrong asymptotic behaviour of the PES of CT states. ΔDFT calculations to compute CT excited states by tweaking the SCF to converge onto the CT solution, for example by using the maximum-overlap methods (MOM)[48], are analogous to including the response of the system to infinite order, thus not having any problem with long-range CT states.

3. An additive long-range correction potential

For the development of our long-range separated potential we adopt a very pragmatic point of view. We distinguish between the failure for the asymptotic behavior of the CT potential energy surfaces and the one for the too low excitation

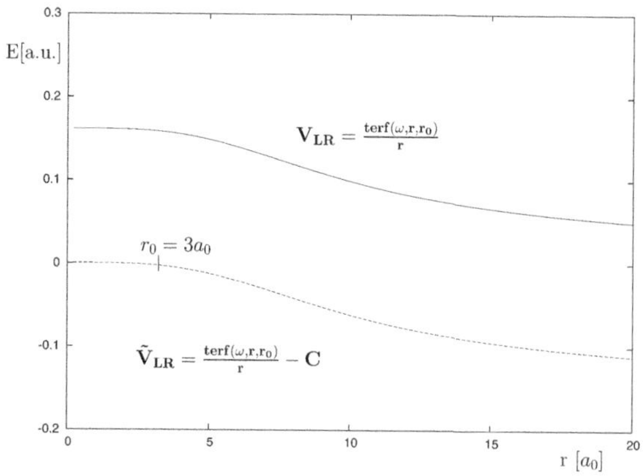

Fig. 1. Additive long-range potentials V_{LR} and $\widetilde{V}_{LR}$. While the first does exhibit significant contributions at short range ($r < r_0$), the latter shifted potential does not, leaving the short-range DFT potential practically unchanged.

energies of charge-transfer excited states at infinite separation, since they have been demonstrated above to have different origins. For each of these failures we introduce a separate, independent correction. Moreover, we want to avoid the refitting of the original xc-potential, such that the correcting potential should be additive and applicable with every existing functional, preserving the well-known accuracy of TDDFT for local excitations.

Let us begin with the correction for the wrong asymptotic behavior of the potential energy surfaces. To correct for this failure in a conceptually simple and straightforward way, we perform a long-range separation of the Coulomb operator as described above and displayed in eq. (9) In contrast to almost all previous long-range corrected xc-potentials, we have chosen to use the *terf*-function, a linear combination of two error-functions, as separator function of the Coulomb operator, as proposed by Dutoi and Head-Gordon [49,50]

$$terf(\omega, r, r_0) = \frac{1}{2}\left[erf(\omega(r + r_0)) + erf(\omega(r - r_0))\right]. \tag{11}$$

In analogy to the *erf*-function, the *terf*-function can also be employed to divide the Coulomb potential into a singular short-range piece and a non-singular long-range piece, according to eq.(9), where ω controls the sharpness of the short-range/long-range division at r_0. With $r_0 = 0$, the terf-function reduces to $erf(\omega r)$, but the additional parameter allows for a more flexible adjustment of the potential. Because the standard local and GGA xc-potentials are well known to converge to zero too quickly with increasing electron-electron distance, we just switch on the long-range Hartree-Fock exchange portion from the split Coulomb operator without explicitly switching off the exchange-correlation potential from DFT. In other words, we just add long-range Coulomb to the exchange operator

$$V_{LR} = \frac{terf(\omega, r, r_0)}{r}, \tag{12}$$

which is displayed in Fig. 1. However, this potential also has contribution from electron-electron distances smaller than r_0, which will influence the short-range part, i.e. the DFT xc-potential. To circumvent this, we simply subtract the value of the long-range potential at $r = 0$ as a constant C from the long-range part. The obtained potential $\tilde{V}_{LR}$ is also shown in Fig. 1, and it is readily apparent that the influence of the long-range potential on the short-range part is negligible for $r < r_0$, where we practically add zero. Here it is important to recognize that the flatness of $\tilde{V}_{LR}$ around $r = 0$ is achieved by setting $\omega r_0 = 1/\sqrt{2}$, as has been shown by Dutoi and Head-Gordon [49]. Since exchange is short-ranged in general, the contributions of the long-range potential to the expectation value of the exchange operator can thus be expected to be generally small. In TDDFT, on the other hand, the added long-range tail of the exchange operator will become Coulombic in linear response, describing the attraction between the transferred electron and the remaining positive charge [3]. Note, that in linear response, only the derivative $\partial\tilde{V}_{LR}/\partial\rho$ is required, thus the additive constant C vanishes and does not play any role here.

As an illustrative example to demonstrate the usefulness of the additive long-range potential $\tilde{V}_{LR}$, we have chosen the pedagogical ethylene-tetrafluoroethylene complex (Fig. 2), which served us already to demonstrate the failure of linear-response TDDFT for charge-transfer states [23,25]. As first step, the geometries of the monomers have been independently optimized employing DFT [51] with the standard GGA xc-functional PBE [52,53] and the 6–31G* basis set. Then, the complex has been composed, for which the vertical excited states have been calculated along the intermolecular distance coordinate R using TDDFT/PBE/6–31G*. The obtained potential energy curves are shown in the upper part of Fig. 3, and it is easily recognized that the curves of the CT excited states do not exhibit the correct asymptotic $1/R$ distance dependence. Let us now use the additive $\tilde{V}_{LR}$ (Fig. 1) within the calculation to improve the asymptotic behavior. With the constraint $\omega r_0 = 1/\sqrt{2}$, we have one additional parameter to determine which is r_0. To find the optimal value, we have repeated the above calculations with r_0-values ranging from 5 Å to 0.5 Å. The obtained excitation energies for the three lowest locally excited (LE) $\pi\pi^*$ states and the lowest CT excited state at an intermolecular separation of 4 Å are given in Table 1 and compared to the values obtained with original TDDFT/PBE ($r_0 = \infty$). While the physically correct asymptotic $1/R$ behavior of the potential energy curves is gradually recovered with decreasing value of r_0 and the correct shape is obtained when it reaches a value of at most r_0 = 1.5Å (Fig. 3), the locally excited states become more and more corrupted when r_0 decreases. While at r_0 = 1.5Å the vertical excitation energies of the locally excited states are shifted upwards only by at most 0.04 eV as compared to the pure TDDFT calculation, for smaller r_0 values the introduced errors become larger and may make a re-fitting of the original xc-potential unavoidable. Note that $r_0 = 0$ would correspond to

Fig. 2. Structure of the coplanar ethylene-tetrafluoroethylene complex and definition of the intermolecular distance coordinate *R*.

an *erf*-splitting of the Coulomb-operator demonstrating the substantial influence of the long-range part of an *erf*-splitting according to eq.(9) also onto the short-range properties of the xc-potential. Although the long-range correction does practically not change the excitation energies of the locally excited states, i.e. the relative energies of these states, the total energies are affected and shifted to higher values. The influence of the long-range correction on the ground state has not been explicitely studied yet, but one can expect it to leave short-ranged properties like geometries and frequencies practically unchanged.

In Figure 3 the obtained potential energy curves of the ethylene-tetrafluoroethylene complex using TDDFT/PBE_{terf} with an r_0-value of 1.5 Å are displayed. Comparison of these curves with the ones obtained at TDDFT/PBE level reveals the advantage of the *terf*-approach. While the locally excited states remain practically unchanged, the excited CT states exhibit now the physically correct $1/R$ shape with respect to the intermolecular separation coordinate, and in addition, their excitation energies are already shifted slightly upwards compared to pure TDDFT/PBE. This owes to the fact that we include the *terf*-functional at the two-electron integral level also already in the ground state DFT calculation. What we observe is a small increase of the HOMO-LUMO gap, a typical effect even when only small portions of exact non-local HF exchange are included. However, this shift is not large enough to compensate for the missing discontinuity in the xc-potential resulting in the too low excitation energies. This error persists also with the LC_{terf} correction.

To further correct the TDDFT/PBE_{terf} approach for the missing discontinuity and thus to shift the charge-transfer excited states to higher and improved excitation energies, we again adopt a very practical approach and introduce this discontinuity as a correction to the orbital energy difference in eq. 6. Before going into detail, one has to recognize that the extent of the error in the excitation energies is related to the spatial overlap of the electron donating and electron accepting orbitals in the CT excited state. Recently, it has been demonstrated that this is in fact the case. Peach et al. [54] suggested a diagnostic for the CT failure of TDDFT, which directly measures the overlap of the Kohn-Sham orbitals involved in the examined transition. The diagnostic is defined as

$$\Lambda = \frac{\sum_{i,a} \kappa_{ia}^2 O_{ia}}{\sum_{i,a} \kappa_{ia}^2}, \tag{13}$$

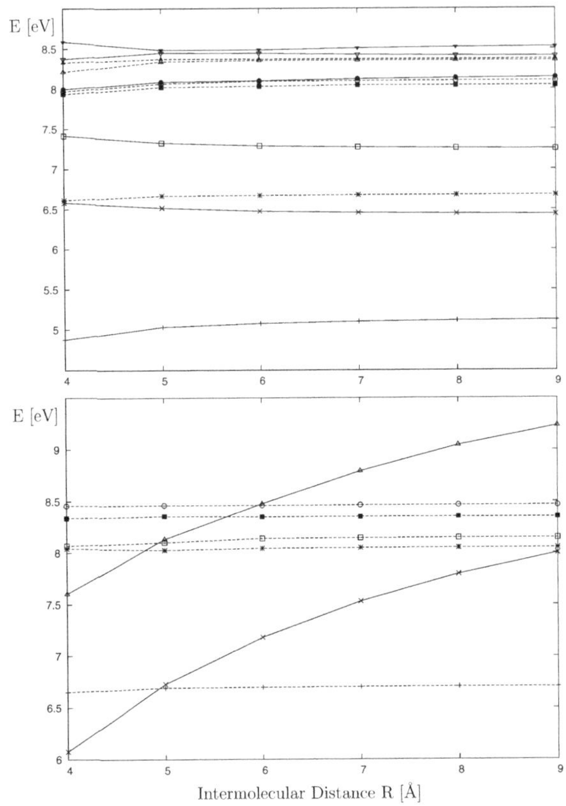

Fig. 3. Potential energy curves of the ethylene-tetrafluoroethylene complex along the intermolecular distance coordinate R as computed with TDDFT/PBE/6–31G* (upper panel) and TDDFT/PBE$_{terf}$/6–31G* (lower panel). Charge transfer states exhibit solid lines, while locally excited states are drawn with dashed lines.

with

$$\kappa_{ia} = X_{ia} + Y_{ia}, \quad (14)$$

being the sum of the X- and Y-amplitudes of the corresponding Slater determinant in the transition vector obtained from solving eg.(2). O_{ia} is the overlap of the orbital moduli

$$O_{ai} = \int |\phi_a(r)||\phi_i(r)|dr. \quad (15)$$

Relating the criterion Λ to the error in the excitation energies obtained with TDDFT/PBE for various molecules exhibiting CT excited states revealed that excited states with $\Lambda > 0.6$ do not exhibit too low excitation energies, while the error in the excitation energies increases with decreasing Λ [54].

Table 1. Dependence of the excitation energies [eV] of the three lowest local excited (LE) states and the lowest excited CT state of the ethylene-tetrafluoroethylene complex on the *terf*-parameter r_0 at an intermolecular distance of 4 Å.

r_0	1 LE	2 LE	3 LE	1 CT
0.5	7.30	8.20	8.43	12.29
1	6.76	8.08	8.26	9.72
1.25	6.70	8.04	8.16	8.72
1.5	6.67	8.02	8.11	7.97
1.75	6.66	8.01	8.09	7.39
2	6.65	8.00	8.08	6.9
3	6.65	8.00	8.07	5.85
5	6.64	8.01	8.07	5.2
∞	6.64	8.01	8.07	5.0

The measure Λ of the spatial overlap of the electron donating and accepting orbitals can now be employed to shift the excitation energies of the calculated excited states individually. This overlap dependent shift (ODS) of the excitation energies is calculated as

$$ODS = k(1 - c\Lambda)\,[(IP - EA) + (\epsilon_{HOMO} - \epsilon_{LUMO})]\,, \qquad (16)$$

where k and c are adjustable parameters, which need to be fitted, and *IP* and *EA* are the ionization potential and electron affinity of the molecules, which need to be calculated prior to the TDDFT calculation. Since Λ is calculated individually for each excited state, the ODS is a state dependent correction, however very easy to implement. Nevertheless, this shift is introduced at each iteration of the Davidson diagonalization of the TDDFT matrix (eq. (2)) and not as an a posteriori shift of the excitation energies. This procedure offers the great advantage that spurious mixing of the wavefunctions of CT excited states and locally excited states is avoided.

For the calculation of improved potential energy curves of the ethylene-tetrafluoroethylene complex, the required parameters have been adjusted to k = 1.0 and c = 1.67 for $\Lambda < 0.6$. For states with $\Lambda \geq 0.6$ no shift is applied. The choice of these parameters has been motivated by the previous analysis of the performance of the PBE xc-functional for CT excited states by Peach et al. (see above) [54], and the fact that the error in the excitation energies seems to increase linearly with decreasing Λ. The shift $(IP{-}EA){+}(\varepsilon_{HOMO}{-}\varepsilon_{LUMO})$ is calculated once to be 1.4 eV at an intermolecular distance of 9 Å and is kept constant over the complete range of the intermolecular distance coordinate *R*. As can be seen by comparison of Fig. 4 with Fig. 3, the potential energy curves of the lowest excited CT states are shifted upwards compared to locally excited states. Employing the ODS, the lowest CT state now clearly lies above the lowest locally excited state. These calculations serve as a proof-of-principle and not to achieve quantitative results, since the parameters and the calculation of the shift need to be improved thoroughly in the future.

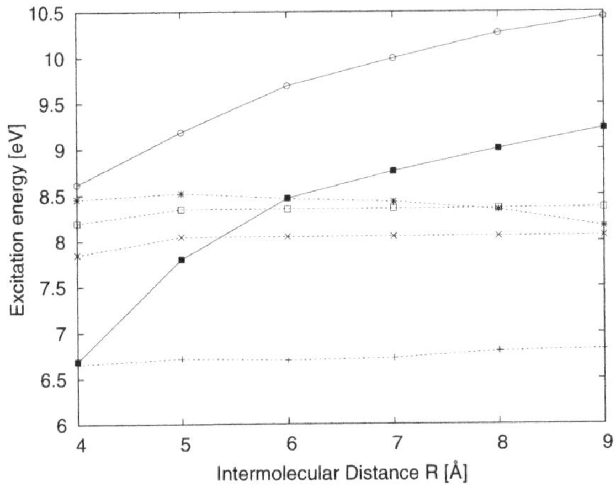

Fig. 4. Potential energy curves of the lowest CT (solid lines) and locally excited (dashed lines) states of the ethylene-tetrafluoroethylene complex along the intermolecular distance coordinate R as computed with TDDFT/$PBE_{terf+ODS}$/6–31G*.

4. Brief summary and conclusions

We have developed a new additive long-range correction potential as well as an overlap dependent shift (ODS) which together correct for the two drastic failures of time-dependent density functional theory for charge-transfer excited states, i.e. their wrong asymptotic behavior with respect to charge separation coordinates as well as the too low excitation energies. The long-range correction rests on the splitting of the Coulomb operator employing a recently suggested *terf*-function as partition function, which offers greater flexibility than the widely used *erf*-function. In particular, the additive long-range Coulomb potential hardly affects the short-range properties of the original xc-potential. The ODS relies on the a priori calculation of the ionization potential and the electron affinity of the molecule under consideration and a state specific shifting of the vertical excitation energies based on the spatial overlap of the electron donating and accepting molecular orbitals. First test calculations for the ethylene-tetrafluoroethylene complex reveal the potential of our new additive potential, since it requires no refitting of the original DFT xc-potential.

In the future, further tests and improvements are required. We are optimistic that universal parameters for the *terf*-splitting can be derived, since it describes electron-electron interaction. However, it remains to be explored whether it is really sufficient to only switch on non-local HF exchange without switching off the xc-potential, or whether one should also switch the latter off using the complementary (1-*terf*)-function. For the ODS, one may need an individual parameter set for each xc-functional, since the dependence of error of the excitation energies of CT state on the spatial overlap varies with xc-functional. This is particularly true for hybrid xc-potentials like e.g. B3LYP. In our first tests, a constant shift and parameters have been employed, derived from TDDFT/PBE calculations. This can be substantially improved by fitting the parameters of the ODS to the errors of long-range corrected TDDFT/PBE_{terf} calculations.

Acknowledgement

This work has been supported by the Deutsche Forschungsgemeinschaft within the Priority Programme SPP 1145 "First Principle Methods". AD holds a Heisenberg-Professorship funded by the Deutsche Forschungsgemeinschaft. ADD is a fellow of the Humboldt foundation. Financial support from the Director of the Office of Basic Energy Sciences, Chemical Sciences Division of the US Department of Energy, under Contract DE-AC03–76SF0098 is gratefully acknowledged.

References

1. M. E. Casida, Recent Advances in Density-Functional. Methods World Scientific, Singapore (1995) pp 155–192.
2. E. U. K. Gross, J. F. Dobson, M. Petersilka, Density Functional Theory II. Springer Heidelberg (1996).
3. A. Dreuw, J. Plötner, L. Lorenz, J. Wachveitl, J. E. Djanhan, J. Brüning, M. Bolte, M. U. Schmidt, Ang. Chem. Int. Ed. **44** (2005) 7783.
4. D. Sundholm, Chem. Phys. Lett. **302** (1999) 480.
5. F. Furche, R. Ahlrichs, C. Wachsmann, E. Weber, A. Sobanski, F. Vögtle; , S. Grimme, J. Am. Chem. Soc. **122** (2000) 1717.
6. A. Dreuw, B. D. Dunietz, M. Head-Gordon, J. Am. Chem. Soc. **124** (2002) 12070.
7. A. Dreuw, G. R. Fleming, M. Head-Gordon, J. Phys. Chem. B **107** (2003) 6500–6503.
8. A. Dreuw, M. Head-Gordon, Chem. Rev. **105** (2005) 4009.
9. M. Wormit, A. Dreuw, J. Phys. Chem. B **110** (2006) 24200.
10. M. Diercksen, S. Grimme, J. Chem. Phys. **120** (2004) 3544.
11. P. A. M. Dirac, Proc. Cam. Phil. Soc. **26** (1930) 376.
12. S. H. Vosko, L. Wilk, M. Nusair, Can. J. Phys. **58** (1980) 1200.
13. A. D. Becke, Phys. Rev. A **38** (1988) 3098–3100.
14. C. Lee, W. Yang, R. G. Parr, Phys. Rev. B **37** (1988) 785.
15. A. D Becke. J. Chem. Phys. **98** (1993) 5648.
16. M. W. Casida, C. Jamorski, K. C. Casida, D. R. Salahub, J. Chem. Phys. **108** (1998) 4439.
17. R. van Leeuwen, E. J. Baerends, Phys. Rev. A **49** (1994) 2421.
18. Z.-L. Cai, K. Sendt, J. R. Reimers, J. Chem. Phys. **117** (2002) 5543.
19. S. Grimme, M. Parac, Chem.Phys.Chem. **3** (2003) 292.
20. D. J. Tozer, R. D. Amos, N. C. Handy, B. J. Roos, L. Serrano-Andres, Mol. Phys. **97** (1999) 859.
21. A. L. Sobolewski, W. Domcke, Chem. Phys. (2003) in press.
22. D. J. Tozer, J. Chem. Phys. **119** (2003) 12697.
23. A. Dreuw, M. Head-Gordon, J. Am. Chem. Soc. **126** (2004) 4007–4016.
24. A. Dreuw, G. R. Fleming, M. Head-Gordon, Phys. Chem. Chem. Phys. **5** (2003) 3247–3256.

25. A. Dreuw, J. L. Weisman, M. Head-Gordon, J. Chem. Phys. **119** (2003) 2943–2946.
26. E. Runge, E. K. U. Gross, Phys. Rev. Lett. **52** (1984) 997.
27. J. Schirmer, A. Dreuw, Phys. Rev. A **75** (2007) 022513.
28. J. Schirmer, A. Dreuw, Phys. Rev. A **78** (2008) 056502.
29. S. Hirata, M. Head-Gordon, Chem. Phys. Lett. **302** (1999) 375.
30. J. P. Perdew, R. G. Parr, M. Levy, J. L. Balduz, Jr. Phys. Rev. Lett. **49** (1982) 1691.
31. Y. Tawada, T. Tsuneda, S. Yanagisawa, T. Yanai, K. Hirao, J. chem. Phys. **1210** (2004) 8425.
32. T. Yanai, D. P. Tew, N. C. Handy, Chem. Phys. Lett. **393** (2004) 51.
33. R. Baer, D. Neuhauser, Phys. Rev. Lett. **94** (2005) 043002.
34. H. Stoll, A. Savin, in Density Functional Methods in Physics. R. M. Dreizler, J. da Providencia (Eds.), Plenum, New York (1985) p 177.
35. A. Savin, in Recent Developments and Applications of Modern Density Functional Theory, J. M. Seminario (Ed.) Elsevier, Amsterdam (1996).
36. T. Leininger, H. Stoll, H.-J. Werner, A. Savin, Chem. Phys. Lett. **275** (1997) 151.
37. H. Iikura, T. Tsuneda, T. Yanai, K. Hirao, J. Chem. Phys. **115** (2001) 5340.
38. T. Sato, T. Tsuneda, K. Hirao, J. Chem. Phys. **126** (2007) 234114.
39. M. Chiba, T. Tsuneda; K. Hirao. J. Chem. Phys. **124** (2006) 144106.
40. M. A. Rohrdanz, J. M. Herbert, J. Chem. Phys. **129** (2008) 034107.
41. M. A. Rohrdanz, K. M. Martins, J. M. Herbert, J. Chem. Phys. **130** (2009) 054112.
42. O. Gritsenko, E. J. Baerends, J. Chem. Phys. **121** (2004) 655.
43. T. Stein, L. Kronik, R. Baer, J. Am. Chem. Soc. **131** (2009) 2818.
44. A. Görling, Phys. Rev. Lett. **83** (1999) 5459.
45. S. Ivanov, S. Hirata, R. J. Bartlett, Phys. Rev. Lett. **83**, 5(1999) 455.
46. T. Gimon, A. Ipatov, A. Heßelmann, A. Görling, J. Chem. Theo. Comp. **5** (2009) 781.
47. T. Ziegler, M. Seth, M. Krykunov, J. Autschbach, J. Chem. Phys. **129** (2009) 184114.
48. A. T. B. Gilbert, N. A. Besley, P. M. W. Gill, J. Phys. Chem. A **112** (2008) 13164.
49. A. D. Dutoi, M. Head-Gordon, J. Phys. Chem. A **112** (2008) 2110.
50. A. D. Dutoi, *Ph. D. Thesis: Novel Approaches to Solving the Electronic Schrödinger Equation for Molecules.* University of California, Berkeley (2006).
51. R. G. Parr, W. Yang, Density-Functional Theory of Atoms and Molecules. Oxford Science Publication New York (1989).
52. J. P. Perdew, K. Burke, M. Enzerhof, Phys. Rev. Lett. **77** (1996) 3865.
53. J. P. Perdew, K. Burke, M. Enzerhof, Phys. Rev. Lett. **78** (1997) 1396.
54. M. J. G. Peach, P. Benfield, T. Helgaker, D. J. Tozer, J. Chem. Phys. **128** (2008) 044118.

Density-Functional Theory with Orbital-Dependent Functionals: Exact-exchange Kohn-Sham and Density-Functional Response Methods

By Andreas Görling*, Andrey Ipatov, Andreas W. Götz, and Andreas Heßelmann

Lehrstuhl für Theoretische Chemie, Egerlandstraße 3, 91058 Erlangen, Germany

(Received July 20, 2009; accepted September 25, 2009)

Density-functional Theory / Time-dependent Density-functional Theory / Orbital-dependent Functionals / Exact Exchange

Kohn-Sham methods with orbital-dependent exchange-correlation functionals, in particular exact-exchange methods, are discussed. The numerical stability of optimized-effective-potential methods to construct the exchange-correlation potential in the case of orbital-dependent energy functionals is considered. A numerically stable exact-exchange Kohn-Sham method is briefly presented. A new ansatz for time-dependent density-functional methods in the response regime is considered and an exact-exchange time-dependent density-functional method based on it is discussed and it is demonstrated that this methods is able to treat qualitatively correctly charge-transfer excitations.

1. Introduction

Conventional density-functional theory (DFT) methods like those based on the local density approximation (LDA) [1–3] or generalized gradient approximations (GGAs) [1–5] employ approximate functionals for the exchange-correlation (xc) energy that are integrals over functions of the electron density or, in the case of GGAs, of the electron density and its gradient. GGA methods nowadays are widely and successfully employed routine methods to investigate electronic ground states and their properties in chemistry and solid state physics [3,6,7]. Electronic excitations and the energies of excited states can be treated via time-dependent density-functional theory (TDDFT) [8,9] relying also on functionals

* Corresponding author. E-mail: goerling@chemie.uni-erlangen.de

© by Oldenbourg Wissenschaftsverlag, München

within the LDA or GGA. However, despite their success, conventional DFT methods are not accurate enough for many questions of interest. For example energies of transition states of chemical reactions often are unreliable, Van-der-Waals interactions can not be described, or charge-transfer excitations are described qualitatively wrong [10–15]. These problems have their origin in shortcomings of the employed GGA functionals. In recent years, therefore, a new generation of DFT methods emerged that uses functionals that not only depend on the electron density and its gradients but also on Kohn-Sham (KS) orbitals. Such functionals shall be referred to as orbital-dependent functionals (ODFs) [82]. Examples are functionals that contain the kinetic energy density, meta-GGA functionals [16–23], or functionals that contain the exchange energy density, hyper-GGA functionals [17,24–32] and local hybrid functionals [28,33–35]. Because the orbitals are functionals of the electron-density, ODFs lie within the realm of DFT.

The evaluation of the xc energy for ODFs poses no principal difficulty. However, within DFT methods, more precisely within KS methods, not only the xc energy but also the xc potential, the functional derivative of the xc energy with respect to the electron density, is required during the self-consistency process to obtain the orbitals. The evaluation of this functional derivative is straightforward for GGA functionals because these functionals are known in terms of the electron density and their gradients. For ODFs, however, this step is problematic because the dependence of the orbitals on the density is not known and the functional derivative of an ODF with respect to the electron density cannot simply be taken via the chain rule. In practice, for meta- and hyper-GGA functionals, instead of taking the functional derivative of the xc energy with respect to the electron density the derivative with respect to the KS orbitals is taken [20,36] in order to obtain an xc potential. An xc potential obtained in this way, however, is not the correct KS xc potential and the resulting approach is, strictly speaking, no longer a KS method.

The KS exchange energy is exactly known in terms of the KS orbitals, i.e., as ODF. For a closed shell system the KS exchange energy in terms of the orbitals has exactly the same form as the Hartree-Fock (HF) exchange energy, i.e., is given by the well-known expression for the exchange energy of a Slater determinant. The value of the KS and HF exchange energy is different because the expression is evaluated for different orbitals, i.e., for KS or for HF orbitals. If the exchange potential is determined as functional derivative of the exchange energy with respect to orbitals the nonlocal HF exchange potential is obtained, while if the exchange potential is determined as functional derivative of the exchange energy with respect to the electron density the local multiplicative KS exchange potential is obtained. This means the difference between the HF method and an exact exchange-only KS method, which shall be denoted EXX (exact exchange) method, is the type of the functional derivative of the exchange energy taken to obtain the exchange potential. While HF and EXX total energies differ very little, the KS orbitals and eigenvalues, in particular, the unoccupied orbitals and their eigenvalues are completely different. Since KS orbitals and eigenvalues are the input quantities for the calculation of response properties, in particular, excitation energies via TDDFT, it is crucial that the correct functional derivative with respect to the electron density is taken. Similar as for EXX and HF methods, the common practice of taking the 'wrong' functional derivative within meta- and hyper-GGA methods in most cases probably has relatively little effect for the calculation of total energies, whereas it is less clear whether this holds true also for the resulting KS orbitals and eigenvalues.

Unoccupied HF orbitals mostly have positive eigenvalues and therefore little physical meaning. The eigenvalue spectra of conventional LDA or GGA methods is qualitatively wrong due to the presence of unphysical Coulomb self-interactions in these methods and, e.g., does not exhibit Rydberg series. EXX methods, on the other hand, yield physical eigenvalue spectra that contain Rydberg series [37,38,83]. As a result TDDFT methods based on LDA or GGA functionals yield very poor excitation energies for transitions in states with Rydberg character [39], which, e.g., often are present as low-lying excited states in organic molecules. EXX orbitals and TDDFT methods based on them, on the other hand, describe Rydberg states very well [37,38]. This demonstrates that it is of importance to employ correct KS xc potentials, i.e., functional derivatives of the xc energy with respect to the electron density. Moreover it seems unsatisfying from a formal point of view to devise density-functional methods which use an xc potential which is wrong in the sense that it is not the xc potential required according to the underlying formalism.

The functional derivative of an ODF with respect to the electron density can, in principle, be obtained via the optimized effective potential (OEP) approach [40–43,82]. Numerically stable OEP methods for atoms based on numerical representations of orbitals and potentials on grids [40,41] as well as for periodic systems based on plane wave basis sets are known [44–46]. OEP methods for molecules employing Gaussian basis sets were proposed [47,48] shortly after plane wave OEP methods were introduced. However, such Gaussian basis set OEP methods suffered from numerical instabilities [47,49–53], the origin of which is closely related to the KS response function which has to be effectively inverted when solving the OEP equation.

Numerically stable OEP methods are a prerequisite for using orbital-dependent xc functionals in self-consistent KS methods, which arguably represent the future of DFT methods. Various attemps to develop numerically stable OEP methods for molecules were made in recent years [37,38,47–50,54–58,60–62,84–87]. Some of these employ regularizations of the basis set representation of the KS response functions [55] or add smoothing terms to the total energy [56]. Other approaches introduce approximations in the one-particle Green's functions occuring in the OEP equation [37,38,61,62]. Within the DFG priority project 1145 we analysed in detail the origin of the numerical instabilities of OEP methods [52,53] and developed a numerically stable Gaussian basis set OEP method [54].

TDDFT methods, using ODFs, in particular time-dependent exact-exchange (TDEXX) methods for molecules [63–65] employing Gaussian basis sets, are numerically even more demanding than EXX KS methods or OEP methods in general. The reason is that in TDEXX methods not only one but two KS response functions occur which have to be effectively inverted. Moreover the frequency-dependent instead of the static KS response function is required. First straightforward implementations of TDEXX methods employing Gaussian basis sets suffer from numerical instabilities [66,67]. Within the DFG priority project 1145 we therefore analysed density-functional response methods [68,75,76] and reformulated the basic TDDFT response equation in such a way that inverse KS response functions no longer occur in TDEXX methods, or generally in TDDFT methods employing ODFs [68]. In this way the source of the numerical instabilities could be removed. The resulting TDEXX approach is numerically stable and can correctly describe charge-transfer excitations [68].

The manuscript is organized as follows: In the next Section OEP methods, in particular EXX methods, their relation to the HF method, and the sources of their numerical instabilities are discussed. Moreover the numerically stable EXX method developed within the DFG priority project 1145 is presented. In Section 3 we present the reformulation of the basic TDDFT response equation and the TDEXX approach based on it. Furthermore the relation between TDEXX and time-dependent HF is discussed. In Section 4 a brief summary and outlook is given. A point that is crucial for OEP and TDDFT methods, but which often is not considered sufficiently, is that the nature of these methods is closely related to the fact that products of occupied and unoccupied KS orbitals form a linearly dependent set. We will therefore put special emphasis to this point in the following.

2. OEP and EXX methods

2.1 Basic formulas and relation between EXX and HF methods

The OEP equation for the KS xc potential $v_{\rm xc}$ corresponding to an ODF for the xc energy is given by

$$\int d\mathbf{r}'\,\chi_s(\mathbf{r},\mathbf{r}')\,v_{\rm xc}(\mathbf{r}') \;=\; t(\mathbf{r}) \tag{1}$$

or

$$v_{\rm xc}(\mathbf{r}) \;=\; \int d\mathbf{r}'\,\chi_s^{-1}(\mathbf{r},\mathbf{r}')\,t(\mathbf{r}') \tag{2}$$

with the static KS response function

$$\chi_s(\mathbf{r},\mathbf{r}') \;=\; 4\sum_i^{occ.}\sum_a^{unocc.}\frac{\varphi_i^{\rm KS}(\mathbf{r})\,\varphi_a^{\rm KS}(\mathbf{r})\,\varphi_a^{\rm KS}(\mathbf{r}')\,\varphi_i^{\rm KS}(\mathbf{r}')}{\varepsilon_i^{\rm KS}-\varepsilon_a^{\rm KS}} \tag{3}$$

containing the occupied and unoccupied KS orbitals φ_i^{KS} and φ_a^{KS}, respectively, and their eigenvalues $\varepsilon_i^{\mathrm{KS}}$ and $\varepsilon_a^{\mathrm{KS}}$. Non-spin-polarized systems with real-valued orbitals shall be considered throughout. Spin is taken into account by appropriate prefactors. The right hand side t of the OEP equation (1) depends on the xc functional. From now on we will concentrate on the exact exchange energy as ODF. For simplicity correlation shall be neglected and we will deal exclusively with the EXX approach. Addition of a conventional GGA correlation functional, however, is straightforward. Parts of the following discussion, in particular questions of the numerical stability of EXX methods, are valid for OEP methods in general, i.e., are valid also for other orbital-dependent xc functionals. The EXX method, in this case, serves as a specific example.

In the exact exchange-only case the right hand side t of the OEP equation (1) has the form

$$t(\mathbf{r}) = 4 \sum_i^{occ.} \sum_a^{unocc.} \frac{\varphi_i^{\mathrm{EXX}}(\mathbf{r})\, \varphi_a^{\mathrm{EXX}}(\mathbf{r})\, \langle \varphi_a^{\mathrm{EXX}} | \hat{v}_{\mathrm{x}}^{\mathrm{NL}} | \varphi_i^{\mathrm{EXX}} \rangle}{\varepsilon_i^{\mathrm{EXX}} - \varepsilon_a^{\mathrm{EXX}}} \tag{4}$$

with $\hat{v}_{\mathrm{x}}^{\mathrm{NL}}$ being a nonlocal exchange operator of the form of the HF exchange operator, however, constructed from KS orbitals. Because we consider the special case of an EXX approach the superscript KS at the orbitals and eigenvalues is replaced by the superscript EXX.

In the EXX case the OEP equation (1) can be rearranged in the form

$$0 = \mathbf{Q}^T(\mathbf{r})\, \mathbf{\Delta}_{\mathrm{x}} \tag{5}$$

with the vectors $\mathbf{Q}(\mathbf{r})$ and $\mathbf{\Delta}_{\mathrm{x}}$ defined by their elements

$$Q_{ia}(\mathbf{r}) = \varphi_i^{\mathrm{EXX}}(\mathbf{r})\, \varphi_a^{\mathrm{EXX}}(\mathbf{r}) \tag{6}$$

and

$$\Delta_{\mathrm{x},ia} = \frac{\langle \varphi_a^{\mathrm{EXX}} | \hat{v}_{\mathrm{x}}^{\mathrm{NL}} | \varphi_i^{\mathrm{EXX}} \rangle - \langle \varphi_a^{\mathrm{EXX}} | \hat{v}_{\mathrm{x}} | \varphi_i^{\mathrm{EXX}} \rangle}{\varepsilon_i^{\mathrm{EXX}} - \varepsilon_a^{\mathrm{EXX}}} . \tag{7}$$

The dimension of the vectors $\mathbf{Q}(\mathbf{r})$ and $\mathbf{\Delta}_{\mathrm{x}}$ equals the number of products of occupied times unoccupied EXX orbitals. The indices i and a of the occupied and unoccupied orbitals, respectively, form a superindex ia labelling the elements of the vectors $\mathbf{Q}(\mathbf{r})$ and $\mathbf{\Delta}_{\mathrm{x}}$. Throughout, indices i or j will label occupied, indices a or b will label unoccupied orbitals.

If the products of occupied times unoccupied orbitals, i.e., the functions $Q_{ia}(\mathbf{r})$, would form a linearly independent set then from Eq. (5) would immediately follow that the vector $\mathbf{\Delta}_{\mathrm{x}}$ equals the zero vector, i.e., that the identity

$$\langle \varphi_a^{\mathrm{EXX}} | \hat{v}_{\mathrm{x}} | \varphi_i^{\mathrm{EXX}} \rangle = \langle \varphi_a^{\mathrm{EXX}} | \hat{v}_{\mathrm{x}}^{\mathrm{NL}} | \varphi_i^{\mathrm{EXX}} \rangle \tag{8}$$

holds for the matrix elements of the KS exchange potential v_{x}. The latter identity, however, would imply that the exchange-only KS and the HF determinants would be identical. To show this we add the operator

$$\hat{\Delta}_{\mathrm{x}} = \hat{v}_{\mathrm{x}}^{\mathrm{NL}} - \hat{v}_{\mathrm{x}} \tag{9}$$

to the exchange-only KS Hamiltonian operator

$$\hat{H}^{\text{EXX}} = \hat{T} + \hat{v}_{\text{ext}} + \hat{v}_{\text{H}} + \hat{v}_{\text{x}} \tag{10}$$

to obtain the Hamiltonian operator

$$\hat{H}^{\text{HF}} = \hat{T} + \hat{v}_{\text{ext}} + \hat{v}_{\text{H}} + \hat{v}_{\text{x}}^{\text{NL}} \tag{11}$$

which has the form of the HF Hamiltonian operator, the Fock operator, but is constructed from EXX orbitals. In the definitions (10) and (11) the operators $\hat{T}$, $\hat{v}_{\text{ext}}$, and $\hat{v}_{\text{H}}$ are the operators of the kinetic energy, the operator corresponding to the external potential v_{ext}, usually the electrostatic potential of the nuclei, and the operator corresponding to the Hartree potential $\hat{v}_{\text{H}}$, respectively.

In the basis of the EXX orbitals the Hamiltonian operator (11) is represented by the matrix

$$\mathbf{H}^{\text{HF}} = \begin{pmatrix} \boldsymbol{\varepsilon}^{\text{EXX,occ}} & \mathbf{0} \\ \mathbf{0} & \boldsymbol{\varepsilon}^{\text{EXX,unocc}} \end{pmatrix} + \begin{pmatrix} \Delta\mathbf{V}_{\text{x}}^{\text{occ}} & \mathbf{0} \\ \mathbf{0} & \Delta\mathbf{V}_{\text{x}}^{\text{unocc}} \end{pmatrix} \tag{12}$$

with the diagonal matrices $\varepsilon^{\text{EXX,occ}}$ and $\varepsilon^{\text{EXX,unocc}}$ containing the EXX eigenvalues $\varepsilon_i^{\text{EXX}}$ and $\varepsilon_a^{\text{EXX}}$ of the occupied and unoccupied EXX orbitals, respectively. The matrix elements of the matrices $\Delta\mathbf{V}_{\text{x}}^{\text{occ}}$ and $\Delta\mathbf{V}_{\text{x}}^{\text{unocc}}$ are given by

$$\Delta V_{\text{x},ij}^{\text{occ}} = \langle \varphi_i^{\text{EXX}} | \hat{v}_{\text{x}}^{\text{NL}} - \hat{v}_{\text{x}} | \varphi_j^{\text{EXX}} \rangle \tag{13}$$

$$\Delta V_{\text{x},ab}^{\text{occ}} = \langle \varphi_a^{\text{EXX}} | \hat{v}_{\text{x}}^{\text{NL}} - \hat{v}_{\text{x}} | \varphi_b^{\text{EXX}} \rangle \tag{14}$$

There are no matrix elements coupling occupied and unoccupied KS orbitals, i.e., no matrices $\Delta\mathbf{V}_{\text{x}}^{\text{occ,unocc}}$ or $\Delta\mathbf{V}_{\text{x}}^{\text{unocc,occ}}$, because the elements of those matrices are the ones collected in $\boldsymbol{\Delta}_x$ which were zero if the products of occupied times unoccupied orbitals were linearly independent. This means if we diagonalize the above Hamiltonian operator (12) then the eigenstates will be obtained by unitary transformations within the occupied and within the unoccupied EXX orbitals, respectively, and the Slater determinant formed by those eigenstates resulting from the unitary transformation of occupied EXX orbitals is the same as the one resulting from the occupied EXX orbitals itself. The Hamiltonian operator obtained by adding $\hat{\Delta}_{\text{x}}$ to the exchange-only KS Hamiltonian operator has the form of the Fock operator, however, is constructed from EXX orbitals. Because all terms in this Hamiltonian operator are either independent of the orbitals, $\hat{T}$ and $\hat{v}_{\text{ext}}$, or invariant with respect to unitary transformations of the occupied orbitals among themselves, $\hat{v}_{\text{H}}$ and $\hat{v}_{\text{x}}^{\text{NL}}$, the Hamiltonian operator remains unchanged if we replace the EXX orbitals by those orbitals that diagonalize the Hamiltonian operator (12). This means we have obtained orbitals that are eigenstates of a Hamiltonian operator of the form of the Fock operator and this Fock operator can be constructed from the orbitals. In other words, the orbitals are HF orbitals and the Hamiltonian operator (12) is the Fock operator represented in the basis of the EXX orbitals.

The products of occupied times unoccupied EXX orbitals and generally of KS orbitals are linearly dependent and therefore Eq. (8) does not hold true ex-

Table 1. Deviation from unity of the overlap between the HF determinant and the EXX and PBEx (Perdew, Burke, Ernzerhof exchange density functional) [79] determinants in units of 10^{-3}. The uncontracted triple zeta basis sets from Ref. [54] were used.

System	EXX	PBEx
Ne	0.01	1.94
H_2O	0.12	2.72
NH_3	0.20	2.51
CH_4	0.36	1.83
CO	0.45	4.11
Ethylene	0.69	3.19
Formaldehyde	0.69	4.73

actly and the EXX and HF determinants are different. However, in practice the differences are very small. In Table 1 the deviation from unity of the overlap of EXX and HF determinants is displayed for a number of molecules. The finding that the differences between the EXX and HF determinants are small does not mean that the canonical EXX and HF orbitals and their eigenvalues are almost identical. Even if the EXX and the HF determinant were exactly identical the EXX and HF orbitals and their eigenvalues could strongly differ from each other. In fact, the spectra of the unoccupied EXX and HF orbitals are completely different. The former is physically meaningful whereas the latter has little physical meaning. Indeed there are usually only very few if any bound unoccupied HF orbitals. On the other hand, the similarity of EXX and HF determinants means that neglecting their differences is a good approximation. This approximation leads to the localized Hartree-Fock approximation [37] for the KS exchange potential which proved to be quite accurate and which can alternatively be derived via a common energy denominator approximation [62].

2.2 Numerical stability of basis set OEP methods, Gaussian basis set EXX method

OEP methods for molecules [47–50,54–56,60,82] and solids [44–46] usually employ two basis sets, the orbital basis set to represent the KS orbitals and an auxiliary basis set to represent the KS exchange potential, the KS response functions, and the right hand side of the OEP equation (1). The latter then turns into a matrix equation [47,48] which is readily solved. However, methods that straightforwardly solve the OEP equation in this way and are based on Gaussian basis sets, in contrast to methods based on plane waves, turn out to be numerically instable [47,49–53]. The resulting xc potentials may exhibit unphysical features and oscillations [51–54]. The main source of the numerical problems is that the basic theorem of DFT, the Hohenberg-Kohn theorem, does not hold for finite orbital basis sets [52,53]. This can easily be seen: If the orbital basis set is localised in a certain region of space, usually the region around the molecule, then we can add an auxiliary function to the effective KS potential which is localized elsewhere without changing the electron density because the auxiliary function added to the effective KS potential will not overlap with the orbital basis set and therefore not couple to the latter. Even if the auxiliary basis set is localized in the same region of space as the orbital basis set it may happen that certain linear combinations of auxiliary basis functions couple either too weakly or not at all to the orbital basis set. As a result the KS response matrix will have eigenvalues with an erroneously small magnitude. Ultimately this means that for a given auxiliary basis set the orbital basis set has to be chosen large enough such that the KS response matrix corresponding to the auxiliary basis set is converged with respect to the orbital basis set. If for a given orbital basis set the auxiliary basis set is enlarged this will inevitably lead to numerical instabilities [52–54].

One might think that the problem could be solved by a regularization of the response matrix through a singular value decomposition or, e.g., by a Tikhonov regularization [50,55]. This, however, does not solve the problem. If the KS response matrix contains corrupted matrix elements which are not converged with respect to the orbital basis set, then these matrix elements will also affect the regularization and the resulting regularized KS matrix. Another seemingly self-evident solution of the problem is to use products of occupied times unoccupied KS orbitals as auxiliary basis set to represent the KS exchange potential [59,60]. Because the KS response function is constructed from exactly these products one would think that the coupling between auxilary basis set and response function is perfect and the KS response function is fully converged. However, this is not the case as shown in Ref. [52,53]. The problem is also not solved if the same basis set is used for the orbitals and for the exchange potential, the response function, and the right hand side of the OEP equation [45,46,52,54,88].

The strategy towards a numerically stable basis set OEP method is to choose an auxiliary basis set that is as small as possible but still sufficiently large to represent the exchange potential accurate enough. Then an orbital basis set has to be chosen large enough to obtain a converged response matrix. For periodic systems with plane wave orbital and auxiliary basis sets such a balancing of the basis sets is straightforward. The energy cutoff for the plane waves of the orbital basis set has to be larger by a certain factor (usually a factor of about 1.75) than the cutoff for the plane waves of the auxiliary basis set [45,46]. For Gaussian orbital basis sets such a balancing is not as simple. In Ref. [54,53] it was found that the EXX orbitals corresponding to the continuum part of the KS eigenvalue spectrum, i.e., KS orbitals with positive eigenvalues, give an important contribution to the KS response matrix. In order to represent this part of the KS spectrum it turned out [54,53] that the basis set needs flexibility within the regions close to the nucleus. This means standard Gaussian orbital basis sets have to be decontracted or augmented by tight basis functions, i.e., Gaussian functions with high exponents. At first it seems strange to represent KS orbitals in the continuum with Gaussian basis sets. Indeed due to their boundary conditions individual KS orbitals within the continuum, and in particular their asymptotic behavior, cannot be represented by Gaussian basis functions. However, we only need to represent sums over KS orbitals in the continuum in the region where the occupied orbitals are localized, see Eqs. (3),(4) and (7). This is possible with suitable Gaussian basis sets, that is with Gaussian basis sets containing functions with high exponents [54].

Instead of Gaussian functions itself, electrostatic potentials of Gaussian functions turned out to be a suitable choice for the auxilary basis functions [47,54]. This means the actual auxiliary basis functions $f_k(\mathbf{r})$ are given by

$$f_k(\mathbf{r}) \;=\; \int d\mathbf{r}' \, \frac{g_k(\mathbf{r}')}{|\mathbf{r}-\mathbf{r}'|} \tag{15}$$

with $g_k(\mathbf{r}')$ denoting a Gaussian basis function. The functions $f_k(\mathbf{r})$ are atom-centered like the functions $g_k(\mathbf{r})$ but unlike the latter have a long-range Coulombic tail. The exchange potential is then given by the linear combination

$$v_x(\mathbf{r}) \;=\; \sum_k v_{x,k}\, f_k(\mathbf{r}) \;=\; \sum_k v_{x,k} \int d\mathbf{r}' \, \frac{g_k(\mathbf{r}')}{|\mathbf{r}-\mathbf{r}'|} \tag{16}$$

with coefficients $v_{x,k}$. The above representation of the KS exchange potential can alternatively be interpreted as follows. We define an exchange charge density ρ_x as a density whose electrostatic potential yields the KS exchange potential

$$v_x(\mathbf{r}) \;=\; \int d\mathbf{r}' \, \frac{\rho_x(\mathbf{r}')}{|\mathbf{r}-\mathbf{r}'|} \tag{17}$$

and we then represent the exchange charge density ρ_x by Gaussian basis functions $g_k(\mathbf{r})$

$$\rho_x(\mathbf{r}) \;=\; \sum_k v_{x,k}\, g_k(\mathbf{r})\,. \tag{18}$$

The exchange charge density ρ_x is of interest in itself. We can similarly define a correlation charge density whose electrostatic potential yields the KS correlation potential. The sum of the electron density plus the exchange and correlation charge densities is an effective electron density that takes into account exchange and correlation effects, i.e., the many-body effects.

By calculating the exchange potential via the exchange charge density ρ_x it is straightforward to enforce the correct asymptotic behavior of the KS exchange potential. The KS exchange potential has to approach $-1/r$ at large distances r away from the molecule [70,71], except on nodal surfaces of the highest occupied molecular orbital [38,69]. The $(-1/r)$ asymptotic can easily be enforced via the condition [47,54]

$$\int d\mathbf{r}\, \rho_x(\mathbf{r}) \;=\; \int d\mathbf{r} \sum_k v_{x,k}\, g_k(\mathbf{r}) \;=\; -1 \tag{19}$$

Enforcement of a second condition,

$$\langle \varphi_{\mathrm{HOMO}}^{\mathrm{EXX}} | \hat{v}_x | \varphi_{\mathrm{HOMO}}^{\mathrm{EXX}} \rangle \;=\; \langle \varphi_{\mathrm{HOMO}}^{\mathrm{EXX}} | \hat{v}_x^{\mathrm{NL}} | \varphi_{\mathrm{HOMO}}^{\mathrm{EXX}} \rangle \tag{20}$$

with $\varphi_{\mathrm{HOMO}}^{\mathrm{EXX}}$ denoting the highest occupied molecular orbital leads to a further increase in the numerical stability [54].

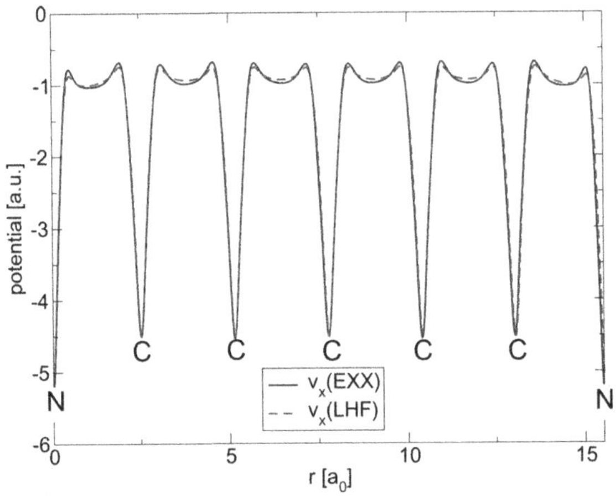

Fig. 1. EXX (solid line) and localized Hartree-Fock (dashed line) exchange potentials for the pyridine molecule along the carbon nitrogen ring calculated with the uncontracted triple zeta orbital and auxiliary basis sets from Ref. [54].

In summary, the approach of Ref. [54] not only turned out to be numerically stable but also to be quite efficient (comparable to HF methods with the same orbital basis set) and easy to implement. Moreover it is a purely analytical approach, no integration grids are needed. As an example Figure 1 displays the KS exchange potential obtained with the OEP method of Ref. [54] for the pyridine molecule. For comparison also the localized Hartree-Fock exchange potential [37] is shown in Figure 1.

3. TDDFT and TDEXX methods

The calculation of excitation energies of molecules within TDDFT or more precisely with density-functional response methods in most cases is carried out on the basis of the eigenvalue equation [72–75]

$$\left[\boldsymbol{\varepsilon}^2 + 4\,\boldsymbol{\varepsilon}^{1/2}\mathbf{K}(\omega)\boldsymbol{\varepsilon}^{1/2}\right]\mathbf{z}_n(\omega) \;=\; \omega_n^2\,\mathbf{z}_n(\omega)\,. \tag{21}$$

The dimension of Eq. (21) is given by the product of the number of occupied times the number of unoccupied KS orbitals. In Eq. (21) the eigenvalue ω_n^2 equals the square of the excitation energy of the n-th excited state. The eigenvectors $\mathbf{z}_n(\omega)$ determine the density-density (potential-density) response function and are commonly used to analyse excitations in terms of the involved occupied and unoccupied orbitals. The elements $K_{ia,jb}(\omega)$ of the matrix $\mathbf{K}(\omega)$ are given by

$$K_{ia,jb}(\omega) \;=\; \int d\mathbf{r}d\mathbf{r}'\;\varphi_i(\mathbf{r})\varphi_a(\mathbf{r})f_{\text{uxc}}(\mathbf{r},\mathbf{r}',\omega)\,\varphi_j(\mathbf{r}')\varphi_b(\mathbf{r}') \tag{22}$$

and the elements of the diagonal matrix ε by $\varepsilon_{ia,jb} = \delta_{ia,jb}\ (\varepsilon_a - \varepsilon_i)$. The sum of the Coulomb kernel $1/|\mathbf{r} - \mathbf{r}'|$ and the xc kernel f_{xc}, the frequency-dependent functional derivative of the KS xc potential with respect to the electron density, is denoted by f_{uxc}. In standard TDDFT methods the adiabatic approximation is made, i.e., the frequency-dependence of the xc kernel is neglected. The kernel then turns into the second derivative of the xc energy functional with respect to the electron density which can easily be evaluated for LDA or GGA functionals.

A crucial drawback of TDDFT methods within the adiabatic LDA or GGA approximation is that charge transfer (CT) excitations cannot be described correctly [10,11,13–15]. CT excitations are characterized in a one-electron picture by an excitation from an occupied orbital φ_i localized in one region of space into an unoccupied orbital φ_a localized in another region of space. If the overlap of the involved orbitals φ_i and φ_a approaches zero then, for a finite xc kernel, the matrix elements $K_{kl,jb}(\omega)$ with $kl = ia$ and/or $jb = ia$ equal zero. This leads to an eigenvector $\mathbf{z}_n(\omega)$ that is a unit vector with the entry of 1 at the component ia and an accompanying eigenfrequency ω equal to $\varepsilon_a - \varepsilon_i$. If the two involved orbitals are the highest occupied molecular orbital (HOMO) and the lowest unoccupied molecular orbital (LUMO) then the CT excitation energy should approach I–A–$1/R$ for increasing distance R of the fragments and for I and A denoting the ionization energy and the electron affinity, respectively. While the negative of the KS eigenvalue ε_{HOMO} in the KS formalism equals the ionization energy the negative of the eigenvalue ε_{LUMO} does not equal the electron affinity. For conventional LDA or GGA functionals even $-\varepsilon_{HOMO}$ is not a very good approximation for the ionization energy I. Moreover the $(-1/R)$ distance behavior is not found in conventional TDDFT methods.

In order to obtain CT excitation energies with density-functional response methods that differ from the HOMO-LUMO eigenvalue difference the xc kernel has to approach infinity in such a way that the vanishing overlap of HOMO and LUMO is compensated and the matrix elements of the xc kernel remain finite. One might assume that this generally requires frequency-dependent kernels. However, we could show [76] that in the special case of CT excitations in the system He H^+ the qualitatively correct CT excitation energy is obtained without any frequency-dependent kernel. In fact in this simple two electron case the exact-exchange kernel equals half the Coulomb kernel and is frequency-independent. In the He H^+ case the correct behavior of the CT energy with the distance R is obtained via the distance behavior of the difference $\varepsilon_{LUMO} - \varepsilon_{HOMO}$ of the involved orbital eigenvalues. This demonstrates that not only the xc kernel but also the eigenvalues may be of importance in order to correctly describe CT excitation energies.

For systems with more than two electrons the xc kernel plays a crucial role for describing CT excitations. A first step to go beyond conventional adiabatic LDA and GGA kernels in density-functional response methods is to use orbital-dependent xc kernels, in particular, the exact-exchange kernel $f_x(\mathbf{r},\mathbf{r}',\omega)$, which is frequency-dependent. The exchange kernel obeys the equation [63–65]

$$f_x(\mathbf{r},\mathbf{r}',\omega) = \int d\mathbf{r}''d\mathbf{r}'''\chi_s^{-1}(\mathbf{r},\mathbf{r}'',\omega)h_x(\mathbf{r}'',\mathbf{r}''',\omega)\chi_s^{-1}(\mathbf{r}''',\mathbf{r}',\omega)\,. \tag{23}$$

In Eq. (23) h_x is an orbital-dependent function given in the Appendix, see also Refs. [63–65] and χ_s now is the frequency-dependent KS response function. Eq. (23) is numerically even more demanding than the OEP equation (2) because it contains not only one but two inverse KS response functions. Moreover the involved KS response function is not the static one but the frequency-dependent one. Indeed even first attempts to employ the simpler adiabatic exact-exchange kernel in density-functional response methods suffered from numerical instabilities [66,67].

In order to avoid the occurence of inverse KS response functions in density-functional response methods we reformulated the underlying basic equations [68]. The basic equation for the first order reponse $\rho^{(1)}(\mathbf{r}',\omega)$ of the electron density on a frequency-dependent external perturbation $v^{(1)}(\mathbf{r}',\omega)$ is [8]

$$\begin{aligned}\int d\mathbf{r}' \left[\delta(\mathbf{r}-\mathbf{r}') \;- \int d\mathbf{r}''\,\chi_s(\mathbf{r},\mathbf{r}'',\omega)\,f_{uxc}(\mathbf{r}'',\mathbf{r}',\omega)\right]\rho^{(1)}(\mathbf{r}',\omega)\\ = \int d\mathbf{r}'\,\chi_s(\mathbf{r},\mathbf{r}',\omega)\,v^{(1)}(\mathbf{r}',\omega)\end{aligned} \tag{24}$$

Eq. (24) can be derived [73] directly from the Runge-Gross theorem [78] without refering to the formally problematic [77] action formalism as in the original derivation. The response of the electron density in TDDFT can be expressed by

$$\rho^{(1)}(\mathbf{r},\omega) \;=\; \int d\mathbf{r}'\,\chi_s(\mathbf{r},\mathbf{r}',\omega)\,v_s^{(1)}(\mathbf{r}',\omega) \tag{25}$$

with $v_s^{(1)}$ denoting the linear response of the effective KS potential that corresponds according to the Runge-Gross theorem [78] to the perturbation $v^{(1)}$. If we neglect the correlation contribution to f_{uxc}, separate the Coulomb and exchange contributions, and insert Eqs. (23) and (25) into the basic equation (24) we obtain [68]

$$\begin{aligned}\int d\mathbf{r}'\Bigg[\chi_s(\mathbf{r},\mathbf{r}',\omega) - \int d\mathbf{r}''d\mathbf{r}'''\frac{\chi_s(\mathbf{r},\mathbf{r}''',\omega)\chi_s(\mathbf{r}'',\mathbf{r}',\omega)}{|\mathbf{r}'''-\mathbf{r}''|}\\ -h_x(\mathbf{r},\mathbf{r}',\omega)\Bigg]v_s^{(1)}(\mathbf{r}',\omega) = \int d\mathbf{r}'\chi_s(\mathbf{r},\mathbf{r}',\omega)v^{(1)}(\mathbf{r}',\omega)\,,\end{aligned} \tag{26}$$

Eq. (26) no longer contains the kernel f_x but the function h_x, which is known [63–65], see Appendix, and in contrast to f_x can be evaluated without solving an integral equation, like Eq. (23), that contains inverse response functions. Eq. (26) is an equation for the response $v_s^{(1)}$ of the effective KS potential instead of the electron density. On the basis of Eq. (26) a TDEXX approach, i.e., a density-functional response method employing the exact frequency-dependent exchange kernel could be implemented [68].

Within the TDEXX approach qualitatively correct CT excitation energies were obtained. In this way it was demonstrated that CT excitation energies can

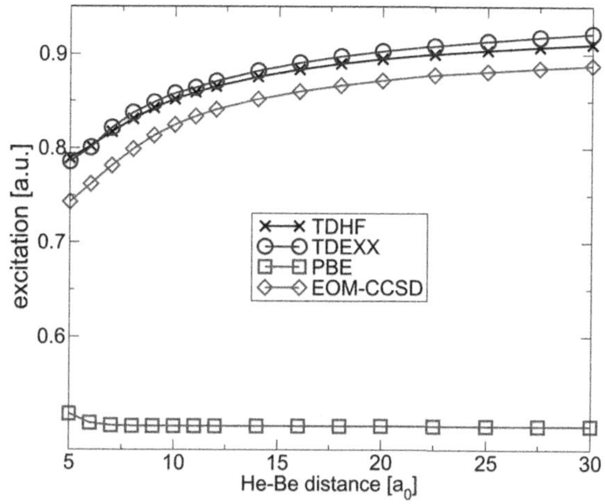

Fig. 2. He(1s) → Be($2p_{x,y}$) charge-transfer excitation energies calculated using time-dependent Hartree-Fock (TDHF), time-dependent exact exchange (TDEXX), time-dependent DFT with Perdew-Burke-Ernzerhof xc potential and kernel [79] (PBE), and equation-of-motion coupled cluster singles and doubles (EOM-CCSD). Calculations were done using the aug-cc-pVQZ basis sets from Dunning *et al.* [80,81].

be treated within TDDFT and that ODFs are of interest in TDDFT. As an example Figure 2 displays the distance behaviour of the excitation energy for the CT transition characterized in a one-electron picture as excitation from the He(1s) to the Be($2p_{x,y}$) orbital in the He Be dimer.

When reconsidering the basic equations of TDDFT and their derivation in order to develop a TDEXX approach we made the unexpected observation [75] that there exist simple valence-valence single-electron excitations in atoms or diatomics that principally cannot be treated in density-functional response methods because these excitations are not accompanied by a first order response of the electron density. For these excitations the usually employed equation (21) does not represent a proper density-functional response equation and yields as excitation energy a difference of KS eigenvalues [75].

4. Summary and outlook

With the development of a numerically stable and efficient Gaussian basis set OEP method [54] it is now possible to perform self-consistent KS calculations with orbital-dependent xc functionals for molecules. So far the exact treatment of exchange was implemented. For the future the development of correlation functionals that are well suited for a combination with an exact treatment of exchange is desirable. Present xc functionals benefit from an error cancellation between their exchange and correlation energy, which no longer is present if the exchange energy is treated exactly. However, note that this error cancellation does not occur between present exchange and correlation potentials, both of which in almost all cases are short-range. The correct asymptotic behavior of the effective KS potential is only obtained with an exchange-correlation potential that exhibits the asymptotic behavior of the exact-exchange potential.

With the response equation for the effective KS potential (26) ODFs can also be employed in TDDFT methods. The finding that with a TDEXX approach charge-transfer excitation energies can be treated qualitatively correctly in contrast to conventional LDA or GGA TDDFT methods shows that ODFs are of interest in TDDFT. The TDEXX method, in contrast to the TDHF method, is a nonlinear method due to the frequency-dependence of the exact-exchange kernel. In how far this leads to additional excitations beyond the number of occupied times unoccupied KS orbitals, in particular whether this enables a description of two-electron excitations, is an interesting question to be investigated in the future. Moreover, also in the case of TDDFT the development of orbital-dependent correlation kernels accompanying the exact-exchange kernel is desirable.

The authors gratefully acknowledge the funding from the priority program 1145 'Modern and universal first-principles methods for many-electron systems in chemistry and physics' of the Deutsche Forschungsgemeinschaft (DFG) and funding of the DFG, which, within the framework of its 'Excellence Initiative', supports the Cluster of Excellence 'Engineering of Advanced Materials' (www.eam.uni-erlangen.de) at the University of Erlangen-Nuremberg.

Appendix

In this appendix, in contrast to the main text, we consider the somewhat more general case that the molecular orbitals may be complex. The function $h_x(\mathbf{r},\mathbf{r}',\omega)$ defining the exchange kernel of equation (23) can be decomposed into four terms $h_x^p(\mathbf{r},\mathbf{r}',\omega)$ with $p = 1,2,3,4$:

$$h_x(\mathbf{r},\mathbf{r}',\omega) = \sum_{p=1,4} h_x^p(\mathbf{r},\mathbf{r}',\omega)\,. \tag{27}$$

The four terms are given by:

$$\begin{aligned} h_x^1(\mathbf{r},\mathbf{r}',\omega) = -2\sum_{ij}^{\text{occ}}\sum_{ab}^{\text{unocc}}\Bigg[&\frac{\varphi_i^*(\mathbf{r})\varphi_a(\mathbf{r})\langle aj|bi\rangle\varphi_b^*(\mathbf{r}')\varphi_j(\mathbf{r}')}{(\varepsilon_i-\varepsilon_a+\omega+\mathrm{i}\eta)(\varepsilon_j-\varepsilon_b+\omega+\mathrm{i}\eta)} \\ &+\frac{\varphi_a^*(\mathbf{r})\varphi_i(\mathbf{r})\langle ib|ja\rangle\varphi_j^*(\mathbf{r}')\varphi_b(\mathbf{r}')}{(\varepsilon_i-\varepsilon_a-\omega-\mathrm{i}\eta)(\varepsilon_j-\varepsilon_b-\omega-\mathrm{i}\eta)}\Bigg], \end{aligned} \tag{28}$$

$$\begin{aligned} h_x^2(\mathbf{r},\mathbf{r}',\omega) = -2\sum_{ij}^{\text{occ}}\sum_{ab}^{\text{unocc}}\Bigg[&\frac{\varphi_i^*(\mathbf{r})\varphi_a(\mathbf{r})\langle ab|ji\rangle\varphi_j^*(\mathbf{r}')\varphi_b(\mathbf{r})}{(\varepsilon_i-\varepsilon_a+\omega+\mathrm{i}\eta)(\varepsilon_j-\varepsilon_b-\omega-\mathrm{i}\eta)} \\ &+\frac{\varphi_a^*(\mathbf{r})\varphi_i(\mathbf{r})\langle ij|ba\rangle\varphi_b^*(\mathbf{r}')\varphi_j(\mathbf{r})}{(\varepsilon_i-\varepsilon_a-\omega-\mathrm{i}\eta)(\varepsilon_j-\varepsilon_b+\omega+\mathrm{i}\eta)}\Bigg], \end{aligned} \tag{29}$$

$$
\begin{aligned}
h_{\mathrm{x}}^{3}(\mathbf{r},\mathbf{r}',\omega) = -2\sum_{ij}^{\text{occ}}\sum_{a}^{\text{unocc}}\Bigg[&\frac{\varphi_i^*(\mathbf{r})\varphi_a(\mathbf{r})\langle j|\hat{v}_{\mathrm{x}}^{\mathrm{NL}}-\hat{v}_{\mathrm{x}}|i\rangle\varphi_a^*(\mathbf{r}')\varphi_j(\mathbf{r})}{(\varepsilon_i-\varepsilon_a+\omega+\mathrm{i}\eta)(\varepsilon_j-\varepsilon_a+\omega+\mathrm{i}\eta)}\\
&+\frac{\varphi_a^*(\mathbf{r})\varphi_i(\mathbf{r})\langle i|\hat{v}_{\mathrm{x}}^{\mathrm{NL}}-\hat{v}_{\mathrm{x}}|j\rangle\varphi_j^*(\mathbf{r}')\varphi_a(\mathbf{r})}{(\varepsilon_i-\varepsilon_a-\omega-\mathrm{i}\eta)(\varepsilon_j-\varepsilon_a-\omega-\mathrm{i}\eta)}\Bigg]\\
+2\sum_{i}^{\text{occ}}\sum_{ab}^{\text{unocc}}\Bigg[&\frac{\varphi_i^*(\mathbf{r})\varphi_a(\mathbf{r})\langle a|\hat{v}_{\mathrm{x}}^{\mathrm{NL}}-\hat{v}_{\mathrm{x}}|b\rangle\varphi_b^*(\mathbf{r}')\varphi_i(\mathbf{r})}{(\varepsilon_i-\varepsilon_a+\omega+\mathrm{i}\eta)(\varepsilon_i-\varepsilon_b+\omega+\mathrm{i}\eta)}\\
&+\frac{\varphi_a^*(\mathbf{r})\varphi_i(\mathbf{r})\langle b|\hat{v}_{\mathrm{x}}^{\mathrm{NL}}-\hat{v}_{\mathrm{x}}|a\rangle\varphi_i^*(\mathbf{r}')\varphi_b(\mathbf{r})}{(\varepsilon_i-\varepsilon_a-\omega-\mathrm{i}\eta)(\varepsilon_i-\varepsilon_b-\omega-\mathrm{i}\eta)}\Bigg],
\end{aligned}
\tag{30}
$$

and

$$
\begin{aligned}
h_{\mathrm{x}}^{4}(\mathbf{r},\mathbf{r}',\omega) = -2\sum_{ij}^{\text{occ}}\sum_{a}^{\text{unocc}}\Bigg[&\frac{\varphi_j^*(\mathbf{r})\varphi_a(\mathbf{r})\langle a|\hat{v}_{\mathrm{x}}^{\mathrm{NL}}-\hat{v}_{\mathrm{x}}|i\rangle\varphi_i^*(\mathbf{r}')\varphi_j(\mathbf{r})}{(\varepsilon_j-\varepsilon_a+\omega+\mathrm{i}\eta)(\varepsilon_i-\varepsilon_a)}\\
&+\frac{\varphi_j^*(\mathbf{r})\varphi_i(\mathbf{r})\langle i|\hat{v}_{\mathrm{x}}^{\mathrm{NL}}-\hat{v}_{\mathrm{x}}|a\rangle\varphi_a^*(\mathbf{r}')\varphi_j(\mathbf{r})}{(\varepsilon_i-\varepsilon_a)(\varepsilon_j-\varepsilon_a+\omega+\mathrm{i}\eta)}\\
&+\frac{\varphi_a^*(\mathbf{r})\varphi_j(\mathbf{r})\langle i|\hat{v}_{\mathrm{x}}^{\mathrm{NL}}-\hat{v}_{\mathrm{x}}|a\rangle\varphi_j^*(\mathbf{r}')\varphi_i(\mathbf{r})}{(\varepsilon_j-\varepsilon_a-\omega-\mathrm{i}\eta)(\varepsilon_i-\varepsilon_a)}\\
&+\frac{\varphi_i^*(\mathbf{r})\varphi_j(\mathbf{r})\langle a|\hat{v}_{\mathrm{x}}^{\mathrm{NL}}-\hat{v}_{\mathrm{x}}|i\rangle\varphi_j^*(\mathbf{r}')\varphi_a(\mathbf{r})}{(\varepsilon_i-\varepsilon_a)\,(\varepsilon_j-\varepsilon_a-\omega-\mathrm{i}\eta)}\Bigg]\\
+2\sum_{i}^{\text{occ}}\sum_{ab}^{\text{unocc}}\Bigg[&\frac{\varphi_i^*(\mathbf{r})\varphi_b(\mathbf{r})\langle a|\hat{v}_{\mathrm{x}}^{\mathrm{NL}}-\hat{v}_{\mathrm{x}}|i\rangle\varphi_b^*(\mathbf{r}')\varphi_a(\mathbf{r})}{(\varepsilon_i-\varepsilon_b+\omega+\mathrm{i}\eta)(\varepsilon_i-\varepsilon_a)}\\
&+\frac{\varphi_a^*(\mathbf{r})\varphi_b(\mathbf{r})\langle i|\hat{v}_{\mathrm{x}}^{\mathrm{NL}}-\hat{v}_{\mathrm{x}}|a\rangle\varphi_b^*(\mathbf{r}')\varphi_i(\mathbf{r})}{(\varepsilon_i-\varepsilon_a)(\varepsilon_i-\varepsilon_b+\omega+\mathrm{i}\eta)}\\
&+\frac{\varphi_b^*(\mathbf{r})\varphi_i(\mathbf{r})\langle i|\hat{v}_{\mathrm{x}}^{\mathrm{NL}}-\hat{v}_{\mathrm{x}}|a\rangle\varphi_a^*(\mathbf{r}')\varphi_b(\mathbf{r})}{(\varepsilon_i-\varepsilon_b-\omega-\mathrm{i}\eta)(\varepsilon_i-\varepsilon_a)}\\
&+\frac{\varphi_b^*(\mathbf{r})\varphi_a(\mathbf{r})\langle a|\hat{v}_{\mathrm{x}}^{\mathrm{NL}}-\hat{v}_{\mathrm{x}}|i\rangle\varphi_i^*(\mathbf{r}')\varphi_b(\mathbf{r})}{(\varepsilon_i-\varepsilon_a)(\varepsilon_i-\varepsilon_b-\omega-\mathrm{i}\eta)}\Bigg].
\end{aligned}
\tag{31}
$$

Matrix elements of the type $\langle aj|bi\rangle$ are defined by

$$
\langle aj|bi\rangle = \int \mathrm{d}^3r \int \mathrm{d}^3r' \; \varphi_a^*(\mathbf{r})\varphi_j^*(\mathbf{r}')\varphi_b(\mathbf{r})\varphi_i(\mathbf{r}')/|\mathbf{r}-\mathbf{r}'|\,. \tag{32}
$$

References

1. R. G. Parr and W. Yang, Density-Functional Theory of Atoms and Molecules. Oxford University Press, Oxford (1989).
2. R. M. Dreizler and E. K. U. Gross, Density Functional Theory. Springer, Heidelberg (1990).

3. W. Koch and M. C. Holthausen, A Chemist's Guide to Density Functional Theory. Wiley-VCH, New York (2000).
4. B. G. Johnson, P. M. W. Gill, and J. A. Pople, J. Chem. Phys. **98** (1993) 5612.
5. K. Burke, J. P. Perdew, and Y. Wang, Electronic density functional theory: Recent progress and new directions. Plenum Press, New York (1998).
6. C. Fiolhais, F. Nogueira, and M. A. L. Marques, editors, A Primer in Density Functional Theory. Volume 620 of Lecture Notes in Physics. Springer (2003).
7. P. Geerlings, F. De Proft, and W. Langenaeker, Chem. Rev. **103** (2003) 1793.
8. M. A. L. Marques et al. (Ed.), Time-Dependent Density Functional Theory. Volume 706 of Lecture Notes in Physics. Springer, Heidelberg, (2006).
9. P. Elliot, F. Furche, and K. Burke, Reviews in Computational Chemistry **26** (2009) 91.
10. A. Dreuw, J. L. Weisman, and M. Head-Gordon, J. Chem. Phys. **119** (2003) 2943.
11. D. Tozer, J. Chem. Phys. **119** (2003) 12697.
12. A. Dreuw, and M. Head-Gordon, Chem. Rev. **105** (2005) 4009.
13. W. Hieringer and A. Görling, Chem. Phys. Lett. **419** (2006) 557; Chem. Phys. Lett. **426** (2006) 234.
14. A. Dreuw and M. Head-Gordon, Chem. Phys. Lett. **426** (2006) 231.
15. M. J. G. Peach, P. Benfield, T. Helgaker, and D. J. Tozer, J. Chem. Phys. **128** (2008) 044118.
16. S. K. Ghosh and R. G. Parr, Phys. Rev. A **34** (1986) 785.
17. M. D. Glossman, L. C. Balbás, A. Rubio, and J. A. Alonso, Int. J. Quant. Chem. **49** (1994) 171.
18. R. Neumann, R. H. Nobes and N. C. Handy, Mol. Phys. **87** (1996) 1.
19. T. Van Voorhis and G. E. Scuseria, J. Chem. Phys. **109** (1998) 400.
20. R. A. King and N. C. Handy, Mol. Phys. **99** (2001) 1005.
21. A. D. Boese and N. C. Handy, J. Chem. Phys. **116** (2002) 9559.
22. A. V. Arbuznikov and M. Kaupp, Chem. Phys. Lett. **381** (2003) 495.
23. J. Tao, J. P. Perdew, V. N. Staroverov, and E. Scuseria, Phys. Rev. Lett. **91** (2003) 146401.
24. A. D. Becke, J. Chem. Phys. **98** (1993) 5648.
25. A. D. Becke, J. Chem. Phys. **98** (1993) 1372.
26. C. Adamo and V. Barone, J. Chem. Phys. **110** (1999) 6158.
27. P. J. Wilson, T. J. Bradley, and D. J. Tozer, J. Chem. Phys. **115** (2001) 9233.
28. J. Jaramillo, G. E. Scuseria, and M. Ernzerhof, J. Chem. Phys. **118** (2003) 1068.
29. J. P. Perdew, V. N. Staroverov, J. Tao, and G. E. Scuseria, Phys. Rev. A **78** (2008) 052513.
30. A. D. Becke, J. Chem. Phys. **119** (2003) 2972.
31. A. D. Becke, J. Chem. Phys. **122** (2005) 064101.
32. P. Mori-Sanchez, A. J. Cohen, and W. Yang, J. Chem. Phys. **124** (2006) 091102.
33. M. Kaupp, H. Bahmann, and A. V. Arbuznikov, J. Chem. Phys. **127** (2007) 194102.
34. B. G. Janesko, A. V. Krukau, and G. E. Scuseria, J. Chem. Phys. **129** (2008) 124110.
35. A. V. Arbuznikov and M. Kaupp, J. Chem. Phys. **128** (2008) 214107.
36. R. Neumann and N. C. Handy, Chem. Phys. Lett. **252** (1996) 19.
37. F. Della Sala and A. Görling, J. Chem. Phys. **115** (2001) 5718.
38. F. Della Sala and A. Görling, J. Chem. Phys. **116** (2002) 5374.
39. M. E. Casida, C. Jamorski, K. C. Casida, and D. R. Salahub, J. Chem. Phys. **108** (1998) 4439.
40. R. T. Sharp and G. K. Horton, Phys. Rev. A **14** (1976) 36.
41. J. D. Talman and W. F. Shadwick, Phys. Rev. A **14** (1976) 36.
42. A. Görling and M. Levy, Phys. Rev. A **50** (1994) 196.
43. A. Görling and M. Levy, Int. J. Quantum Chem. Symp. **29** (1995) 93.
44. A. Görling, Phys. Rev. B 53 (1996) 7024; Phys. Rev. B **59** (1999) 10370(E).

45. M. Städele, J. A. Majewski, P. Vogl, and A. Görling, Phys. Rev. Lett. **79** (1997) 2089.
46. M. Städele, M. Moukara, J. A. Majewski, P. Vogl, and A. Görling, Phys. Rev. B **59** (1999) 10031.
47. A. Görling, Phys. Rev. Lett. **83** (1999) 5459.
48. S. Ivanov and S. Hirata, and R. J. Bartlett, Phys. Rev. Lett. **83** (1999) 5455.
49. S. Hamel, M. E. Casida, and D. R. Salahub, J. Chem. Phys. **116** (2002) 8276.
50. S. Hirata, S. Ivanov, I. Grabowski, R. Bartlett, K. Burke, and J. D. Talman, J. Chem. Phys. **115** (2001) 1635.
51. V. K. Staroverov, G. E. Scuseria, and E. R. Davidson, J. Chem. Phys. **124** (2006) 141103.
52. A. Görling, A. Heßelmann, M. Jones, and M. Levy, J. Chem. Phys. **128** (2008) 104104.
53. A. Heßelmann and A. Görling, Chem. Phys. Lett. **455** (2008) 110.
54. A. Heßelmann, A. W. Götz, F. Della Sala, and A. Görling, J. Chem. Phys. **127** (2007) 054102.
55. Q. Wu and W. Yang, J. Theo. Comp. Chem. **2** (2003) 627.
56. T. Heaton-Burgess, F. A. Bulat, and W. Yang, Phys. Rev. Lett. **98** (2007) 256401.
57. D. R. Rohr, O. V. Gritsenko, and E. J. Baerends, J. Mol. Struct. (THEOCHEM) **762** (2006) 193.
58. S. Kümmel and J. P. Perdew, Phys. Rev. Lett. **90** (2003) 043004.
59. C. Kollmar and M. Filatov, J. Chem. Phys. **127** (2007) 114104.
60. C. Kollmar and M. Filatov, J. Chem. Phys. **128** (2008) 064101.
61. J. B. Krieger, Y. Li, and G. J. Iafrate, Phys. Rev. A **46** (1992) 5453.
62. O. V. Gritsenko and E. J. Baerends, Phys. Rev. A **64** (2001) 042506.
63. A. Görling, Int. J. Quantum Chem. **69** (1998) 265.
64. A. Görling, Phys. Rev. A **57** (1998) 3433.
65. Y.-H. Kim and A. Görling, Phys. Rev. B **66** (2002) 035144.
66. S. Hirata, S. Ivanov, I. Grabowski, and R. J. Bartlett, J. Chem. Phys. **116** (2002) 6468.
67. Y. Shigeta, K. Hirao, and S. Hirata, Phys. Rev. A **73** (2006) 010502(R).
68. A. Heßelmann, A. Ipatov, and A. Görling, Phys. Rev. A **80** (2009) 012507.
69. F. Della Sala and A. Görling, Phys. Rev. Lett. **89** (2002) 033003.
70. M. Levy, J. P. Perdew, and V. Sahni, Phys. Rev. A **31** (1984) 3231.
71. C.-O. Almbladh and U. von Barth, Phys. Rev. B **31** (1985) 3231.
72. M. E. Casida, Recent Advances in Density Functional Methods. D. P. Chong (Ed.), volume I, 155, World Scientific, Singapore (1995).
73. A. Görling, H. H. Heinze, S. Ph. Ruzankin, M. Staufer, and N. Rösch, J. Chem. Phys. **110** (1999) 2785.
74. F. Furche, J. Chem. Phys. **114** (2001) 5982.
75. A. Heßelmann and A. Görling, Phys. Rev. Lett. **102** (2009) 233003.
76. T. Gimon, A. Ipatov, A. Heßelmann, and A. Görling, J. Chem. Theory Comput. **5** (2009) 781.
77. R. van Leeuwen, Phys. Rev. Lett. **80** (1998) 1280.
78. E. Runge and E.K.U. Gross, Phys. Rev. Lett. **52** (1984) 997.
79. J. P. Perdew, K. Burke, and M. Ernzerhof, Phys. Rev. Lett. **77** (1996) 3865.
80. T. H. Dunning, J. Chem. Phys. **90** (1989) 1007.
81. Kendall, T.H. Dunning, and Harrison, J. Chem. Phys. **96** (1992) 6769.
82. A. Görling, J. Chem. Phys. **123** (2005) 062203.
83. F. Della Sala, A. Görling, Int. J. Quant. Chem. **91** (2003) 131.
84. V. N. Staroverov, G. E. Scuseria, E. R. Davidson, J. Chem. Phys. **125** (2006) 081104.
85. A. F. Izmaylov, V. N. Staroverov, G. E. Scuseria, E. R. Davidson, G. Stoltz, E. Cances, J. Chem. Phys. **126** (2007) 084107.

86. A. F. Izmaylov, V. N. Staroverov, G. E. Scuseria, E. R. Davidson, J. Chem. Phys. **127** (2007) 084113.
87. R. Colle, R. K. Nesbet, J. Phys. B **34** (2001) 2475.
88. S. Ivanov, S. Hirata, and R. J. Bartlett, J. Chem. Phys. **116** (2002) 1269.

Electron Structure Quantum Monte Carlo

By Arne Lüchow*, René Petz, and Annett Schwarz

Institut für Physikalische Chemie, RWTH Aachen, Landoltweg 2, 52056 Aachen, Germany

(Received July 30, 2009; accepted September 17, 2009)

Quantum Monte Carlo / QMC / DMC

The diffusion quantum Monte Carlo method (DMC) is capable of calculating accurately the electronic energy for molecules and molecular aggregates. An overview is given on recent developments for the optimization of the guide function that determines the accuracy of the method. Furthermore, the versatility of DMC is shown with applications to Rydberg states, transition metal compounds, and weakly interacting systems.

1. Introduction

The term quantum Monte Carlo is used in different contexts where quantum processes are described with Monte Carlo methods. In this paper we focus on how the electronic many-body problem can be solved numerically using random walk processes in the diffusion Monte Carlo method (DMC). In DMC, all electrons together constitute a random walker that is guided by a stochastic process through position space, and its probability density describes the electronic wave function for all electrons simultaneously. This stochastic process is equivalent to the projection of the exact ground state wave function ψ_0 from an almost arbitrary initial function ψ according to

$$e^{-Ht}\psi \longrightarrow \psi_0, \quad \text{for} \quad t \to \infty \tag{1}$$

such that the stationary solution of the stochastic process represents the exact ground state wave function. Therefore, the correlated movement of the electrons that leads to the electronic correlation energy is built into the DMC method from the very beginning. This is in contrast to most other electronic structure methods that start with a mean field approach and then incorporate the correlation energy either implicitly with a correlation functional as in DFT or explicitly with a perturbative or coupled cluster approach. Because of the Monte Carlo nature of

* Corresponding author. E-mail: luechow@rwth-aachen.de

© by Oldenbourg Wissenschaftsverlag, München

the process a favorable scaling of the DMC method with system size is achieved if the so-called fixed node approximation is employed. This leads to an error denoted node location error which is very small in absolute terms, but significant in terms of chemical accuracy when calculating energy differences such as reaction or activation energies. In this paper we focus on progress in the last few years about how to minimize the node location error.

After an introduction to the DMC method we discuss several recent methods to reduce the node location error in chemical applications and present many recent results for difficult problems in quantum chemistry such as excited states, transition metal compounds, and weakly interacting molecules.

2. The DMC method

The DMC method has been presented in several review articles [1,2] and a monograph [3], so here only an overview is given. In the following we use x for the collection of the cartesian coordinates of n electrons: $x \in \mathbb{R}^{3n}$, Δ denotes the $3n$-dimensional Laplacian $\Delta = \Sigma_{i=1}^{3n}[\partial^2/\partial x_i^2]$, and $V = V(x)$ the potential energy. $\psi_k(x)$ and E_k are the eigenfunctions and eigenvalues, respectively, of the electronic Hamiltonian $H = -[1/2]\Delta + V$.

The so-called importance sampling transformation is the basis of numerically stable and efficient diffusion Monte Carlo algorithms[4,5]. Let $\phi(x)$ be an approximation to the ground state wave function $\psi_0(x)$ that has been obtained from other methods. The operator $\tilde{H} = \phi(x)H\phi(x)^{-1} - E_{ref}$ is non-Hermitian and has the eigenfunctions $f_k(x) = \psi_k(x)\phi(x)$ and the shifted eigenvalues $\tilde{E}_k = E_k - E_{ref}$. The partial differential equation

$$\frac{\partial f}{\partial t} = -\tilde{H} f \tag{2}$$

has the long-time stationary solution

$$f_\infty(x) = \lim_{t\to\infty} f(t,x) = \phi(x)\psi_0(0) \tag{3}$$

containing the ground state solution $\psi_0(x)$. After evaluation of $\phi\Delta\phi^{-1}$ one obtains

$$\frac{\partial f}{\partial t} = -Lf - (E_L(x) - E_{ref})f \tag{4}$$

with the Fokker-Planck operator

$$L = -\frac{1}{2}\Delta + \nabla b(x), \quad b(x) = \frac{\nabla\phi}{\phi} \tag{5}$$

and the local energy $E_L(x) = \phi(x)^{-1}H\phi(x)$.

It is well known that partial differential equations of the Fokker-Planck type

$$\frac{\partial f}{\partial t} = -Lf \tag{6}$$

are closely related to stochastic differential equations which can be simulated in time-discretized form with a drift-diffusion process

$$X_{k+1} = X_k + b(X_k)\tau + \Delta W_\tau \tag{7}$$

where X_k, $k = 0,1,2,\ldots$ is a family of random variables. The time t is discretized into N steps of time step $\tau = t/N$. ΔW_τ is a normal variable with $\mu = 0$ and $\sigma^2 = \tau$. The drift term $b(x) = \nabla\phi/\phi$ causes the random walk to move away from small values of ϕ to large ones, therefore sampling the wave function where it is important. Since ϕ "guides" the random walk toward the important regions it is usually called guide function.

The drift-diffusion process simulates the Fokker-Planck equation because the probability density $p_{X_N}(x)$ converges toward $f(t,x)$ with $\tau \to 0$ if the initial density is the initial function: $p_{X_0}(x) = f(0,x)$.

The importance sampled Schrödinger equation Eq. (2) or (4) is simulated with the drift-diffusion process Eq. (7) with additional weights that are accumulated along the process

$$W_{k+1} = W_k \cdot e^{-\left[\frac{1}{2}(E_L(X_k)+E_L(X_{k+1}))-E_{ref}\right]t}, \tag{8}$$

$$W_0 = 1 \tag{9}$$

A sample of the random variables is usually called *walkers* and constructed by realizing the drift-diffusion process Eq. (7) many times. Let $\{x_k^i\}_{i=1,\ldots M}$ denote a sample of size M of the random variable X_k. The corresponding time is $k\tau$ such that $k = N$ corresponds to time t. Each *walker* x_k^i has a weight w_k^i obtained from the random variable W_k using Eq. (8). Therefore, $\{x_N^i\}_{i=1,\ldots M}$ and $\{w_N^i\}_{i=1,\ldots M}$ denote samples of size M of the random variables X_N and W_N at time t. At time t the weighted probability density represents the normalized solution $\bar{f}(t,x)$ in the sense that

$$\int dx\, g(x)\bar{f}(t,x) = \frac{\int dx\, g(x)f(t,x)}{\int dx\, f(t,x)} = \lim_{M,N\to\infty} \frac{\sum_{i=1}^M w_N^i \cdot g(x_N^i)}{\sum_{i=1}^M w_N^i} \tag{10}$$

for any function $g(x)$. The limit $M \to \infty$ is required by the central limit theorem while $N \to \infty$ for given t is equivalent to $\tau \to 0$ with $t = N\tau$. The stationary solution is obtained at large t:

$$\frac{\int dx\, g(x)\phi(x)\psi_0(x)}{\int dx\, \phi(x)\psi_0(x)} = \lim_{t\to\infty} \lim_{M,N\to\infty} \frac{\sum_{i=1}^M w_N^i \cdot g(x_N^i)}{\sum_{i=1}^M w_N^i} \tag{11}$$

Stationarity is quickly reached due to the exponential convergence, allowing to average not only over different realizations of the stochastic process but also over different times. In practice the weight distribution is stabilized by branching or reconfiguration steps[5,6].

The right-hand side of Eq. (10) is called a mixed estimator, and the most important one is the mixed energy estimator

$$\frac{\int dx\, E_L(x)\phi(x)\psi_0(x)}{\int dx\, \phi(x)\psi_0(x)} = \frac{\int dx\, \phi(x)H\psi_0(x)}{\int dx\, \phi(x)\psi_0(x)} = E_0 \tag{12}$$

where the hermicity of H has been employed to demonstrate that it yields the exact ground state energy E_0. We are thus able to calculate the true ground state energy of H provided that $f(t,x)$ can be represented by a weighted density.

The electronic ground state wave function obeys the Pauli principle which requires the wave function to change sign under position exchange of two electrons while the true mathematical ground state is nodeless and belongs to a bosonic state. The DMC method is best suited for the true ground state such as in multidimensional vibrational problems where the zero point energy is of interest. In order to treat electronic systems, the nodes are usually enforced on the simulation in the fixed-node approximation[7]. The nodes form a $3n$–1-dimensional hypersurface that partitions $\mathbb{R}^{3n}$ into J domains D_j, $j = 1\ldots,J$ with $J \geq 2$. The boundaries of the domains (i.e. the nodes) are denoted ∂D_j. The Schrödinger equation can be defined for each domain D_j:

$$H\psi^j(x) = E^j\psi^j(x),\ x \in D_j \quad \psi^j(x) = 0,\ x \in \partial D_j,\ j = 1, \ldots J \tag{13}$$

If the nodes coincide with the nodes of the true eigenfunction ψ_k all lowest eigenvalues E_0^j, corresponding to nodeless eigenfunctions ψ_0^j, are obviously identical to E_k: $E_0^j = E_k$, $\forall j$. The exact nodes are of course not known in general, but can be approximated by the nodes of an antisymmetric trial function $\phi(x)$. Most or all E_0^j are identical even for inaccurate nodes due to permutation symmetry of the nodes and the Hamiltonian. Electronically excited states have additional nodes such that not all domains D_j are related by permutation and different values for E_0^j are possible.

Therefore, solving the Schrödinger equation in a domain D_j defined by the nodes of an antisymmetric approximate ground state function ϕ leads to $\tilde{E}_0 := E_0^j$ independent of j, unless the domain is due to spurious nodes. The difference $E_0 - \tilde{E}_0$ is denoted node location error. If ϕ approximates an electronically excited state, different values E_0^j are expected that converge toward each other and toward E_k as the nodes of ϕ converge toward the nodes of ψ_k. It can be shown that $\tilde{E}_0$ is an upper bound $\tilde{E}_0 \geq E_0$ [3,8]. Since the Schrödinger equation can be solved exactly (numerically with statistical error) for non-negative functions with the DMC method we can state the fixed-node DMC (FN-DMC) method as a stochastic technique solving the Schrödinger equation with an additional boundary condition (and importance sampling)

$$\frac{\partial f}{\partial t} = -\tilde{H}f, \quad x \in D^j,\ f(x) = 0,\ x \in \partial D^j \tag{14}$$

The boundary condition is naturally enforced with an antisymmetric guide function ϕ because the drift $b(x) = \nabla\phi/\phi$ becomes singular at the nodes. In the limit $\tau \rightarrow 0$ the random walker cannot cross the boundary.

The drift-diffusion algorithm with weighting is denoted pure diffusion Monte Carlo. Its efficiency can be improved with a branching step and a Metropolis-Hastings step. More details can be found in the literature [6,9,10].

3. Optimization of guide functions

Since the FN-DMC method solves the Schrödinger equation exactly for given nodes the guide function plays a fundamental role. The fluctuations or variance of its local energy $E_L(x)$ determines the efficiency of the method and its nodes determine the accuracy. For the use in quantum chemistry it is not necessarily important that the node location error is small it is sufficient that it cancels to a large extent when calculating energy differences. The product of one or several Slater determinants Ψ_n and a Jastrow correlation function e^U is the standard form for guide function in DMC, often called Slater-Jastrow function

$$\phi(x) = e^U \cdot \sum_n c_n \Psi_n . \tag{15}$$

The nodes obviously depend only on the Slater determinant part but are independent of the Jastrow term.

In standard FN-DMC calculations the orbitals (and CI coefficients) are taken from HF, DFT calculations or MCSCF/CASSCF calculations. The correlation function U is parametrized often with the Schmidt-Moskowitz form [11]. The parameters p of U can be determined efficiently by minimization of the variance

$$V(p) = \frac{1}{M} \sum_{i=1}^{M} (E_L(x_i; p) - E_{ref})^2 \tag{16}$$

with respect to p over a fixed sample $\{x_i\}_{i=1,\dots,M}$ with the density $\phi(x;p_0)^2$ obtained with the Metropolis-Hastings method[11]. It has been investigated several times how well the node location errors cancel when dissociation energies are calculated, e.g. [12,13].

Recently, the interest in optimizing the guide wave function has been renewed because the optimization of parameters p that modify the nodes may reduce the node location error and thus finally increase the accuracy of FN-DMC. So far, no method is known to efficiently calculate derivatives of the fixed-node energy $\tilde{E}_0$ with respect to the parameters.

Umrigar and collaborators have recently developed energy minimization schemes that allow optimization of CI, orbital, and Jastrow parameters of the guide function with respect to the variational energy[14,15]. With this method the first-row diatomics and their dissociation energy have been calculated with fully optimized parameters to very high accuracy[16].

The ultimate goal in FN-DMC is the direct optimization of the FN-DMC energy rather than the variational energy. This would require optimal nodal surfaces. Unfortunately, not much is known about the exact nodes although exact nodes are known for a few systems. The first ^3S state of the He atom has the excitation node $r_1 = r_2$ which follows easily from the symmetry of the system. Mitas and coworkers have recently found exact nodes based on symmetry arguments for a number of high spin systems[17,18]. These nodes are very smooth and is has been suggested that the exact nodes are in general simpler and of higher symmetry than the corresponding wave functions[19,20]. Bressanini and

Reynolds investigated in detail the the nodes of the 2^1S state of He and found a very simple node structure independent of the interelectronic angle θ_{12} [21]. Very accurate Hylleraas calculations confirmed the simple structure, but a very small, converged cosine-like dependence on θ_{12} was indeed found[22].

A step toward optimization of the nodes has been taken by Reboredo and Kent who suggest node optimization based on effective functionals[23]. Bressanini et al. investigated the influence of configuration state function (CSF) on the FN-DMC energy and where able to optimize guide functions by carefully selecting individual CSFs[20].

A different approach toward optimization of the nodes has been employed by the current authors. The basic idea of this approach is the development of a criterion that is sensitive to the local accuracy of the node of $\phi(x)$. A successful criterion is the distance η between the nodes of ϕ and $H\phi$. The distance η vanishes everywhere for exact wave functions $\psi_k(x)$ because the nodal surfaces of ψ_k and $H\psi_k$ coincide due to

$$H\psi_k = E_k\psi_k \tag{17}$$

For an approximate wave function such as the guide function we have instead

$$H\phi = E\phi + \chi \tag{18}$$

where the function $\chi(x)$ is small. While it is possible to construct guide functions with $\chi = 0$ at $\phi = 0$, i.e. with distance $\eta = 0$, good guide functions have in general small but non-zero distance η. As the guide function is improved toward the exact wave function the distance η decreases and finally vanishes. The distance η is particularly well suited as a local measure of the quality of the nodes because the unit normal vector $n = \nabla\phi/\|\nabla\phi\|$ at the surface of ϕ is readily available in quantum Monte Carlo codes. Recently, a simple formula for an estimate η_0 to η has been derived[24]

$$\eta_0 = \left.\frac{\Delta\phi}{2E_T\|\nabla\phi\|}\right|_{\phi(x)=0} \tag{19}$$

that required only a rough estimate E_T to the exact energy E_0. All quantities on the right hand side are readily available in a DMC code at no extra cost. Higher order estimates have been derived as well in Ref. [24]. For the model systems H atom (excited 2s, 3s states), Hooke's atom, and He atom ($2\ ^1S$ state) it has been shown that these formulae are excellent approximations to the true distance η as long as the trial function ϕ is a reasonable approximation to the exact eigenfunction. Furthermore, iterative minimization of η_0 lead to quick convergence to the exact nodes of the wave function.

For more chemical systems, the distance η_0 is directly minimized with respect to the parameters of ϕ. Similar to the variance minimization with a fixed sample a random sample of points on the nodal hypersurface is created. The mean sample distance for a sample of size M is

$$\bar{\eta}_0 = \sqrt{\frac{1}{M}\sum_{j=1}^{M}\eta_0^2(j)}. \tag{20}$$

Table 1. FN-DMC energies in E_h with one determinant (single det) and with optimized CI coefficients using $\overline{\eta}_0$ optimization ($\overline{\eta}_0$ opt) compared to the exact non-relativistic energy (exact).

	CSFs	single det	$\overline{\eta}_0$ opt	exact
Be	2	–14.6573(1)	–14.6672(1)	–14.66735[56]
BeH	2	–15.2460(2)	–15.2464(1)	–15.2468[12]
CH	2	–38.4634(4)	–38.4685(4)	–38.4787[12]
C_2	2	–75.8671(6)	–75.8855(5)	–75.923(5)[57]
Li_2	3	–14.9915(1)	–14.9929(1)	–14.9954[57]
C_2	4	–75.8671(6)	–75.8894(5)	–75.923(5)[57]

$\overline{\eta}_0$ is minimized with the Levenberg-Marquardt method or by simply varying the parameters[24]. The minimization of $\overline{\eta}_0$ is fast compared to a FN-DMC calculation.

Initial application of the direct node optimization by minimization of $\overline{\eta}_0$ is the optimization of CI coefficients in Be, BeH, Li_2, CH, and C_2, all small systems with nondynamical electron correlation that are well described by few configuration state functions (CSF).

The coefficient of the first CSF is fixed at $c_1 = 1$, all other coefficients are optimized. The orbitals are calculated with the MCSCF method using STO basis functions [25–27]. The CSFs are ordered with respect to their CI coefficient, and the first two, three (Li_2) or four (C_2) CSFs are used for the optimization. Details on the optimization are given in [28] where first results for the Be atom and C_2 are discussed. The FN-DMC energies with the unoptimized ($c_2 = c_3 = c_4 = 0$), the optimized wavefunction, and the exact nonrelativistic energies are shown in Table 1.

The data show that the $\overline{\eta}_0$ optimization of the nodal hypersurface leads in all cases to significantly improved FN-DMC energies although our value for Li_2 –14.9929(1) E_h is higher than the DMC energy –14.9939(2) E_h that Bressanini et al. obtained [20] with the same number of CSFs. While the node distance η_0 depends on the quality of the wave function at the nodes it is not guaranteed that the minimum of $\overline{\eta}_0$ with respect to the parameters is exactly the minimum of $\tilde{E}_0$, the FN-DMC energy. To test the reliability of the $\overline{\eta}_0$ minimization we have determined the FN-DMC minimum in all cases where only one parameter was optimized, i.e. Be, BeH, CH, C_2, by calculating a few points of the curve $\tilde{E}(c_2)$. The FN-DMC minimum coincides in all four cases with the optimized energy within statistical error bars.

Additionally, we have determined the dependence of the variational energy on the parameter c_2 for Be and C_2. The c_2 values at the minima coincide for variational and FN-DMC energies for the Be atom while values at the minima differ significantly for the carbon dimer (c_2 = –0.52 and –0.31, respectively). The FN-DMC energy with c_2 at the variational minimum is –75.8812(4) E_h or more than 4 mE_h higher than the FN-DMC minimum –75.8855(5) E_h which emphasizes the need for a FN-DMC parameter minimization.

4. Applications of the FN-DMC method

Recent calculations on difficult quantum chemical problems with the FN-DMC method by the authors and by other research groups will be presented. The difficulties of FN-DMC due to the additional nodes in excited states are discussed. Substantial progress has been achieved in calculations of transition metal compounds as well as weakly interacting systems.

4.1 Excited states

The FN-DMC method is suitable for electronically excited states as well as for ground states as long as an appropriate nodal hypersurface, given by a guide function, is available. Although the quantum simulation is in principle unchanged a few problems arise because the FN-DMC energy may depend on the nodal domain of the simulation as discussed above. Additionally, the FN-DMC energy is not necessarily variational[8].

In the last few years, a number of QMC papers have appeared for excited states. While MCSCF or CASSCF wave functions should result in accurate nodes these functions do not scale well with the system size and result often in a large number of determinants which is unfavorable for QMC. Porter et al. showed that CIS functions can be efficiently summed in QMCs[29]. Schautz and Filippi demonstrated that accurate DMC excited state calculations are possible when CSFs are constructed from orbitals obtained with the ROKS open shell DFT method[30,31]. Accuracy of about 0.1 eV has been achieved with QMC for selected excitations such as n–π^*, π–π^*, and σ excitations. Bande et al. constructed excited state guide functions from open shell localized Hartree-Fock (OSLHF) orbitals by Della Sala and Görling[32,33]. This DFT method scales as well with system size as other DFT methods with exact exchange, but is, in addition, able to describe open shell systems correctly. The OSLHF approach allows for an efficient one or two-determinant guide function supplemented with a standard Jastrow correlation factor for Rydberg states. The approach has first been applied to low lying Rydberg state of the C atom and the CO molecule[34]. The accuracy of FN-DMC excitation energies is better than 0.1 eV for CO and the singlet triplet splittings are in agreement with experimental results.

The excited states of C are simple enough to allow a thorough investigation of possible spurious node domains and node domains with different FN-DMC energy as discussed above. Spurious nodes have recently been found by Hachmann et al. in systems like the Be atom and the question has been raised whether these nodes effect the FN-DMC energy[35]. Analysis of domains and their probability distribution and energy contribution revealed that the spurious nodes do not influence the FN-DMC energy as long as the basis set is suitably chosen. It was also demonstrated that accurate FN-DMC calculations of excited states require averaging over the domain energies weighted by the contribution to the wave function[36].

Table 2. Ground state wave function variances in atomic units for Sc - Fe, with Stuttgart (S) and Burkatzki (B) pseudopotential.

	Sc	Ti	V	Cr	Mn	Fe
B	0.69	0.86	0.87	1.33	1.78	1.86
S	0.91	1.09	1.59	2.15	2.74	4.14

4.2 Transition metal compounds with FN-DMC

In general, the FN-DMC method scales favorably with the system size (n^{3-4}). Nevertheless there is a strong dependence on the atomic number Z with $Z^{5.5-6.5}$ for current QMC methods [3]. The application of pseudopotentials for transition metals is thus required for efficient calculations.

Recently "soft" pseudopotentials for QMC calculations have been developed by the Lester and the Needs groups[37–40]. The latest set of soft pseudopotentials for QMC was published by Burkatzki, Filippi and Dolg with optimized basis sets for the main group elements [41] and subsequently for 3d transition metals [42].

With FN-DMC and the Lester soft pseudopotentials on C and O and Stuttgart pseudopotentials on the transition metal, Diedrich et al. calculated accurately, within the experimental error bar of 2 kcal/mol, the dissociation energy of one CO in the carbonyls $Cr(CO)_6$ and $Ni(CO)_4$[43] while $Fe(CO)_5$ was less accurate. With a similar approach Bande and Lüchow obtained accurate dissociation and ionization energies for a number of vanadium oxide species including VO and VO_2. In cases where experimental data were available, the FN-DMC results were at least as accurate as CCSD(T)[44]. The largest species calculated with FN-DMC in this study was V_2O_5. Accurate dissociation energies and bond lengths have been obtained with FN-DMC by Wagner and Mitas[45,46] for the transition metal oxides ScO to MnO with the exception of CrO. In these studies only one determinant has been used because single-determinant FN-DMC is most efficient.

The availability of the new soft pseudopotentials by Burkatzki, Filippi, and Dolg [41,42] allow more efficient FN-DMC calculations. We have investigated the variances of atomic wave functions using either the standard Stuttgart or the soft QMC-adapted new pseudopotentials for the 3d transition metals. The results on the atoms shown in Table 2 demonstrate a decrease of the variance by 30 - 50 % with the new pseudopotentials. The decreased variances lead directly to shorter cpu times for a given standard error of the energy. In addition, the FN-DMC calculations are more stable.

Initial results for the oxides ScO and TiO have been published by Burkatzki and collaborators with the new pseudopotentials. They calculated ionization potentials and the dissociation energies and obtained agreement with experiment. In both cases, a multideterminant guide function was necessary to obtain dissociation energies with an accuracy better than 0.1 eV [42].

First results on the performance of these pseudopotentials on 3d transition metal sulfides is shown here using single determinant guide functions built from

Table 3. Dissociation energies D_e in eV for ScS, TiS, and VS.

	ScS	TiS	VS
BP86/PPI	5.30	5.59	5.23
BP86/PPII	5.29	5.58	5.25
DMC/PPI	4.743(5)	4.891(4)	4.408(7)
DMC/PPII	4.903(5)	4.943(5)	4.395(7)
CCSD(T)[48]	4.62	4.36	4.02
Expt	4.92	4.75	4.62

BP86 orbitals. Experimental bond lengths are employed[47,48], and the triple zeta basis set provided with the pseudopotential is used. Two guide functions are used for each molecule. In the first function (denoted PPI), the Stuttgart pseudopotential [49] is used on the transition metal for comparison while the new soft pseudopotential is used in the second one (PPII). The soft pseudopotential is used in both cases on the sulfur atom[42]. g functions have been deleted from the basis sets because they serve to describe electron correlation and are thus unnecessary when a Jastrow correlation function is present. All FN-DMC energies are extrapolated to zero time step. The results in Table 3 show that it is advantageous to use the new soft pseudopotentials. The deviations of the values obtained with PPII from the experimental values are overall smaller. The DMC dissociation energies are in better agreement with experiment than both DFT calculations and CCSD(T). The FN-DMC dissociation energies could be systematically improved with multideterminant guide functions. Finally, the single determinant approach without special optimization of the Kohn Sham orbitals works surprisingly well and encourages us to pursue FN-DMC calculations for larger transition metal systems.

4.3 Weak interaction with diffusion Monte Carlo

Structure and energetics in chemistry are determined not only by covalent bonds but also by non-covalent bonds, and in many areas van-der-Waals and hydrogen bonding interactions are more important for structure than the covalent bonds. Well known examples are protein and DNA structure or solvation. While hydrogen bonding is already quite well described at the Hartree-Fock level neither Hartree-Fock nor most DFT methods are able to correctly treat dispersive interactions. Since Møller-Plesset calculations at the MP2 level tend to overestimate the van-der-Waals energy the accurate calculation of larger weakly interacting species remains a challenge for quantum chemistry. The FN-DMC method is expected to perform well in these cases because the electron correlation is very accurately obtained and finite basis sets leading to basis set super position errors are not used. FN-DMC has been successfully applied to the He_2 and the LiH-He interaction by Mella and Anderson[50]. In an study on larger systems Diedrich, Lüchow, and Grimme demonstrated that FN-DMC can equally well describe pure van-der-Waals dimers (methane), pure hydrogen-bonded dimers (ammonia, water) as well as the difficult case of the benzene dimer where a T-shaped conformer competes with a parallel displaced conformer[51]. The benzene dimer is a particularly difficult problem because the interaction in T-shaped conformer is dominated by electrostatics while the parallel one is dominated by π interaction. Only when both interaction a equally well described a reliable estimation of the relative energy can be expected. The parallel displaced benzene dimer has been investigated with FN-DMC also be Sorella et al. These authors used a resonating valence bond wave function with all parameters carefully optimized[52]. Gurtubay and Needs published recently a very thorough study of the water dimer using all-electron and pseudo potential guide functions. They obtained binding energies in agreement with the most accurate experimental and theoretical data[53]. Even with HF orbitals and large diffuse basis sets FN-DMC works reliably over a range of systems. In a recent study Korth, Lüchow, and Grimme calculated the S22 set of 22 weakly bound reference systems with different types of interactions where accurate extrapolated CCSD(T) data are available [54]. FN-DMC binding energies agreed with a mean absolut deviation of only 0.68 kcal/mol from the extrapolated ab initio values [55]. The authors compared also the binding energies of the base pairs adenine/thymine and guanine/cytosine in the hydrogen bonded and the π-stacking conformation. The results are again in excellent agreement with extrapolated CCSD(T) data.

5. Conclusion and outlook

In recent years, the fixed-node diffusion quantum Monte Carlo method (FN-DMC) has shown its capability of calculating accurately binding or dissociation energies even for difficult systems such as transition metal compounds and weakly interacting systems or excitation energies. The accuracy is comparable to CCSD(T) in all investigations so far even with a single-determinant guide function in DMC. The accuracy of FN-DMC can be further improved with optimized guide functions where considerable progress has been achieved. The development of optimal guide functions for FN-DMC is being investigated by several groups.

Acknowledgement

The authors gratefully acknowledge sustained support by the Deutsche Forschungsgemeinschaft (DFG) in the priority program SPP 1145. Computational resources have been provided by the Rechenzentrum of RWTH Aachen university.

References

1. W. M. C. Foulkes, L. Mitas, R. J. Needs, G. Rajagopal, Rev. Mod. Phys. **73** (2001) 33.

2. A. Lüchow, J. B. Anderson, Annu. Rev. Phys. Chem. **51** (2000) 501–526.
3. B. L. Hammond, W. A. Lester Jr., P. J. Reynolds, Monte Carlo Methods in Ab Initio Quantum Chemistry. World Scientific: Singapore (1994).
4. R. C. Grimm, R. G. Storer, J. Comp. Phys. **7** (1971) 134.
5. P. J. Reynolds, D. M. Ceperley, B. Alder, W. A. Lester, J. Chem. Phys. **77** (1982) 5593.
6. R. Assaraf, M. Caffarel, A. Khelif, Phys. Rev. E **61**(4) (2000) 4566.
7. J. B. Anderson, J. Chem. Phys. **65** (1976) 4121.
8. W. M. C. Foulkes, R. Q. Hood, R. J. Needs, Phys. Rev. B. **60** (1999) 4558–4570.
9. R. N. Barnett, P. J. Reynolds; W. A. Lester, J. Chem. Phys. **96** (1991) 258.
10. C. J. Umrigar, M. P. Nightingale, K. J. Runge, J. Chem. Phys. **99** (1993) 2865.
11. K. E. Schmidt, J. W. Moskowitz, J. Chem. Phys. **93** (1990) 4172.
12. A. Lüchow, J. B. Anderson, J. Chem. Phys. **105** (1996) 7573–7578.
13. J. C. Grossman, J. Chem. Phys. **117** (2002) 1434–1440.
14. C. J. Umrigar, J. Toulouse, C. Filippi, S. Sorella, R. G. Henning, Phys. Rev. Lett. **98** (2007) 110201.
15. J. Toulouse, C. J. Umrigar, J. Chem. Phys. **126** (2007) 084102.
16. J. Toulouse, C. J. Umrigar, J. Chem. Phys. **128** (2008) 174101.
17. M. Bajdich, L. Mitas, G. Drobny, L. K. Wagner, Phys. Rev. B**72**(7) (2005) 075131.
18. L. Mitas, Phys. Rev. Lett. **96** (2006) 240402.
19. D. Bressanini, D. M. Ceperley, P. J. Reynolds, in Recent advances in Quantum Monte Carlo methods, Part II; W. A. Lester, Jr., S. M. Rothstein, S. Tanaka (Eds.), World Scientific: Singapore, (2002) pages 3–11.
20. D. Bressanini, G. Morosi, S. Tarasco, J. Chem. Phys. **123** (2005) 204109–1-204109–11.
21. D. Bressanini, P. J. Reynolds, Phys. Rev. L. **95** (2005) 110201–1-110201–4.
22. T. C. Scott, A. Lüchow, D. Bressanini; J. D. Morgan III, Phys. Rev. A **75** (2007) 060101R.
23. F. A. Reboredo, P. R. C. Kent, Phys. Rev. B **77** (2008) 245110.
24. A. Lüchow, T. C. Scott, J. Phys. B **40** (2007) 851–867.
25. P. E. Cade, W. Huo, Atomic. Data Nucl. Tables **12** (1973) 415.
26. P. E. Cade, A. C. Wahl, Atomic. Data Nucl. Tables **13** (1974) 339.
27. E. Clementi, C. Roetti, Atomic. Data Nucl. Tables **14** (1974) 177.
28. A. Lüchow, R. Petz, T. C. Scott, J. Chem. Phys. **126**(2007) 144110.
29. A. R. Porter, O. K. Al-Mushadani, M. D. Towler, R. J. Needs, J. Chem. Phys. **114** (2001) 7795–7804.
30. F. Schautz, C. Filippi, J. Chem. Phys. **120** (2004) 10931.
31. F. Schautz, F. Buda, C. Filippi, J. Chem. Phys. **121** (2004) 5836–5844.
32. F. Della Sala, A. Görling, J. Chem. Phys. **118**(23) (2003) 10439.
33. F. Della Sala, A. Görling, Phys. Rev. Lett. **89** (2002) 033003.
34. A. Bande, A. Lüchow, F. Della Sala, A. Görling, **124**(11) (2006) 114114.
35. J. Hachmann, P. T. A. Galek, T. Yanai, G. Chan, N. C. Handy, Chem. Phys. Lett. **392** (2004) 55–61.
36. A. Bande; A. Lüchow, in J. B. Anderson, S. Rothstein (Eds.), Recent Advances in Quantum Monte Carlo. ACS Proceedings (2007).
37. C. W. Greeff; W. A. Lester, Jr., J. Chem. Phys. **109**(5) (1998) 1607.
38. I. Ovcharenko, A. Aspuru-Guzik; W. A. Lester, Jr., J. Chem. Phys. **114**(18), (2001) 7790.
39. J. R. Trail, R. J. Needs, J. Chem. Phys. **122**(1) (2005) 014112.
40. J. R. Trail, R. J. Needs, J. Chem. Phys. **122**(17) (2005) 174109.
41. M. Burkatzki, C. Filippi, M. Dolg, J. Chem. Phys. **126** (2007) 234105.
42. M. Burkatzki, C. Filippi, M. Dolg, J. Chem. Phys. **129** (2008) 164115.
43. C. Diedrich, A. Lüchow, S. Grimme, J. Chem. Phys. **122** (2005) 21101.

44. A. Bande, A. Lüchow, Phys. Chem. Chem. Phys. **10** (2008) 3371–3376.
45. L. Wagner, L. Mitas, Chem. Phys. Lett. **370** (2003) 412–417.
46. L. K. Wagner, L. Mitas, J. Chem. Phys. **126** (2007) 034105.
47. Z. J. Wu, M. Y. Wang, Z. M. Su, Journal of Computational Chemistry **28**(3) (2007) 703–714.
48. C. W. Bauschlicher, Jr., P. Maitre, Theor Chim. Acta **90** (1995) 189.
49. A. Bergner, M. Dolg, W. Kuechle, H. Stoll, H. Preuss, Mol. Phys. **80** (1993) 1431.
50. M. Mella, J. B. Anderson, J. Chem. Phys. **119** (2003) 8225–8228.
51. C. Diedrich, A. Lüchow, S. Grimme, J. Chem. Phys. **123** (2005) 184106.
52. S. Sorella, M. Casula, D. Rocca, J. Chem. Phys. **127** (2007) 014105.
53. I. G. Gurtubay, R. J. Needs, J. Chem. Phys. **127** (2007) 124306.
54. P. Jurecka, J. Sponer, J. Cerny, P. Hobza, Phys. Chem. Chem. Phys. **8** (2006) 1985.
55. M. Korth, A. Lüchow, S. Grimme, J. Phys. Chem. A **112** (2008) 2004–2009.
56. E. R. Davidson, S. A. Hagstrom, S. J. Chakravorty, V. M. Umar, C. F. Fischer, Phys. Rev. A **44** (1991) 7071.
57. C. Filippi, C. J. Umrigar, J. Chem. Phys. **105** (1996) 213.

First-Principles Calculation of Electronic Excitations in Solids with SPEX

By Arno Schindlmayr[1,*], Christoph Friedrich[2], Ersoy Şaşıoğlu[2], and Stefan Blügel[2]

[1] Department Physik, Universität Paderborn, 33095 Paderborn, Germany
[2] Institut für Festkörperforschung and Institute for Advanced Simulation, Forschungszentrum Jülich, 52425 Jülich, Germany

(Received August 17, 2009; accepted September 29, 2009)

Many-Body Perturbation Theory / Self-Energy / Quasiparticle Band Structure / Dielectric Function / Spin Waves

We describe the software package SPEX, which allows first-principles calculations of quasiparticle and collective electronic excitations in solids using techniques from many-body perturbation theory. The implementation is based on the full-potential linearized augmented-plane-wave (FLAPW) method, which treats core and valence electrons on an equal footing and can be applied to a wide range of materials, including transition metals and rare earths. After a discussion of essential features that contribute to the high numerical efficiency of the code, we present illustrative results for quasiparticle band structures calculated within the *GW* approximation for the electronic self-energy, electron-energy-loss spectra with inter- and intraband transitions as well as local-field effects, and spin-wave spectra of itinerant ferromagnets. In all cases the inclusion of many-body correlation terms leads to very good quantitative agreement with experimental spectroscopies.

1. Introduction

First-principles computations, which do not rely on any empirical input parameters, have become an important tool in materials science. Ideally, such calculations use only the fundamental laws of nature together with the specified elemental composition in order to predict the structural and electronic properties of a material. As chemical bonding and the response to external fields are determined by the microscopic dynamics of the ions and electrons inside the solid, the relevant laws in this case are those of quantum mechanics. The complete Hamiltonian can, in fact, be readily written down, because the interaction potentials

* Corresponding author. E-mail: Arno.Schindlmayr@uni-paderborn.de

Z. Phys. Chem. **224** (2010) 357–368
© by Oldenbourg Wissenschaftsverlag, München

between all constituent particles, given by Coulomb's law, are known exactly. However, a direct solution of the Schrödinger equation or its relativistic counterpart, the Dirac equation, is not feasible for extended solids, because the number of electrons N is of the order of $10^{23}/\text{cm}^3$ and hence too large for a numerical treatment of the correlated many-electron wave function $\Psi(\mathbf{r}_1,\ldots,\mathbf{r}_N)$. Instead, practical applications rely on alternative approaches that are formally equivalent to the Schrödinger equation but do not employ the many-electron wave function as the basic variable. A prominent example is density-functional theory (DFT), which is based on the ground-state electron density $n(\mathbf{r})$, a real quantity that depends only on a single spatial position vector. In spite of this vast simplification, the Hohenberg-Kohn theorem [1] asserts that knowledge of the ground-state density alone is sufficient, at least in principle, to determine all observables of a stationary system in equilibrium. The density itself can be calculated from the Kohn-Sham scheme [2], which introduces an auxiliary system of non-interacting electrons with the same ground-state density as the real interacting system and requires the solution of a single-particle Schrödinger-like equation, the Kohn-Sham equation, with a self-consistent local potential.

Despite the enormous scope of the Hohenberg-Kohn theorem, practical applications are limited because for most observables no explicit formulas for the actual dependence on $n(\mathbf{r})$ are known. A notable exception, besides the density itself, is the ground-state total energy, where the dominant contributions are given exactly and the remaining exchange-correlation functional may be replaced by the local-density approximation (LDA) [2] or generalized gradient approximations like PBE [3]. By comparing the total energies for different atomic configurations it is thus possible to predict crystal structures and related quantities like the elastic moduli without empirical parameters. The huge success of density-functional theory stems from the good agreement with crystallographic measurements for a wide range of different materials. In contrast, electronic excitation spectra often show substantial deviations from experimental spectroscopies. The most famous example is the severe underestimation of the band gaps of semiconductors. As a large part of the error in this case is systematic and arises from the pervasive but incorrect identification of the Kohn-Sham eigenvalues with the true quasiparticle energies, better approximations for the exchange-correlation functional will not solve the problem. Therefore, quantitative methods for electronic excitations in solids are now chiefly sought outside density-functional theory. Perhaps the simplest extension are hybrid functionals like PBEh [4] or HSE [5], which replace a fraction of the local exchange potential by non-local Hartree-Fock exchange. As Hartree-Fock in turn tends to overestimate band gaps, a linear combination with suitably chosen coefficients can improve the agreement with experiments. Unfortunately, the optimal relative weights depend on the material: While an admixture of 25% Hartree-Fock yields good results for many typical semiconductors, a higher fraction is required for large-gap insulators, where screening is much weaker [6]. For metals any inclusion of Hartree-Fock exchange even increases the already too large valence-band widths further, so that hybrid functionals are in fact in worse agreement with experimental band structures than pure LDA results [6].

A better-founded scheme without adjustable parameters is many-body perturbation theory [7], which is based on the Green function $G(\mathbf{r},\mathbf{r}';\omega)$. As it contains no inbuilt systematic errors, the accuracy is only limited by functional approximations like the *GW* approximation for the electronic self-energy [8], which can be improved if necessary. As a consequence, this approach yields excitation energies in significantly better agreement with experiments than density-functional theory. The main drawback is the high computational cost, which is related to the more complicated mathematical form of the non-local and frequency-dependent Green function in comparison to the density. For this reason, practical applications have so far been limited to moderately complex systems with up to about one hundred atoms per unit cell. Although this suffices to study, for example, carbon nanotubes [9] or point defects at semiconductor surfaces [10], simulations with tens of thousands of atoms are now possible using density-functional theory with linear-scaling algorithms [11]. Furthermore, many actual *GW* calculations contain a number of additional simplifications.

Besides individual quasiparticle properties, collective excitations of the electron system in solids constitute another major challenge for electronic-structure calculations. Prominent examples are plasmons and excitons, which are associated with resonances in the dielectric function and can be probed by electron-energy-loss or optical spectroscopies, but also spin-wave modes in magnetic materials. As a single-particle picture cannot describe such collective excitations, many-body perturbation theory has become the method of choice for quantitative simulations [12].

Here we describe a new software package, SPEX [13], which contains an implementation of many-body perturbation theory and can be used to simulate various spectroscopic techniques. The code is designed to avoid many shortcomings of previous implementations and to keep additional approximations at a minimum. Most importantly, it is not based on the prevalent pseudopotential concept but uses the full-potential linearized augmented-plane-wave (FLAPW) method [14], which opens the door to a wider range of materials, including elements with localized *d* or *f* orbitals. In the following we first discuss some essential features of the code. Then we present selected results for several spectroscopies that probe electronic excitations, such as quasiparticle band structures, electron energy loss or spin waves, in order to illustrate the possible applications. Finally, we briefly summarize and comment on planned future developments.

2. The SPEX code

Although many-body perturbation theory is a self-contained mathematical framework, actual applications are most efficient when combined with density-functional theory. The Kohn-Sham eigenvalues, which are often in qualitative agree-

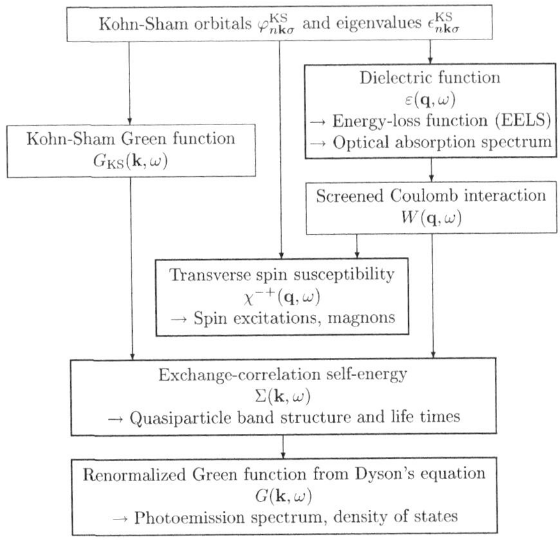

Fig. 1. Flowchart for the perturbative calculation of different electronic excitations and spectroscopies as implemented in SPEX.

ment with the true band structure, then serve as the starting point for a diagrammatic expansion in terms of the dynamically screened Coulomb interaction $W(\mathbf{r},\mathbf{r}';\omega)$. The screening arises from the formation of exchange-correlation holes around charged fermions and leads to the concept of quasiparticles, which incorporate the polarization of the local environment and constitute one class of elementary excitations of the electron system. As the screened interaction between the quasiparticles is much weaker than the bare Coulomb potential, a perturbative treatment is justified. A first-order expansion of the self-energy leads to the *GW* approximation, which is just the linear term in $W(\mathbf{r},\mathbf{r}';\omega)$. In accordance with standard perturbation theory, its matrix elements must be evaluated with the original unperturbed Kohn-Sham orbitals.

Following this philosophy, SPEX is designed as a separate module that computes electronic excitation spectra from a given set of Kohn-Sham orbitals and eigenvalues. It can be combined with any density-functional code whose output data is convertible to the FLAPW form. We use our own package FLEUR [13], but other choices are equally possible. The flowchart in Fig. 1 illustrates the course of the calculation.

A premier goal during the code development was to avoid additional simplifications wherever possible, so that the results depend exclusively on controllable convergence parameters and the choice of functional approximations like the *GW* approximation for the self-energy or the random-phase approximation (RPA) for the dielectric function. Besides, emphasis was placed on high efficiency, especially in terms of CPU time, so that complex materials with large unit cells can be treated. Finally, the code should be versatile and easily adaptable to new physical problems, for instance in the emerging areas of nanomagnetism and spintronics. To satisfy these requirements SPEX contains a number of important features that are summarized below.

(i) Consistent employment of FLAPW, which treats core and valence electrons on the same footing. Unlike the atomic-sphere approximation used in early all-electron codes, the full-potential treatment is also suitable for surfaces or defects. The linearization error caused by expanding the muffin-tin functions around fixed energy parameters, which is negligible for states close to the Fermi level but becomes important for high unoccupied bands that contribute to the self-energy, can be eliminated by including higher energy derivatives as additional local orbitals [15].

(ii) The FLAPW basis set is optimized for the Kohn-Sham orbitals but does not span the complete Hilbert space. Representing products of wave functions, for example in the polarizability or matrix elements of the Coulomb potential, in this basis hence implies a loss of accuracy. Instead, we use a mixed product basis [16] that is constructed from products of the basis functions and allows exact projections. After linear dependencies are removed, this set may be truncated in a controlled fashion, if desired.

(iii) The dielectric function is constructed either in the random-phase or the time-dependent local-density approximation, without recourse to plasmon-pole models. The full frequency dependence as well as local-field effects are thus properly included. Besides, it gives access to the imaginary part of the self-energy, which leads to complex quasiparticle energies and describes the finite lifetime of the excited states due to scattering.

(iv) Frequency convolutions are normally evaluated by means of contour integrations in the complex plane, although the analytic continuation of functions calculated on the imaginary axis to real frequencies, which is faster but less well controlled, is also possible.

(v) Symmorphic and non-symmorphic spatial symmetries are exploited wherever possible. For systems with inversion symmetry we also use the fact that the Coulomb matrix and response functions on the imaginary frequency axis may be chosen real and thus processed in compact form.

(vi) Following a procedure developed in Ref. [17] for plane waves, the singularity of the Coulomb matrix at $\mathbf{k} = \mathbf{0}$ in reciprocal space is treated exactly through an expansion that separates the divergent and regular parts [18]. A subsequent diagonalization confines the singularity to a single divergent eigenvalue, which can be processed analytically. At the same time, the transformation from the mixed product basis to the eigenvectors of the Coulomb matrix allows a very efficient truncation by eliminating the least significant scattering channels with the smallest eigenvalues.

(vii) The spin degree of freedom is fully supported. We make no simplifying assumptions about the orbitals in the two spin channels and allow a completely spin-unrestricted treatment of magnetic materials.

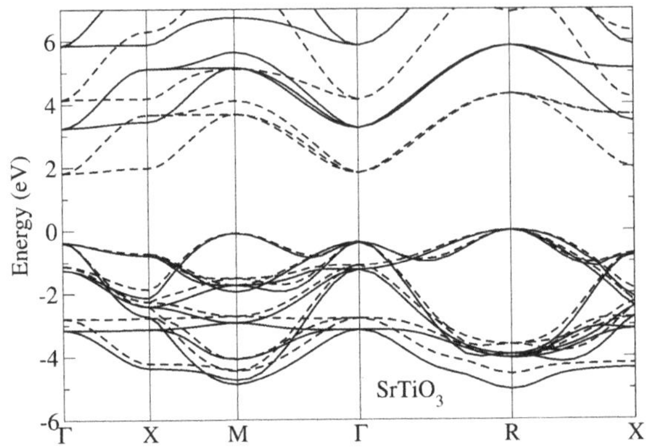

Fig. 2. Quasiparticle band structure of cubic $SrTiO_3$ calculated in DFT-PBE (dashed lines) and the *GW* approximation (solid lines).

(viii) The code is applicable both to insulators and to metals. Where intraband transitions require a special treatment, appropriate provisions are made. In particular, a Drude term is included in the dielectric function in this case. All calculations are done at zero temperature.

(ix) Relativistic corrections are treated at the scalar-relativistic level for valence states and the full Dirac equation for core states.

As a performance test we conducted *GW* band-structure calculations for diamond supercells of various sizes, using a **k**-point sampling equivalent to $4 \times 4 \times 4$ mesh points in the full Brillouin zone corresponding to the elementary diatomic unit cell and with optimized parameters that otherwise guarantee a convergence of the quasiparticle shifts to within 0.01 eV [19]. Our results demonstrate that simulations even with 128 atoms per cell are perfectly feasible on a standard single-processor work station, and a complete quasiparticle band structure requires less than one and a half days of CPU time in this case. We find a scaling behavior that lies between quadratic and cubic with the number of atoms in this size range.

3. Results

In this section we present illustrative results for different spectroscopies. All materials discussed here contain transition-metal elements and would be difficult to treat with conventional pseudopotential codes, because the localized *d* orbitals require a large number of plane waves.

3.1 Quasiparticle band structures

In the perturbative approach adopted here the quasiparticle energies are derived from the Kohn-Sham eigenvalues and the self-energy according to

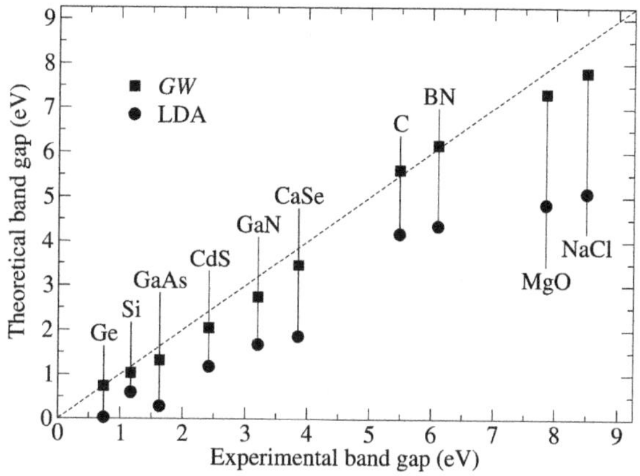

Fig. 3. Calculated band gaps of selected materials in the LDA (circles) and the *GW* approximation (squares) compared to experimental values.

$$\epsilon_{n\mathbf{k}\sigma} = \epsilon_{n\mathbf{k}\sigma}^{\mathrm{KS}} + \langle \varphi_{n\mathbf{k}\sigma}^{\mathrm{KS}} | \Sigma_\sigma(\epsilon_{n\mathbf{k}\sigma}) - V_\sigma^{\mathrm{xc}} | \varphi_{n\mathbf{k}\sigma}^{\mathrm{KS}} \rangle \,. \tag{1}$$

The self-energy is evaluated in the *GW* approximation, and the matrix elements of the local exchange-correlation potential $V_\sigma^{\mathrm{xc}}(\mathbf{r})$ are subtracted to avoid double counting. Although *GW* calculations for real materials have been feasible since the 1980s [20,21], the prevailing reliance on plane waves and pseudopotentials meant that applications were long restricted almost exclusively to *sp*-bonded semiconductors and simple metals. No such restrictions apply to the FLAPW method. As an example, in Fig. 2 we show the band structure of strontium titanate ($SrTiO_3$), a high-κ dielectric insulator from the perovskite family, in the cubic phase observed at room temperature. The lattice constant within DFT-PBE [3] is 7.46 Bohr. For the self-energy construction we use 64 mesh points in the full Brillouin zone, 550 unoccupied bands, and cutoff parameters G_{max} = 5 Bohr^{-1} and $L_{\mathrm{max}} = 4$ for the mixed product basis in the interstitial region and in the muffin-tin spheres, respectively. Second energy derivatives are included as additional local orbitals. Although DFT-PBE gives the correct qualitative picture, both the indirect band gap of 1.81 eV between R and Γ and the direct gap of 2.18 eV at Γ are significantly too low. In contrast, the *GW* approximation yields 3.23 eV and 3.61 eV in very good agreement with the experimental values 3.25 eV and 3.75 eV [22]. The size of the band gap is quite important in this case, as $SrTiO_3$ has long been technically used as an optically transparent synthetic diamond simulant. The comparison with a previously published *GW* value of 5.07 eV for the indirect gap [23], obtained with a parametrized model dielectric function, underlines the need for an accurate evaluation of the self-energy correction.

In Fig. 3 we display results for a wider range of materials from small-gap to large-gap insulators. In all cases the *GW* approximation corrects virtually the entire error of the LDA. The data support our previous conclusion, based on an in-depth study of silicon [15], that carefully performed all-electron *GW* calculations yield very good agreement with experiments. This had initially been in doubt after an early FLAPW calculation found unexpectedly large deviations both from experiments and from established pseudopotential values [24]. However, this observed discrepancy is now understood to have arisen from incomplete convergence, especially with respect to the number of unoccupied conduction bands in the self-energy [15,25]. Our own results are meanwhile confirmed by other independent all-electron implementations [26,27].

Incidentally, these studies also revealed that well converged all-electron *GW* band gaps do not coincide with pseudopotential values after all. The remaining difference stems from the pseudization of the wave functions and the linearized core-valence interaction in the latter approach [28].

3.2 Electron-energy-loss spectroscopy

While the single-particle Green function and self-energy describe the final states of photoemission experiments, where the particle number changes due to electron emission or injection, low-energy spectroscopies involving intraband or interband transitions are related to the dielectric function $\varepsilon(\mathbf{r},\mathbf{r}',\omega)$, which is linked to the density-density correlation function and characterizes the linear response to an external electric field. In particular, electron-energy-loss spectroscopy (EELS) measures the imaginary part of the inverse macroscopic dielectric function, the so-called energy-loss function $-\mathrm{Im}\ \varepsilon^{-1}(\mathbf{q},\omega)$. In Fig. 4 we display the EELS spectrum of ferromagnetic nickel calculated in the RPA. The response function is constructed with $40\times40\times40$ mesh points in the full Brillouin zone, $G_{\mathrm{max}} = 5\ \mathrm{Bohr}^{-1}$, $L_{\mathrm{max}} = 4$ and 118 unoccupied bands [15]. Second and third energy derivatives are included as local orbitals. The resulting curve is in very good agreement with experimental measurements [29], especially if transitions from the 3*s* and 3*p* core orbitals are taken into account; in this case the step in the energy-loss function around 64 eV, which corresponds to the onset of transitions from these states, is also well reproduced. The RPA formally corresponds to a complete neglect of dynamic exchange-correlation effects in the linear density response function. The latter can be approximately included by the adiabatic local-density approximation [30], but in most cases the results change very little compared to the RPA.

3.3 Spin-wave spectra

Electric fields couple to the charge of the electrons and induce characteristic excitations, such as interband optical transitions or collective plasmon modes, which correspond to resonances in the dielectric function and can be measured with frequency-resolved spectroscopies. Likewise, magnetic fields couple to the

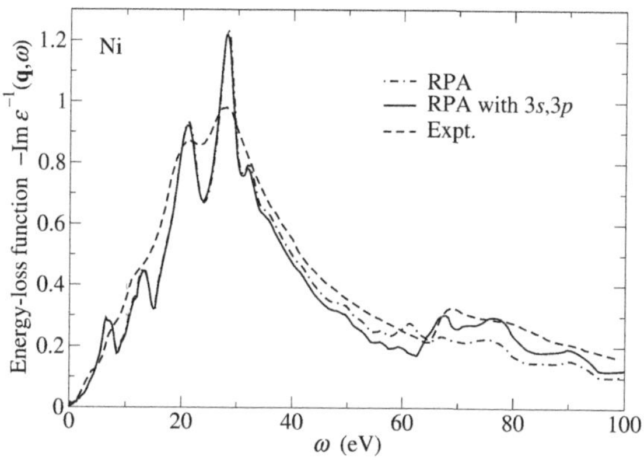

Fig. 4. Electron-energy-loss spectrum (EELS) of ferromagnetic Ni for $\mathbf{q}$ = (0.25,0,0) $2\pi/a$ calculated in the RPA with (solid line) and without (dot-dashed line) transitions from the $3s$ and $3p$ core states compared to experimental data (dashed line) from Ref. [29].

spin of the electrons and give rise to excitations of the spin system. These also fall into two groups, spin-flip Stoner excitations of individual electrons and collective spin waves, and can be identified as resonances in the dynamic transverse spin susceptibility.

Most theoretical studies of spin waves are based on the Heisenberg model, although the assumption of localized spins makes its justification doubtful for ferromagnetic metals with itinerant electrons. In contrast, very few attempts at first-principles calculations were reported so far [31,32]. The difficulty is twofold: As magnetic behavior originates in localized d or f orbitals of transition-metal and rare-earth elements, an all-electron scheme is mandatory. Furthermore, the treatment of spin waves requires dynamic exchange-correlation effects not contained in the RPA. We have solved the latter by explicitly including the screened Coulomb interaction between electrons and holes in the two different spin channels [33]. The resulting two-particle problem is analogous to the Bethe-Salpeter equation used for excitons in semiconductors. As the dominant part of the dynamic correlation in ferromagnets is caused by multiple scattering of electron-hole pairs at the same atomic site, we transform the screened interaction to a basis of maximally localized Wannier functions [34], which allows a very efficient truncation. Here we take only matrix elements with four Wannier functions that are all localized at the same site into account, but systematic extensions are of course possible to ensure full convergence [35].

As an example we show results for iron, obtained with 30×30×30 mesh points in the full Brillouin zone, G_{max} = 4.5 Bohr^{-1}, L_{max} = 4 and 100 unoccupied bands. Figure 5 shows the transverse spin susceptibility of the non-interacting Kohn-Sham system. As there is no dynamic correlation in this case, only single-particle Stoner excitations exist. As a consequence, the spectral function Im $\chi_{\mathrm{KS}}^{-+}(\mathbf{q},\omega)$ exhibits a peak at around 2 eV, which equals the exchange splitting

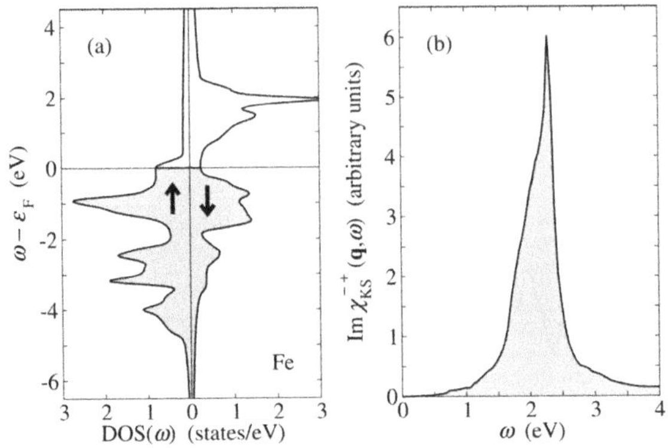

Fig. 5. (a) Density of states DOS(ω) for the majority (up) and minority (down) spin channel of ferromagnetic Fe. (b) Imaginary part of the non-renormalized Kohn-Sham spin susceptibility for $\mathbf{q}$ = (0.1,0.1,0) $2\pi/a$.

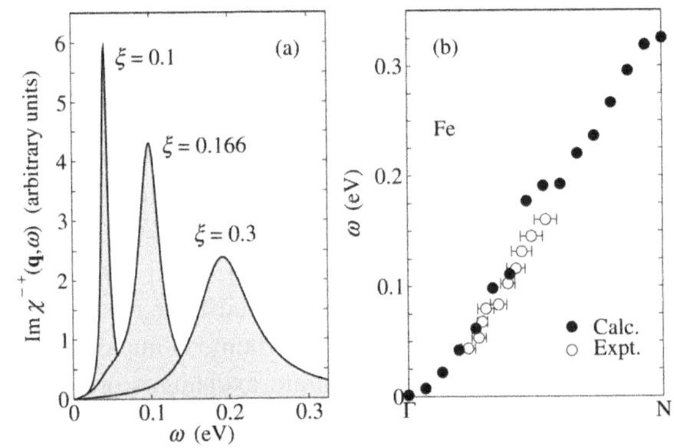

Fig. 6. (a) Imaginary part of the renormalized spin susceptibility of Fe for wave vectors $\mathbf{q}$ = (ξ,ξ,0) $2\pi/a$ and (b) calculated spin-wave dispersion along the [110] direction compared to experimental data from Ref. [36].

visible in the density of states and corresponds to spin-flip transitions between occupied majority and unoccupied minority states. When dynamic correlation is included in the form of the screened Coulomb interaction, an additional spin-wave peak appears at low energies as illustrated in Fig. 6(a). Plotting the peak positions as a function of the wave vector yields the spin-wave energies as displayed in Fig. 6(b) for the [110] direction in iron. The dispersion obtained in this way, with no empirical parameters, is in excellent agreement with experimental data [36].

4. Summary and outlook

We have discussed a new implementation of many-body perturbation theory within the FLAPW method. A number of features, such as the copious use of symmetries, the treatment of the Coulomb singularity or expedient basis transformations that allow efficient truncations, which are described in detail elsewhere [18,19], ensure a high computational efficiency. Our results for electronic excitations and associated spectroscopies in solids are in excellent quantitative agreement with experiments and show that full-potential calculations are now feasible even without plasmon-pole models or other far-reaching simplifications. As the FLAPW method is applicable to transition metals and other complex materials that cannot be easily treated with pseudopotentials, this opens up new prospects for theoretical investigations. We have already started out on this path by exploring spin-wave excitations in ferromagnets. Without any empirical parameters, we obtained spin-wave dispersions in very good agreement with experiments. The rich physics of spin-dependent phenomena means that many further developments are still necessary, however. For example, the relativistic spin-orbit coupling and non-collinear magnetism require appropriate extensions, as do spin dynamics at finite temperature or the inclusion of spin-dependent scattering in the electronic self-energy. Outside the field of nanomagnetism, the linear and nonlinear optical properties of modern optoelectronic materials provide another timely challenge.

Acknowledgement

We benefitted from useful discussions with G. Bihlmayer, M. Niesert, A. Gierlich, F. Freimuth, T. Kotani and T. Miyake. Financial support from the Deutsche Forschungsgemeinschaft through the Priority Programme 1145 and from the EU's Sixth Framework Programme through the Nanoquanta Network of Excellence (NMP4-CT-2004–500198) is gratefully acknowledged.

References

1. P. Hohenberg and W. Kohn, Phys. Rev. **136** (1964) B864.
2. W. Kohn and L. J. Sham, Phys. Rev. **140** (1965) A1133.
3. J. P. Perdew, K. Burke and M. Ernzerhof, Phys. Rev. Lett. **77** (1996) 3865; **78** (1997) 1396(E).
4. J. P. Perdew, M. Ernzerhof and K. Burke. J. Chem. Phys. **105** (1996) 9982.
5. J. Heyd, G. E. Scuseria and M. Ernzerhof, J. Chem. Phys. **118** (2003) 8207; **124** (2006) 219906(E).
6. J. Paier, M. Marsman, K. Hummer, G. Kresse, I. C. Gerber and J. G. Ángyán, J. Chem. Phys. **124** (2006) 154709; **125** (2006) 249901(E).
7. G. D. Mahan, *Many-Particle Physics*. Plenum, New York (1990).
8. L. Hedin, Phys. Rev. **136** (1965) A796.
9. C. D. Spataru, S. Ismail-Beigi, L. X. Benedict and S. G. Louie, Appl. Phys. A **78** (2004) 1129.

10. M. Hedström, A. Schindlmayr and M. Scheffler, Phys. Status Solidi B **234** (2002) 346.
11. S. Goedecker, Rev. Mod. Phys. **71** (1999) 1085.
12. G. Onida, L. Reining and A. Rubio, Rev. Mod. Phys. **74** (2002) 601.
13. http://www.flapw.de/
14. D. J. Singh, *Planewaves, Pseudopotentials and the LAPW Method*. Kluwer, Dordrecht (1994).
15. C. Friedrich, A. Schindlmayr, S. Blügel and T. Kotani, Phys. Rev. B **74** (2006) 045104.
16. T. Kotani and M. van Schilfgaarde, Solid State Commun. **121** (2002) 461.
17. C. Freysoldt, P. Eggert, P. Rinke, A. Schindlmayr, R. W. Godby and M. Scheffler, Comput. Phys. Commun. **176** (2007) 1.
18. C. Friedrich, A. Schindlmayr and S. Blügel, Comput. Phys. Commun. **180** (2009) 347.
19. C. Friedrich, S. Blügel and A. Schindlmayr, Phys. Rev. B **81** (2010) 125102.
20. M. S. Hybertsen and S. G. Louie, Phys. Rev. Lett. **55** (1985) 1418.
21. R. W. Godby, M. Schlüter and L. J. Sham, Phys. Rev. Lett. **56** (1986) 2415.
22. K. van Benthem, C. Elsässer and R. H. French, J. Appl. Phys. **90** (2001) 6156.
23. G. Cappellini, S. Bouette-Russo, B. Amadon, C. Noguera and F. Finocchi, J. Phys.: Condens. Matter **12** (2000) 3671.
24. W. Ku and A. G. Eguiluz, Phys. Rev. Lett. **89** (2002) 126401.
25. M. L. Tiago, S. Ismail-Beigi and S. G. Louie, Phys. Rev. B **69** (2004) 125212.
26. M. Shishkin and G. Kresse, Phys. Rev. B **74** (2006) 035101.
27. M. van Schilfgaarde, T. Kotani and S. V. Faleev, Phys. Rev. B **74** (2006) 245125.
28. R. I. Gómez-Abal, X. Li, M. Scheffler and C. Ambrosch-Draxl, Phys. Rev. Lett. **101** (2008) 106404.
29. L. A. Feldkamp, M. B. Stearns and S. S. Shinozaki, Phys. Rev. B **20** (1979) 1310.
30. S. Botti, A. Schindlmayr, R. Del Sole and L. Reining, Rep. Prog. Phys. **70** (2007) 357.
31. S. Y. Savrasov, Phys. Rev. Lett. **81** (1998) 2570.
32. K. Karlsson and F. Aryasetiawan, Phys. Rev. B **62** (2000) 3006.
33. F. Aryasetiawan and K. Karlsson, Phys. Rev. B **60** (1999) 7419.
34. F. Freimuth, Y. Mokrousov, D. Wortmann, S. Heinze and S. Blügel, Phys. Rev. B **78** (2008) 035120.
35. E. Şaşıoğlu, A. Schindlmayr, C. Friedrich, F. Freimuth, and S. Blügel, Phys. Rev. B **81** (2010) 054434.
36. C.-K. Loong, J. M. Carpenter, J. W. Lynn, R. A. Robinson and H. A. Mook, J. Appl. Phys. **55** (1984) 1895.

Development of a Wavefunction-based *Ab Initio* Method for Metals Applying the Method of Increments

By Elena Voloshina and Beate Paulus*

Institut für Chemie und Biochemie - Physikalische und Theoretische Chemie, Freie Universität Berlin, Takustraße 3, 14195 Berlin, Germany

(Received July 30, 2009; accepted September 17, 2009)

Ab Initio Method / Method of Increments / Metals / Wavefunction-based Methods

Ab initio electron correlation calculations based on quantum-chemical methods are successfully applied to metallic systems via the method of increments, which is an expansion of the correlation energy in terms of one-body, two-body, and higher-order contributions from localized orbital groups. To deal with the two distinct problems that occur in metals, the difficulty of localization of the orbitals and the generation of clusters with neutral atoms in the center, we proposed an embedding scheme which has itself no metallic character but can mimic the metal in the internal region, where the atoms are correlated. The first application was made for solid mercury, where a very good agreement with experimental ground-state properties was achieved. Further the approach has been extended to other group 2 and 12 metals (Be, Mg, Zn, and Cd) where the metallic character is more pronounced than in mercury. Different variants of the embedding and the localized orbitals used for the incremental scheme have been developed and tested. Application of the method of increments to the investigated metallic systems allows us not only to obtain values close to experimental data but also to understand the influence of individual correlation-energy increments on cohesive properties and to clarify thereby some aspects of the structural features of the group 12 metals.

1. Introduction

Ab initio electronic structure methods are designed to solve the full electronic Hamiltonian. One very successful method to do this is the density-functional theory (DFT) [1]. Within a DFT-based formalism one avoids constructing the many-body wavefunction of the system and instead computes directly ground-state properties from its charge density. However, the exact form of the func-

* Corresponding author. E-mail: beate@chemie.fu-berlin.de

© by Oldenbourg Wissenschaftsverlag, München

tional embodied in the Hohenberg-Kohn theorem is unknown. Many, sometimes sophisticated, approximations for the exchange-correlation functional have been developed. However, their applicability depends on the system involved. Generally, the present DFT-based approaches are not amenable to systematic improvements.

The alternative approach to the problem of electron correlations determines the many-body wavefunction of the system by approximately solving the corresponding Schrödinger equation. One can systematically improve this approach by including more terms in the expansion of the wavefunction of the system. However such wavefunction-based approaches scale quite unfavorably with the size of the system. Therefore in their *ab initio* form they have been applied mainly to quite small molecules. In condensed-matter physics, wavefunction-based many-body approaches are generally used in conjugation with model Hamiltonians. Usually various approximations have to be made, and the extension of *ab initio* quantum-chemistry-type methods to infinite systems did not seem feasible for a long time.

However, it has been shown recently that it is possible to apply quantum-chemical correlation methods to extended systems when using the idea of local excitations (the property, that the correlation hole, i.e. the region, where the electrons interact instantaneously, is fairly local). For a solid it is useful, therefore, to switch from the description with delocalized Bloch states to localized Wannier orbitals [2]. This is the starting-point for the so-called local correlation methods. Historically, the first method of this type is the Local Ansatz (LA) developed by Stollhoff and Fulde [3]. Whereas e.g. the coupled cluster approach [4] uses all possible single and double excitations (CCSD), the LA uses only a subset of these excitations, relying on a physical picture of the material examined. Therefore this method can easily yield an insight into the mechanism of the correlations, but an error estimation is difficult. Another possibility to restrict the excitation space is the local correlation method of Pulay [5], where for each occupied localised orbital an excitation domain is constructed. A further method, the method of increments [6] developed by Stoll, combines Hartree-Fock (HF) calculations for periodic systems with correlation calculations for finite embedded clusters, and the total correlation energy per unit cell of a solid is written as a cumulant expansion in terms of contributions from localized orbital groups of increasing size. In contrast to the method of Pulay, here an *a priori* truncation of the virtual space is not necessary due to the partitioning of the correlation energy into correlation-energy increments. But of course a combination of the method of increments with the local correlation method of Pulay is possible to reduce the computational effort even further [7]. Due to the computational savings with the restriction of the excitation space it is now possible to extend the local correlation method according to Pulay to solids. First attempts are made by Pisani and Schütz to extend the program package CRYSTAL [8] to correlations obtained with the local MP2 method resulting in the program code CRYSCOR [9]. Note, however, that up to now CRYSCOR treatment is limited to non-metals.

The focus of this project lies on determining ground-state properties of metals with wavefunction-based correlation methods using the method of increments. Calculations based upon the method of increments have been performed on a variety of solids with an energy gap (For a review see Ref. [10]). In all cases the ground-state properties obtained agree well with the experimental values.

Metals require a special treatment because of two distinct features. Since the conduction bands are only partially filled, one cannot construct well-localized orbitals from them. Furthermore one must deal properly with charge screening which obviously is a correlation effect. One would expect that the metallic solid should be well described within DFT. Unfortunately, this seems to be not the case. For instance, in the case of solid mercury the method of increments was successfully applied [11], whereas various DFT-functionals fail.

2. Method of increments

2.1 General formalism

Quantum-chemical correlation methods, developed for finite systems, can be applied to periodic systems using the method of increments [6]. In this approach, the total energy is written as $E = E^{HF} + E^{\text{corr.}}$, where E^{HF} is the HF energy of the system, and $E^{\text{corr.}}$ is the contribution of correlation effects to the total energy per unit cell. The correlation contribution is computed using the incremental expansion:

$$E^{\text{corr.}}_{\text{solid}} = \sum_{A \in u.c.} \varepsilon_A + \frac{1}{2!} \sum_{\substack{A \neq B \\ A \in u.c. \\ B \in \text{solid}}} \Delta\varepsilon_{AB} + \frac{1}{3!} \sum_{\substack{A \neq B \neq C \\ A \in u.c. \\ B,C \in \text{solid}}} \Delta\varepsilon_{ABC} + \dots. \tag{1}$$

The ε_A (one-body increment) is computed by considering excitations only from the A-orbitals, freezing the rest of the solid at the HF level. The two-body increment is defined as $\Delta\varepsilon_{AB} = \varepsilon_{AB} - (\varepsilon_A + \varepsilon_B)$,where ε_{AB} is the correlation energy of the joint orbital system AB. Higher-order increments are defined in an analogous way. Thus, for the three-body term we get: $\Delta\varepsilon_{ABC} = \varepsilon_{ABC} - (\varepsilon_A + \varepsilon_B + \varepsilon_C) - (\Delta\varepsilon_{AB} + \Delta\varepsilon_{AC} + \Delta\varepsilon_{BC})$. Finally, summing up all increments, with the proper weight factors (according to their occurrence in the solid), one obtains the correlation energy per unit cell of the infinite system. In order to get reliable results a size-extensive correlation method must be used. Of course, the incremental expansion only makes sense if it is well convergent, i.e., if $\Delta\varepsilon_{AB}$ rapidly decreases with increasing distance between the positions A and B, and if the three-body terms are significantly smaller than the two-body ones and so on.This means that only a few increments need to be calculated, yet a nearly full account of the short range correlations is achieved in this way.

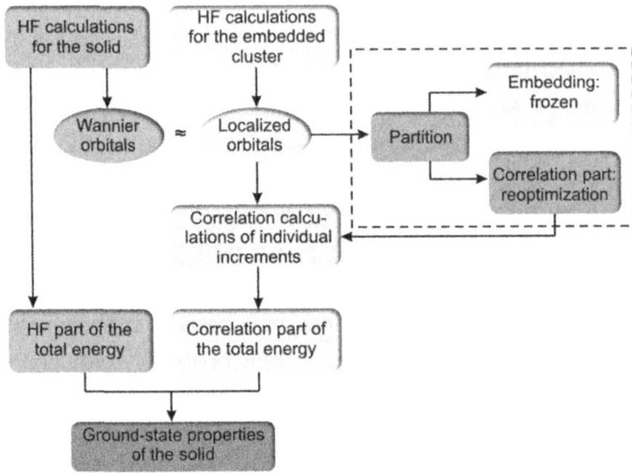

Fig. 1. Schematic overview of the ground-state calculations with the method of increments. Part which is specific for metals is marked by dashed line. This Fig. is taken from Ref. [13].

Due to the need to calculate the correlation energy increments in finite embedded clusters, we have an additional requirement that the energy must converge with the size of the embedding. As an additional advantage, we can calculate individual terms of the expansion from suitable embedded finite clusters of reasonable size. A schematic overview of the procedure is given in Fig. 1.

A direct transfer of this approach to metallic systems is not possible since localized orbitals become very long-range entities. Therefore, a many-body expansion in terms of such orbitals cannot be expected to have useful convergence characteristics.

2.2 Modifications for applying the method of increments for metals

To deal with the two distinct problems that occur in metals, the difficulty of localization of the orbitals and the generation of clusters with neutral atoms in the center, we suggest the following procedure [12,13].

Selection of fragments of the solid: The correlation-energy increments are calculated for selected fragments which reflect the geometry of the crystal. These fragments have two components, firstly the atoms A, B, C, ... to be correlated (in the center of the fragment) and secondly the embedding atoms. We select the positions of the atoms to be correlated and surround each of these atoms by embedding shells of various sizes. This corresponds to a spherical cut-off, which best simulates the Madelung forces in the center. As an example, Fig. 2 shows the one-body cluster built on the basis of *hcp* structure of Mg.

Calculation of the cluster with a minimal valence basis set: The embedding atoms are required to simulate the environment of an atom of the infinite solid, without allowing the electronic charge to diffuse towards the surface of the cluster. Additionally the basis set should be chosen in such a way to enable

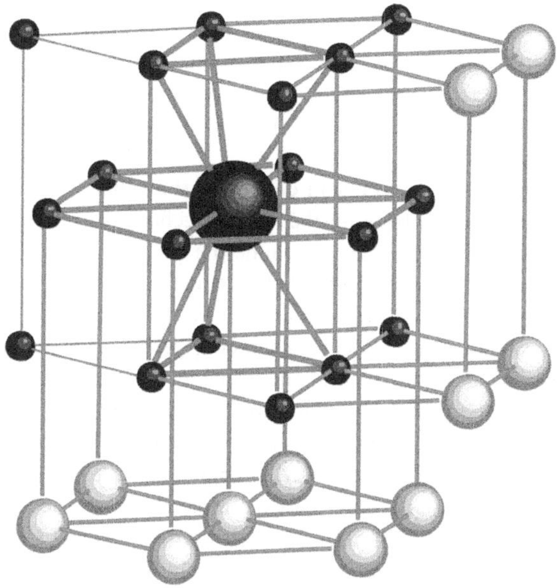

Fig. 2. Crystallographic structure of *hcp* Mg. Dark spheres show the studied one-body cluster: the large dark sphere is the central atom (described with cc-pVTZ basis set) and the small dark spheres are 18 embedding atoms (described with large-core pseudopotential and [*s*] basis set). The light atoms are shown in order to complete the *hcp* cell of Mg. This Fig. is taken from Ref. [18].

localization of the orbitals. Both requirements can be satisfied, if we supply only *s*-basis functions for the valence electrons. This way no metallic character can be described, but each atom still has its correct crystal surroundings with respect to the electrostatic interaction.

Localization of the orbitals: Within the initial description of only valence *s*-basis functions, we perform a HF calculation of the cluster and due to the neglect of any metallic character we can use standard procedures to localize the orbitals (e.g. the Foster-Boys criterion or that of Pipek-Mezey [14]). This set of localized orbitals contains both the embedding orbitals which are centered at the embedding atoms and also the orbitals located at the atoms which are to be correlated.

Allowing for delocalization in the central part: While the embedding orbitals are kept frozen, we supply on the atoms to be correlated an extended basis set of at least double-ζ quality with polarization functions, recalculate the integrals and perform an HF calculation for the orbitals to be correlated. During this HF calculation the central orbitals are reoptimized and can develop a metallic character. But the long-range tails are absent, because the embedding region is kept frozen.

Correlation calculation of the central part: The reoptimized orbital set may be correlated with any size-extensive method. Within this correlation calculation it is possible to freeze part of the semi-core orbitals, (e.g. the *d*- or underlying *sp*-shell in the case of the group 12 metals) to analyze the influence of the valence only correlations [11,15].

Determination of the correlation energy of the solid: After evaluating the individual increments and checking their convergence with increasing distance of the atoms involved, we have to determine the weight factors with which the individual increments occur in the solid state structure and sum them up for the correlation energy of the solid per unit cell (or per atom) as described in eqn.1.

One more problem that can occur when considering metallic systems is the quasi-degenerate bands. This render questionable a single-reference approach like the coupled-cluster (CC) [16] which we have used in many of our previous applications [10]. In such a case one has to check whether it is necessary to treat the electrons within a multi-reference (MR) approach [17]. In such an approach, the correlation energy is partitioned into a complete active space self-consistent field (CASSCF) part which describes the static correlations and a MR configuration interaction part for the dynamic correlations. In the case of metals we select an active space for the CASSCF calculation including the important bands around the Fermi level. The dynamic correlations are treated on top of the CASSCF wavefunction with an approximately size-extensive MR correlation method, i.e. an MR averaged coupled pair functional (MR-ACPF).

3. Influence of the embedding on the correlation energy

The correlation effects which influence the ground-state properties of metals are quite local. Therefore, the question can arise, whether it is really necessary to embed the clusters for the correlation calculation. Could we not take the correlation part from free fragments of the metals, i.e. the two-body part from dimers, the three-body parts from trimers and so on. Still the long range HF part of the ground-state properties would be calculated in the periodic system. In order to answer this question, the correlation contributions extracted from free and embedded fragments of the *hcp* structure of magnesium have been compared [18].

In Fig. 3 the various contribution to the cohesive energy of magnesium are presented. Up to the three-body correlation correction the both series show rather good convergence ($\Sigma\Delta\varepsilon_{AB} >> \Sigma\Delta\varepsilon_{ABC}$) and can lead to more or less similar results. The situation changes drastically when considering 4-body correlation-energy terms. While one can clearly see the good convergence behavior of the incremental scheme when using the embedded clusters ($\Sigma\Delta\varepsilon_{AB} >> \Sigma\Delta\varepsilon_{ABC} >> \Sigma\Delta\varepsilon_{ABCD}$), free Mg-clusters yield $\Sigma\Delta\varepsilon_{ABCD}$ which is even 1.5 times higher than the corresponding two-body contribution. Of course, such data yield a great overestimation of expected correlation energy.

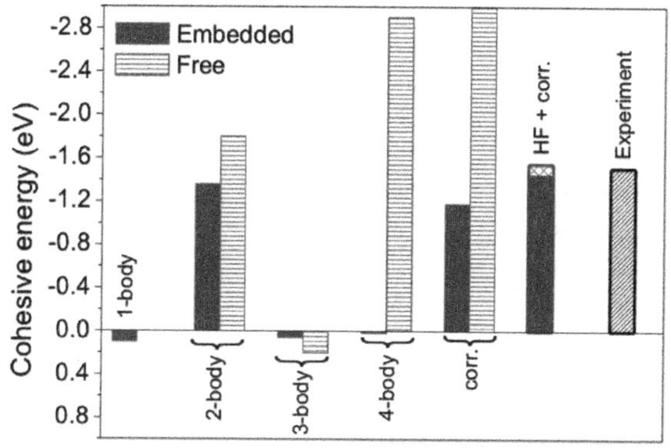

Fig. 3. Cohesive energy of the solid magnesium: 1-, 2-, 3-, and 4-body contributions obtained with cc-pVTZ basis set for the two different approaches (cf. text) and calculated as compared with experimental value. The shading of the final result (HF + corr.) indicates the uncertainty in the HF value. This Fig. is taken from Ref. [13].

In order to understand the reason for such behaviour one has to consider the nature of binding in free Mg-clusters in detail. Whereas the equilibrium distance of the dimer is about 20 % larger than the distance in the solid, for Mg_3 (equilateral triangle) the interatomic distance is only 5 % larger, and for the Mg-tetramer (tetrahedron) it is even smaller by 3 %. Also the binding energy per atom increases drastically from Mg_3 to Mg_4. That is also seen in a DFT/BP-86 study for Mg clusters up to 22 atoms [19], where the cohesive energy per atom has a bend at $n = 4$. This gives a hint, that for Mg_4 the electronic structure is changed from purely van der Waals-like to binding with covalent and metallic contributions.

Thus, by modifying the nature of binding, the embedding influences also the correlation energy, that can be extracted from the corresponding clusters. Our attempt to get good agreement with experiment for Mg-bulk correlation energy, using the free clusters, fails due to the mixing of species of different nature. At the same time when using the embedded scheme one reaches nearly 100 % of the expected value with the method of increments [18]. An analogous study concerning the use of non-embedded species was performed for mercury [11]. In that case, in contrast to the embedded clusters, the convergence of the incremental scheme also failed, although the result is not so dramatic since the metallic character in solid Hg is less pronounced. Another reason is that approximately half of the cohesion of mercury is due to *d*-correlations which have very local character.

Table 1. Selected local increments (in Hartree per atom) extracted from Be clusters of different size (the most compact structures). The calculation have been performed with CCSD(T) method; only 2*s* electrons are correlated. For the correlated atoms a cc-pVTZ basis set was employed. A cut-off of $1.4a_0$ was used for embedding with [2*s*] basis set.

Cluster	Be_4	Be_8	Be_{13}
ε_A	–0.032164	–0.031200	–0.031123
$\Delta\varepsilon_{AB}$ [1]	–0.013791	–0.013203	–0.013065
$\Delta\varepsilon_{ABC}$ [2]	+0.002231	+0.002004	+0.001833

[1] r_{AB} = 2.23 Å; [2] r_{AB} = 2.23 Å, r_{AC} = 2.23 Å, r_{BC} = 2.29 Å.

4. Particular case of beryllium

The beryllium crystal has the same *hcp* structure as magnesium (they differ in their c/a ratio with respect to the ideal one); the free atoms are characterized by closed ns^2 shells. At the same time, Mg is almost free-electron-like whereas beryllium, although in general metallic, shows a lower density of states of almost exclusively *p* character at the Fermi energy. Therefore, in order to get reliable result one has to model this physical picture. Our first attempt to do this via an embedding scheme, namely, to divide the embedding into two regions: a true embedding region, where the atoms are described with a minimal basis [*s*], and an intermediate region, where the occupied orbitals are kept frozen, but [2*s*1*p*] basis sets are supplied for the re-optimization of the orbitals in the central cluster (in the middle) [20] failed since we did not get convergence with respect to the order of increments ($\Sigma\Delta\varepsilon_{AB} > \Sigma\Delta\varepsilon_{ABC}$, but $\Sigma\Delta\varepsilon_{ABC} < \Sigma\Delta\varepsilon_{ABCD}$) although when summing up all calculated contributions an excellent agreement with experiment was reached. Taking into account our observation regarding Mg-clusters described in previous section (see Sec. 3 and Ref. [18]), for the further investigation we decided to use Be entities to be correlated of at least 4 atoms.

In Tab. 1 we list selected correlation-energy increments for Be extracted from different clusters. From these results one can see the good transferability of the increments. This is not always the case when using smaller clusters and can be reached mostly through more complicated embedding scheme [20]. Another advantage of using larger clusters is the possibility to estimate the contribution of higher-order increments. For example, in the case of Be_8-cluster, contributions from four-, five-, six-, seven-, and even one eight-body species can be expressed as $E^{corr}_{\text{total}}–\Sigma\varepsilon_A–\Sigma\Delta\varepsilon_{AB}–\Sigma\Delta\varepsilon_{ABC}$ = 0.75 mHartree (i.e. less than 1 % from the total correlation energy). Here, $\Sigma\varepsilon_A$, $\Sigma\Delta\varepsilon_{AB}$, and $\Sigma\Delta\varepsilon_{ABC}$ are sums of the 8 one-body, 28two-body, and 56 three-body contributions, which can be obtained from 8-body Be-cluster.

Work is under way in our laboratory to settle the incremental expansion beryllium using larger entities like Be_4 pyramids as one-body increments.

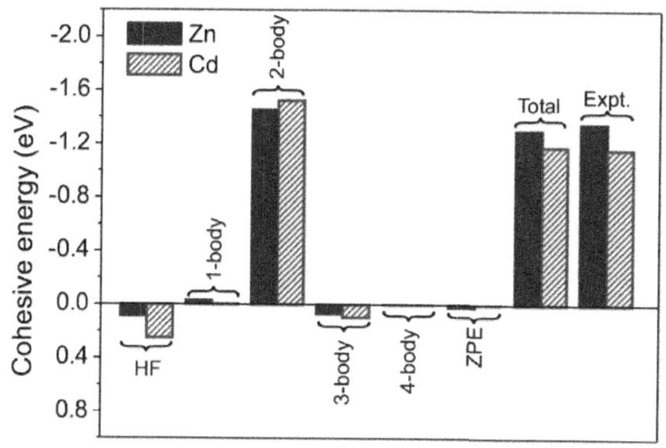

Fig. 4. The different contributions to the total binding energy from the method of increments are plotted, for the experimental lattice parameters, and compared to the experimental value. The zero-point energy (ZPE) correction is estimated from the Debye temperature [21]. This Fig. is taken from Ref. [13].

5. Metallic binding due to correlations

For the group 12 metals, zinc and cadmium, which condense in the *hcp* structure [21] we perform a wavefunction-based correlation treatment on top of periodic HF calculations for these materials applying the method of increments [15]. Although the HF treatment yields no binding and no equilibrium geometry for zinc and cadmium [22], the binding of the ground-state structure is fully described by electronic correlations. The performance of the method of increments for Zn and Cd at the experimental lattice parameters is summarized in Fig. 4. The correlation contribution to the cohesive energy from one-atom increments, where the correlation contribution of the free atom is subtracted, is small. The two-body correlation energy is clearly the dominant part of the total binding energy: The two-body contributions yield about 105 % of the total correlation part to the cohesive energy. It is interesting to note that the two-body correlation energy of cadmium is larger than that of zinc, and the decreased binding energy of cadmium relative to zinc is due to the increased repulsion in the solid at the HF level. The three-body terms are small again (8 % of the correlation part to the cohesive energy) and repulsive. *d* correlation is 42 % of the total correlation for zinc and 48 % for cadmium. Therefore an uncorrelated treatment of the *d* shell will yield unrealistic results. Overall we reach a very good agreement with the experimental values. The largest error in determining the cohesive energy is not in the correlation part (neglected far-away two- and three-body terms and four body terms are of the order of the zero-point vibrational energy) but in the HF part due to the difficult balance between the basis sets describing the crystal and the free atom [23].

Table 2. Lattice constants (Å), cohesive energies (eV), and cell volumes (Å^3) for minima on the potential energy surface of zinc.

	a	c/a	E_{coh}	V
Expt.[a]	2.67	1.86	−1.35	30.4
expt.-like	2.61	1.91	−1.35	29.3
ideal-like	2.72	1.60	−1.39	27.8
min-3	2.61	1.80	−1.35	27.7

[a]Lattice constants [21], cohesive energy [24]

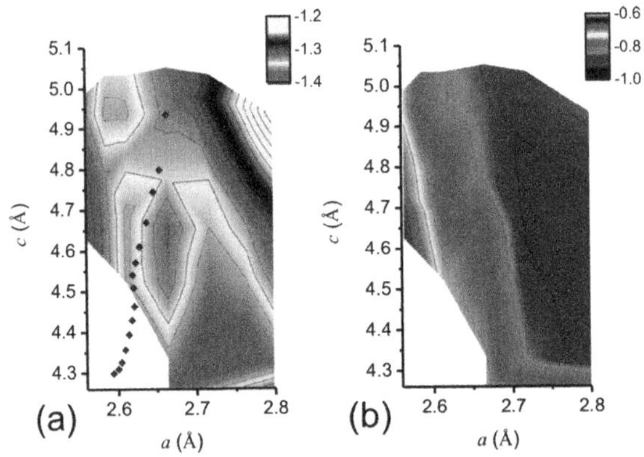

Fig. 5. Total binding energy of Zn with respect to the lattice constants (in eV), with (a) and without (b) d correlation. Note the different scales of the contour plots. The blue diamonds in the left diagram mark the lattice constants measured by Takemura [25] under high pressure. This Fig. is taken from Ref. [15].

In contrast to all other *hcp* elements metallic Zn and Cd constitute conspicuous exceptions, showing extreme values of c/a = 1.86 and 1.89 for Zn and Cd, respectively, deviating by 14 % and 16 % from the ideal value [21]. Using the methods of increments we have been able to draw a detailed picture of the *hcp* structure of zinc. Whereas the two-body terms determine the cohesive energy, for the lattice parameters the three-body terms are essential [15]. The obtained lattice parameters, cohesive energies, and cell volumes for minima on the PES are summarized in Tab. 2.

The cohesive energy at the minimum called "expt.-like" in Tab. 2 matches the experimental one. The corresponding a parameter deviates by −2 %, the c-parameter by less than 1 %. One reason for the over-binding in the a-b plane may be the neglect of the correlation contributions of the semicore orbitals (3s, 3p). They were computed for a few points on the PES and lead to a reduction of the binding energy by 0.01 to 0.02 eV. This effect is smaller at larger lattice constants.

Behind a barrier of 0.02 eV height towards smaller c parameters it features two valleys separated by a ridge. The corresponding minima are called "ideal-like" and "min-3" in Tab. 2. The ideal-like minimum with a nearly ideal c/a ratio even exhibits a slightly larger binding energy than the expt.-like one. This fact raises the question why the ideal-like minimum thus far has not been accessible via high-pressure experiments. The reason is given by the special shape of the energy landscape. Because of the hydrostatic conditions in the experiments that prevent a signiÞcant expansion of the *a*-*b* plane, the system moves through the left valley. Taking into account that our calculations slightly overestimate the binding in the hexagonal plane, the high- pressure data of Takemura [25] fit very well to our PES (Fig. 5). The additional local minimum in the left valley (min-3 in Tab. 2) is another indication for the strong bond in the hexagonal planes and may further explain the anomalies observed in high-pressure experiments under certain conditions.

6. Conclusion

Since 1992, when H. Stoll has applied the idea of local excitations in the electronic theory of solids for considering the correlation energy of diamond, silicon and graphite, the method of increments has been succesfully applied for various systems with an energy gap. The problem to extend the method to metallic systems is twofold. The first problem consists in the modelling of a metallic solid by a finite cluster, and connected with this the question about the localisation of orbitals in a metal arises. As a solution we suggest an embedding scheme which has itself no metallic character but can mimic the metal in the internal region, where the atoms are correlated. Within the incremental scheme we allow for a delocalisation of the orbitals and therefore account for the metallic character of the system. With this approach the convergence rate, especially with the order of increments, is not as good as for insulators, but still three- and higher-body terms account for only about 15 % of the correlation contribution to the binding. Up to now we have carefully described closed shell metals [13]. However, it is possible to extend this approach also to opened-shell cases. Thus, the correlation energy of a metallic Li-ring has been considered by applying a projection technique for non-orthogonal occupied and virtual orbitals within the incremental scheme [17]. Even an embedding scheme was suggested for 3-dimensional lithium relying on pairs of atoms [26]. In the case of lithium the band is crossing the Fermi level and is half occupied. In this regard the second problem occuring in metals has to be mentioned, namely the quasi-degeneracy of the orbitals around the Fermi surface. Then a single-reference correlation method fails and the incremental scheme has to be extended to the use of multi-reference correlation treatments. This can be the case not only in open-shell systems, but also, for example, in barium [27], where the *d*-bands are crossing the Fermi level.

Thus, with the method of increments we have a tool, which allows the evaluation of the ground-state properties of metals. Starting from the HF solution for the extended systems, which is a routine calculation nowadays, we can localize the orbitals and perform an incremental expansion for the correlation energy. An asset of the incremental scheme is that applying it allows one not only to obtain systematically improvable results, but also to investigate the influence of individual contributions to the correlation energy. This opens up a large field of applications. With such a method at hand we can try to understand unusual structures of some metals. Preliminary work has been done in our laboratory for gallium [28], which forms Ga-dimers within the metallic phase. One more class of very interesting systems is intermetallic compounds, especially Laves phases. Here our approach could be useful for understanding their stability and phase formation as a function of components.

Acknowledgement

Financial support through the priority program SPP 1145 ("Modern and universal first-principles methods for many-electron systems in chemistry and physics") of the DFG is acknowledged. The authors want to thank Prof. Dr. Herrmann Stoll (Stuttgart, Germany), Prof. Dr. Krzysztof Rosciscewski (Krakow, Poland), Dr. Nicola Gaston (Lower Hutt, New Zealand), Dr. Ulrich Wedig (Stuttgart, Germany), Prof. Dr. Martin Jansen (Stuttgart, Germany), and PD Dr. Dirk Andrae (Berlin, Germany) for the intensive collaboration in various joined projects.

References

1. W. Kohn and L. J. Sham, Phys. Rev. **140** (1965) A1133; P. Hohenberg and W. Kohn, Phys. Rev. **136** (1964) B864.
2. G. H. Wannier, Phys. Rev. **52** (1937) 191.
3. G. Stollhoff, P. Fulde, Z. Phys. B **26** (1977) 251; G. Stollhoff, P. Fulde, J. Chem. Phys. **73** (1980) 4548; G. Stollhoff, J. Chem. Phys. **105** (1996) 227.
4. F. Coester, Nucl. Phys. **7** (1958) 421; F. Coester, H. Kümmel, Nucl. Phys. **17** (1960) 477; H. Kümmel, Nucl. Phys. **22** (1961) 177; J. Cizek, Adv. Chem. Phys. **14** (1969) 35.
5. P. Pulay, Chem. Phys. Lett. **100** (1983) 151.
6. H. Stoll, Phys. Rev. B **46** (1992) 6700.
7. S. Kalvoda, B. Paulus, M. Dolg, H. Stoll, H.-J.Werner, Phys. Chem. Chem. Phys. **3** (2001) 514. H. Stoll, B. Paulus, P. Fulde, J. Chem. Phys. **123** (2005) 144108.
8. R. Dovesi, V. R. Saunders, C. Roetti, *et al., CRYSTAL06 Users Manual.* University of Torino, Torino (2006); see also, http://www.crystal.unito.it.
9. C. Pisani, M. Busso, G. Capecchi, S. Casassa, R. Dovesi, L. Maschio, C. Zicovich-Wilson and M. Schütz, J. Chem. Phys. **122** (2005) 094113; see also CRYSCOR home page: http://www.cryscor.unito.it/home.html.
10. B. Paulus, Phys. Rep. **428** (2006) 1.
11. B. Paulus and K. Rosciszewski, Chem. Phys. Lett. **394** (2004) 96; B. Paulus, K. Rosciszewski, N. Gaston, P. Schwerdtfeger, and H. Stoll, Phys. Rev. B **70** (2004) 165106; N. Gaston, B. Paulus, K. Rosciszewski, P. Schwerdtfeger, and H. Stoll, Phys. Rev. B **74** (2006) 094102.
12. E. Voloshina, N. Gaston, and B. Paulus, J. Chem. Phys. **126** (2007)134115.
13. E. Voloshina and B. Paulus: Wavefunction-based ab initio correlation method for metals: Application of the incremental scheme to Be, Mg, Zn, Cd, and Hg, in *Chemi-*

cal Modelling: Applications and Theory, Vol. 6, M. Springborg (ed.), R.S.C., Cambridge, 2009.
14. J. M. Foster and S. F. Boys, Rev. Mod. Phys. **32** (1960) 296; J. Pipek and P. G. Mezey, J. Chem. Phys. **90** (1989) 4916.
15. N. Gaston and B. Paulus, Phys. Rev. B **76** (2007) 214116; N. Gaston, B. Paulus, U. Wedig, and M. Jansen, Phys. Rev. Lett. **100** (2008) 226404.
16. C. Hampel, K. Peterson, and H.-J. Werner, Chem. Phys. Lett. **190** (1992) 1; M. J. O. Deegan and P. J. Knowles, Chem. Phys. Lett. **227** (1994) 321. J. D. Watts, J. Gauss and R. J. Bartlett, J. Chem. Phys. **98** (1995) 8718.
17. H. Stoll, Ann. Physik, **5** (1997) 355; B. Paulus, Chem. Phys. Lett., **371** (2003) 7.
18. E. Voloshina and B. Paulus, Mol. Phys. **105** (2007) 2849; E. Voloshina and B. Paulus, Phys. Rev. B **75** (2007) 245117.
19. A. Köhn, F. Weigend, and R. Ahlrichs, Phys. Chem. Chem. Phys. **3** (2001) 711.
20. E. Voloshina, B. Paulus, and H. Stoll, J. Phys.: Conf. Ser. **117** (2008) 012029.
21. C. Kittel, *Introduction to Solid State Physics*, 7th edition. John Wiley and Sons, New York, Chichester (1996).
22. U. Wedig, M. Jansen, B. Paulus, K. Rosciszewski, and P. Sony, Phys. Rev. B **75** (2007) 205123.
23. B. Paulus and K. Rosciszewski, J. Phys.: Condens. Matter **19** (2007) 346217.
24. CRC Handbook of Chemistry and Physics. CRC Press, New York (1997).
25. K. Takemura, Phys. Rev. B **56** (1997) 5170.
26. H. Stoll, B. Paulus, and P. Fulde, Chem. Phys. Lett. **469** (2009) 90.
27. B. Paulus and A. Mitin, Lecture series on computer and computational science **7** (2006) 935.
28. E. Voloshina, K. Rosciszewski, and B. Paulus, Phys. Rev. B **79** (2009) 045113.

Recent Advances in Explicitly Correlated Coupled-Cluster Response Theory for Excited States and Optical Properties

By Jun Yang and Christof Hättig*

Lehrstuhl für Theoretische Chemie, Ruhr-Universität Bochum, Universitätstrasse 150, 44801 Bochum, Germany

(Received July 27, 2009; accepted September 24, 2009)

Explicitly Correlated Coupled-cluster / Response Theory / Equation of Motion / Dipole Moment / Optical Property / Excited State

The derivation of coupled-cluster response theory for explicitly correlated wavefunctions with terms linear in the interelectronic distance r_{12} (CC-R12) or a function f_{12} thereof (CC-F12) is reviewed and an implementation at the CCSD(R12) and CCSD(F12) for excitation energies and the linear, quadratic and cubic reponse functions is reported for ansätze 1 and 2. We discuss the results of first applications for vertical excitation energies, excited state equilibrium structures and vibrational frequencies, polarizabilities and first and second hyperpolarizabilities with focus on present limitations and future prospects of the approach. We show in particular that the standard choice of the geminal functions and wavefunction expansion can in certain cases lead to a bias towards the unperturbed ground state and how this can be avoided by extending the geminal space. If taken care of this, in particular CCSD(F12)/ansatz 2 shows for the potential energy curves of excited states and for optical properties a similar enhanced basis set convergence of the correlation contribution due to connected double excitations as for ground state energies. For diffuse excited states and hyperpolarizbilities the correlation contribution converges with CCSD(F12)/ansatz 2 often faster than orbital relaxation contributions. The convergence of the latter can be improved by allowing single excitations into a space spanned by a complementary auxiliary basis set.

1. Introduction

Explicity correlated coupled-cluster methods have experienced during the last decade a tremendous increase in efficiency and accuracy and are becoming now a standard tool of computational chemistry for the calculation of accurate ground state energies [1]. The slow convergence of the correlation contribution to the

* Corresponding author. E-mail: christof.haettig@rub.de

Z. Phys. Chem. **224** (2010) 383–395
© by Oldenbourg Wissenschaftsverlag, München

electronic wavefunction is also a severe bottleneck in the calculation of linear and nonlinear optical properties. The extension of explicitly correlated techniques to coupled-cluster response theory thus proffers the possibility to extend considerably the applicability of coupled-cluster response theory towards larger systems by lowering the arduous basis demands of such calculations.

The application of the techniques and approximations developed within R12 (or F12) theory for ground state calculations in the framework of coupled-cluster response theory to allow for the calculation of excited states or optical properties is, however, not as straight forward as one might at first expect. As is known, for example from local correlation methods [2,3], the requirements for the excitation vectors, which describe in response theory the excited states, or for responses of the cluster amplitudes to e.g. external electric fields are quite different from those for the ground state cluster amplitudes or wavefunction. While the cluster amplitudes for the unperturbed ground state only account for correlation effects, the excitation vectors describe the excitation process beside differential correlation and orbital relaxation effects. In coupled-cluster response theory usually the unperturbed Hartree-Fock wavefunction is used as reference state to avoid in the frequency-dependent case spurious unphysical poles and orbital relaxation effects are described through the cluster operator. Thus the response of the cluster amplitudes to external fields describes the polarization not only of the correlation contribution to the wavefunction but also of the orbitals. As will be shown below, the balanced desription of these contributions is a challenge for R12 theory which requires at least in some situations an extension of the wavefunction ansatz which is usually used in CC-R12 or CC-F12 calculations.

2. Theory

CC-R12 and CC-F12 methods include in the coupled-cluster wavefunction ansatz

$$|\mathrm{CC}\rangle = \exp(\hat{T})|\mathrm{HF}\rangle \tag{1}$$

in addition to connected single ($\hat{T}_1$), double ($\hat{T}_2$), triple ($\hat{T}_3$), etc. excitations into the virtual space of the molecular orbital basis also double excitations into explicitly correlated pair functions (geminals):

$$\chi_{xy}(1,2) = \hat{Q}_{12} f_{12}(r_{12}) |\varphi_x(1)\varphi_y(2)\rangle\,, \tag{2}$$

where φ_x and φ_y are taken from a suitable chosen subset of the molecular orbitals (for ground state energies usually the active occupied MOs) and $\hat{Q}_{12}$ is a projection operator ensuring strong orthogonality which for ansatz 1 and 2 is defined as, respectively,

$$\hat{Q}_{12}^{(1)} = (1-\hat{P}_1)(1-\hat{P}_2) \quad \text{and} \quad \hat{Q}_{12}^{(2)} = (1-\hat{O}_1)(1-\hat{O}_2) - \hat{V}_1\hat{V}_2\,. \tag{3}$$

In the last equation $\hat{O} = \Sigma_I |\varphi_I\rangle\langle\varphi_I|$ is a projector onto the space spanned by all occupied MOs, $\hat{V} = \Sigma_a |\varphi_a\rangle\langle\varphi_a|$ projects onto the space spanned by the virtual MOs, and $\hat{P} = \hat{O}+\hat{V}$ onto the space of all MOs. The function $f_{12}(r_{12})$ is in R12 methods just r_{12}, in F12 methods usually a slater function $\exp(-\gamma r_{12})$ with a preoptimized exponent γ. If we introduce a formally complete one-electron basis $\{\varphi_\alpha\}$, the explicitly correlated connected double excitations can (assuming that $|\mathrm{HF}\rangle$ is closed-shell) be written as

$$\hat{T}_{2'} = \sum_{ix \geq yj} c_{ij}^{xy} \sum_{\alpha\beta} F_{xy}^{\alpha\beta} E_{\alpha i} E_{\beta j} = \sum_{ix \geq jy} c_{ij}^{xy} \tau_{xiyj} \, , \tag{4}$$

with $F_{xy}^{\alpha\beta} = \langle \alpha\beta | \hat{Q}_{12} f_{12} | xy \rangle$. In the following we refer for brevity to the amplitudes t_{ai}, t_{aibj}, ... and c_{ij}^{xy} collectively as t_μ and to the single (E_{ai}), double ($E_{ai}E_{bj}$), etc. excitation operators and τ_{xiyj} as τ_μ, so that we can write the cluster operator as $\hat{T} = \Sigma_\mu \, t_\mu \, \tau_\mu$.

We can now combine this wavefunction ansatz with coupled-cluster response theory to derive expressions for linear and nonlinear response functions and excitation energies. For this we introduce the time-dependent Lagrange function:

$$L_{\mathrm{CC}} = \langle \mathrm{HF} | (H - i\tfrac{d}{dt}) e^{\hat{T}} | \mathrm{HF} \rangle + \sum_\mu \bar{t}_\mu \langle \mu | e^{-\hat{T}} (H - i\tfrac{d}{dt}) e^{\hat{T}} | \mathrm{HF} \rangle \tag{5}$$

Response functions are obtained as derivatives of the real part of L_{CC} with respect to strengths ε_i of external (e.g. electromagnetic) fields:

$$\langle\langle X_1, X_2, \ldots, X_n \rangle\rangle_{\omega_2,\ldots,\omega_n} = \left(\frac{d^n \mathrm{Re}(L_{\mathrm{CC}})}{d\epsilon_1(\omega_1) \ldots \epsilon_n(\omega_n)} \right)_{\epsilon_i = 0} , \tag{6}$$

and excitation energies are obtained as solutions of the eigenvalue problem

$$\sum_\mu A_{\nu\mu} E_\mu = \omega \sum_\mu S_{\nu\mu} E_\mu \, , \tag{7}$$

where $A_{\mu\nu}$ and $S_{\nu\mu}$ are, respectively, the Jacoby and metric matrices:

$$A_{\nu\mu} = \frac{\partial^2 L_{\mathrm{CC}}}{\partial \bar{t}_\nu \partial t_\mu} \quad \text{and} \quad S_{\nu\mu} = \langle \nu | \tau_\mu | \mathrm{HF} \rangle \, . \tag{8}$$

By restricting now the cluster operator to at most double excitations one obtains the CCSD-R12 or CCSD-F12 models. If one accounts in addition for connected triple excitations, e.g. as in the CC3 model, one arrives at CC3-R12 or CC3-F12. However, it turned out that contributions which are higher-order in the explicitly correlated double excitations are negligible. Thus the computational costs of the explicitly correlated CCSD variants can without significant loss in accuracy be reduced by approximating the doubles equations as,

$$0 = \langle \mu_2 | e^{-\hat{T}_1} (H + [H, \hat{T}_2 + \hat{T}_2'] + \tfrac{1}{2}[[H, \hat{T}], \hat{T}_2 + 2\hat{T}_2']) e^{\hat{T}_1} | \mathrm{HF} \rangle \, , \tag{9}$$

$$0 = \langle \mu_{2'} | [F, \hat{T}_{2'}] + e^{-\hat{T}_1} (H + [H, \hat{T}_2]) e^{\hat{T}_1} | \mathrm{HF} \rangle \, , \tag{10}$$

which leads to the CCSD(F12) model [4,5], where – similar as for CC2 and CC3 – all similarity transformations with $\hat{T}_1$ are kept since the single excitations play an important role in response theory. For the CCSD(F12) model the computational costs are approximately those of an MP2-F12 calculation plus 5–7 times those of a conventional CCSD calculation. Since they give typical results of quintuple-ζ quality already in triple-ζ basis sets, this reduces tremendously the computational costs, since for a (conventional) CCSD calculation in a quintuple-ζ basis they are typically by more than a factor of 50 higher than in a triple-ζ basis. For example with the present implementation in the Dalton program used for the present study we find for the BF molecule in the t-aug-cc-pVTZ basis set that one iteration for a conventional CCSD calculation takes 211 seconds. For CCSD(F12) an iteration takes 1314 seconds, i.e. roughly 6 times longer. But to reach the same accuracy as CCSD(F12)/t-aug-cc-pVTZ with conventional CCSD the basis set must be improved to t-aug-cc-pV5Z which increases the required CPU time per iteration to 14784 seconds, roughly an order magnitude more than for CCSD(F12)/t-aug-cc-pVTZ.

For the calculation of ground state energies further approximations can be made which reduce the computational costs of the explicitly correlated CCSD approaches essentially to the costs of a conventional CCSD and an MP2-F12 calculation [6,7]. For computational convenience CC3(F12) and CCSD(F12)(T) neglect in addition all coupling terms between connected triple excitations and double excitations into the explicitly correlated pair functions. In the following we will in addition present some results for the CC2-R12 or CC2-F12 models, which are obtained by approximating the equations for $\hat{T}_2$ and $\hat{T}_{2'}$ to first order in the fluctuation potential in exactly the same manner as it is done in the conventional CC2 model.

3. Excited states

Excitation energies play a decisive role in response theory as they are not only important for the description of excited states but determine also the dispersion of frequency-dependent optical properties. As indicated in the introduction, a balanced description of ground and excited states can be a challenge for R12 methods since the usual construction of the geminal space with φ_x and φ_y in Eq. (1) restricted to occupied orbitals is tuned for ground state energies. For small highly symmetric molecules this choice can bias the results for excitation energies and response properties.

Two illustrative examples which highlight this problem are the vertical excitation energies of Be and BH for which some results are reported in Table 1. For the ^{1}P state of Be and the $^1\Pi$ state of BH CC2-R12 overestimates in the smaller basis sets considerably the excitation energies and converges in fact slower to the basis set limit than the corresponding conventional CC2 results. The convergence of CC2-R12 is evidently slower than that of the conventional CC2 model. CC2-R12 improves the basis set convergence only for the $^1\Sigma^+$ state of BH.

Table 1. Vertical excitation energies (eV) for Be and BH with CC2 (R12 results obtained with ansatz 1). *X* denotes the cardinal number of orbital basis sets. For details about basis and bond lengths see Ref. [8].

	Be ^1P			BH $^1\Pi$			BH $^1\Sigma^+$		
X	conv	R12	R12+	conv	R12	R12+	conv	R12	R12+
2	5.54	5.70	5.30	2.87	2.99	2.92	6.40	6.49	6.49
3	5.21	5.27	5.13	2.83	2.89	2.86	6.44	6.49	6.49
4	5.15	5.17	5.12	2.82	2.85	2.83	6.47	6.49	6.49
5	5.13	5.14	5.11	2.82	2.84	2.82	6.47	6.49	6.48
∞	5.11±0.005			2.81±0.005			6.48±0.005		

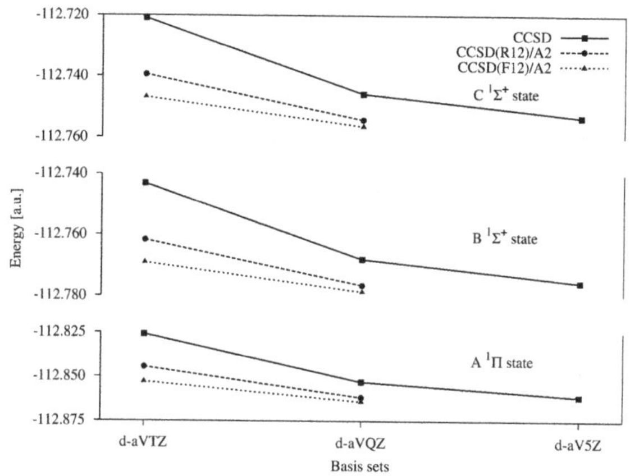

Fig. 1. Basis set dependence of total energies for excited states of CO with ansatz 2.

The reason for the unbalanced description of the Be ^1P and BH $^1\Pi$ states is that the geminal space constructed by multiplying f_{12} with products of only occupied orbitals does not contain any pair functions with, respectively, P or Π symmetry which can correlated the excited state. A remedy for this problem is to extend the set of orbitals used in the construction of the geminal functions with a few additional virtual orbitals [9], which is indicated in the following with R12+. Typically, the inclusion of a few additional valence orbitals is sufficient to correlate the excited states with a similar accuracy as the ground state and to obtain a fast basis set convergence. This is demonstrated by the CC2-R12+ results in Table 1. It should, however, pointed out that such an extreme loss of balance as for the Be ^1P and BH $^1\Pi$ states is a pecularity of small highly symmetric molecules. For example, without including virtual orbitals in the construction of the pair functions, the total energies of several singlet excited states for CO (c.f. Fig. 1) show with CCSD(R12) and CCSD(F12) a remarkably improved convergence which is similar to what is observed for ground state energies.

Accurate total energies for excited states are important for their equilibrium structures and vibrational frequencies. Some results for equilibrium bond lengths

Table 2. Bond lengths (R_e in pm) with CCSD(R12) and CCSD(F12) using ansatz 2 and d-aug-cc-pVXZ basis sets (indicated as XZ). Results parentheses include a d-aV5Z correction for the Hartree-Fock value. For further technical details see Ref. [13].

		CCSD			CCSD(R12)		CCSD(F12)	
		TZ	QZ	∞	TZ	QZ	TZ	QZ
N_2	$a'^1\Sigma_u^-$	125.4	125.1	124.9	125.2	124.9	125.1	124.9
		(125.3)			(125.0)		(125.0)	
	$a^1\Pi_g$	120.7	120.3	120.1	120.4	120.2	120.4	120.2
		(120.6)			(120.3)		(120.2)	
	$w^1\Delta_u$	124.9	124.5	124.4	124.6	124.4	124.6	124.4
		(124.7)			(124.5)		(124.4)	
CO	$A^1\Pi$	123.1	122.5	122.3	122.7	122.3	122.6	122.3
		(122.8)			(122.4)		(122.4)	
	$B^1\Sigma^+$	111.6	111.2	111.0	111.3	111.0	111.2	111.0
		(111.4)			(111.1)		(111.0)	
	$C^1\Sigma^+$	111.5	111.0	110.8	111.2	110.9	111.1	110.9
		(111.2)			(111.0)		(110.9)	
BF	$A^1\Pi$	131.1	130.5	130.2	130.7	130.3	130.6	130.3
		(130.7)			(130.3)		(130.2)	
	$B^1\Sigma^+$	121.2	120.7	120.5	121.0	120.6	120.9	120.6
		(120.9)			(120.6)		(120.5)	
	$C^1\Sigma^+$	122.5	122.0	121.8	122.3	121.9	122.2	121.9
		(122.1)			(121.9)		(121.8)	

obtained in Ref. [12] with CCSD(R12) and CCSD(F12) using ansatz 2 (cmp. Eq. 3) are shown in Table 2. With conventional CCSD the bond lengths of most states deviate even with the 5Z basis by 0.1 pm from the basis set limit. The CCSD(R12) and CCSD(F12) results in the QZ basis are for all investigated states close to the basis set limits, which have been extrapolated from the conventional QZ and 5Z results, and surpass the conventional 5Z values. With the triple-ζ basis set the improvements are less satisfactory. The reason for this is, however, not the correlation treatment, but the uncorrelated wavefunction which limits now the overall basis set convergence. This becomes obvious when the TZ potential energy curves are approximately corrected for this error by adding the differences between HF response results in the d-aug-cc-pV5Z and d-aug-cc-pVTZ basis sets. With this correction (c.f. results given in Table 2 in parentheses) the CCSD(R12) and CCSD(F12) results in the triple-ζ basis agree for all states with the extrapolated basis set limits within the accuracy of the latter.

Results for harmonic frequencies are presented in Table 3. The conventional CCSD results deviate for most states even with the 5Z basis still by several wavenumbers from the extrapolated basis set limit. With CCSD(R12) and CCSD(F12) the convergence is significantly improved and all frequencies are already in the QZ basis very close to the basis set limit. With the triple-ζ basis

Table 3. Harmonic vibrational frequencies (ω_e in cm^{-1}) with CCSD(R12) and CCSD(F12) using ansatz 2 and d-aug-cc-pVXZ basis sets (indicated as XZ). Results parentheses include a d-aV5Z correction for the Hartree-Fock value. For further technical details see Ref. [13].

		CCSD			CCSD(R12)		CCSD(F12)	
		TZ	QZ	∞	TZ	QZ	TZ	QZ
N_2	$a'^1\Sigma_u^-$	1714	1725	1730	1725	1729	1724	1729
		(1711)			(1722)		(1721)	
	$a^1\Pi_g$	1830	1848	1854	1846	1854	1847	1854
		(1836)			(1851)		(1852)	
	$w^1\Delta_u$	1733	1744	1749	1744	1748	1744	1748
		(1734)			(1745)		(1745)	
CO	$A^1\Pi$	1561	1581	1592	1579	1589	1580	1589
		(1566)			(1583)		(1584)	
	$B^1\Sigma^+$	2221	2244	2253	2241	2252	2246	2253
		(2230)			(2248)		(2254)	
	$C^1\Sigma^+$	2259	2281	2289	2277	2289	2282	2289
		(2267)			(2284)		(2289)	
BF	$A^1\Pi$	1258	1271	1279	1271	1278	1275	1278
		(1260)			(1273)		(1277)	
	$B^1\Sigma^+$	1704	1715	1720	1714	1720	1717	1720
		(1708)			(1717)		(1720)	
	$C^1\Sigma^+$	1620	1632	1637	1630	1637	1633	1637
		(1625)			(1635)		(1638)	

the CCSD(R12) and CCSD(F12) results have very much the same accuracy as the conventional results in the QZ basis. Again, the accuracy is at this level limited by the basis set error in the uncorrelated part and can be improved by applying the above described correction scheme. With this correction the CCSD(F12) results for all Σ^+ states agree with the extrapolated basis set limit within its accuracy and also the frequencies for the $a^1\Pi_g$ and $w^1\Delta_u$ states of N_2 are close to quintuple-ζ quality. Slightly larger deviations are found for the $^1\Pi$ states of BH, CO and BF.

4. Linear and nonlinear optical properties

In this section, we turn to the assessment of recents results [13] obtained with the CCSD(R12) and CCSD(F12) approaches with ansatz 2 for static and frequency-dependent higher-order properties.

Before discussing polarizabilities and hyperpolarizabilities we briefly comment on results for dipole moments as an example of first-order properties to show that the RI approximation used in the CCSD(R12) and CCSD(F12) methods represents accurately the important matrix elements with the perturbation

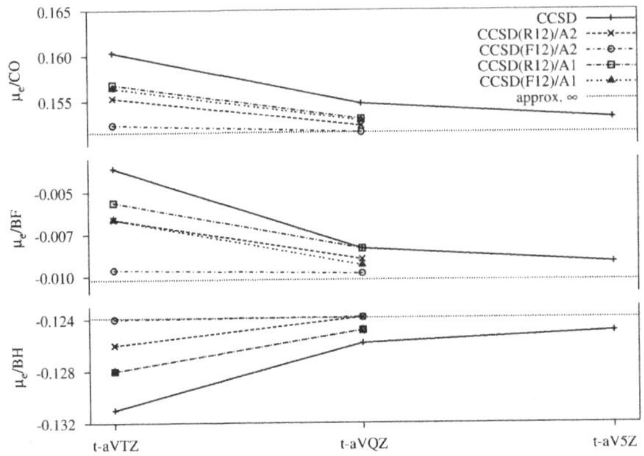

Fig. 2. Basis set convergence of the correlation contributions to total eletrcic dipole moments μ_e for BH (bottom), BF (middle) and CO (top) diatomic molecules (in a.u.) in the t-aug-cc-pVXZ basis sets. A1 and A2 denote, respectively ansatz 1 and 2.

operators. Figure 2 shows some results for dipole moments of BH, BF and CO. This comparison reveals that the CCSD(F12) model with ansatz 2 gives in the t-aug-cc-pVTZ basis dipole moments which are better converged than the t-aug-cc-pV5Z results with conventional CCSD. With the quadruple-ζ basis the CCSD(F12)/ansatz 2 calculations agree with the extrapolated basis set limits and may even surpass them in accuracy. CCSD(R12) and also CCSD(F12) with ansatz 1 are less accurate and improve the basis set convergence compared to conventional CCSD only by one cardinal number.

The CCSD, CCSD(R12) and CCSD(F12) results for the polarizability (α) and the second hyperpolarizability ($\gamma_\parallel$) for Ne are shown Fig. 3. The explicitly correlated models reach in a basis with cardinal number X for both α and $\gamma_\parallel$ about the same accuracy as standard CCSD calculations with X+1 for ansatz 1 and X+2 for ansatz 2. In variance to total energies and first-order properties we observe for (hyper-)polarizabilities with the present approach no advantage of CCSD(F12)/ansatz 2 compared to CCSD(R12)/ansatz 2. In particular for α the CCSD(R12) results are closer to the basis set limit than the CCSD(F12) values, which underestimate the polarizability. A possible cause of this problem might be that the exponent in the F12 geminal function, which has been optimized for ground state correlation energies, leads in the response calculations to a bias towards the unperturbed state.

The BH molecule has been a critical case for assessing the performance of R12 or F12 response theory as the standard choice for the geminal space contains only pair functions of σ symmetry, which biases the results for excitations into the A $^1\Pi$ state [8,10]. A similar bias is observed for the xx component of the polarizability α unless a few virtual orbitals with π symmetry are included in the construction of the geminal functions. However, this does not necessarily affect

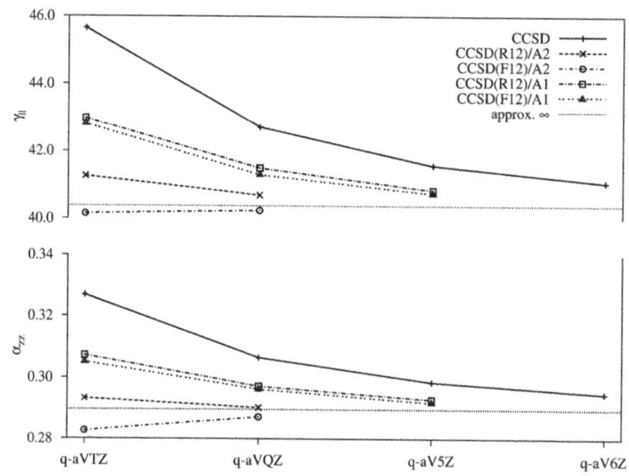

Fig. 3. Convergence of the correlation contributions to the static polarizabilities α_{zz} and second hyperpolarizabilities $\gamma_{\|}$ of Ne (in a.u.) with the q-aug-cc-pVXZ basis sets. A1 and A2 denote ansatz 1 and 2, respectively.

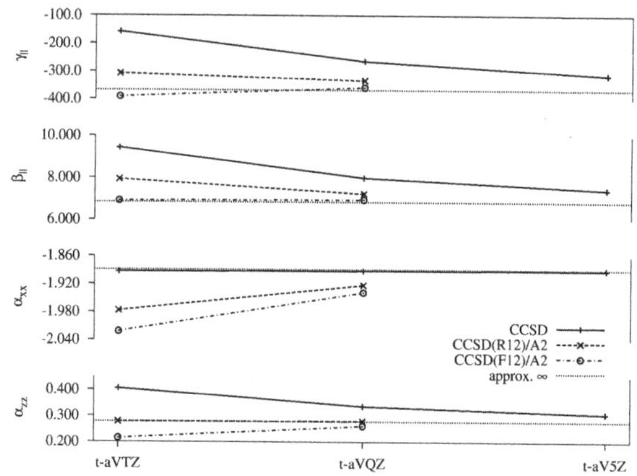

Fig. 4. Basis set convergence of the electronic correlation contributions to static polarizabilities α_{zz} and α_{xx}, first and second hyperpolarizabilities $\beta_{\|}$ and $\gamma_{\|}$ for BH (in a.u.) using the t-aVXZ basis sets with triply augmented diffuse functions. A2 stands for ansatz 2.

the parallel components of the polarizability (α_{zz}) and the first and second hyperpolarizabilities ($\beta_{\|}$ and $\gamma_{\|}$) presented in Table 4 and Fig. 4. With CCSD(R12)/ansatz 2 and the t-aug-cc-pVTZ basis the correlation contribution to α_{zz} deviates by less then 0.001 a.u. from the basis set limit. CCSD(F12) gives again a slightly to small polarizability, while for $\beta_{\|}$ and $\gamma_{\|}$ the CCSD(F12) results are closest to the extrapolated basis set limits.

Table 4. Static polarizabilities α_{zz} and α_{xx} and second hyperpolarizabilities $\gamma_{\|}$ for BH, BF, CO and N_2 (in a.u.) with conventional CCSD and CCSD(R12) and CCSD(F12) with ansatz 2 and t-aug-cc-pVXZ basis sets.

BH		α_{zz}	α_{xx}	$\beta_{\|}$	$\gamma_{\|}$
CCSD	TZ	23.01	21.00	64.66	10764
	QZ	22.90	21.01	63.39	10642
	5Z	22.87	21.02	62.91	10593
	∞	22.84	21.02	62.31	10540
CCSD(R12)	TZ	22.88	20.91	63.16	10615
CCSD(F12)	TZ	22.81	20.87	62.14	10531
BF		α_{zz}	α_{xx}	$\beta_{\|}$	$\gamma_{\|}$
CCSD	TZ	18.15	20.05	53.53	6246
	QZ	18.04	20.03	52.41	6172
	5Z	18.00	20.01	52.13	6148
	∞	17.97	20.00	51.83	6111
CCSD(R12)	TZ	17.99	19.96	52.67	6156
CCSD(F12)	TZ	17.93	19.93	51.83	6087
CO		α_{zz}	α_{xx}	$\beta_{\|}$	$\gamma_{\|}$
CCSD	TZ	15.71	11.88	25.25	1410
	QZ	15.64	11.84	25.03	1382
	5Z	15.61	11.82	24.98	1370
	∞	15.58	11.81	24.90	1359
CCSD(R12)	TZ	15.58	11.80	25.07	1376
CCSD(F12)	TZ	15.53	11.76	24.81	1357
N_2		α_{zz}	α_{xx}		$\gamma_{\|}$
CCSD	TZ	14.68	10.19		908
	QZ	14.63	10.15		891
	5Z	14.61	10.15		883
	∞	14.58	10.12		874
CCSD(R13)	TZ	14.57	10.11		883
CCSD(F12)	TZ	14.54	10.08		872

For BF, CO and N_2, which have occupied σ and π orbitals, already the standard construction scheme for the geminal space includes pair functions of σ, π and δ symmetry. For these molecules we do not observe a similar bias in the results for α_{xx} (cmp. Table 4). With ansatz 2 both CCSD(R12) and CCSD(F12) accelerate the convergence with the cardinal number of the basis set considerably. However, as already observed above for Ne, the CCSD(F12) results tend to underestimate the polarizability and hyperpolarizabilities. This might partially be a consequence of the fact that the exponent for the pair functions was optimized for the unperturbed ground state energy. But at this level of accuracy also the limited convergence of the underlying uncorrelated value in the t-aug-cc-pVTZ basis contributes significantly to the remaining basis set incompleteness errors.

Table 5. ESHG second hyperpolarizabilities $\gamma_{\parallel}^{\mathrm{ESHG}}(\omega)$ for N_2 in a.u. at the CCSD as well as CCSD(R12) and CCSD(F12) levels with ansatz 2 and d-aug-cc-pVXZ basis sets.

	CCSD				CCSD(R12)	CCSD(F12)
λ (nm)	TZ	QZ	5Z	est. ∞	TZ	TZ
1055.0	956.5	943.4	937.2	926.7	929.2	918.2
694.3	1038.8	1023.0	1015.8	1004.3	1007.9	995.6
457.9	1272.3	1248.1	1237.8	1223.8	1230.6	1214.0

For CCSD(R12) the polarizabilities are with the t-aug-cc-pVTZ basis similar close to the extrapolated basis set limit as the conventional CCSD/t-aug-cc-pV5Z results, which is partially a result of error cancellation between the correlation contribution and the uncorrelated result. Encouraging is that also the static second hyperpolarizabilities $\gamma_{\parallel}$ confirm the strength of CCSD(R12) and CCSD(F12) with ansatz 2: already the calculations in the triple-ζ basis set give results that are rather close to the estimated basis set limits.

As example for a frequency-dependent higher-order property we investigated the parallel average of the electric field induced second harmonic generation (ESHG) hyperpolarizability for N_2. In Table 5 we listed results for some selected wavelengths. The basis set limits for $\gamma_{\parallel,ESHG}(\omega)$ were estimated using a multiplicative scheme:

$$\gamma^{\infty}_{\parallel,ESHG}(\omega) \approx \gamma^{5Z}_{\parallel,ESHG}(\omega)\left(\frac{\gamma_{\parallel}^{\infty}(0)}{\gamma_{\parallel}^{5Z}(0)}\right), \qquad (11)$$

where the $\gamma_{\parallel}^{\infty}(0)$ is obtained by the extroplation scheme described in Refs. [11, 12].

The results demonstrate that both CCSD(R12) and CCSD(F12) with ansatz 2 tremendously improve the basis set convergence of the dispersion, which is for both approaches already in the t-aug-cc-pVTZ basis better than in a conventional CCSD/t-aug-cc-pV5Z calculation. The remaining deviations from the extrapolated basis set limits are less than 10 a.u. (i.e. less than 1%) and the largest deviations occur for the shortest wavelengths and are probably mainly due to errors of the approximation in Eq. (11) which increase with increasing frequency ω, i.e. decreasing wavelengths.

5. Summary and outlook

The equations for explicitly correlated CC-R12 and CC-F12 response theory with ansatz 1 and 2 have been derived and implemented in the program DALTON [14] for the CC2-R12, CCSD(R12) and CC3(R12) models and the corresponding F12 variants. This has allowed first applications of these approaches to excited states and optical response properties as dipole polarizabilities and hyperpolarizabilities.

The results show that for small and highly symmetric molecules the standard construction scheme for the explicitly correlated pair functions can lead to a bias towards the electron ground state. Including a few virtual orbitals (e.g. some valence orbitals) rectifies this problem and a considerably improved basis set convergence is observed for energies and structures of excited states and linear and nonlinear response properties.

At the example of equilibrium bond lengths and harmonic vibrational frequencies of singly excited states of N_2, CO, BF and BH we showed that in CCSD(R12) and CCSD(F12) calculations with ansatz 2 it is already in the d-aug-cc-pVTZ basis no longer the correlation contribution but the underlying uncorrelated contribution which limits the overall basis set convergence. If one corrects for these errors (e.g. by using a basis set correction obtained from Hartree-Fock response results) the explicitly correlated results reach in the d-aug-cc-pVTZ basis an accuracy which in conventional CCSD calculations is only obtained with quintuple-ζ basis sets, i.e. much larger computational costs. An alternative way to correct for the incompleteness of the one-electron space in explicitly correlated coupled-cluster response calculations would be to allow single excitations from the occupied space into the space spanned by the complementary auxiliary basis. If we compare CCSD(R12) and CCSD(F12), we see e.g. for equilibrium bond lengths and harmonic vibrational frequencies not such large improvements with the exponential correlation factor of CCSD(F12) as have been observed for ground state energies. A reason for this might be that the exponents optimized for ground state correlation energies are less suitable for excited states.

For ground state properties we find that the CCSD(F12) method with ansatz 2 gives already in a triple-ζ basis correlation contributions to dipole moments and second hyperpolarizabilities with an accuracy which conventional CCSD achieves first in quintuple-ζ or larger basis sets. Here the improvements are comparable to those observed with this approach for ground state correlation energies.

CCSD(R12) with ansatz 2 gives also correlation contributions to polarizabilities in triple-ζ basis sets with quintuple-ζ quality, while CCSD(F12) with ansatz 2 tends to slightly underestimate the polarization. The source of this effect might again be that the exponent for the pair functions is optimized for the unperturbed ground state, but also the remaining basis set incompleteness error in the one-electron space limits at this level the overall basis set convergence. A remedy for the latter problem would again to include in the CC response calculation single excitations into the complementary auxiliary basis. For first hyperpolarizabilities the performance of CCSD(R12) and CCSD(F12) is intermediate between those observed for α and γ.

In conclusion, we find that already with the standard setup CCSD(R12) and in particular CCSD(F12) response theory are very valuable tools to reduce the computational costs for accurate coupled-cluster response calculations. A certain challenge is, however, to obtain a balanced description of the unperturbed ground state and excited states or a polarization of the ground state wavefunction and also a balance between the remaining basis set incompleteness errors in the one-electron and the pair function space. Concerning the first problem, the influence of the exponents of the Slater-type correlation factor used in CCSD(F12) on the results for excitation energies and (hyper-)polarizabilities needs a more careful examination. For both CCSD(R12) and CCSD(F12) the inclusion of a few unoccupied valence orbitals in the construction of the explicitly correlated pair functions facilitates in difficult cases a balanced description of excited states or the polarizability and hyperpolarizability components. With the accuracy of the correlation treatment reached with this, the errors caused by the incompleteness of the one-electron space limit now in several cases the overall basis set convergence of explicitly correlated coupled-cluster response calculations. This error can, however, be reduced with minor additional costs by including single excitations into the CABS space. This will probably also allow to reduce the large number of diffuse basis functions which is presently needed for accurate calculations of excitation energies and (hyper-)polarizabilities and are a bottleneck in the integral evaluation.

Summing up, with these improvements of both accuracy and computational efficiency the explicitly correlated CCSD(F12) and CC3(F12) response models reduce considerably the computational costs and thereby extent the applicability of accurate coupled-cluster response calculations.

References

1. W. Klopper, F. R. Manby, S. Ten-No, E. F. Valeev, Int. Rev. Phys. Chem. **24** (2006) 427.
2. T. Korona, H. J. Werner, J. Chem. Phys. **118** (2003) 3006.
3. D. Crawford, R. A. King, Chem. Phys. Lett. **336** (2002) 611.
4. H. Fliegl, W. Klopper, C. Hättig, J. Chem. Phys. **122** (2005) 084107.
5. D. P. Tew, W. Klopper, C. Neiss, C. Hättig, Phys. Chem. Chem. Phys. **9** (2007) 1.
6. G. Knizia, T. B. Adler, H. J. Werner, Phys. Chem. Chem. Phys. **10** (2008) 3400.
7. D. P. Tew, W. Klopper, C. Hättig, Chem. Phys. Lett. **452** (2008) 326.
8. H. Fliegl, C. Hättig, W. Klopper, J. Chem. Phys. **124** (2006) 044112.
9. C. Neiss, C. Hättig, W. Klopper, J. Chem. Phys. **125** (2006) 064111.
10. C. Neiss, C. Hättig: J. Chem. Phys. **126** (2007) 154101.
11. T. Helgaker, W. Klopper, H. Koch, J. Noga: J. Chem. Phys. **106** (1997) 9639.
12. J. Yang, C. Hättig: J. Chem. Phys. **130** (2009) 124101.
13. J. Yang, C. Hättig: J. Chem. Phys., submitted for publication.
14. DALTON, a molecular electronic structure program, release 2.0, http://www.kjemi.uio.no/software/dalton/dalton.html, 2005.

A Linear-Scaling MP2 Method for Large Molecules by Rigorous Integral-Screening Criteria

By Bernd Doser, Jan Zienau, Lucien Clin, Daniel S. Lambrecht, and Christian Ochsenfeld*

Universität Tübingen, Theoretical Chemistry, Auf der Morgenstelle 8, 72076 Tübingen, Germany

(Received July 21, 2009; accepted October 12, 2009)

Linear-scaling / Electron-correlation / Perturbation-theory / MP2 / Large Molecules / Integral-screening

A brief review of our linear-scaling method for atomic-orbital (AO) second-order Møller-Plesset perturbation theory (MP2) is given. The key feature of our method is the rigorous preselection of numerically significant four-center two-electron integrals based on multipole-based integral estimates (MBIE) that do not only account for the exponential coupling between the Gaussian-type basis functions forming charge distributions, but also for the $1/R$ coupling between the charge distributions. This coupling turns for the required integral products in AO-MP2 into at least a $1/R^4$ or even a $1/R^6$ decay behavior. Using MBIE we attain linear scaling, which is illustrated for DNA fragments with up to 1052 atoms and 10 674 basis functions as computed on a single processor. The largest molecule calculated in our present work at the scaled-opposite spin (SOS-) AO-MP2 level is an RNA system comprising 1664 atoms and 19 182 basis functions. Furthermore, we present results for the use of Cholesky-decomposed pseudo-density matrices in Laplace-based MP2, that offers the advantage of exploiting occupied/virtual blocking both with and without auxiliary basis sets.

1. Introduction

For the reliable ab-initio description of molecular systems it is crucial to account for electron-correlation effects that are neglected in the Hartree-Fock (HF) mean-field approach to the Schrödinger equation [1]. A suitable pathway to describe correlation effects is provided by wavefunction-based correlation methods which allow to systematically approach the exact solution of the Schrödinger equation.

* Corresponding author. E-mail: christian.ochsenfeld@uni-tuebingen.de

Z. Phys. Chem. **224** (2010) 397–412
© by Oldenbourg Wissenschaftsverlag, München

The corresponding hierarchy of approximations to the exact solution is extremely useful for the reliable and systematic study of molecular systems. However, a central drawback of wave-function based correlation methods is the limited applicability to larger molecules due to the steep increase of the computational cost with molecular size M: even the simplest approach, second-order Møller-Plesset perturbation theory (MP2), scales as $\mathcal{O}(M^5)$. Therefore, it is crucial to reduce the computational scaling to linear, which is the main goal of our present work.

It is worthwhile to note that such linear-scaling quantum-chemical (QM) methods for treating large molecules remain important even if one employs approximative QM/MM (QM/molecular mechanics) schemes that offer interesting possibilities for the study of local changes in molecular systems such as, e. g., reactions in active sites of enzymes. Recent studies showed that, although QM/MM approximations typically accelerate the convergence of properties with respect to the chosen QM region size, even for simple molecular systems often QM regions with more than 300–1000 atoms are required for reliable accuracies [2]. The necessity for choosing large QM regions becomes even more important if not only local changes such as in enzymatic active sites occur, but expanded interaction areas as, e.g., in DNA/RNA-protein or cell-virus complexes [3] need to be described.

For reducing the scaling behavior of the MP2 approach, much work has been done by many research groups: in particular the seminal work by Pulay and Saebø in 1985 [4–7] and subsequent work by Werner, Schütz, and other groups (see e.g. Refs. [8–27]) has led to much success in reducing the computational scaling. Nevertheless, a multitude of difficulties remain: If distance-based criteria are employed for preselecting significant contributions, they may depend strongly on the molecular system and can affect the reliability adversely, so that for example discontinuous potential energy surfaces may result [28]. Other approaches lead to strongly increased prefactors as compared to efficient MO-based implementations, which makes it difficult to reach the linear-scaling regime, unless severe approximations are introduced. Hence, none of these linear-scaling methods has so far fully replaced the traditional fifth-order scaling MO-based MP2 method.

In order to tackle this challenge, we recently devised an alternative approach for attaining a linear-scaling computational effort for the MP2 method [27, 29, 30], where we aim for:

– numerically fully controllable results,
– continuous potential energy surfaces,
– linear scaling for systems with local electronic structures,
– early crossovers to traditional schemes,
– system sizes of 500–1000 atoms to be computable on today's workstation computers.

In order to fulfill these criteria for a linear-scaling MP2 method, the key element was to introduce rigorous integral bounds, denoted as multipole-based integral estimates (MBIE) [26, 27, 29, 30], that allow to preselect numerically significant contributions to the MP2 energy within the atomic orbital (AO), the Cholesky, or other local representations. This becomes possible, since MBIE provides rigorous upper bounds to four-center two-electron integrals ubiquitous in quantum chemistry, while accounting not only for the exponential coupling between the basis functions forming the charge distributions, but also for the $1/R$ coupling between interacting charge distributions. It is this $1/R$ coupling within conventional two-electron integrals which turns into a $1/R^4$ or even a $1/R^6$ coupling for the transformed integral products occuring within AO-based MP2 theory (the coupling for exchange-type products is even exponential).

Here, we briefly review our MBIE-based AO-MP2 method and demonstrate the linear-scaling behavior for alkanes and DNA systems. The largest system computed so far comprises 1052 atoms and 10 674 basis functions using a single processor [30]. In our present work, we were able to calculate an RNA system comprising 1664 atoms and 19 182 basis functions using a parallel version of our code [29]. Furthermore, we discuss the use of Cholesky-decomposed pseudo-density matrices in Laplace-transformed MP2.

2. Laplace Formalism

The canonical closed-shell MP2 correlation energy is given in the MO basis by

$$E_{\mathrm{MP2}} = -\sum_{ijab} \frac{(ia|jb)\,[2(ia|jb) - (ib|ja)]}{\varepsilon_a + \varepsilon_b - \varepsilon_i - \varepsilon_j} \tag{1}$$

with the MO integrals

$$(ia|jb) = \iint d\mathbf{r}_1 d\mathbf{r}_2\, \varphi_i^*(\mathbf{r}_1)\varphi_a(\mathbf{r}_1)\, \frac{1}{r_{12}}\, \varphi_j^*(\mathbf{r}_2)\varphi_b(\mathbf{r}_2)\,. \tag{2}$$

Indices i, j denote occupied orbitals, whereas a, b are virtuals. In order to avoid canonical, delocalized MOs Almlöf suggested in 1991 [13] to use a Laplace transformation for eliminating the disturbing denominator $x_q = \varepsilon_a + \varepsilon_b - \varepsilon_i - \varepsilon_j$:

$$\frac{1}{x_q} = \int_0^{\infty} \exp\left(-x_q t\right) dt \approx \sum_{\alpha=1}^{\tau} \omega^{(\alpha)} \exp\left(-x_q t^{(\alpha)}\right)\,. \tag{3}$$

In this way, the orbital energies in the exponential can be decoupled and combined with the MO coefficient matrix to form occupied and virtual pseudo-density matrices, so that an AO-based MP2 formulation becomes possible as described in the next section. We note in passing, that such a Laplace-based scheme can also be employed for a linear-scaling reformulation of coupled-perturbed self-consistent field theory in context of the calculation of molecular response properties [31].

It is important to note that for the Laplace quadrature (Eq. (3)) just a few summation points α are required to attain μHartree accuracy. For determining the Laplace coefficients $\omega^{(\alpha)}$ and $t^{(\alpha)}$, we employ the least-square algorithm of Levenberg-Marquardt [32]:

$$\sum_{q}^{N_{grid}} f(q)\left[\frac{1}{x_q} - \sum_{\alpha}^{\tau} \omega^{(\alpha)} \exp(-x_q t^{(\alpha)})\right]^2 \stackrel{!}{=} \min \tag{4}$$

with the weighting function *f*(*q*) and the number of supporting points N_{grid} dividing the fitting region into subintervals.

Recently, Takatsuka *et al.* [33] proposed the minimax technique which minimizes the Chebyshev norm of the quadrature error and is considered to be the best possible approximation. Kats *et al.* [34] have shown that in the case of MP2 the accuracy of the least-square method is comparable to the minimax technique and that the influence of the weighting function is much more important. The best possible weighting function is given by the MO integral values which, however, obviously have to be avoided due to the computational effort [14,30,34]. The most trivial choice is an equal weighting of the supporting points (f(q) = 1), which often provides good results and is also used in the minimax scheme. However, fluctuating errors in dependence of the number of quadrature points are a problem. An alternative is to define the weighting function over the number of integral terms which reduces error fluctuations but also increases the necessary number of Laplace coefficients. As a good compromise between error fluctuations and the number of Laplace coefficients we chose a weighting function that accounts only for subintervals which contain integral contributions and weights them equally, whereas empty subintervals are discarded [30].

The required number of Laplace coefficients is reduced by employing the frozen-core approach since the size of the fitting region is reduced. Here, not only the frozen core approximation, but also the frozen virtual approximation is of help. This is particularly important for the calculation of counterpoise corrections [35] to interaction energies since the core orbitals of the ghost basis would lead to high-lying virtual orbitals [30].

3. AO-MP2

As discussed above, the Laplace transformed denominator allows decoupling of the orbital energies, so that they can be combined with the MO coefficients to form occupied and virtual pseudo-density matrices, which are sparse for electronically local systems:

$$\begin{aligned} \underline{P}_{\mu\nu}^{(\alpha)} &= \left|\omega^{(\alpha)}\right|^{1/4} \sum_{i}^{N_{occ}} C_{\mu i}\, C_{\nu i}\, e^{\varepsilon_i t^{(\alpha)}} \\ \overline{P}_{\mu\nu}^{(\alpha)} &= \left|\omega^{(\alpha)}\right|^{1/4} \sum_{a}^{N_{virt}} C_{\mu a}\, C_{\nu a}\, e^{-\varepsilon_a t^{(\alpha)}} \quad , \end{aligned} \tag{5}$$

although canonical MO-based quantities can also be entirely avoided by employing matrix exponentials over potentially sparse quantities [16,36,37]. With these pseudo-density matrices the AO-MP2 energy expression results as

$$\begin{aligned} E_{\text{AO-MP2}} &= -\sum_{\alpha}^{\tau} 2e_J^{(\alpha)} - e_K^{(\alpha)} \\ &= -\sum_{\alpha}^{\tau} \sum_{\mu\nu\lambda\sigma}^{\text{N}} (\underline{\mu}\bar{\nu}|\underline{\lambda}\bar{\sigma})^{(\alpha)} \; [\, 2\,(\mu\nu|\lambda\sigma) - (\mu\sigma|\lambda\nu)\,] \ , \end{aligned} \tag{6}$$

where α stems from the Laplace quadrature replacing the energy denominator of canonical MO-MP2 and both untransformed AO-based four-center two-electron integrals $(\mu\nu|\lambda\sigma)$ and fully-transformed integrals enter the formulation. The latter are obtained by four consecutive transformations

$$(\underline{\mu}\bar{\nu}|\underline{\lambda}\bar{\sigma})^{(\alpha)} = \sum_{\mu'\nu'\lambda'\sigma'} (\mu'\nu'|\lambda'\sigma') \, \underline{P}_{\mu\mu'}^{(\alpha)} \, \overline{P}_{\nu\nu'}^{(\alpha)} \, \underline{P}_{\lambda\lambda'}^{(\alpha)} \, \overline{P}_{\sigma\sigma'}^{(\alpha)} \ . \tag{7}$$

The order of summations within the contraction (Eq. (6)) and the transformation (Eq. (7)) can be interchanged, so that an alternative formulation of the AO-MP2 energy involving only half-transformed integrals is obtained [27,29]:

$$e_J^{(\alpha)} = \sum_{\mu\nu\lambda\sigma} (\mu\nu|\underline{\lambda}\bar{\sigma})^{(\alpha)} (\underline{\mu}\bar{\nu}|\lambda\sigma)^{(\alpha)} \tag{8}$$

$$e_K^{(\alpha)} = \sum_{\mu\nu\lambda\sigma} (\mu\nu|\underline{\lambda}\sigma)^{(\alpha)} (\mu\bar{\sigma}|\lambda\bar{\nu})^{(\alpha)} \ , \tag{9}$$

with the half-transformed integrals (HTI)

$$(\underline{\mu}\bar{\nu}|\lambda\sigma)^{(\alpha)} = \sum_{\mu'\nu'} (\mu'\nu'|\lambda\sigma) \, \underline{P}_{\mu\mu'}^{(\alpha)} \, \overline{P}_{\nu\nu'}^{(\alpha)} \ . \tag{10}$$

4. Preselection of Significant Contributions

The AO formulation described above avoids canonical MOs so that the pathway towards a linear-scaling method is open. However, for this it is crucial to preselect the numerically significant integral products prior to their calculation. Therefore, rigorous upper bounds to the quantities contributing to the AO-MP2 energy are necessary.

As mentioned already in the introduction, it can be shown that the size of the two-electron integral products decays at least as $1/\text{R}^4$ or even as $1/\text{R}^6$ with the distance R between the corresponding charge distributions [27]. For the exchange-part this coupling between charge distributions is even exponential, so that in this sense the Coulomb-type part is the more complicated one, although the exchange-type part is technically involved. Therefore, we focus in the following solely on the Coulomb-type contractions (denoted by J subscript).

It is worthwhile to note that the counted number of numerically significant transformed products indicates an extremely early onset of linear scaling [27]: e.g., the number of significant Coulomb-type products for an alkane with four to five carbon atoms scales as $M^{1.5}$ instead of $\mathcal{O}(M^4)$; for one to two stacked DNA base pairs the scaling is even $M^{1.0}$ (6–31G* basis). This holds also for larger basis sets, for which the scaling behavior is roughly unchanged (see discussion in Ref. [29]) and only the prefactors increase. Such an early onset of linear scaling is not known from SCF theory (see e.g. Refs. [38–40]).

In order to allow for preselection of the decaying integral products it is necessary to employ upper bounds to four-center two-electron integrals that incorporate the $1/R$ coupling between the charge distributions. Therefore, the simple Schwarz estimates commonly used in quantum chemistry cannot be employed for reaching linear scaling. A solution is provided by our MBIE upper bounds recently introduced for two-electron integrals, which are based on a presummation of multipole expansions [26]. The only requirement for such expansions is that the charge distributions are well separated, which can be guaranteed. Since the presummation can be done starting from any multipole order, tightness and cost of our bounds can be balanced as desired, while in any order an upper bound to the two-electron integral is obtained. A positive side effect is that these bounds do not deteriorate with an increasing ratio of the two exponents characterizing the charge distributions, which has been found to be a problem in other integral estimates [41,42]. Most importantly, MBIE represents the first bound accounting for the $1/R$ coupling between charge distributions.

Employing MBIE it becomes possible to preselect numerically significant HTIs (Eq. 10) within AO-MP2 [26,27,29,30]. Since these bounds are valid only for well-separated charge distributions, we combine the strengths of Cauchy-Schwarz-type screening (QQZZ) [42,15] and MBIE [26,27,29,30].

5. Accuracy and Efficiency

The accuracy of AO-MP2 is controlled numerically by two factors: the number of quadrature points and the integral screening threshold. In our experience, both criteria are only weakly dependent on the chosen system, so that a reliable choice is straightforward.

For an accuracy better than 10 μHartree (chemical accuracy), five quadrature points are required for the 6–31G** basis. The required number increases for the aug-cc-pVTZ basis set to eight Laplace points [29].

The dependence upon the numerical integral screening threshold is illustrated in Fig. 1 for the methane molecule using different basis sets up to aug-cc-pVQZ. Here, the energy difference (ΔE) with respect to unscreened AO-MP2$_J$ is given, so that the quadrature error is canceled out. For all basis sets, an exponential decay of the screening error upon lowering the numerical screening threshold is obtained. For a given integral threshold (ϑ) the error increases with the basis set

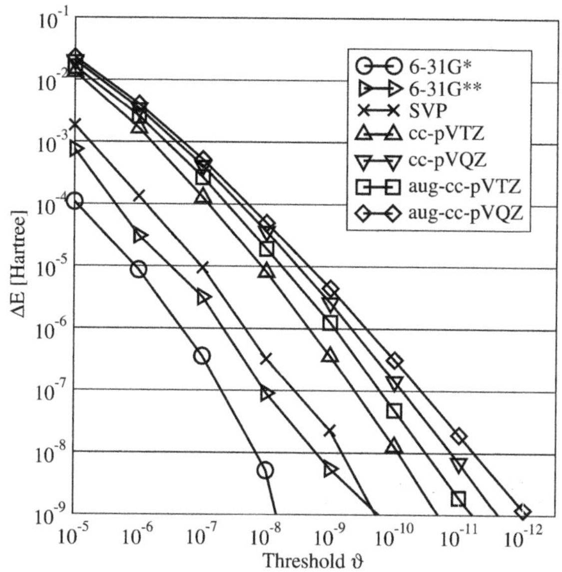

Fig. 1. Numerical accuracy ΔE of AO-MP2$_J$ (in Hartree) in dependence of the integral screening threshold ϑ for methane using the 6–31G*, 6–31G**, SVP, cc-pVXZ, and aug-cc-pVXZ (X = T, Q) basis sets.

size. While for achieving chemical accuracy for the 6–31G* basis set an integral threshold of 10^{-6} is sufficient, an integral threshold of 10^{-9} is necessary for the aug-cc-pVQZ basis.

The performance of MBIE in preselecting numerically significant integral products is illustrated in Fig. 2, where similiar accuracies are compared. Here, the number of significant products is preselected by MBIE, the conventional QQZZ screening, and by exact counting (in the latter scheme all products are calculated first and then the number of significant products is counted). The data clearly shows that MBIE reduces the number of preselected products asymptotically to linear, while the true number is slightly overestimated. It has to be stressed that no numerically significant products are discarded in either preselection scheme.

The efficiency of our MBIE-based AO-MP2 approach is illustrated in Fig. 3 for DNA fragments (6–31G* basis). Here, we compare the total time required for AO-MP2$_J$, conventional MO-MP2$_J$, and RI-MO-MP2$_J$. Since no auxiliary basis sets are available for the 6–31G* basis, we employ the aux-SVP basis [43] for the RI calculations.

The crossover to MO-MP2$_J$ and to RI-MO-MP2$_J$ is achieved around DNA$_1$ and DNA$_2$, respectively. Here, the AO-MP2 threshold was chosen such as to provide a numerical error of about 50 μHartree for DNA$_1$ and about 250 μHartree for DNA$_2$. The reduced AO-MP2 scaling opens the way to the largest system

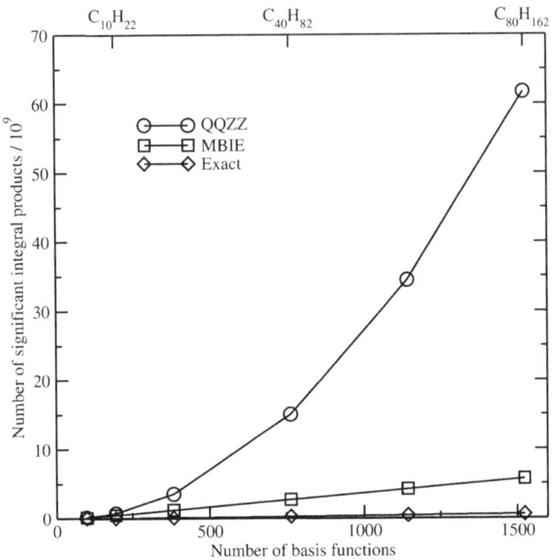

Fig. 2. Number of numerically significant half-transformed integral products ($\vartheta = 10^{-6}$) as preselected by QQZZ and MBIE for linear alkanes using the 6–31G* basis (5 Laplace points) as taken from Ref. [29]. In addition the exact number of products is listed, where a threshold of $\vartheta = 10^{-10}$ was selected for providing comparable accuracy.

computed so far at the MP2 level using a single core (Intel Xeon EM64T dual-core 1.86 GHz, 16 GB RAM): a DNA system with 16 stacked base pairs, DNA_{16}, comprising 1052 atoms and 10 674 basis functions (6–31G*). For this system a conventional MO-MP2 calculation would have required more than 100 years on the same computer (this value being extrapolated conservatively with $N^{4.4}$ from DNA_2).

6. AO-MP2 Calculations for a Ribozyme system

As a first application, we employed our linear-scaling AO-MP2 method to study systematically the catalytic activity of a Diels-Alder ribozyme, which had been synthesized in the group of A. Jäschke [44] and characterized using X-ray analysis by Serganov *et al.* [45]. In our present work, we focus on the calculation of the interaction energy between the Diels-Alder product and the ribozyme at the SOS-AO-MP2/6–31G** level (see below for a discussion of the method). Details of our study (including the notation) and influences upon the catalytic activity will be given elsewhere [46]. Here, we focus on the dependence of the interaction energy upon the chosen environment of the active site (compare Fig. 4 for the different region sizes) and the influence of numerical preselection thresholds

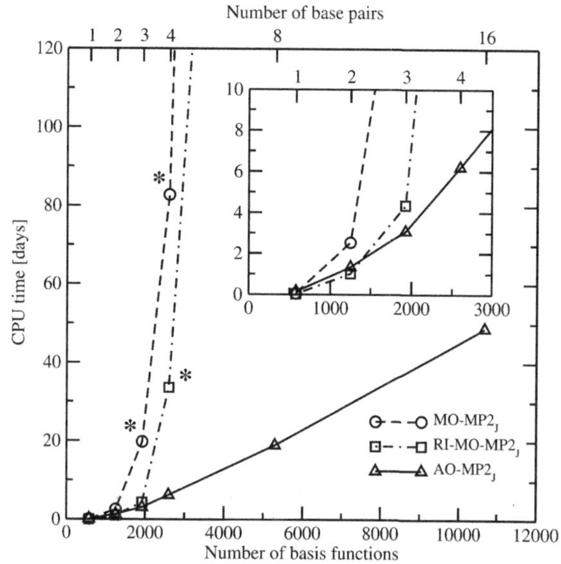

Fig. 3. Total Wall time of AO- and MO-MP2$_J$ for DNA(A-T) fragments using the 6–31G* basis ($\vartheta = 10^{-6}$). In addition, RI-MO-MP2$_J$ timings are shown employing the aux-SVP auxiliary basis. The data points (*) are extrapolated conservatively with the scaling behavior of previous points. It has to be stressed that the calculation for DNA_{16} was split into parts (due to limited computational resources), so that the time is artificially overestimated (see Ref. [30]). Reprinted with permission from B. Doser *et al.*, Journal of Chemical Physics, 130, 064107 (2009). Copyright 2009, American Institute of Physics.

upon the accuracy of the interaction energies. While it is clear that the numerical accuracies are fully controlled in our MBIE-based AO-MP2 scheme, our available computational resources limited our threshold selection for the largest computed systems. The interaction energies for the different ribozyme fragments and various thresholds are listed in Tab. 1. It is also worthwhile to note that despite the remaining inaccuracies in the interaction energies for the loose thresholds, the trend upon enlarging the system seems adequately described. Once a size of the ribozyme fragment with 999 atoms is chosen, the computed interaction energies become largely converged with the system size as indicated by comparison to the interaction energy of the entire system. The latter is, to the best of our knowledge, with 1664 atoms and 19 182 basis functions the largest MP2 calculation without empirical restrictions performed so far.

7. Basis-Set Dependence of Scaling Behavior

It is well known that the use of larger basis sets is often crucial for obtaining reliable electron-correlation energies. Here, a slow convergence of results with

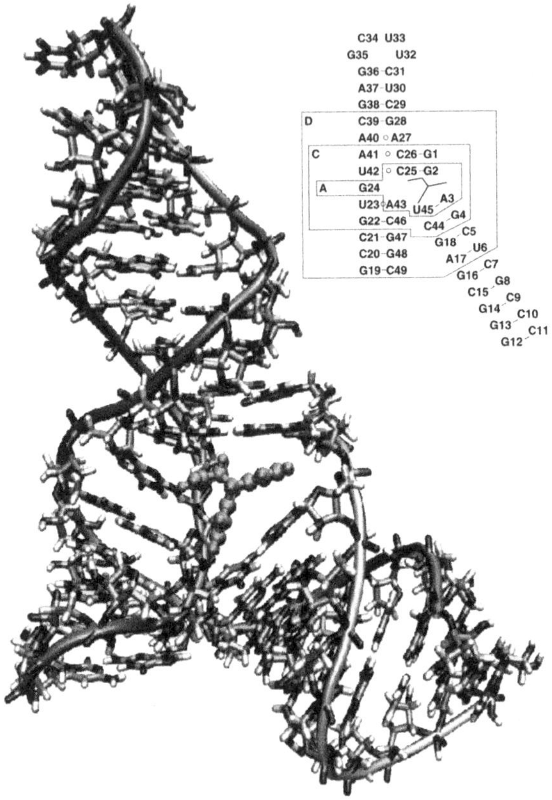

Fig. 4. Ribozyme with the Diels-Alder product of maleimide and anthracene in the active site. In the upper picture the environment selections A, C, and D are illustrated. The entire system consists of 1664 atoms.

basis-set size is usually encountered. The introduction of so-called R12 or even more so F12 methods (see, e.g. Refs. [47, 48]), however, allows an accelerated convergence by including explicit interelectronic distances into the wavefunction. In this way, even basis sets of triple-zeta quality provide results close to the basis set limit [48].

Another, however very pragmatic pathway to circumvent the basis set convergence problem in MP2 theory can in some cases be obtained by using so-called spin-component scaled (SCS) methods introduced by Grimme [49]. Here, often a double-zeta basis set is sufficient to provide accuracies comparable to triple- or even quadruple-zeta bases. A simplified version of SCS-MP2 by restriction to the Coulomb-type part of the MP2 expression has been introduced by Jung *et al.* and has been denoted as scaled opposite-spin (SOS) method [22]. It offers appealing features although for the S22 test set of Hobza *et al.* [50] slightly larger errors than those of conventional MP2 are observed.

Table 1. Interaction energy ΔE (SOS-AO-MP2/6–31G**) in kJ/mol between Diels-Alder ribozyme and substrate using different selections for the active site environment (see Fig. 4, in parentheses the number of atoms are given). The basis set superposition error (BSSE) was accounted for using the counterpoise correction.

ϑ	A (227)	C (531)	D (999)	All (1664)
10^{-6}	-124.7	-159.8	-163.4	-163.2
10^{-7}	-129.2	-162.6	-	-
10^{-8}	-129.7	-	-	-

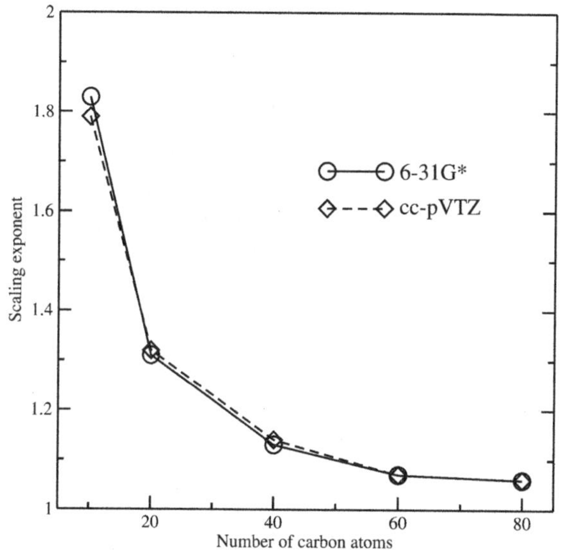

Fig. 5. Scaling exponent M^x with respect to the next smaller system (starting with C_5H_{12}) for significant halftransformed integral products of linear alkanes using 6–31G* ($\vartheta = 10^{-6}$) and cc-pVTZ ($\vartheta = 10^{-8}$) comparing similar accuracies.

Despite this progress in improved basis convergence, it remains important to allow for larger basis-set calculations also at the AO-MP2 level. The dependence of the scaling behavior upon increasing the basis set is illustrated in Fig. 5 for linear alkanes: For providing similar accuracies in the AO-MP2 energy, the scaling behavior of numerically significant integral products is roughly the same for 6–31G* and cc-pVTZ basis sets. Only the prefactor is of course much larger for the cc-pVTZ basis, so that such calculations are extremely costly at this stage. Although the basis-set independence of the scaling behavior offers promising perspectives, we focus in the following section on how to reduce the computational prefactors of AO-MP2 for larger basis sets.

8. Use of Cholesky-decomposed pseudo-densities

The main drawback in moving from canonical MO-MP2 to AO-MP2 theory arises from the need for formally using the total basis dimension in the AO basis as compared to the virtual and occupied dimensions in MO-MP2. While this disadvantage is less severe for smaller basis sets, the factor becomes more and more important for larger basis sets.

A possible solution for reducing this prefactor arising in AO-MP2 is provided by a Cholesky decomposition of the positive semi-definite pseudo-density matrices:

$$\underline{P}^{(\alpha)}_{\mu\mu'} = \sum_{i}^{N_{occ}} \underline{L}^{(\alpha)}_{\mu i}\underline{L}^{(\alpha)}_{\mu' i} \quad \text{and} \quad \overline{P}^{(\alpha)}_{\nu\nu'} = \sum_{a}^{N_{virt}} \overline{L}^{(\alpha)}_{\nu a}\overline{L}^{(\alpha)}_{\nu' a}, \tag{11}$$

where the Cholesky factors $\mathbf{L}$ have dimensions $N \times N_{occ}$ and $N \times N_{virt}$, respectively [51]. While this decomposition preserves the locality of the standard AO-based pseudo-density matrices, it reintroduces the restriction of orbital indices to occupied and virtual dimensions.

In terms of Cholesky decomposed pseudo-densities, the Coulomb-type part of the MP2 energy results as

$$e_J^{(\alpha)} = \sum_{ij}^{N_{occ}} \sum_{ab}^{N_{virt}} \left(\underline{i}\overline{a}|\underline{j}\overline{b}\right)^{(\alpha)} \left(\underline{i}\overline{a}|\underline{j}\overline{b}\right)^{(\alpha)} \tag{12}$$

with fully transformed integrals (FTIs) obtained by transforming the two-electron integrals with the Cholesky factors

$$\left(\underline{i}\overline{a}|\underline{j}\overline{b}\right)^{(\alpha)} = \sum_{\mu\nu\lambda\sigma}^{N} (\mu\nu|\lambda\sigma)\underline{L}^{(\alpha)}_{\mu i}\overline{L}^{(\alpha)}_{\nu a}\underline{L}^{(\alpha)}_{\lambda j}\overline{L}^{(\alpha)}_{\sigma b} \tag{13}$$

(application to the exchange part is straightforward, but not in focus of our present work).

The resulting so-called Cholesky-decomposed density-based (CDD-) $\mathrm{MP2}_J$ method has restricted orbital indices i, a, j, b, while at the same time the linear-scaling behavior of AO-MP2 theory is retained [51]. A crucial prerequisite for attaining a linear-scaling behavior of CDD-$\mathrm{MP2}_J$ is again the possibility to preselect numerically significant quantities by MBIE [26,51].

Although the CDD-$\mathrm{MP2}_J$ method performs better than AO-$\mathrm{MP2}_J$ for larger basis sets, we have not yet fully explored its capabilities which is subject to future work [52]. Instead we have focused in a first step to reduce the prefactors further by combining CDD-$\mathrm{MP2}_J$ with the well known resolution of the identity (RI) technique [19,20,53] for calculating the FTIs

$$\left(\underline{i}\overline{a}|\underline{j}\overline{b}\right) \approx \sum_{k}^{N_{aux}} B_{\underline{i}\overline{a}k} B_{\underline{j}\overline{b}k}, \tag{14}$$

where the intermediary matrix $\mathbf{B}$ is constructed as

$$B_{\underline{i}\overline{a}l} = \sum_{\mu\nu} \underline{L}_{\mu i} \overline{L}_{\nu a} B_{\mu\nu l} \qquad \text{with} \tag{15}$$

$$B_{\mu\nu l} = \sum_{k} (\mu\nu|k)\,(k|l)^{-\frac{1}{2}}. \tag{16}$$

Indices k, l denote indices of the auxiliary basis with size N_{aux}, and $(k|l)^{-1/2}$ elements of the inverse square root of the matrix of two-center integrals over auxiliary basis functions. The advantages of the RI-based CDD-MP2$_J$ (RI-CDD-MP2$_J$) method are twofold: Firstly, the prefactors for the construction of **B** (Eq. 16) and transformation (Eq. 15) are small as they involve only three indices for the target matrices. Secondly, the rate-determining step, namely the transformed integrals assembly (Eq. 14) profits from restriction of the indices to the occupied/virtual subspaces, respectively and represents a simple matrix multiplication that is amenable to the benefits of highly efficient linear-algebra libraries.

The major drawback of using RI is, however, that employing the full auxiliary basis set leads to quadratically and cubically scaling steps (where the former are rate-determining in the systems we investigated so far), so that the RI-CDD-MP2$_J$ method is not linear scaling. Here, possible improvements, as suggested in different context, might be obtained by restricting the auxiliary basis set to local domains [54,55] or by working in an attenuated Coulomb-metric [56,57]. We will not explore these approaches in our present work and will use the full auxiliary basis set throughout. Despite the loss of linear scaling for the RI-version of CDD-MP2$_J$, we expect the method to be useful for a range of medium-sized molecules.

The performance comparison of RI-CDD-MP2$_J$ to linear-scaling AO-MP2$_J$ for the two molecules C_5H_{12} and DNA$_1$ and various basis sets (6–31G*, SVP, and cc-pVTZ) shows gains of one to two orders of magnitude for the RI-CDD-based method [51], where the gain increases with increasing basis set size. Here it is important to note that the given benefits for C_5H_{12} [51] can be considered to be close to the maximum obtainable ones, as for larger systems screening will become more effective and thus linear-scaling AO-MP2$_J$ will become more competitive compared to the higher-order scaling RI-CDD-MP2$_J$. This is already seen for DNA$_1$, where the gains are smaller than for C_5H_{12}.

Furthermore, a comparison of the total timings (Wall times) for linear alkanes and DNA fragments shows that RI-CDD-MP2$_J$ has small prefactors as compared to RI-MO-MP2 even for small-sized systems. It also exhibits an early crossover point located between $C_{20}H_{42}$ and $C_{40}H_{82}$ and DNA$_2$ and DNA$_4$, respectively. A more detailed account of our RI-CDD-MP2$_J$ method and its performance can be found in Ref. [51] and is also subject to future work.

9. Conclusion

We have reviewed our linear-scaling AO-MP2 method which is based on the MBIE-preselection of numerically significant integral products that decay at least as $1/R^4$ with the distance of the corresponding charge distributions. The linear scaling is illustrated for both alkanes and DNA fragments, the latter with up to 1052 atoms and 10 674 basis functions. Furthermore, the so far largest SOS-AO-MP2 calculation is presented for an RNA system comprising 1664 atoms and 19 182 basis functions.

Whereas pure AO-MP2 theory sacrifices the occupied/virtual blocking of conventional fifth-order scaling MO-MP2 theory, we recently introduced a Laplace-based MP2 method which employs Cholesky-decomposed pseudo-density matrices. In this way, the advantageous occupied/virtual rank can be exploited, while the sparsity of the pseudo-density matrices is preserved. The advantages are obvious, in particular for larger basis sets, while the linear-scaling behavior of our AO-MP2 method is retained. Furthermore, we combined our linear-scaling CDD-MP2 approach with the RI approximation in order to reduce prefactors. Although the RI approach destroys the linear scaling if the full auxiliary basis is employed, it offers pragmatic advantages for medium-sized molecules.

While the calculation of molecules such as the RNA system with 1664 atoms clearly shows the new possibilities of our MBIE-based AO-MP2 approach, much work remains to be done in the future. Our Cholesky-based approach provides a first step towards the usability of larger basis sets, but also other prefactors in the calculation offer much room for improvements.

Acknowledgement

C. O. acknowledges financial support (OC35/3) by SPP 1145 'Modern and universal first-principles methods for many-electron systems in chemistry and physics' of the DFG ('Deutsche Forschungsgemeinschaft'). J. Z. thanks the 'Landesgraduiertenförderung' for a graduate fellowship.

References

1. E. Schrödinger, Ann. Phys. **79** (1926) 361.
2. C. V. Sumowski and C. Ochsenfeld, J. Phys. Chem. A **113** (2009) 11734.
3. B. D. Persson, S. Müller, D. M. Reiter, B. B. T. Schmitt, M. Marttila, C. V. Sumowski, S. Schweizer, U. Scheu, C. Ochsenfeld, N. Arnberg, and T. Stehle, J. Virol. **83** (2009) 673.
4. P. Pulay, Chem. Phys. Lett. **100** (1983) 151.
5. S. Saebø and P. Pulay, Chem. Phys. Lett. **113** (1985) 13.
6. P. Pulay and S. Saebø, Theor. Chem. Acc. **69** (1985) 357.
7. S. Saebø and P. Pulay, J. Chem. Phys. **86** (1987) 914.
8. C. Hampel and H.-J. Werner, J. Chem. Phys. **104** (1996) 6286.
9. M. Schütz, G. Hetzer, and H.-J. Werner, J. Chem. Phys. **111** (1999) 5691.
10. G. Hetzer, M. Schütz, H. Stoll, and H.-J. Werner, J. Chem. Phys. **113** (2000) 9443.
11. P. E. Maslen and M. Head-Gordon, Chem. Phys. Lett. **283** (1998) 102.
12. M. S. Lee, P. E. Maslen, and M. Head-Gordon, J. Chem. Phys. **112** (2000) 3592.
13. J. Almlöf, Chem. Phys. Lett. **181** (1991) 319.
14. M. Häser and J. Almlöf, J. Chem. Phys. **96** (1992) 489.
15. M. Häser, Theor. Chim. Acta **87** (1993) 147.

16. P. Y. Ayala and G. E. Scuseria, J. Chem. Phys. **110** (1999) 3660.
17. R. A. Friesner, R. B. Murphy, M. D. Beachy, M. N. Ringnalda, W. T. Pollard, B. D. Dunietz, and Y. Cao, J. Phys. Chem. A **103** (1999) 1913.
18. D. Walter, A. B. Szilva, K. Niedfeldt, and E. A. Carter, J. Chem. Phys. **117** (2002) 1982.
19. M. Feyereisen, G. Fitzgerald, and A. Komornicki, Chem. Phys. Lett. **208** (1993) 359.
20. F. Weigend, M. Häser, H. Patzelt, and R. Ahlrichs, Chem. Phys. Lett. **294** (1998) 143.
21. H.-J. Werner, F. Manby, and P. J. Knowles, J. Chem. Phys. **118** (2003) 8149.
22. Y. Jung, R. C. Lochan, A. D. Dutoi, and M. Head-Gordon, J. Chem. Phys. **121** (2004) 9793.
23. M. Kobayashi and H. Nakai, Chem. Phys. Lett. **420** (2006) 250.
24. T. Nakajima and K. Hirao, Chem. Phys. Lett. **427** (2006) 225.
25. F. Neese, F. Wennmohs, and A. Hansen, J. Chem. Phys. **130** (2009) 114108.
26. D. S. Lambrecht and C. Ochsenfeld, J. Chem. Phys. **123** (2005) 184101.
27. D. S. Lambrecht, B. Doser, and C. Ochsenfeld, J. Chem. Phys. **123** (2005) 184102.
28. N. J. Russ and T. D. Crawford, J. Chem. Phys. **121** (2004) 691.
29. B. Doser, D. S. Lambrecht, and C. Ochsenfeld, Phys. Chem. Chem. Phys. **10** (2008) 3335.
30. B. Doser, D. S. Lambrecht, J. Kussmann, and C. Ochsenfeld, J. Chem. Phys. **130** (2009) 064107.
31. M. Beer and C. Ochsenfeld, J. Chem. Phys. **128** (2008) 221102.
32. W. H. Press, B. P. Flannery, S. A. Teukolsky, and W. T. Vetterling, *Numerical Recipes in C: The Art of Scientific Computing*, Cambridge University Press, 2nd edition, 2002.
33. A. Takatsuka, S. Ten-no, and W. Hackbusch, J. Chem. Phys. **129** (2008) 044112.
34. D. Kats, D. Usvyat, S. Loibl, T. Merz, and M. Schütz, J. Chem. Phys. **130** (2009) 127101.
35. S. B. Boys and F. Bernardi, Mol. Phys **19** (1970) 553.
36. P. R. Surján, Chem. Phys. Lett. **406** (2005) 318.
37. S. Schweizer, B. Doser, and C. Ochsenfeld, J. Chem. Phys. **128** (2008) 154101.
38. C. Ochsenfeld and M. Head-Gordon, Chem. Phys. Lett. **270** (1997) 399.
39. P. E. Maslen, C. Ochsenfeld, C. A. White, M. S. Lee, and M. Head-Gordon, J. Phys. Chem. A **102** (1998) 2215.
40. C. Ochsenfeld, J. Kussmann, and D. S. Lambrecht, *Linear-Scaling Methods in Quantum Chemistry*, in Reviews in Computational Chemistry, edited by K. B. Lipkowitz and T. L. Cundari, volume 23, pages 1–82, VCH Publishers, New York, 2007.
41. P. M. W. Gill, B. G. Johnson, and J. A. Pople, Chem. Phys. Lett. **217** (1994) 65.
42. M. Häser and R. Ahlrichs, J. Comput. Chem. **10** (1989) 104.
43. K. Eichkorn, F. Weigend, O. Treutler, and R. Ahlrichs, Theor. Chem. Acc. **97** (1997) 119.
44. B. Seelig and A. Jäschke, Chem. Biol. **6** (1999) 167.
45. A. Serganov, S. Keiper, L. Malinina, V. Tereshko, E. Skripkin, C. Höbartner, A. Polonskaia, A. T. Phan, R. Wombacher, R. Micura, Z. Dauter, A. Jäschke, and D. J. Patel, Nat. Struct. Mol. Biol. **12** (2005) 218.
46. B. Doser and C. Ochsenfeld, in preparation (2010).
47. W. Klopper, F. R. Manby, S. Ten-no, and E. F. Valeev, Int. Rev. Phys. Chem. **25** (2006) 427.
48. H.-J. Werner, J. Chem. Phys. **129** (2008) 101103.
49. S. Grimme, J. Chem. Phys. **118** (2003) 9095.
50. P. Jurecka, J. Sponer, J. Cerný, and P. Hobza, Phys. Chem. Chem. Phys. **8** (2006) 1985.

51. J. Zienau, L. Clin, B. Doser, and C. Ochsenfeld, J. Chem. Phys. **130** (2009) 204112.
52. L. Clin, J. Zienau, B. Doser, and C. Ochsenfeld, in preparation (2010).
53. J. L. Whitten, J. Chem. Phys. **58** (1973) 4496.
54. H.-J. Werner, F. Manby, and P. J. Knowles, J. Chem. Phys. **118** (2003) 8149.
55. A. Sodt, J. E. Subotnik, and M. Head-Gordon, J. Chem. Phys. **125** (2006) 194109.
56. Y. Jung, A. Sodt, P. M. W. Gill, and M. Head-Gordon, Proc. Natl. Acad. Sci. U.S.A. **102** (2005) 6692.
57. S. Reine, E. Tellgren, A. Krapp, T. Kjærgaard, T. Helgaker, B. Jansik, S. Høst, and P. Salek, J. Chem. Phys. **129** (2008) 104101.

A Quasirelativistic Two-component Density Functional and Hartree-Fock Program

By Christoph van Wüllen*

Fachbereich Chemie and Forschungszentrum OPTIMAS, Technische Universität Kaiserslautern, Erwin-Schrödinger Straße 52, 67663 Kaiserslautern, Germany

(Received June 10, 2009; accepted September 25, 2009)

Relativistic / Spin-orbit Coupling / Two-component / Hartree-Fock / Density Functional Theory / Magnetic Anisotropy

This work describes the implementation of an efficient two-component quasirelativistic density functional and Hartree-Fock program. The fact that the basis functions are real can be exploited if a special internal representation of operators and density matrices is used. This also leads to a considerable reduction of the effort in the closed shell case. While in most applications to open shell molecules, the noncollinear approach to define a relativistic spin density is preferable, the collinear approach finds its application in the calculation of magnetic anisotropy energies. Linear algebra steps in the SCF procedure have a higher relative weight compared to the nonrelativistic case, therefore some care was necessary to make them fast when parallelizing the code. The quasirelativistic Hamiltonians that have been implemented are the 'zeroth-order regular approximation' (ZORA) Hamiltonian, the Douglas-Kroll-Hess Hamiltonian up to the sixth order, together with an accurate approximation to treat the picture change effect of the electron interaction, and effective core potential (ECP) matrix elements. Geometry gradients are available for the ZORA and ECP methods.

1. Introduction

The Schrödinger equation describes a nonrelativistic world as its classical limit corresponds to Newtonian mechanics. In classical physics, this is appropriate as long as the relative velocities of the particles are small compared to the speed of light, or, in other words, the kinetic energy of a particle is small compared to its rest energy mc^2. It is easily seen that this condition is not fulfilled for electrons moving in the potential of heavy (highly charged) nuclei, and since the mid-1970s it is also clear that this does not only affect electrons in the inner cores of heavy elements, but has an impact on the chemical (valence shell) properties of

* Corresponding author. E-mail: vanwullen@chemie.uni-kl.de

© by Oldenbourg Wissenschaftsverlag, München

these elements [1,2]. During the 1990s, almost all molecular electronic structure codes have been extended to take relativistic effects into account, since otherwise applications to heavy-element chemistry would have remained rather limited. For a small review of this period, the reader is referred to Ref. [3]. However, most implementations have focused on so-called scalar relativistic approximations which neglect the spin-orbit interaction. Without the spin-orbit interaction, the quasirelativistic Hamiltonian has the same symmetry properties as in the nonrelativistic case. Therefore, adding scalar relativitiy to a nonrelativistic quantum chemical code is not particularly difficult, a modification of the one-electron integrals is sufficient in most cases, and most quantum chemistry codes have been extended along these lines.

For most molecules, in particular *closed shell* molecules, spin-orbit coupling has only a small effect on valence ground-state properties like bond lengths and vibrational frequencies. Excitation energies on the other hand are much more affected, especially if the spin-orbit interaction splits states which are (nearly) degenerate in a scalar-relativistic description. A standard method used in such situations is spin-orbit CI starting with relatively few highly correlated wave functions from a scalar-relativistic calculation and adding a spin-orbit configuration interaction calculation in this space [4–9]. To investigate spin-orbit effects for most molecular ground-state properties spin-orbit CI is however less suitable, and one should include the spin-orbit interaction already in the orbital optimization procedure. This not only takes care of higher-order spin-orbit effects such as orbital relaxation (atomic $2p_{1/2}$ and $2p_{3/2}$ orbitals may have quite different radial parts) but also includes spin-orbit effects in a single-configuration description that is inherent in the density functional methodology. The drawback of such an approach is that the orbitals are no longer spin eigenfunctions (either *spin up* or *spin down*) and can no longer be chosen real, such that they form complex two-component quantities. Although this is a more technical rather than a conceptual problem, it means that an existing quantum chemical computer code cannot simply be extended to run with two-component orbitals but requires a major re-design. During the past years, a number of groups have developed density functional codes with two-component spinors [10–16]. There are older two-component Hartree-Fock implementations [17,18] which seemingly have never been applied to molecules with more than a few atoms. A more recent implementation of two-component Hartree-Fock [19,20] is mainly used to generate two-component spinors for a subsequent electron correlation treatment as this implementation is restricted to the *closed shell* case.

In this article we report our implementation of the two-component computational method which differs from other implementations we are aware of, and which offers some advantages in terms of storage and number of floating point operations needed as well as an easier interface to existing non-relativistic or scalar relativistic computer code. It is shortly discussed how to define a 'spin density' in a relativistic description. The next sections give some notes on the performance and the parallelization of the code and on the quasirelativistic Hamiltonians we are currently using.

2. Program structure

Since the Kohn-Sham density functional and Hartree-Fock method are essentially the same from the viewpoint of a computer implementation, we will treat both cases together. For an n-electron system, one starts from a single Slater determinant (either the Hartree-Fock wave function itself of the Kohn-Sham reference wave function)

$$\Psi = |\psi_1 \psi_2 \dots \psi_n| \tag{1}$$

built from two-component complex molecular spinors (orbitals)

$$\psi_i = \begin{pmatrix} \phi_i^{(\alpha)} \\ \phi_i^{(\beta)} \end{pmatrix} \tag{2}$$

The basis set used to expand the scalar functions ϕ_i^{α}, ϕ_i^{β} consists of N conventional (real and scalar) Gaussian basis functions B_μ. Formally, this makes a basis of $2N$ real but two-component basis functions to expand the molecular spinors

$$\psi_i = \sum_{\mu=1}^{N} C_{\mu i}^{(\alpha)} \begin{pmatrix} B_\mu \\ 0 \end{pmatrix} + \sum_{\mu=1}^{N} C_{\mu i}^{(\beta)} \begin{pmatrix} 0 \\ B_\mu \end{pmatrix} \tag{3}$$

We will make use of the fact that our $2N$ two-component basis functions are real and eigenfunctions of σ_z, the third of the Pauli spin matrices

$$\sigma_x = \begin{pmatrix} 0 & 1 \\ 1 & 0 \end{pmatrix}, \quad \sigma_y = \begin{pmatrix} 0 & -\mathrm{i} \\ \mathrm{i} & 0 \end{pmatrix}, \quad \sigma_z = \begin{pmatrix} 1 & 0 \\ 0 & -1 \end{pmatrix}. \tag{4}$$

The matrix representation $\mathbf{H}$ of a one-particle operator such as the one-electron part of the Hamiltonian $\hat{h}$ is a complex $2N \times 2N$ matrix which can be written in terms of four complex $N \times N$ blocks

$$\mathbf{H} = \begin{pmatrix} \mathbf{H}^{(\alpha\alpha)} & \mathbf{H}^{(\alpha\beta)} \\ \mathbf{H}^{(\beta\alpha)} & \mathbf{H}^{(\beta\beta)} \end{pmatrix} \tag{5}$$

with, for example (the other blocks are defined likewise)

$$H_{\mu\nu}^{(\alpha\beta)} = \left\langle \begin{pmatrix} B_\mu \\ 0 \end{pmatrix} \middle| \hat{h} \middle| \begin{pmatrix} 0 \\ B_\nu \end{pmatrix} \right\rangle \tag{6}$$

Most two-component density functional programs seem to use such a representation of spin-dependent operators through the matrix elements with two-component basis functions, but our implementation is different as we express the spin-dependent operator $\hat{h}$

$$\hat{h} = \hat{h}^{(0)} + \sigma_x \hat{h}^{(x)} + \sigma_y \hat{h}^{(y)} + \sigma_z \hat{h}^{(z)} \tag{7}$$

through *spin-free* operators $\hat{h}^{(0)}$, $\hat{h}^{(x)}$, $\hat{h}^{(y)}$ and $\hat{h}^{(z)}$ and the Pauli spin matrices. This can be done for any spin-dependent one-particle operator. In the absence of magnetic fields, the Hamiltonian $\hat{h}$ is invariant w.r.t. time reversal, that is

$$\hat{h} = \hat{T}\hat{h}\hat{T}^{-1} \tag{8}$$

$$\hat{T} = -\mathrm{i}\,\sigma_y \hat{C} \tag{9}$$

($\hat{C}$ is the complex conjugation operator), and a little algebra shows then that

$$\hat{h}^{(0)} = \hat{C}\hat{h}^{(0)}\hat{C} \tag{10}$$

$$\hat{h}^{(S)} = -\hat{C}\hat{h}^{(S)}\hat{C}, \quad S \in \{x, y, z\} \tag{11}$$

Internally, we represent the operator $\hat{h}$ through the matrix elements of these spin-free operators, and this gives four $N \times N$ matrices $\mathbf{H}^{(0)}$, $\mathbf{H}^{(x)}$, $\mathbf{H}^{(y)}$ and $\mathbf{H}^{(z)}$ with, for example (the other matrices are defined likewise)

$$H^{(x)}_{\mu\nu} = \left\langle B_\mu \middle| \hat{h}^{(x)} B_\nu \right\rangle \tag{12}$$

Switching to this representation has several advantages. For example, from Eqs.(10,11) follows that for real basis functions ($B_\mu = \hat{C}\, B_\mu$) the matrix $\mathbf{H}^{(0)}$ is real while the matrices $\mathbf{H}^{(x)}$, $\mathbf{H}^{(y)}$ and $\mathbf{H}^{(z)}$ are purely imaginary, and only their imaginary part need be stored. Compared to the full matrix $\mathbf{H}$ (or its four blocks), our representation needs only half of the storage. The full complex $2N \times 2N$ matrix can easily be constructed through

$$\begin{aligned} \mathbf{H}^{(\alpha\alpha)} &= \mathbf{H}^{(0)} + \mathbf{H}^{(z)}, \quad \mathbf{H}^{(\alpha\beta)} = \mathbf{H}^{(x)} - \mathrm{i}\mathbf{H}^{(y)} \\ \mathbf{H}^{(\beta\beta)} &= \mathbf{H}^{(0)} - \mathbf{H}^{(z)}, \quad \mathbf{H}^{(\beta\alpha)} = \mathbf{H}^{(x)} + \mathrm{i}\mathbf{H}^{(y)} \end{aligned} \tag{13}$$

The only occasion where this is actually done in our program is for the matrix representation of the Fock (Kohn-Sham) operator before it is transformed to the molecular orbital basis as part of the diagonalization, because this enables us to use BLAS routines for the matrix multiplications that are part of this transformation (see Sec. 3).

The one-electron part of the energy expectation value of the Slater determinant Ψ is

$$\begin{aligned} E_{1e} &= \sum_{i=1}^{n} < \psi_i | \hat{h} \psi_i > \\ &= \sum_{\mu,\nu=1}^{N} \left\{ D^{(0)}_{\mu\nu} H^{(0)}_{\mu\nu} + D^{(x)}_{\mu\nu} H^{(x)}_{\mu\nu} + D^{(y)}_{\mu\nu} H^{(y)}_{\mu\nu} + D^{(z)}_{\mu\nu} H^{(z)}_{\mu\nu} \right\} \end{aligned} \tag{14}$$

This defines the density matrices $\mathbf{D}^{(0)}$, $\mathbf{D}^{(x)}$, $\mathbf{D}^{(y)}$ and $\mathbf{D}^{(z)}$ which can be computed from the molecular orbital coefficients as

$$\begin{aligned}
D_{\mu\nu}^{(0)} &= \sum_{i=1}^{n} \left\{ \left(C_{\mu i}^{(\alpha)}\right)^{\star} C_{\nu i}^{(\alpha)} + \left(C_{\mu i}^{(\beta)}\right)^{\star} C_{\nu i}^{(\beta)} \right\} \\
D_{\mu\nu}^{(x)} &= \sum_{i=1}^{n} \left\{ \left(C_{\mu i}^{(\alpha)}\right)^{\star} C_{\nu i}^{(\beta)} + \left(C_{\mu i}^{(\beta)}\right)^{\star} C_{\nu i}^{(\alpha)} \right\} \\
D_{\mu\nu}^{(y)} &= -\mathrm{i} \sum_{i=1}^{n} \left\{ \left(C_{\mu i}^{(\alpha)}\right)^{\star} C_{\nu i}^{(\beta)} - \left(C_{\mu i}^{(\beta)}\right)^{\star} C_{\nu i}^{(\alpha)} \right\} \\
D_{\mu\nu}^{(z)} &= \sum_{i=1}^{n} \left\{ \left(C_{\mu i}^{(\alpha)}\right)^{\star} C_{\nu i}^{(\alpha)} - \left(C_{\mu i}^{(\beta)}\right)^{\star} C_{\nu i}^{(\beta)} \right\}
\end{aligned} \tag{15}$$

To compute E_{1e}, only the real part of $\mathbf{D}^{(0)}$ and the imaginary parts of $\mathbf{D}^{(x)}$, $\mathbf{D}^{(y)}$ and $\mathbf{D}^{(z)}$ are needed and thus requires $4N^2$ (real) floating point multiplications (not counting the hermiticity of the $\mathbf{D}$ and $\mathbf{H}$ matrices, which leads to the same savings in both representations). The computational effort to contract the (complex) $2N \times 2N$ matrix $\mathbf{H}$ with the corresponding (complex) density matrix is at least twice as high.

So far we have not imposed any constraints on the molecular orbitals ψ_i except being normalized and orthogonal. This is known as as the *Kramers unrestricted* procedure, it corresponds to spin unrestricted Kohn-Sham and Hartree-Fock formalisms. Since spin restricted methods are rarely used in nonrelativistic density functional theory, so we did not implement the two-component *Kramers restricted* case either. We do however obtain a reduction both in storage and CPU time for the *closed shell* case, where all the orbitals ψ_i form Kramers pairs (n is necessarily even in this case). Two orbitals ψ_i and ψ_j form a Kramers pair if (except for a phase factor) they are related through

$$\psi_j = \begin{pmatrix} \phi_j^{(\alpha)} \\ \phi_j^{(\beta)} \end{pmatrix} = \begin{pmatrix} -\left(\phi_i^{(\beta)}\right)^{\star} \\ \left(\phi_i^{(\alpha)}\right)^{\star} \end{pmatrix} = -\mathrm{i}\,\sigma_y \hat{C} \psi_i \tag{16}$$

Because the basis functions are real, one can verify that in the *closed shell* case, $\mathbf{D}^{(0)}$ is real while $\mathbf{D}^{(x)}$, $\mathbf{D}^{(y)}$ and $\mathbf{D}^{(z)}$ are purely imaginary. This does not only halve the storage needed for the density matrices but also leads to considerable computational savings when calculating the contributions from the electron interaction to the Kohn-Sham (Fock) matrix, and implies that the spin density vanishes everywhere (see below).

The electron interaction part of the energy expectation value is

$$\begin{aligned}
E_{2e} &= \frac{1}{2} \sum_{i,j=1}^{n} \left\{ (\psi_i \psi_i | \psi_j \psi_j) - a_X (\psi_i \psi_j | \psi_j \psi_i) \right\} \\
&\quad + \int F_{xc}\left(\rho(\vec{r}), s(\vec{r})\right) \mathbf{d}\vec{r}
\end{aligned} \tag{17}$$

In the Hartree-Fock case, $a_X = 1$ and F_{xc} vanishes, while for pure density functionals, $a_X = 0$. For the so-called hybrid (or exact exchange) density functionals first suggested by Becke [21,22] F_{xc} is non-vanishing and a_X assumes a non-zero value (frequently around 0.2). As written here, the function $F_{xc}(\vec{r})$ depends locally on the density $\rho(\vec{r})$ and the spin density $s(\vec{r})$ (local density approximation). Modern 'gradient-corrected' functionals additionally depend on the derivatives $\vec{\nabla}\rho \cdot \vec{\nabla}\rho$, $\vec{\nabla}\rho \cdot \vec{\nabla} s$ and $\vec{\nabla} s \cdot \vec{\nabla} s$. Although this case has been implemented as well (some details can be found in Ref. [23]), it will not be discussed here since it is a straightforward albeit tedious extension well known from the nonrelativistic case.

The first contribution to E_{2e} involves two-electron integrals with two-component orbitals in Mulliken notation, which can be expressed in conventional two-electron integrals (involving only scalar orbitals) for arbitrary spinors ψ_p

$$\begin{aligned}(\psi_p\psi_q|\psi_r\psi_s) = (\phi_p^{(\alpha)}\phi_q^{(\alpha)}|\phi_r^{(\alpha)}\phi_s^{(\alpha)}) + (\phi_p^{(\alpha)}\phi_q^{(\alpha)}|\phi_r^{(\beta)}\phi_s^{(\beta)}) \\ + (\phi_p^{(\beta)}\phi_q^{(\beta)}|\phi_r^{(\alpha)}\phi_s^{(\alpha)}) + (\phi_p^{(\beta)}\phi_q^{(\beta)}|\phi_r^{(\beta)}\phi_s^{(\beta)})\end{aligned} \tag{18}$$

To facilitate the notation we use the conventional definition of the Coulomb and exchange operators $\hat{J}$ and $\hat{K}$ which are defined through their matrix elements

$$\begin{aligned}<\psi_p|\hat{J}\psi_q> &= \sum_{j=1}^{n}(\psi_p\psi_q|\psi_j\psi_j) \\ <\psi_p|\hat{K}\psi_q> &= \sum_{j=1}^{n}(\psi_p\psi_j|\psi_j\psi_q)\end{aligned} \tag{19}$$

In the matrix representation in terms of our real basis functions, $\mathbf{J}^{(0)}$ is a real matrix and $\mathbf{J}^{(x)}$, $\mathbf{J}^{(y)}$, $\mathbf{J}^{(z)}$ vanish (because $\hat{J}$ is a spin-free operator). The two-electron energy can be re-expressed as

$$\begin{aligned}E_{2e} = &\frac{1}{2}\sum_{\mu,\nu=1}^{N} D_{\mu\nu}^{(0)}J_{\mu\nu}^{(0)} \\ &- \frac{a_X}{4}\sum_{\mu,\nu=1}^{N}\left\{D_{\mu\nu}^{(0)}K_{\mu\nu}^{(0)} + D_{\mu\nu}^{(x)}K_{\mu\nu}^{(x)} + D_{\mu\nu}^{(y)}K_{\mu\nu}^{(y)} + D_{\mu\nu}^{(z)}K_{\mu\nu}^{(z)}\right\} \\ &+ \int F_{xc}\left(\rho(\vec{r}), s(\vec{r})\right) \mathbf{d}\vec{r}\end{aligned} \tag{20}$$

The $\mathbf{J}$ and $\mathbf{K}$ matrices can be calculated from the conventional two-electron integrals using standard techniques

$$\begin{aligned}J_{\mu\nu}^{(0)} &= \sum_{\rho,\sigma=1}^{N} D_{\rho\sigma}^{(0)}(B_\mu B_\nu|B_\rho B_\sigma) \\ K_{\mu\nu}^{(S)} &= \sum_{\rho,\sigma=1}^{N} D_{\rho\sigma}^{(S)}(B_\mu B_\sigma|B_\rho B_\nu), \quad S \in \{0, x, y, z\}\end{aligned} \tag{21}$$

Note that only the real part of $\mathbf{D}^{(0)}$ is needed to construct $\mathbf{J}^{(0)}$, such that the algorithm from a nonrelativistic program can be used without any change. The exchange operator is spin-dependent even for *closed shell* systems, but in this special case $\mathbf{K}^{(0)}$ is real and $\mathbf{K}^{(x)}$, $\mathbf{K}^{(y)}$, $\mathbf{K}^{(z)}$ are purely imaginary. This follows from the properties of the density matrices and the fact that the two-electron integrals $(B_\mu\ B_\sigma \mid B_\rho\ B_\nu)$ are real. In fact we store the real and imaginary parts of the density matrices separately and thus also separately construct the real and imaginary parts of the $\mathbf{K}$ matrices using well-tested code from our nonrelativistic program. In addition, half of the work can just be skipped in the *closed shell* case where the imaginary part of $\mathbf{K}^{(0)}$ and the real parts of $\mathbf{K}^{(x)}$, $\mathbf{K}^{(y)}$ and $\mathbf{K}^{(z)}$ vanish and need not be computed. Comparing the computational effort of a scalar relativistic and a two-component program, there is no difference in the calculation of the matrix elements of $\hat{J}$, but for the $\mathbf{K}$ matrices the *processing* of the two-electron integrals is four times more expensive (the *calculation* of the two-electron integrals is the same of course). The exchange-correlation energy is evaluated using numerical quadrature as in the nonrelativistic case. At each grid point of the integration grid, one needs to evaluate the density and the spin density. The density is given by

$$\rho(\vec{r}) = \sum_{i=1}^{n} \psi_i^\dagger(\vec{r})\psi_i(\vec{r}) = \sum_{\mu,\nu=1}^{N} D_{\mu\nu}^{(0)} B_\mu(\vec{r}) B_\nu(\vec{r}) \tag{22}$$

Only the real part of the density matrix $\mathbf{D}^{(0)}$ contributes, such that there is no difference to the nonrelativistic implementation. In a quasirelativistic procedure is however not clear *a priori* how to define the spin density. This has been investigated in some detail in Ref. [24]. Basically, one defines a (spin) magnetization vector $\vec{m} = (m_x, m_y, m_z)$ with

$$\begin{aligned}
m_x(\vec{r}) &= \sum_{i=1}^{n} \psi_i^\dagger(\vec{r})\sigma_x\psi_i(\vec{r}) = \sum_{\mu,\nu=1}^{N} D_{\mu\nu}^{(x)} B_\mu(\vec{r}) B_\nu(\vec{r}) \\
m_y(\vec{r}) &= \sum_{i=1}^{n} \psi_i^\dagger(\vec{r})\sigma_y\psi_i(\vec{r}) = \sum_{\mu,\nu=1}^{N} D_{\mu\nu}^{(y)} B_\mu(\vec{r}) B_\nu(\vec{r}) \\
m_z(\vec{r}) &= \sum_{i=1}^{n} \psi_i^\dagger(\vec{r})\sigma_z\psi_i(\vec{r}) = \sum_{\mu,\nu=1}^{N} D_{\mu\nu}^{(z)} B_\mu(\vec{r}) B_\nu(\vec{r})
\end{aligned} \tag{23}$$

Note that again only the real parts of the density matrices contribute. These are zero in the *closed shell* case where of course $\vec{m} = 0$ everywhere (and need not be computed). There are two ways to define the spin density in terms of $\vec{m}$, and both have been implemented in our program. The older one is the *collinear approach*, where one chooses a unit vector $\vec{e} = (e_x, e_y, e_z)$ (frequently called the spin alignment direction or the spin quantization axis) pointing in an arbitrary but fixed direction, and defines $s(\vec{r}) = \vec{e} \cdot \vec{m}(\vec{r})$. It has been pointed out [24] that the collinear approach together with spin-orbit coupling lets the total energy of a molecule depend on its orientation which is not desirable in most cases (an exception is the calculation of magnetic anisotropy energies, see below). An energy expression that is rotationally invariant results from the *non-collinear approach* where one defines $s(\vec{r}) = |\vec{m}(\vec{r})|$. The non-collinear approach requires a larger computational effort since one has to calculate the three quantities m_x, m_y, m_z at each grid point before evaluating the spin density there, while in the collinear approach one first constructs a single effective spin density matrix $\mathbf{D}^{(e)}$ and evaluates the spin density at each point as

$$\mathbf{D}^{(e)} = e_x \mathbf{D}^{(x)} + e_y \mathbf{D}^{(y)} + e_z \mathbf{D}^{(z)}$$

$$s(\vec{r}) = \sum_{\mu,\nu=1}^{N} D_{\mu\nu}^{(e)} B_\mu(\vec{r}) B_\nu(\vec{r}) \tag{24}$$

To optimize the orbitals, matrix elements of the Fock (or Kohn-Sham) operator

$$\hat{F} = \hat{h} + \hat{J} - a_x \hat{K} + V_{xc} \tag{25}$$

are needed, all of them have been discussed except for the exchange-correlation potential V_{xc}. For real basis functions all the matrix representations $\mathbf{V}_{xc}^{(0)}$, $\mathbf{V}_{xc}^{(x)}$, $\mathbf{V}_{xc}^{(y)}$ and $\mathbf{V}_{xc}^{(z)}$ are real, and $\mathbf{V}_{xc}^{(0)}$ is calculated as in the nonrelativistic case

$$\left(V_{xc}^{(0)}\right)_{\mu\nu} = \int B_\mu B_\nu \frac{\delta F_{xc}}{\delta \rho} \mathrm{d}\vec{r} \tag{26}$$

This means that in the closed shell case, where the spin density vanishes and thus $\mathbf{V}_{xc}^{(x)} = \mathbf{V}_{xc}^{(y)} = \mathbf{V}_{xc}^{(z)} = 0$, the exchange-correlation part of a two-component density functional calculation needs nothing beyond that already available from the nonrelativistic implementation. In the collinear approach, one calculates a single effective spin-dependent matrix $\mathbf{V}_{xc}^{(e)}$

$$\left(V_{xc}^{(e)}\right)_{\mu\nu} = \int B_\mu B_\nu \frac{\delta F_{xc}}{\delta s} \mathrm{d}\vec{r} \tag{27}$$

and constructs the matrices $V_{xc}^{(x,y,z)}$ easily through

$$\mathbf{V}_{xc}^{(x)} = e_x \mathbf{V}_{xc}^{(e)}, \quad \mathbf{V}_{xc}^{(y)} = e_y \mathbf{V}_{xc}^{(e)}, \quad \mathbf{V}_{xc}^{(z)} = e_z \mathbf{V}_{xc}^{(e)} \tag{28}$$

In contrast, one has to construct three independent matrices in the non-collinear approach (all quantities in the integrands depend on $\vec{r}$)

$$\begin{aligned}
\left(V_{xc}^{(x)}\right)_{\mu\nu} &= \int B_\mu B_\nu \frac{m_x}{s} \frac{\delta F_{xc}}{\delta s} \mathrm{d}\vec{r} \\
\left(V_{xc}^{(y)}\right)_{\mu\nu} &= \int B_\mu B_\nu \frac{m_y}{s} \frac{\delta F_{xc}}{\delta s} \mathrm{d}\vec{r} \\
\left(V_{xc}^{(z)}\right)_{\mu\nu} &= \int B_\mu B_\nu \frac{m_z}{s} \frac{\delta F_{xc}}{\delta s} \mathrm{d}\vec{r}
\end{aligned} \tag{29}$$

Thus, for *closed shell* systems and also *open shell* systems with the collinear approach, the computational effort for the two-component exchange-correlation part is the same as for the nonrelativistic case (actually the same code is used), while the effort is doubled for the *open shell* case in the non-collinear approximation.

Geometry gradients for this formalism have been implemented [23] using well-established techniques, provided that the geometrical derivatives of the matrix elements of the quasirelativistic Hamiltonian $\hat{h}$ are available (see Sec. 4). There is little difference in CPU time betweeen the nonrelativistic and the two-component gradient since both are dominated by the calculation of the derivative two-electron integrals.

Early implementations of the two-component density functional method were restricted the collinear approach with a single 'hard-wired' spin alignment direction $\vec{e} = (0,0,1)$. This has largely been abandoned because of the loss of rotational invariance caused by singling out a spin alignment direction [24]. For a special application, namely the calculation of magnetic anisotropy energies (or zero field splittings) the collinear approach is however the method of choice [25–27]. Because of the spin-orbit interaction (together with relativistic corrections to the electron interaction), the ground state energy of a (laboratory-fixed) molecule in a strong homogeneous external magnetic field depends on the relative orientation of the molecule and the magnetic field. This dependence is called the magnetic anisotropy energy and is closely related to zero field splitting. Note that the relation between these two quantities has not always been fully understood, such that too small zero field splitting parameters have been extracted from magnetic anisotropy calculations [27]. To calculate magnetic anisotropy energies with density functional methods, one could expose the molecule an external magnetic field strong enough to align the spins but that does not mix Zeeman levels of different spin states (that is, increase the overall spin polarization). For this special purpose it is better to use the collinear definition of the spin density, because this will create an effective magnetic field that aligns the spins where there is spin density, but vanishes elsewhere. With this application in mind we have implemented the collinear approach in its full generality, that is, for any quantization axis $\vec{e}$ given by the user in the program input. Starting from scalar relativistic orbitals (where $\vec{m}(\vec{r})$ is parallel to the z axis everywhere) we first rotate the spins such that $\vec{m}$ is parallel to $\vec{e}$ everywhere, and iterate until self-consistency using the collinear approach. This is repeated for several different directions $\vec{e}$ and a magnetic anisotropy energy profile thus generated (for an application to the extensively studied molecular magnet $Mn_{12}O_{12}$ acetate, a system with nearly 150 atoms, see Ref. [27]).

3. Efficiency and Parallelization

In a typical calculation on a molecular system of about 100 atoms, the calculation of the **J**, **K** (for $a_x \neq 0$) and $\mathbf{V}_{xc}$ (for $F_{xc} \neq 0$) matrices requires the largest part of the CPU time. While the calculation of **J** is the same for a nonrelativistic (or scalar-relativistic) and a two-component calculation with the same basis set (actually the same code is used), calculating $\mathbf{V}_{xc}$ in the two-component non-collinear approach is twice as expensive, and *processing* the two-electron integrals to construct **K** even requires four times more floating point operations compared to the nonrelativistic case. Since the effort to *calculate* the two-electron integrals is the same in either case, we estimate that for Hartree-Fock and exact exchange containing density functionals, the effort associated with two-electron integrals is twice as large for the two-component calculations. After having properly implemented Pulay's "FDS-SDF" convergence accelerator [28] for the two-component case, the number of SCF iterations is similar for scalar-relativistic and two-component calculations on the same molecule, such that a two-component calculation is about twice as expensive as a scalar-relativistic calculation. The construction of the Fock matrix (calculation/processing of the two-electron integrals, numerical integration) can be parallelized using the same techniques as in the nonrelativistic case and need not be discussed here, but it is important to realize that the remaining, non-parallelized matrix algebra steps (the transformation of the Fock matrix to the molecular orbital basis, the diagonalization of the resulting matrix, the MO orthogonalization every few iterations, and the Pulay procedure) have a higher relative cost in the two-component case. This stems from the fact that matrix dimensions are twice as large (this gives a factor of eight because the effort for matrix multiplications and diagonalizations scales cubically with the dimension) and complex (this gives another factor of four). This analysis is supported by computer experiments [16]. To obtain good parallelization efficiency up to 32 parallel CPUs, all matrix algebra is done by BLAS and LAPACK routines (notably, subroutines zgemm for matrix multiplication, zheevd for matrix diagonalization, and zpotrf as well as ztrsm for MO orthogonalization via Cholesky decomposition). As an additional advantage, multi-threaded BLAS and LAPACK libraries are available that give a certain degree of automatic parallelization of these matrix routines on modern multicore CPUs. This together brings down the wall-clock time for the matrix steps well below that for the construction of the Fock matrix even in a parallel calculation. The overall efficiency of the two-component program is thus such that almost every scalar-relativistic calculation that can be done is also possible using the correspondig spin-dependent (two-component) quasirelativistic Hamiltonian.

In Table 1, CPU times (in seconds, single 2.8 GHz Xeon CPU core) for matrix operations (building the density matrices, similarity transformations, diagonlizations), the calculation and processinig of two-electron integrals, and the numerical quadrature are reported for a single SCF iteration on the *open shell* gold cluster Au_{21} and the *closed shell* system obtained by adsorbing a Tl atom onto it. ECPs were used in these calculations, and a $7s7p7d1f$ primitive basis set, contracted to $5s5p4d1f$ was used for Au, while $7s8p7d1f$ contracted to $5s6p4d1f$ was used for Tl. For Au_{21} there were 987 contracted scalar basis functions and our (fairly accurate) integration grid had ~ 457000 grid points with nonzero weight (1037 basis functions and ~ 476000 grid points for $Au_{21}Tl$). The measured CPU times are consistend with the analysis given above. Point-group symmetry has not been used in any of these calculations because it has not been implemented for the two-component case. The proper implementation of double

Table 1. CPU time for a single SCF iteration for scalar-relativistic ("sc") and two-component ("2c") ECP calculation for $Au_{21}Tl$ (*closed shell*) and Au_{21} (*open shell*). Results are given for a gradient-corrected (BP) and an exact-exchange containing (B3LYP) exchange-correlation functional. See text for computational details.

Molecule			matrix operations	two-electron integrals [a]	numerical quadrature
$Au_{21}Tl$	BP	sc	2	1594	90
		2c	45	1602	90
	B3LYP	sc	2	2273	90
		2c	44	5091	90
Au_{21}	BP	sc	3	1406	174
		2c	39	1409	406
	B3LYP	sc	3	2418	174
		2c	39	5259	405

[a] time for *both* calculating and processing the two-electron integrals

group symmetry as necessary for two-component calculations is slightly more complicated than in the scalar-relativistic or nonrelativistic case, on the other hand we expect smaller computational savings because the relativistic symmetry is usually lower. For example, there is only a single fermion irrep in the relativistic (double) point groups C_{2v}^* and D_2^*.

4. Quasirelativistic Hamiltonians

The ZORA Hamiltonian, which goes back to Ref. [29], has been thoroughly investigated (and given the name ZORA) by the Amsterdam group [30–32] and found much use in relativistic quantum chemistry since then. In our implementation, the ZORA matrix elements are evaluated through numerical quadrature, which makes the implementation of the spin-orbit parts and their geometrical derivatives quite facile. There is a fundamental problem with ZORA, namely that it is not gauge invariant. Different implementations of ZORA have introduced different methods to work around this problem, and our approach, the 'model potential' variant of ZORA [33, 34] is the only one that does *not* rely on a mapping of basis functions to nuclear centers. Systematic improvements beyond the ZORA model [35,36] have not found much use.

There is another quasirelativistic all-electron Hamiltonian of comparable popularity, namely the Douglas-Kroll-Hess approach. For a recenct review of its concepts see Ref. [37]. For more than a decade, it has only been used at its second order, extensions beyond [38,39] have revealed that the Douglas-Kroll Hamiltonian is not uniquely defined beyond the fourth order. This was not understood and quite baffling until it was clearly analyzed by the present author [40] on the occasion of formulating sixth-order Douglas-Kroll for the first time. Numerical experiments have shown that for one-electron systems, positive-energy Dirac eigenvalues can be reproduced by Douglas-Kroll to any desired accuracy if one extends the order of the Douglas-Kroll Hamiltonian far enough [41,42]. However, getting numerically exact results for one-electron systems is not sufficient: it is also necessary to include the electron interaction contribution to the effective potential when constructing the Douglas-Kroll Hamiltonian since otherwise large errors for spin-orbit splittings [12] or magnetic anisotropies [27] arise. Second, for high accuracy one has to compensate that the physical (or "true") electron density cannot directly be calculated from the two-component orbitals (the so-called *picture change effect*). A treatment of the electron interaction including full account of the picture change effect would be even more expensive than performing a fully relativistic four-component calculation in the first place, therefore higher-order Douglas-Kroll Hamiltonians only make sense if there is an efficient but still accurate approximation to the electron interaction part. Stimulated by the success of the 'model potential' approximation in the ZORA framework, we have suggested an approximation along the same lines for Douglas-Kroll [43]. It relies on the observation that the largest differences between the "true" electron density and those calculated from the two-component orbitals occur in the inner cores of the atoms where they are not very sensitive to the chemical environment. Information from atomic calculations (performed once and forever) can thus be used to correct the density obtained from the two-component orbitals.

Most scalar relativistic calculations performed today probably use effective core potentials (ECPs). For a recent review focusing on the type of ECPs mostly used in our applications, see Ref. [44]. It is straightforward to include spin-orbit terms in such ECPs, and the techniques to evaluate the molecular integrals that arise are well known [45,46]. We have implemented these integrals in our program and could improve the computational efficiency quite a bit [47]. In gradient calculations geometrical derivatives of the matrix elements of the quasirelativistic Hamiltonian are needed. We extended the techniques known from the scalar-relativistic case to implement geometrical derivatives of the spin-dependent parts of the ZORA and ECP integrals [23]. These become more complicated for Douglas-Kroll, as reported by Nasluzov and Rösch [48], and will also be computationally expensive for large molecules. For ZORA a so-called *strictly atomic* approximation has been suggested [49] where relativistic corrections are applied only to one-center atomic matrix elements. Within such an approximation, that also been suggested in the framework of Douglas-Kroll [50], relativistic correction to the geometrical derivatives of the one-particle Hamiltonian matrix elements vanish and need not be calculated.

Our experience [11,23,47,51] shows that very similar molecular structures and vibrational frequencies (and to a large extent also binding energies) are obtained from fully relativistic four-component and quasirelativistic ZORA or Douglas-Kroll calculations. In the latter case, it is hardly necessary to go beyond the fourth order for these properties. Note that our comparison is limited to diatomic molecules since we did not implement the geometrical derivatives of the Douglas-Kroll matrix elements, as this seems unnecessary in view of the above observation. A statement of similar generality cannot be made for ECPs since some variations between ECPs from different sources have been found. However, for systems with many heavy atoms *any* type of all-electron calculation (let it be nonrelativistic, scalar-relativistic or two-component) is rather expensive and ECPs furnish a quite attractive alternative there.

5. Conclusions

If properly implemented, a quasirelativistic two-component density functional or Hartree-Fock program exhibits the same efficiency we are used to in nonrelativistic calculations, and our strategy to achieve this has been presented in this article. Issues specific to the two-component formalism are how to define the spin density (collinear vs. non-collinear approach) and the higher relative weight of the linear algebra steps in the SCF procedure which needs to be considered when parallelizing the code. Through the availabilty of forces (geometry gradients), automatic geometry optimizations can be carried out which is a prerequisite for obtaining molecular structures and binding energies for polyatomic compounds. We must however admit that spin-orbit effects on molecular structures are rather small for most chemical compounds. The magnetic anisotropy energy on the other hand is a property which not only depends strongly on spin-orbit coupling, but is a genuine relativistic effect as it would vanish in a nonrelativistic or scalar-relativistic world. Therefore, a two-component description is really needed if one is interested in magnetic anisotropy energies of molecules with heavy atoms.

Acknowledgement

The development of the two-component program has been supported over the years by several grants of the Deutsche Forschungsgemeinschaft (DFG, grants No. Wu288/2–1, Wu288/4–1, Wu288/4–2), largely within the priority programme SPP 1145 'Modern and universal first-principles methods for many-electron systems in chemistry and physics'.

References

1. K. S. Pitzer, Accounts Chem. Res. **12** (1979) 271.
2. P. Pyykkö and J. P. Desclaux, Accounts Chem. Res. **12** (1979) 276.
3. B. A. Hess, Ber. Bunsen-Ges. Phys. Chem. **101** (1997) 1.
4. C. Teichteil, M. Pelissier, and F. Spiegelmann, Chem. Phys. **81** (1983) 273.
5. A. B. Alekseyev, H.-P. Liebermann, I. Boustani, G. Hirsch, and R. J. Buenker, Chem. Phys. **173** (1993) 333.
6. T. Fleig and C. M. Marian, Chem. Phys. Lett. **222** (1994) 267.
7. M. Sjøvoll, O. Gropen, and J. Olsen, Theor. Chem. Acc. **97** (1997) 301.

8. S. Yabushita, Z. Y. Zhang, and R. M. Pitzer, J. Phys. Chem. A **103** (1999) 5791.
9. V. Vallet, L. Maron, C. Teichteil, and J. P. Flament, J. Chem. Phys. **113** (2000) 1391.
10. E. van Lenthe, J. G. Snijders, and E. J. Baerends, J. Chem. Phys. **105** (1996) 6505.
11. W. Liu, C. van Wüllen, Y. K. Han, Y. J. Choi, and Y. S. Lee, Adv. Quantum Chem. **39** (2001) 325.
12. M. Mayer, S. Krüger, and N. Rösch, J. Chem. Phys. **115** (2001) 4411.
13. L. Gagliardi, B. Schimmelpfennig, L. Maron, U. Wahlgren, and A. Willetts, Chem. Phys. Lett. **344** (2001) 207.
14. J. E. Peralta and G. E. Scuseria, J. Chem. Phys. **120** (2004) 5875.
15. I. Malkin, O. L. Malkina, V. G. Malkin, and M. Kaupp, J. Chem. Phys. **123** (2005) 244103.
16. M. K. Armbruster, F. Weigend, C. van Wüllen, and W. Klopper, Phys. Chem. Chem. Phys. **10** (2008) 1748.
17. P. Hafner and W. H. E. Schwarz, Chem. Phys. Lett. **65** (1979) 537.
18. Y. S. Lee, W. C. Ermler, and K. S. Pitzer, J. Chem. Phys. **73** (1980) 360.
19. S. Y. Lee and Y. S. Lee, J. Comput. Chem. **13** (1992) 595.
20. Y.-K. Han, C. Bae, Y. S. Lee, and S. Y. Lee, J. Comput. Chem. **19** (1998) 1526.
21. A. D. Becke, J. Chem. Phys. **98** (1993) 1372.
22. A. D. Becke, J. Chem. Phys. **98** (1993) 5648.
23. C. van Wüllen and N. Langermann, J. Chem. Phys. **126** (2007) 114106.
24. C. van Wüllen, J. Comput. Chem. **23** (2002) 779.
25. H. J. F. Jansen, Phys. Rev. B **59** (1999) 4699.
26. D. Fritsch, K. Koepernik, M. Richter, and H. Eschrig, J. Comput. Chem. **29** (2008) 2210.
27. C. van Wüllen, J. Chem. Phys. **130** (2009) 194109.
28. P. Pulay, J. Comput. Chem. **3** (1982) 556.
29. C. Chang, M. Pelissier, and P. Durand, Phys. Scr. **34** (1986) 394.
30. E. van Lenthe, E. J. Baerends, and J. G. Snijders, J. Chem. Phys. **99** (1993) 4597.
31. E. van Lenthe, E. J. Baerends, and J. G. Snijders, J. Chem. Phys. **101** (1994) 9783.
32. E. van Lenthe, R. van Leeuwen, E. J. Baerends, and J. G. Snijders, Int. J. Quantum Chem. **57** (1996) 281.
33. C. van Wüllen, J. Comput. Chem. **18** (1997) 1985.
34. C. van Wüllen, J. Comput. Chem. **20** (1999) 51.
35. K. G. Dyall and E. van Lenthe, J. Chem. Phys. **111** (1999) 1366.
36. W. Klopper, J. H. van Lenthe, and A. C. Hennum, J. Chem. Phys. **113** (2000) 9957.
37. M. Reiher, Theor. Chem. Acc. **116** (2006) 241.
38. T. Nakajima and K. Hirao, J. Chem. Phys. **113** (2000) 7786.
39. A. Wolf, M. Reiher, and B. A. Hess, J. Chem. Phys. **117** (2002) 9215.
40. C. van Wüllen, J. Chem. Phys. **120** (2004) 7307.
41. M. Barysz and A. J. Sadlej, J. Chem. Phys. **116** (2002) 2696.
42. M. Reiher and A. Wolf, J. Chem. Phys. **121** (2004) 10945.
43. C. van Wüllen and C. Michauk, J. Chem. Phys. **123** (2005) 204113.
44. H. Stoll, B. Metz, and M. Dolg, J. Comput. Chem. **23** (2002) 767.
45. L. E. McMurchie and E. R. Davidson, J. Comput. Phys. **44** (1981) 289.
46. R. M. Pitzer and N. W. Winter, Int. J. Quantum Chem. **40** (1991) 773.
47. A. V. Mitin and C. van Wüllen, J. Chem. Phys. **124** (2006) 064305.
48. V. A. Nasluzov and N. Rösch, Chem. Phys. **210** (1996) 413.
49. J. H. van Lenthe, S. Faas, and J. G. Snijders, Chem. Phys. Lett. **328** (2000) 107.
50. J. Thar and B. Kirchner, J. Chem. Phys. **130** (2009) 124103.
51. W. Liu, C. van Wüllen, F. Wang, and L. Li, J. Chem. Phys. **116** (2002) 3626.

Self-interaction Free Relativistic Spin-density Functional Theory

By D. Ködderitzsch[1,*], H. Ebert[1], E. Engel[2], and H. Akai[3]

[1] Ludwig-Maximilians-Universität München, Department Chemie und Biochemie, Butenandtstraße 11, 81377 München, Germany
[2] Johann Wolfgang Goethe University Frankfurt am Main, Center for Scientific Computing, Max-von-Laue-Str. 1, 60438 Frankfurt am Main, Germany
[3] Department of Physics, Graduate School of Science, Osaka University, 1–1 Machikaneyama 1, Osaka 560–0043, Japan

(Received July 1, 2009; accepted September 24, 2009)

DFT / OEP / Relativistic KKR

We review the progress in the recent development of the relativistic optimized effective potential (ROEP) method within spin-density functional theory. The ROEP-equations for spin-polarized systems are derived and their application to open-shell atoms using the exact exchange approximation for the exchange correlation-functional is presented. Further, we expand the ROEP framework to treat extended systems within the KKR-multiple scattering formalism. We illustrate the application of the theory to open-shell free atoms and solids.

1. Introduction

As the success of density functional theory (DFT) [1,2] hinges on the availability of good approximations to the exchange-correlation (xc) functionals much effort has been devoted to the development of this basic ingredient to the theory.

DFT-xc-functionals have evolved from using the density as basic variable to a second generation involving generalized gradient approximations GGAs and, finally, to the so-called third generation which uses orbitals as ingredients to capture the physics of a broad range of systems.[3,4]

Utilizing orbital functionals proceeds in the spirit of what made DFT so successful, namely using orbitals as part of a functional (i.e. their use in the non-interacting kinetic energy functional of the Kohn-Sham mapping). Orbital functionals are, in contrast to earlier density based explicit functionals, implicit

* Corresponding author. E-mail: dkopc@cup.uni-muenchen.de

Z. Phys. Chem. **224** (2010) 427–439
© by Oldenbourg Wissenschaftsverlag, München

density functionals. Therefore, obtaining directly the xc-potential as a functional derivative of the xc-energy with respect to the density is not possible. Instead, the use of orbital dependent functionals requires the optimized effective potential method (OEP).[5,6] This procedure recasts the problem of determining $\delta E_{xc}/\delta n$ into an integral equation (OEP equation) whose ingredients are response functions and further functional derivatives (see section 2.1)

As an immediate result the computational cost involved in the numerical solution of the OEP equations is higher than when using explicit density functionals. Further it turns out, that not only constructing the constituents of the OEP equations (i.e. the response functions and the inhomogeneities of the integral equation) is computationally demanding but also numerically delicate. Up to now the solution of the OEP equations often employ the exact exchange (EXX) approximation to the xc-functional. Few works combine EXX with existing correlation functionals, however the results are often disappointing, as existing correlation functionals are incompatible with EXX (for an overview see reference [7]). Therefore orbital dependent correlation-functionals using the RPA have been studied[8] which lead to even more demanding numerical intricacies. Numerical challenges are also illustrated by the ongoing controversies concerning all-electron-approach versus pseudo-potential methods [7,9,10,11], and the problems in solving the OEP equations when using basis-set expansions of both the potential and orbitals.[12,13]

In spite of the aforementioned difficulties the use of orbitals in tailoring exchange-correlation functionals turned out to be very fruitful as using orbital functionals in the EXX approach cured some of the notorious problems of LDA and GGAs. In particular, this approach is self-interaction-free and the discontinuities of the functional derivative at integer electron numbers, which the exact exchange-correlation functional should have, are recovered (for an overview, see Ref. [14]). Also, the use of orbital functionals has opened a route for a systematic inclusion of correlations into DFT, once the exchange has been dealt with exactly.[15] For the same reasons, orbital dependent xc-functionals play an important role in time-dependent DFT and are currently studied intensively.[16–18]

Various applications of the OEP-method address the band gap problem for solids and find for semiconductors good agreement to experimental band gaps when supplying EXX with LSDA-correlation.[9,19] Even though for other systems EXX still underestimates the band gap, EXX calculations seem to supply a good starting points for subsequent GW-calculations.[20] A recent study [21] reports on the opening of band gap in all of the strongly correlated 3d transition-metal oxides (TMO). This is a landmark, insofar, as it opens the band gap in all the TMO in a pure parameter-free DFT description employing a local orbital-independent potential. This in contrast to various other schemes, e.g. LDA+U [22], LDA+DMFT [23] or the self-interaction corrected schemes [24,25]) which introduce orbital-dependent potentials.

As demonstrated above a lot of work has been invested in developing the OEP in the non-relativistic context (see also the detailed reviews by Grabo[14],

Engel[4] and Kümmel[7]). However, much less work has been devoted to the development of a relativistic OEP.

We formulate the equations and present exact numerical solutions of the relativistic-OEP within spin-density functional theory (SDFT) for free open-shell atoms and solids in the EXX approximation to the xc-functional. Our work is motivated by several aspects. First, studying a spin-polarized relativistic formulation in which spin-orbit coupling effects are naturally included is a starting point for studying magnetic anisotropies in solids and analyzing the impact of relativistic effects in spectroscopy. Also, starting from a Dirac Hamiltonian where these effects are included per se and no recurse to a perturbative treatment is needed is more elegant from a purists point of view. Second, an exact numerical, i.e. basis-set-free solution of the ROEP evades any discussion about the choice of basis sets, cutoffs, etc. In our approach, we will avoid, in particular, the so-called sum over states approach to construct the Green's function. Finally, of course, relativistic effects play an important role in heavy atoms and are notoriously hard to handle within standard quantum chemical schemes.

Let us mention here that prior to the work reviewed here, a non-spin polarized version of a relativistic OEP applied to closed shell atoms has already been published.[26,27] Extending it to the open-shell case with a non-zero exchange field leads to a more involved problem when constructing the ingredients of the ROEP equations (see section 2.1).

In the first part of the paper the theory of a ROEP-method is worked out with emphasis on finite systems. Further the extension of the formalism to a multiple-scattering KKR framework is derived to study extended systems. To treat core states in solids use is made of the formalism for the free atoms. We then review application of the ROEP to open-shell atoms and non-magnetic solids.

2. Formalism

2.1 Finite systems

Our starting point is the Dirac equation within relativistic Kohn-Sham spin density functional theory (KS-SDFT) (see e.g. [28])

$$\mathcal{H}_D \, \phi_k(\mathbf{r}) = \epsilon_k \, \phi_k(\mathbf{r}) \; . \tag{1}$$

The Hamiltonian (in atomic Rydberg units) is given as

$$\mathcal{H}_D = -ic\boldsymbol{\alpha} \cdot \boldsymbol{\nabla} + \frac{c^2}{2}\beta + V_{KS}(\mathbf{r}) \; . \tag{2}$$

The Dirac and spin matrices α_i, β and Σ have the usual meaning.[28–30] The index k comprises the set of quantum numbers to label the atomic, core or band states, respectively. The KS potential can be decomposed into a spin-independent and a spin-dependent part (here the presence of an external magnetic field is excluded):

$$V_{KS}(\mathbf{r}) = \bar{V}_{KS}(\mathbf{r}) + \beta\mathbf{\Sigma}\cdot\mathbf{B}_{xc}(\mathbf{r})$$

$$= V_{ext}(\mathbf{r}) + V_H(\mathbf{r}) + \bar{V}_{xc}(\mathbf{r}) + \beta\mathbf{\Sigma}\cdot\mathbf{B}_{xc}(\mathbf{r})$$

that in turn splits up into an external part, a Hartree term and terms containing the spin averaged and spin dependent part of the exchange correlation potential. The KS-potential is given as a functional of the particle and spin magnetization density

$$n(\mathbf{r}) = \frac{1}{2}\sum_{k}^{occ}\phi_k^\dagger(\mathbf{r})\phi_k(\mathbf{r}) + cc. \tag{3}$$

$$\mathbf{m}(\mathbf{r}) = \frac{1}{2}\sum_{k}^{occ}\phi_k^\dagger(\mathbf{r})\beta\mathbf{\Sigma}\phi_k(\mathbf{r}) + cc. \tag{4}$$

For the sake of convenience from now on we restrict ourselves to the case $\mathbf{B}_{xc} = B_{xc,z}\,\mathbf{e}_z \equiv B_{xc}\mathbf{e}_z$ and a magnetisation density which is collinear to $\mathbf{B}_{xc}$, i. e. $\mathbf{m} = m_z\,\mathbf{e}_z \equiv m\mathbf{e}_z$ such that the KS-potential takes the form

$$V_{KS}(\mathbf{r}) = \bar{V}_{KS}(\mathbf{r}) + \beta\Sigma_z B_{xc}(\mathbf{r})\ . \tag{5}$$

In what follows we will derive the ROEP equations starting from Eq. (1) and recast it in different forms to make contact to other formulations found in the literature [6,14,31,32] and to set the framework for a numerical solution of the problem.

Starting from the definition of the KS-potentials

$$\bar{V}_{xc}(\mathbf{r}) = \frac{\delta E_{xc}[n,m]}{\delta n(\mathbf{r})} \qquad B_{xc}(\mathbf{r}) = \frac{\delta E_{xc}[n,m]}{\delta m(\mathbf{r})} \tag{6}$$

and applying the chain rule for functional derivatives one obtains (neglecting the eigenvalue dependence of E_{xc})

$$\int d^3r' \begin{pmatrix} \bar{V}_{xc}(\mathbf{r}') \\ B_{xc}(\mathbf{r}') \end{pmatrix}^T \chi(\mathbf{r}',\mathbf{r}) = \begin{pmatrix} I_V(\mathbf{r}) \\ I_B(\mathbf{r}) \end{pmatrix}^T , \tag{7}$$

as an coupled integral equation for the relativistic optimized effective KS-potentials $\bar{V}_{xc}$ and B_{xc}. Here we have used the abbreviations

$$I_V(\mathbf{r}) = \int d^3r' \sum_k \frac{\delta\phi_k^\dagger(\mathbf{r}')}{\delta\bar{V}_{KS}(\mathbf{r})}\frac{\delta E_{xc}[n,m]}{\delta\phi_k^\dagger(\mathbf{r}')} + c.c. \tag{8}$$

$$I_B(\mathbf{r}) = \int d^3r' \sum_k \frac{\delta\phi_k^\dagger(\mathbf{r}')}{\delta B_{xc}(\mathbf{r})}\frac{\delta E_{xc}[n,m]}{\delta\phi_k^\dagger(\mathbf{r}')} + c.c. \tag{9}$$

to denote the inhomogeneity of the integral equation (7) and

$$\chi(\mathbf{r},\mathbf{r}') = \begin{pmatrix} \frac{\delta n(\mathbf{r})}{\delta \bar{V}_{KS}(\mathbf{r}')} & \frac{\delta n(\mathbf{r})}{\delta B_{xc}(\mathbf{r}')} \\ \frac{\delta m(\mathbf{r})}{\delta \bar{V}_{KS}(\mathbf{r}')} & \frac{\delta m(\mathbf{r})}{\delta B_{xc}(\mathbf{r}')} \end{pmatrix} = \begin{pmatrix} \chi_{nn}(\mathbf{r},\mathbf{r}') & \chi_{nm}(\mathbf{r},\mathbf{r}') \\ \chi_{mn}(\mathbf{r},\mathbf{r}') & \chi_{mm}(\mathbf{r},\mathbf{r}') \end{pmatrix} \tag{10}$$

is the static non-interacting KS-response function comprised of subblocks corresponding to charge-charge, magnetization-magnetization and off diagonal response. In deriving Eq. (7) use has been made of

$$\sum_{\mu=n,m} \int d^3r \, (\chi^{-1}(\mathbf{r}'',\mathbf{r}))_{\nu\mu}\chi_{\mu\sigma}(\mathbf{r},\mathbf{r}''') = \delta_{\nu\sigma}\delta(\mathbf{r}''-\mathbf{r}''') \,. \tag{11}$$

The KS-response function can be expressed in terms of the orbitals ϕ_k (see below). The functional derivatives of the orbitals ϕ_k with respect to the KS-potentials can be obtained by first order perturbation theory, with the so called orthogonal Green's function involved (see below). Therefore, if an explicit form for E_{xc} is given this integral equation can be solved to obtain the potentials $\bar{V}_{xc}$ and B_{xc}. The potentials $\bar{V}_{xc}(\mathbf{r})$ and $B_{xc}(\mathbf{r})$ as solution of the integral equation (7) are determined up to a constant (see section 3).

First order perturbation theory gives for the various functional derivatives occurring in Eqs. (8) and (9)

$$\begin{aligned} \frac{\delta\phi_k(\mathbf{r}')}{\delta \bar{V}_{KS}(\mathbf{r})} &= G_k(\mathbf{r}',\mathbf{r})\phi_k(\mathbf{r}) \ , \quad \frac{\delta\phi_k(\mathbf{r}')}{\delta B_{xc}(\mathbf{r})} G_k(\mathbf{r}',\mathbf{r})\beta\Sigma_z\phi_k(\mathbf{r}) \\ \frac{\delta\phi_k^\dagger(\mathbf{r}')}{\delta \bar{V}_{KS}(\mathbf{r})} &= \phi_k^\dagger(\mathbf{r})G_k(\mathbf{r},\mathbf{r}') \ , \quad \frac{\delta\phi_k^\dagger(\mathbf{r}')}{\delta B_{xc}(\mathbf{r})} \phi_k^\dagger(\mathbf{r})\beta\Sigma_z G_k(\mathbf{r},\mathbf{r}') \end{aligned} \tag{12}$$

with the orthogonal Green's function

$$G_k(\mathbf{r}',\mathbf{r}) = \sum_{i\neq k} \frac{\phi_i(\mathbf{r}')\phi_i^\dagger(\mathbf{r})}{\epsilon_k - \epsilon_i} = G_k^\dagger(\mathbf{r},\mathbf{r}'), \tag{13}$$

of the Dirac equation (1) projected into the subspace orthogonal to ϕ_k.[6,26,33] It is obtained as solution of

$$(\epsilon_k - \mathcal{H}_D)G_k(\mathbf{r},\mathbf{r}') = \delta(\mathbf{r}-\mathbf{r}') - \phi_k(\mathbf{r})\phi_k^\dagger(\mathbf{r}') \,. \tag{14}$$

For the spin-dependent potential considered in Eq. (5) its construction is much more involved than in the non-relativistic and non-spin polarized relativistic case, respectively. Details on its construction are given in Refs. [34,35].

Using the operators $O^\nu \in \{\mathbb{1}_4, \beta\Sigma_z\}$, where $\nu = n,m$ denotes the association with the charge- and magnetization-density, respectively, the static KS response functions can be written as

$$\chi_{\mu\nu}(\mathbf{r}',\mathbf{r}) = \sum_k^{occ} \phi_k^\dagger(\mathbf{r}')O^\mu G_k(\mathbf{r}',\mathbf{r})O^\nu\phi_k(\mathbf{r}) \,. \tag{15}$$

Further alternative forms of the ROEP equations Eq. (7) have been obtained [35] by using the orbital shift ψ_k^{sft}

$$\psi_k^{sft}(\mathbf{r}) = \int d^3r' \, G_k(\mathbf{r},\mathbf{r}') \left[\left(\bar{V}_{xc}(\mathbf{r}') + \beta\Sigma_z B_{xc}(\mathbf{r}') \right) \phi_k(\mathbf{r}') - \frac{\delta E_{xc}[n,m]}{\delta\phi_k^\dagger(\mathbf{r}')} \right] . \tag{16}$$

Using the latter, Eqs. (7) can be expressed by the equivalent requirements

$$\sum_{k}^{occ} \phi_k^\dagger(\mathbf{r})\psi_k^{sft}(\mathbf{r}) + c.c. = 0\,, \tag{17}$$

$$\sum_{k}^{occ} \phi_k^\dagger(\mathbf{r})\beta\Sigma_z\psi_k^{sft}(\mathbf{r}) + c.c. = 0\,. \tag{18}$$

The orbital shift ψ_k^{sft} has been discussed extensively in the context of non-relativistic formulations of the OPM.[14,31,36] It represents the first order change in the wavefunctions due to the replacement of the orbital-dependent potential $(\delta E_{xc}[n,m])/(\delta\phi_k^\dagger(\mathbf{r}))$ with the common potential $\bar{V}_{xc}(\mathbf{r}) + \beta\Sigma_z\, B_{xc}(\mathbf{r})$ seen by all electrons. In a non-relativistic formulation the counterpart of Eq. (17) has been used by Kümmel and Perdew [31] to construct an iterative scheme to solve the OEP equations bypassing some problems in the conventional formulation caused by the need to invert response functions.

Following earlier ideas of Krieger et al. [36] and Kreibich et al. [37] one can derive yet another explicit form of the ROEP equations, which makes use of a differential equation for the orbital shifts obtained from Eqs. (17) and (18). Using Eq. (14) one obtains

$$\begin{aligned}\left[\mathcal{H}_D - \epsilon_k\right]\psi_k^{sft}(\mathbf{r}) = -\left[\left(\bar{V}_{xc}(\mathbf{r}) + \beta\Sigma_z B_{xc}(\mathbf{r})\right)\phi_k(\mathbf{r}) - \frac{\delta E_{xc}}{\delta\phi_k^\dagger(\mathbf{r})}\right] + \\ \left[\bar{A}_{xck} - \bar{u}_{xck}\right]\phi_k(\mathbf{r})\,,\end{aligned} \tag{19}$$

where

$$\begin{aligned}\bar{A}_{xck} &= \int d^3r\,\phi_k^\dagger(\mathbf{r})\left(\bar{V}_{xc}(\mathbf{r}) + \beta\Sigma_z B_{xc}(\mathbf{r})\right)\phi_k(\mathbf{r})\,,\\ \bar{u}_{xck} &= \int d^3r\,\phi_k^\dagger(\mathbf{r})\,\frac{\delta E_{xc}}{\delta\phi_k^\dagger(\mathbf{r})}\,.\end{aligned}$$

Using Eq. (16) gives

$$\begin{pmatrix}\bar{V}_{xc}(\mathbf{r})\\ B_{xc}(\mathbf{r})\end{pmatrix} = \frac{\mathcal{J}^{-1}(\mathbf{r})}{2}\begin{pmatrix}\sum_k\left\{a_{xck}^V(\mathbf{r}) + \left[\bar{A}_{xck} - \bar{u}_{xck}\right]\phi_k^\dagger(\mathbf{r})\phi_k(\mathbf{r}) + c.c.\right\}\\ \sum_k\left\{a_{xck}^B(\mathbf{r}) + \left[\bar{A}_{xck} - \bar{u}_{xck}\right]\phi_k^\dagger(\mathbf{r})\beta\Sigma_z\phi_k(\mathbf{r}) + c.c.\right\}\end{pmatrix}, \tag{20}$$

where

$$
\begin{aligned}
a^V_{xck}(\mathbf{r}) &= \phi^\dagger_k(\mathbf{r})\frac{\delta E_{xc}}{\delta \phi^\dagger_k(\mathbf{r})} + \phi^\dagger_k(\mathbf{r})\hat{D}_k\psi^{sft}_k(\mathbf{r}) + c.c. \\
a^B_{xck}(\mathbf{r}) &= \phi^\dagger_k(\mathbf{r})\beta\Sigma_z\frac{\delta E_{xc}}{\delta \phi^\dagger_k(\mathbf{r})} + \phi^\dagger_k(\mathbf{r})\beta\Sigma_z\hat{D}_k\psi^{sft}_k(\mathbf{r}) + c.c. \\
\hat{D}_k &= ic\alpha\cdot\nabla - \frac{c^2}{2}(\beta-1) - \beta\Sigma_z B_{xc}(\mathbf{r}) + \epsilon_k \qquad (21)\\
\mathcal{J}(\mathbf{r}) &= \begin{pmatrix} n(\mathbf{r}) & m(\mathbf{r}) \\ m(\mathbf{r}) & n(\mathbf{r}) \end{pmatrix}.
\end{aligned}
$$

Equation (20) represents an exact transformation of the ROEP equation. By omitting the terms in ψ^{sft} in the coefficients a^V_{xck} and a^B_{xck} (Eqs. (21)) the relativistic KLI approximation (RKLI) is obtained which has been extensively discussed in the non-spin polarized relativistic [32,37] as well as the non-relativistic context.[14,38] That the latter approximation is indeed in the spirit of the traditional KLI-approximation can be verified by using the constant denominator approximation for G_k:

$$
\tilde{G}_k(\mathbf{r},\mathbf{r}') = \frac{1}{\Delta\epsilon}\sum_{i\neq k}\phi_i(\mathbf{r})\phi^\dagger_i(\mathbf{r}') = \frac{1}{\Delta\epsilon}\left[\delta(\mathbf{r}-\mathbf{r}') - \phi_k(\mathbf{r})\phi^\dagger_k(\mathbf{r}')\right]. \qquad (22)
$$

Insertion into the orbital shift term Eq. (16) leads to:

$$
\begin{aligned}
\psi^{sft}_k(\mathbf{r}) &= \frac{1}{\Delta\epsilon}\Big\{\left[\bar{V}_{xc}(\mathbf{r}) + \beta\Sigma_z B_{xc}(\mathbf{r})\right]\phi_k(\mathbf{r}) \\
&\quad - \frac{\delta E_{xc}[n,m]}{\delta\phi^\dagger_k(\mathbf{r})} - \phi_k(\mathbf{r})\left[\bar{A}_{xck} - \bar{u}_{xck}\right]\Big\}. \qquad (23)
\end{aligned}
$$

Inserting Eq. (23) into the OEP Eqs. (17) and (18) yields Eq. (20) with the orbital shift term neglected.

The results presented below are obtained using the relativistic exact exchange as an approximation for the xc-functional which is given as an explicit functional of the orbitals

$$
E_x = -\sum_{kl}^{occ}\int d^3r\int d^3r'\,\frac{\phi^\dagger_k(\mathbf{r})\,\phi_l(\mathbf{r})\,\phi^\dagger_l(\mathbf{r}')\,\phi_k(\mathbf{r}')}{|\mathbf{r}-\mathbf{r}'|}. \qquad (24)
$$

We assume a spherically symmetric KS-potential, i. e. $V_{KS}(\mathbf{r}) = V_{KS}(r)$ which implies the same property for $\bar{V}_{xc}$ and B_{xc}. Making the latter assumption and following the notation and coupling scheme given in Ref. [39] results in the following expansion of the wavefunction into four spinors

$$
\phi^\gamma_k(\mathbf{r}) = \begin{pmatrix} g^\gamma_{\Lambda k}(r)\chi_\Lambda(\hat{r}) \\ if^\gamma_{\Lambda k}(r)\chi_{-\Lambda}(\hat{r}) \end{pmatrix} + \begin{pmatrix} g^\gamma_{\bar{\Lambda}k}(r)\chi_{\bar{\Lambda}}(\hat{r}) \\ if^\gamma_{\bar{\Lambda}k}(r)\chi_{-\bar{\Lambda}}(\hat{r}) \end{pmatrix}, \qquad (25)
$$

$\Lambda = (\kappa,\mu)$ is a combined index comprising the spin-orbit and magnetic quantum number.[1] γ labels the physical solutions of Eq. (1) for the case of having a coupling induced by a non-zero B_{xc}. These forms of the wavefunctions are assumed for atomic/core states as well as for the scattering solutions (see section 2.2) when constructing the ingredients of the ROEP equations.

2.2 ROEP method for extended non-magnetic system within multiple scattering KKR

To adapt the formalism to extended systems to be used within a relativistic multiple scattering KKR formalism [40] we partition the ingredients of the ROEP Eqs. (7) into parts associated with the core and valence electronic structure of the system.

Partitioning the expression for the exchange-correlation energy results in four terms associated with core-core (cc), core-valence (cv), valence-core (vc) and valence-valence (vv) contributions:

$$E_{xc} = E_{xc}^{(cc)} + E_{xc}^{(cv)} + E_{xc}^{(vc)} + E_{xc}^{(vv)} \ . \tag{26}$$

The inhomogeneity I_V Eq. (7) splits accordingly, with the expressions for the core-core-part given by Eq. (8) and the (cv,vc) and (vv)-term discussed below. In what follows we use exact-exchange (EXX, E_x) as an approximation to the exchange-correlation functional and give the terms needed for solving Eq. (7) according to the partitioning given in Eqs. (26) and (31) (see below).

The valence properties of the paramagnetic ground state are represented within the relativistic KKR method by the multiple scattering Green's function [41]:

$$\begin{aligned} G^{(v)}(\mathbf{r},\mathbf{r}',E) \;=\; & \sum_{\Lambda\Lambda'} Z^n_\Lambda(\mathbf{r},E)\tau^{nn'}_{\Lambda\Lambda'} Z^{n'\times}_{\Lambda'}(\mathbf{r}',E) \\ & - \sum_{\Lambda} \Big[Z^n_\Lambda(\mathbf{r},E) J^{n\times}_\Lambda(\mathbf{r}',E)\Theta(r'-r) \\ & + J^n_\Lambda(\mathbf{r},E) Z^{n\times}_\Lambda(\mathbf{r}',E)\Theta(r-r') \Big] \delta_{nn'} \, , \end{aligned} \tag{27}$$

where Z^n_Λ and J^n_Λ are four component wavefunctions being regular and irregular solutions, respectively, of the single site Dirac equation at a site n, and $\tau^{nn'}_{\Lambda\Lambda'}$ is the scattering path operator subsuming the multiple scattering events in the solid. Using $G^{(v)}$ the relativistic exact exchange functional can be written as:

$$\begin{aligned} E_x^{(vv)}[n] \;=\; & -\frac{1}{\pi^2}\mathrm{Tr} \int d^3r \int d^3r' \frac{\left[\Im \int dE G^{(v)}(\mathbf{r},\mathbf{r}',E)\right]}{|\mathbf{r}-\mathbf{r}'|} \\ & \times \left[\Im \int dE' G^{(v)}(\mathbf{r}',\mathbf{r},E')\right] , \end{aligned} \tag{28}$$

where use has been made of the expression for the density matrix $n^{(v)}(\mathbf{r},\mathbf{r}')$ in terms of $G^{(v)}$:

[1] We use the following notation: $\bar{\Lambda} = (-\kappa-1,\mu)$ and $-\Lambda = (-\kappa,\mu)$

$$n^{(v)}(\mathbf{r},\mathbf{r}') = -\frac{1}{\pi}\Im \int dE\, G^{(v)}(\mathbf{r},\mathbf{r}',E)\,. \tag{29}$$

To determine the contribution of the valence states to the inhomogeneity in Eq. (7), Eq. (28) is varied with respect to $\bar{V}_{KS}(\mathbf{r})$ which gives:

$$\begin{aligned} I_V^{(vv)}(\mathbf{r}) &= \frac{\delta E_x^{(vv)}}{\delta \bar{V}_{KS}(\mathbf{r})} \\ &= -\frac{2}{\pi^2}\mathrm{Tr}\int d^3r' \int d^3r'' \frac{\left[\Im \int dE G^{(v)}(\mathbf{r}',\mathbf{r}'',E)\right]}{|\mathbf{r}'-\mathbf{r}''|} \\ &\quad \times \left[\Im \int dE' G^{(v)}(\mathbf{r}'',\mathbf{r},E') G^{(v)}(\mathbf{r},\mathbf{r}',E')\right]\,. \end{aligned} \tag{30}$$

The inhomogeneities $I_V^{(vc)}$ and $I_V^{(cv)}$ are obtained by adopting Eq. (28), replacing the appropriate parts by the core density matrix $n^{(c)}(\mathbf{r},\mathbf{r}')$ and varying with respect to $\bar{V}_{KS}(\mathbf{r})$.

The response is given by the sum of core (c) and valence (v) contributions, respectively:

$$\chi(\mathbf{r},\mathbf{r}') = \chi^{(c)}(\mathbf{r},\mathbf{r}') + \chi^{(v)}(\mathbf{r},\mathbf{r}')\,, \tag{31}$$

where $\chi^{(c)}$ is determined using the formalism given in section 2.1. The contribution of the valence states to the response function is determined using $G^{(v)}$ as:[42,43]

$$\chi^{(v)}(\mathbf{r},\mathbf{r}') = -\frac{1}{\pi}\mathrm{Tr}\int_{E_0}^{E_F} dE\, G^{(v)}(\mathbf{r},\mathbf{r}',E) G^{(v)}(\mathbf{r}',\mathbf{r},E)\,, \tag{32}$$

where E_0 is chosen such that the integral includes the valence states, and E_F is the Fermi-energy.

3. Application

3.1 Free atoms

In the numerical procedure the effective potentials as solution of the ROEP equation (7) are obtained by a discretization of the inhomogeneities and the response function on real space meshes and subsequent matrix inversion. Choosing the effective potentials to be spherically symmetric corresponding spherical averaged inhomogeneities and response functions are used in the procedure.

For the atomic calculation the range of the radial mesh depends on the system and is chosen such that numerical stability as well as high accuracy of the result is guaranteed. The meshes in this case typically extend to about 30 to 50 a.u. Outer boundary conditions are imposed using the procedure given in Ref. [35], by using a spin-projected representation of the potentials and using the exact result for $r \rightarrow \infty$.

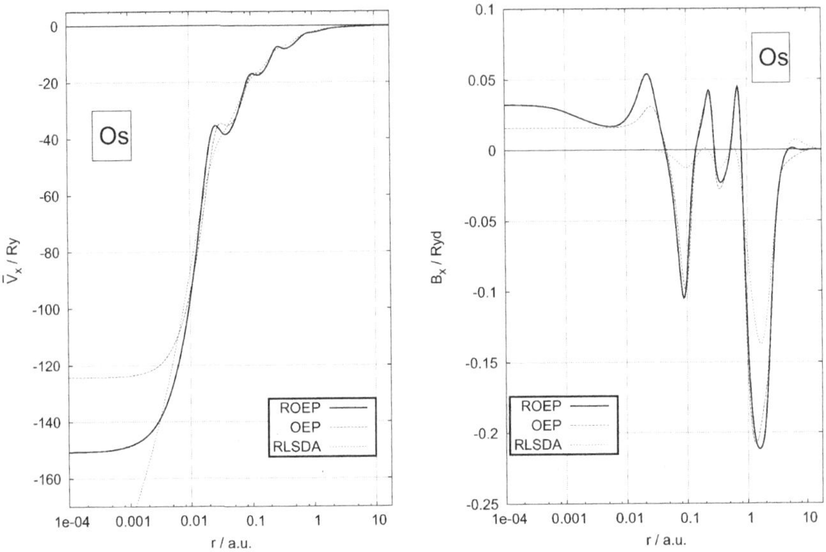

Fig. 1. ROEP potentials for the Os atom (Z = 76) with a d^6 configuration assumed, i. e. [Xe]$4f^{14}5d^6 6s^2$. Left: Spin-averaged potential, comparing ROEP to non-relativistic OEP and RSLDA. Right: The same for the spin-dependent potentials.

In Fig. 1 the spin-averaged and spin-dependent exchange ROPM potentials for the open-shell Os atom as a representative example is shown and compared to the non-relativistic OPM and RLDA. A d^6 configuration has been assumed, [Xe]$4f^{14}5d^6 6s^2$. In this case as well as in others cases studied [35] it turns out that the outermost d-orbitals have almost 100% spin-polarization. Further, with growing ordering number the spin-averaged exchange potential in the ROPM gets pushed down at the nucleus when compared to the non-relativistic OPM. The differences in the spin-dependent exchange do not follow such a systematic trend but show also distinct changes.

As can be seen in Fig. 1 the spin dependent ROPM exchange potentials B_x as compared to the RLDA has a larger amplitude which also extends to the vicinity of the nucleus. This is a manifestation of the non-locality of the ROPM potentials. In a local approximation like the RLDA a non-vanishing exchange field can only arise in spatial regions where a finite magentisation is present. This also has an important impact on the hyperfine fields.[35] A very prominent shortcoming of the local approximation to density functional theory is the underestimation of the core polarization contribution to the magnetic hyperfine field due to spin-polarized valence states.[39] Calculations performed in the ROEP framework presented here show [34] that for most elements the exact exchange leads to better agreement with the experimental hyperfine constants than the conventional xc functionals, the improvement being particularly obvious for the 4d series.

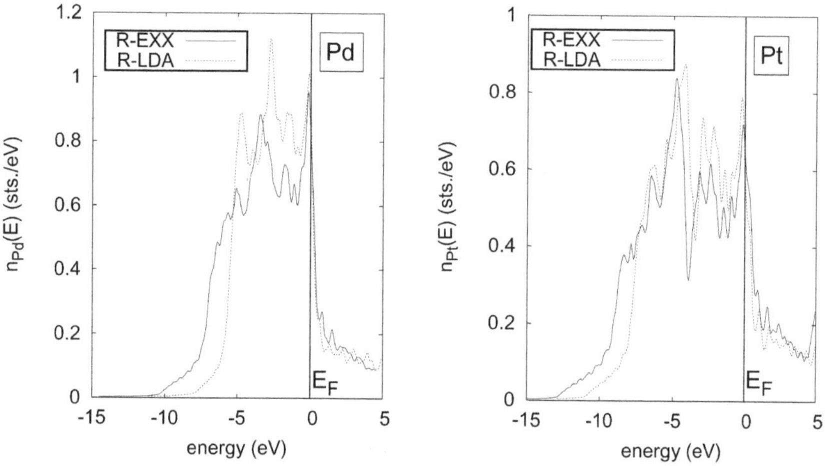

Fig. 2. DOS for the transition metals Pd (left) and Pt (right) calculated using relativistic exact exchange, compared to R-LDA.

3.2 Solids

When determining the effective potentials for the solid, the computationally most demanding part is the evaluation of the Coulomb integrals in Eq. (30) on a double energy mesh (E,E'). In contrast to former non-relativistic LMTO and KKR implementations [44,45], where the variations are computed numerically, we performed the variation in Eq. (30) analytically. More details on the analytical treatment of the functional variation of E_x in terms of the Green's function for the valence states will be published elsewhere [46]. The energy integration is done on a complex energy path. For determining the inhomogeneities $I^{(vv)}$ up to 4-th neighbor shell contributions have been taken into account. The angular cut-off for the expansion in scattering solutions for the valence states has been chosen as $l = 2(= d)$ and $l = 3$ for Pd and Pt, respectively.

We use the atomic sphere approximation (ASA) and the inversion is done on an exponential radial mesh spanning the Wigner-Seitz(WS)-radius. The boundary condition at the WS-radius is chosen as in Ref. [45] and we obtain the δ-like feature of V_x as reported there.

Calculations for alkali metals obtained in the presented framework have been published elsewhere.[47]

In Fig. 2 we show the density of states for the transition-metals Pd ($Z = 46$, $4d$) and Pt ($Z = 78$, $5d$) as obtained in the relativistic EXX-framework compared to RLDA (xc-correlation functional for the RLDA Ref. [48], lattice constants: $a_{\mathrm{Pd}} = 3.89\text{Å}$, $a_{\mathrm{Pt}} = 3.92\text{Å}$, *fcc* crystal lattice structure). For both Pd and Pt the DOS is broadened as compared to the RLDA, with weight shifted downwards. The Fermi energy is still pinned to upper edge of the d-band manifold.

To summarize, we presented the relativistic optimized effective potential formalism worked out in the framework of SDFT. In a first step we derived the ROEP-equations for finite systems and cast them into different forms. Further, we extended the theory to treat extended systems in the framework of multiple scattering KKR-theory. Results for open-shell atoms and non-magnetic solids were presented. Further application of the ROEP for magnetic solids is in progress.

Acknowledgement

This work was supported by the Deutsche Forschungsgemeinschaft within the priority program "Moderne und universelle first-principles-Methoden für Mehrelektronensysteme in Chemie und Physik" (SPP 1145).

References

1. P. Hohenberg and W. Kohn, Phys. Rev. **136** (1964) B 864.
2. W. Kohn and L. J. Sham, Phys. Rev. **140** (1965) A 1133.
3. R. M. Dreizler and E. K. U. Gross, Density Functional Theory. Springer-Verlag, Berlin, Heidelberg, (1990).
4. E. Engel, A Primer in Density Functional Theory. Springer, Berlin (2003), chap. Orbital-dependent functionals for the exchange-correlation energy: A third generation of density functional theory. p. 56.
5. R. T. Sharp and G. K. Horton, Phys. Rev. **90**(2), (1953) 317.
6. J. D. Talman and W. F. Shadwick, Phys. Rev. A **14**(1), (1976) 36.
7. S. Kümmel and L. Kronik, Rev. Mod. Phys. **80** (2008) 3.
8. T. Kotani and H. Akai, J. Magn. Magn. Materials **177–181** (1998) 569.
9. M. Städele, J. A. Majewski, P. Vogl, and A. Görling, Phys. Rev. Letters **79** (1997) 2089.
10. S. Sharma, J. K. Dewhurst, and C. Ambrosch-Draxl, Phys. Rev. Lett. **95** (2005) 136402.
11. E. Engel, Phys. Rev. B (2009), submitted.
12. V. N. Staroverov, G. E. Scuseria, and E. R. Davidson, J. Chem. Phys. **124** (2006) 141103.
13. V. N. Staroverov, G. E. Scuseria, and E. R. Davidson, J. Chem. Phys. **125** (2006) 141103.
14. T. Grabo, T. Kreibich, S. Kurth, and E. K. U. Gross, Strong Coulomb Correlations and Electronic Structure Calculations: Beyond Local Density Approximations. Gordon and Breach (2000), chap. 1.
15. E. Engel, A. Bonetti, S. Keller, I. Andrejkovics, and R. Dreizler, Phys. Rev. A **58** (1998) 964.
16. C. A. Ullrich, U. J. Gossmann, and E. K. U. Gross, Phys. Rev. Lett. **74** (1995) 872.
17. R. van Leeuwen, Phys. Rev. Lett. **80** (1998) 1280.
18. M. Mundt and S. Kümmel, Phys. Rev. A **74** (2006) 022511.
19. D. M. Bylander and L. Kleinman, Phys. Rev. Lett. **74** (1995) 3660.
20. P. Rinke, A. Qteish, J. Neugebauer, C. Freysoldt, and M. Scheffler, New Journal of Physics **7** (2005) 126.
21. E. Engel and R. N. Schmid, Phys. Rev. Lett. **103** (2009) 036404.
22. V. I. Anisimov, F. Aryasetiawan, and A. I. Lichtenstein, J. Phys.: Condensed Matter **9** (1997) 767.

23. J. Kunes, V. I. Anisimov, S. L. Skornyakov, A. V. Lukoyanov, and D. Vollhardt, Phys. Rev. Lett. **99** (2007) 156404.
24. A. Svane and O. Gunnarsson, Phys. Rev. Lett. **65** (1990) 1148.
25. Z. Szotek, W. Temmerman, and H. Winter, Phys. Rev. Lett. **72**(8),(1994) 1244.
26. B. A. Shadwick, J. D. Talman, and M. R. Norman, Comp. Phys. Commun. **54** (1989) 95.
27. E. Engel, S. Keller, A. F. Bonetti, H. Müller, and R. M. Dreizler, Phys. Rev. A **52** (1995) 2750.
28. H. Eschrig, The Fundamentals of Density Functional Theory (B G Teubner Verlagsgesellschaft, Stuttgart, Leipzig (1996).
29. M. E. Rose, Elementary Theory of Angular Momentum. Wiley, New York (1961).
30. M. E. Rose, Relativistic Electron Theory. Wiley, New York (1961).
31. S. Kümmel and J. P. Perdew, Phys. Rev. Lett. **90** (2003) 043004.
32. E. Engel, A. Facco Bonetti, S. Keller, I. Andrejkovics, and R. Dreizler, Phys. Rev. A **58** (1998) 964.
33. J. D. Talman, Comp. Phys. Commun. **54** (1988) 85.
34. E. Engel, D. Ködderitzsch, and H. Ebert, Phys. Rev. B **78** (2008) 235123.
35. D. Ködderitzsch, H. Ebert, and E. Engel, Phys. Rev. B **77** (2008) 045101.
36. J. B. Krieger, Y. Li, and G. J. Iafrate, Phys. Rev. A **46**(9),(1992) 5453.
37. T. Kreibich, E. K. U. Gross, and E. Engel, Phys. Rev. A **57**(1),(1998) 138.
38. J. B. Krieger, Y. Li, and G. J. Iafrate, Phys. Rev. A **45**(1),(1992) 101.
39. H. Ebert, J. Phys.: Condensed Matter **1** (1989) 9111.
40. H. Ebert, in Electronic Structure and Physical Properties of Solids., H. Dreyssé (Ed.), Springer, Berlin (2000), vol. 535 of Lecture Notes in Physics, p. 191.
41. J. B. Staunton, B. L. Györffy, and P. Weinberger, J. Phys. F: Met. Phys. **10** (1980) 2665.
42. M. J. Stott and E. Zaremba, Phys. Rev. A **21** (1980) 12.
43. M. Deng, H. Freyer, J. Voitländer, and H. Ebert, J. Phys.: Condensed Matter **13** (2001) 8551.
44. T. Kotani, Phys. Rev. Lett. **74** (1995) 2989.
45. T. Kotani and H. Akai, Phys. Rev. B **54** (1996) 16502.
46. D. Ködderitzsch, to be published (2009).
47. D. Ködderitzsch, H. Ebert, H. Akai, and E. Engel, J. Phys.: Condensed Matter **21** (2009) 064208.
48. S. H. Vosko, L. Wilk, and N. Nussair, Can. J Phys. **58** (1980) 1200.

Second Order Local Møller-Plesset Perturbation Theory for Periodic Systems: the CRYSCOR Code

By Denis Usvyat[1], Lorenzo Maschio[2], Cesare Pisani[2], and Martin Schütz[1,*]

[1] Institute of Physical and Theoretical Chemistry, University of Regensburg, Universitätsstraße 31, 93040 Regensburg, Germany

[2] Dipartimento di Chimica IFM, and Centre of Excellence NIS (Nanostructured Interfaces and Surfaces), Università di Torino, via Giuria 5, I-10125 Torino, Italy

(Received July 14, 2009; accepted September 7, 2009)

Local Correlation / Density Fitting / Titanium Dioxide

This article reviews the periodic LMP2 method and its implementation in the CRYSCOR code. The main steps of the LMP2 calculations and the techniques employed are briefly described. Illustrative single-point calculations for three TiO_2 polymorphs: rutile, anatase and brookite in their experimental geometry are performed. It is shown that the method scales linearly with respect to the number of atoms per unit cell, and can be applied to relatively complex periodic systems. The LMP2 method in contrast to DFT positions rutile slightly lower in the energy than anatase. Brookite is found to be the most stable. However, since the energy differences are relatively small, a further investigation of the influence of the domain-sizes, geometry, zero-point vibrations, *etc.* on the relative stability of these systems is required.

1. Introduction

Density functional theory (DFT) is presently the most common technique for calculating the electronic structure of extended molecular systems. In its Kohn-Sham reformulation it combines low computational cost with (in most cases) appreciable accuracy. However, DFT, at least in the context of standard exchange-correlation functionals presently used, possesses a few but significant drawbacks as compared to wavefunction-based theories, like the lack of the possibility for a systematic improvement of predictions, the inability to describe weak interactions (dispersive forces), the underestimation of the excitation energies with charge-transfer character, *etc.* Wave-function-based (quantum chemi-

* Corresponding author. E-mail: martin.schuetz@chemie.uni-regensburg.de

© by Oldenbourg Wissenschaftsverlag, München

cal) approaches like MP2, CCSD(T), *etc.* are conceptually free of these problems. However, due to prohibitively steep scaling with respect to the system size they are hardly applicable for systems even with several dozens of atoms without additional approximations. In the last decades though, local approaches to the correlation problem (see for example Ref. [1] and references therein) significantly reduced the scaling of such theories, in some cases even to linear. These techniques, especially when combined with the density fitting approximation [2,3], made quantum chemical calculations even for quite extended systems (say, with more than a hundred atoms) routine.

In crystalline systems, due to their extended nature the scaling problems of the quantum chemical methods become even more pronounced. Therefore, DFT has a dominant position in solid state calculations. Nevertheless, in the last two decades new wave-function based methods, adjusted to the periodic problem have been proposed. Several approaches have been developed specifically for the case of 1D-periodicity (see Ref. [4] and references therein). A molecular orbital invariant formalism for MP2, based on the Laplace transform of the energy denominators has been applied to the periodic case by Scuseria and coworkers [5,6]. Some of approaches to correlation in solids employ local schemes and localized functions instead of canonical fully delocalized crystalline orbitals [7–10]. Among this group of methods we also list the CRYSCOR project and the periodic local MP2 method implemented in it [4,11–13]. This method is the topic of the present paper and will be discussed in detail in the following. Correlation approaches for crystals, reported very recently, include the finite cluster scheme, suggested by Manby and coworkers [14] and the canonical MP2 method using the plane wave basis by Kresse *et al.* [15].

We started the CRYSCOR project several years ago, aiming for a code performing the post-Hartree-Fock correlation treatment for 1D, 2D and 3D-periodic systems with relatively large unit cells (up to few tens of atoms). The main requirements for the code have been the generality (*i.e.* applicability to systems with different types of structure, binding, symmetry, dimensionality, *etc.*), efficiency, a favorable scaling with respect to the cell size, the possibility to use rather rich basis sets (at least, of the triple-zeta quality), automatization and user-friendliness. This paper reviews the current status of the CRYSCOR code and is organized as follows. In section 2 we sketch the general structure of the program and discuss the theoretical approaches used at different stages of the calculation. Section 3 presents as an illustrative example, an LMP2 study on the relative stability of titanium oxide polymorphs. Conclusions follow in section 4.

2. Theory

Consider the general expression for the MP2 (or LMP2) energy:

$$E_2 = \sum_{ij} \sum_{ab} (ia|jb)(2T^{ij}_{ab} - T^{ij}_{ba}), \tag{1}$$

where the amplitudes T^{ij}_{ab} are the solutions of the equation system

$$0 = (ia|jb) + \sum_c (F_{ac}T^{ij}_{cb} + T^{ij}_{ac}F_{cb}) - \sum_k (F_{ik}T^{kj}_{ab} + T^{ik}_{ab}F_{kj}). \tag{2}$$

The indices i, j and k denote orbitals representing the Hartree-Fock (HF) occupied space, a, b and c the virtual space. The $(ia \mid jb)$ are the electron repulsion integrals (ERIs) in the chemical notation and F_{ab} and F_{ij} are the external-external and internal-internal blocks of the Fock matrix, respectively.

The starting point for the LMP2 calculation is the HF reference, computed by the well-known LCAO periodic code CRYSTAL [16]. CRYSTAL also delivers the Fock matrix occurring in eq. (2). The LMP2 energy can be rather sensitive to the accuracy of the Fock matrix. Since construction of the Fock matrix in an infinite periodic system is not straightforward, we give a short description of how this is accomplished in the CRYSTAL code [17]. The Coulomb part of the Fock matrix implies only conditionally converging lattice sums, for which the Ewald summation technique is used. The 4-index integrals in this part are evaluated explicitly only when the interacting densities are overlapping, while for the remaining ones the multipole-moment approximation is employed. For the exchange term no infinite summations occur, since the slow decay of the Coulomb interaction is compensated for by the exponentially decaying density matrix, the 4-index integrals are contracted with. Yet, since the density matrix is initially not known, the range of the relevant ERIs is determined using an orbital overlap criterion, which in some approximate way can represent the result of the contraction with the density matrix.

In our local MP2 approach both occupied and virtual spaces are spanned by local orbitals. For the occupied space the orthogonal localized orbitals - Wannier functions (WFs) are constructed by means of the localization-wannierization scheme [18] as also implemented in CRYSTAL program. Since the subsequent CRYSCOR correlation treatment is essentially exploiting the point symmetry of the system (*vide infra*), the Wannier functions are symmetrized accordingly. Presently an *a posteriori* scheme [19] is applied, *i. e.* the symmetry is imposed onto already localized orbitals. This has the advantage that the localization procedure is not affected by any symmetry constraints and provides the WFs, being the most localized according to the chosen localization criteria. Yet, localization of *a priori* symmetrized functions [20], can be beneficial [4,21], since (i) when the symmetry of a given WF is not unique, the most convenient (regarding savings within the subsequent LMP2 calculations) symmetry can be chosen; (ii) the localization procedure itself is then performed within smaller subspaces (corresponding to the irreducible representations of the site groups), which can substantially reduce its cost.

To span the virtual space projected AOs (PAOs) [1,22] are used. It implies a local representation for the virtual space through projected AOs (PAOs). PAOs are centered on atoms, possess the same symmetry as the AOs, and remain localized (though usually with a somewhat higher spread than that of the corresponding AOs). PAOs are non-orthogonal and linearly dependent. The former leads to inclusion of the PAO overlap matrix in the LMP2 equations (2), while the latter is resolved by removal of the redundant functions (see *e.g.* Refs. [1,4] for details).

Although the PAOs are needed in the direct space (where the spatial locality can be exploited), they are constructed in the reciprocal space, sampled by an appropriate *k*-mesh. The reciprocal space treatment is more efficient since different *k*-points can be processed independently, while all the indices are restricted to the reference cell. We start from the Fock matrix in the AO basis in the direct space representation taken from the HF calculation. Having the Fock matrix in the direct space form allows utilizing such *k*-meshes, which are suitable for our purposes and do not coincide with the HF *k*-mesh. The AO overlap matrix in the direct space is generated directly in CRYSCOR. Then both matrices are Fourier-transformed into the reciprocal space. For each *k*-point the reciprocal image of the Fock matrix, transformed into the orthogonalized basis, is diagonalized in order to construct the density matrix and, optionally, to evaluate the singles contribution in the framework of the dual basis scheme (*vide infra*). The complement to the density matrix so obtained, is a projector onto the virtual space for a given *k*-point. It generates the reciprocal image of the PAOs for this *k*-point. The Fock and overlap matrices in the PAO basis and the Fock matrix in the WF basis, needed for the LMP2 equations (2) are obtained also at this stage. Finally all these objects are back-transformed to direct space within the region consistent with the *k*-mesh chosen for the reciprocal representation of the considered matrices [12,23].

Further, following the molecular local correlation techniques [1,22], for each pair of WFs *ij* in (1) and (2) the virtual excitation space is restricted to orbital pair domains, which contain only PAOs centered not far from the centroids of either of the two WFs. The pair domains are unions of the related orbital domains. The latter can be determined either by the Boughton-Pulay scheme [22] or, alternatively, by explicitly specifying the number of the atomic coordination spheres to be included in the domain. In molecular crystals it is convenient to include the complete molecular units in the domains. Restriction of the virtual space to pair domains automatically truncates the PAO summations in (1) and (2) making them *ij*-pair-specific.

In molecular methods the sum over the occupied indices runs over all local orbitals, whose inter-domain distance is smaller than a certain cut-off value [22]. This truncation is justified by the R^{-6} decay of the incremental pair energies, which at some point become negligibly small within a given accuracy. Truncation of the pair list and specification of domains together reduce the overall scaling of the molecular local MP2 method with the system size to linear. The default value for the cut-off distance in molecules is usually set to 15 Bohr. In crystals, however this issue demands more care. Firstly, in periodic systems the meaningful quantity is obviously not the total MP2 energy (which is infinite for an infinite object), but rather the energy per unit cell. Hence one of the occupied indices is restricted to the unit cell. The second occupied index formally runs over an infinite number of orbitals, which implies an infinite number of equations to be solved. Hence, the truncation of the second occupied index is not just a matter of computational convenience, as in the molecular case, but rather a prerequisite for a periodic LMP2 method. At the same time, in 3D packed systems the R^{-6} decay, due to the quadratic growth of the number of orbital pairs with the inter-orbital distance decreases to R^{-4} [4], which demands that higher cut-off distances must be used in 2D and especially 3D cases. We usually employ the value of 20 Bohr for this parameter.

On the other hand, the 3D periodicity opens a possibility to extrapolate the contribution of all pairs up to infinity [4]. Due to the R^{-6} asymptotical decay of the *ij*-pair energy for a given class of *i* and *j* WFs, the corresponding C_6 Lennard-Jones-like coefficient for this class of pairs can be fitted using those pair energies explicitly included in the LMP2 calculations (by default we use the energies of the pairs in the of range of 12–20 Bohr to perform this fit).

Up to this point all the summation ranges and ingredients of the LMP2 equations (1) and (2) have been evaluated, apart from the ERIs $(ia \mid jb)$. A large fraction of the integrals can be calculated virtually for free by invoking the multipolar expansion [4]. Yet, the multipolar approximation for an ERI $(ia \mid jb) \equiv (\rho_{ia} \mid \rho_{jb})$ is valid only if the interacting product densities ρ_{ia} and ρ_{jb} are non-overlapping. This means that this approximation can be exploited only for distant pairs. In CRYSCOR we employ this technique for pairs with interorbital distance larger than 6 Å. However, when the AO basis set contains diffuse functions, the multipole expansion should be utilized with care, since the PAOs corresponding to the diffuse basis functions and the product densities involving those are diffuse as well. This issue is discussed in more detail in section 3.

If the densities ρ_{ia} and ρ_{jb} are overlapping, the ERIs $(ia \mid jb)$ must be calculated explicitly. Even in molecules the transformation of such 4-index ERIs in the molecular-orbital basis by far dominates the whole LMP2 computation time [22]. In 3D periodic systems this problem is even more severe. In the pilot implementation of the periodic LMP2 method in the CRYSCOR code[11], which did not contain the density fitting approximation (*vide infra*), the cost of evaluation of the 4-index ERIs was prohibitively large [13]. For example, evaluation of the LMP2 energy for a relatively simple system like diamond with a basis set of a double-zeta quality could have taken several weeks of computing. One of the reasons for such a poor performance is that the cost for the computation of the 4-index integrals in 3D systems is proportional to the 12th power of the localization radius of WFs and PAOs. In order to reduce the spatial extent of WFs we developed a method for generating very compact, but highly overlapping WFs [24]. Although this method was able to partly circumvent the problem, the non-orthogonality of the WFs had to be included in the LMP2 equations (2), which substantially complicates the LMP2-equations solving procedure. Furthermore, the spread of the PAOs remained unchanged, leaving this part of the problem unsolved.

A much more powerful technique, which we use in order eliminate the mentioned computational bottleneck, is the density fitting approximation. For decades now this technique is successfully utilized in the context of molecular calculations [2,3,25–27]. It implies a factorization of the 4-index ERIs in terms of products of 2- and 3-index quantities:

$$(ia|jb) \approx \sum_P d_P^{ia}(P|jb), \tag{3}$$

where P indexes the functions from an auxiliary (fitting) basis set, and the fitting coefficients are determined through the linear equations system

$$\begin{aligned} d_P^{ia} &= \sum_{P'}(ia|P')[J^{-1}]_{P'P}\,, \\ J_{P'P} &\equiv (P'|P). \end{aligned} \tag{4}$$

The fitting basis consists of atom centered functions and is usually 3–5 times larger than the orbital basis. The density fitting approximations in molecular LMP2 calculations allows improving the performance by an order of magnitude, while the error in the energy introduced by this approximation remains negligible for a properly chosen fitting basis set.

Yet a straightforward application of the molecular DF technique in the periodic context is non-trivial [6,12,28–30]. Due to the infinite nature of the system the direct space formulation would imply the inversion of a matrix (4) with infinite dimension. The reciprocal treatment, on the other hand, suffers from slow convergence of the lattice summations, occurring in the Fourier transform, and from dense k-meshes, which would be generally needed [12]. We have developed two techniques, which circumvent these difficulties. The first one is a generalization of molecular local fitting [2,3,6], performed completely in the direct space. For each particular WF and all PAOs belonging to a given atom, a local fit-domain is defined, comprising a local subset of fitting functions. This technique requires a slightly more complicated (three-term rather than one) expression for the factorization (3) of the 4-index ERIs. This technique does not exploit the translational symmetry and therefore is expected to be advantageous in systems with a rather large unit cell and not too large basis sets.

The second technique is a combined direct-reciprocal treatment [30]. One part of the fitting basis is restricted to fit-domains and processed in the direct space, while the other part has full coverage and is transformed to the reciprocal space. In order to decouple the two fits the direct-space part of the fitting basis is projected out of the space spanned by the reciprocal counterpart (Direct-reciprocal decoupled density fitting - DRD-DF) [29,30].

A common choice of fitting functions in molecular calculations are the usual Gaussian-type orbitals (GTOs). However, in the periodic case the GTOs can be safely used only in the direct-space fit, since the Fourier-transforms of the Coulomb integrals involving these functions converge very slowly or even might not converge at all [12]. As a convenient alternative, the Poisson-type orbitals (PTOs), the Laplacian of GTOs [31,32], can be employed for this purpose. The

Coulomb integrals involving PTOs reduce to one-electron integrals, which makes the gain from using them twofold: (i) there is a considerable reduction of the cost for the ERI generation, (ii) the lattice sums of such integrals are exponentially converging, the latter being particularly essential for the reciprocal-space techniques. Yet, a fit with an only-PTO fitting basis would provide rather poor results, since the mentioned properties of the PTOs are consequences of their inability to represent any multipole moments [32]. Therefore, in order to fit the actual moments of the densities the PTO basis set should be complemented by at least a few GTOs. In our approach we include always a set of GTOs (one per angular momentum) on each atom of the fit-domain, which are then processed in the direct space. The density fitting approximation tremendously speeds up the evaluation and transformation of the ERIs, providing a gain in the computational time of few orders of magnitude compared to the non-DF calculations [13,30].

Once the 4-index ERIs are assembled the LMP2 equations (2) can finally be solved. The iterative procedure has been discussed in detail in Ref. [4]. The only important point we mention here is that, due to the coupling of individual pairs via the Fock matrix, all the amplitudes are presently kept in memory. For crystals with large unit cells, low symmetry, reasonably good basis, and large orbital-domains this step can constitute actual computational and/or memory bottleneck (see section). A possible general solution to this problem might be the Laplace-transformed local MP2 scheme [27].

In addition to the LMP2 energy the CRYSCOR code can deliver the correction to the HF density matrix [33,34]. This correction can be used to evaluate different first order properties like Compton profiles, X-ray factors, electron density distribution, *etc.* at the correlated level. The LMP2 density matrix at the present stage does not include orbital-relaxation effects and is evaluated solely from the first-order doubles. Neglect of orbital relaxation is much more severe for MP2 than for CC models, which include the $\exp(T_1)$ factor compensating for this deficiency [35]. Work to improve this issue is currently in progress.

An important ingredient of the CRYSCOR LMP2 implementation is the extensive exploitation of point group symmetry, which is also a distinguishing feature of the CRYSTAL code. The symmetry treatment, though being rather cumbersome to implement, leads to essential reduction of the computational cost for the vast majority of studied systems, which in contrast to large molecules, as a rule, possess at least some and in many cases quite high symmetry.

Another important option of the CRYSCOR code is the possibility to expand the orbital basis set at the stage of the correlation calculation. This dual basis set technique, well-known in the molecular context [36], is helpful when the correlation part of the calculation due the approximations introduced becomes less expensive than HF. It exploits the modest requirements of the HF method regarding the basis set quality, compared to correlation methods. In the periodic case, similarly to other aspects of the LMP2 method, this technique yet is not just an optional scheme to save the computing time, but rather a principal way to use extended basis sets in correlation calculations. This is due to the fact that the periodic HF method is very sensitive to the diffuseness of the basis set. Sometimes HF convergence is very difficult to achieve even with a basis of triple-zeta quality. MP2 calculation, on the other hand, demands more extensive basis sets (triple-zeta at least). The dual basis set technique naturally brings together the basis set requirements of the two methods.

If the HF reference corresponds to a smaller basis set, the Brillouin theorem within the expanded basis may no longer be valid. This can be taken into account in a perturbative way by including the first-order singles correction to the energy [4]. The singles equations are completely decoupled from the doubles equations, and the singles energy then is nothing else but a perturbative estimation of the HF first iteration energy increment in the larger basis, if the starting guess is the converged HF with the smaller basis. Yet, the difference between the two basis sets should not be too large, since otherwise the perturbative approach fails. We evaluate the singles energy in the reciprocal space using the canonical representation (*vide supra*).

3. Application

Although the CRYSCOR code is rather new and still at the stage of optimization and testing, several applications using the code have already been performed. These cover mainly weakly bound systems, such as the CO_2 crystal [4,13], noble gas crystals [37], Ice [4,38] and the physisorption of methane on the MgO surface [4]. Several other applications are underway.

In this paper we present an illustratory application of the LMP2 method to the investigation of the relative stability of three polymorphs of the TiO_2 crystal occurring in nature: rutile, anatase and brookite, shown in Fig. 1. Rutile and anatase belong to two tetragonal groups $P4_2/mmm$ and $I4/amd$, respectively, and contain 2 formula units in the cell. Brookite has 8 formula units per cell and the symmetry of the orthorhombic *Pbca* group. The relative stability of the three polymorphs of TiO_2 is still a matter of debate. The experimental data are somewhat scarce (especially for brookite) and to a certain extent contradictory (see *e. g.* Ref. [39] and references therein). Most of the measurements, though, reveal that rutile is a few kJ/mol more stable than anatase, and the stability of brookite is somewhat in between. In several DFT studies the stability of anatase and rutile has been compared (see Refs. [40,41]). DFT, in contrast to the experimental data, commonly predicts anatase to be more stable than rutile with the relative energy depending quite substantially on the chosen functional. The HF calculation exhibits the same tendency, however a semi-empirical estimate for the correlation contribution reverts this order [42]. Yet, a further investigation of this effect at a correlated *ab initio* level has so far not been carried out. The sole existing high quality correlation calculation for the TiO_2 crystal has been done only for the rutile structure, employing the incremental scheme[43]. As for the

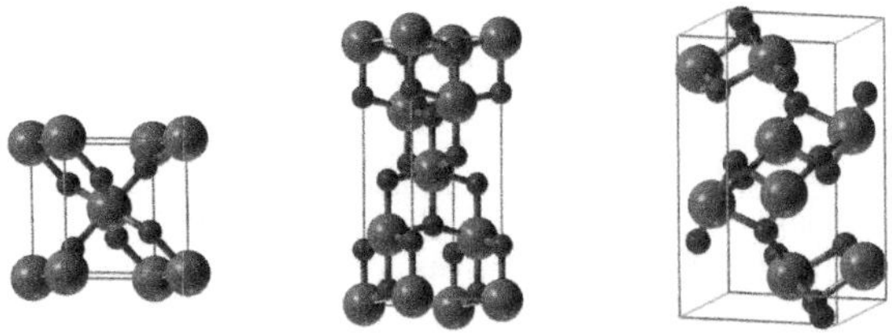

(a) Rutile (b) Anatase (c) Brookite

Fig. 1. Crystallographic unit cells of the studied TiO_2 polymorphs.

brookite crystal, its relative stability to our knowledge has been theoretically studied only in the context of pressure induced transitions [44].

Since the situation with the relative stability of these (and many other) compounds is far from being resolved, correlated methods become important for the study of such systems. The three TiO_2 polymorphs serve as a good test example for investigating the reliability and cost of the LMP2 calculations with the CRYSCOR code in view of its application to more complicated systems. Since brookite has a four times larger unit cell and two times less symmetry elements in the point group than rutile or anatase, a comparative study of these systems can also provide information on the scaling of the different stages of the calculations with respect to the cell size, and on the savings due to symmetry exploitation.

We performed a series of LMP2, and, for comparison, DFT calculations for the three structures at the experimental geometries [45,46]. In order to investigate the subtle energy differences between the polymorphs, it would be preferable to consider the equilibrium geometries. Yet, the scope and the purpose of this paper is not to provide the final MP2 results for the relative stability of these polymorphs, but rather to explore the possibility of the LMP2 method implemented in the CRYSCOR code to be applied to such systems.

As a basis set, we employ the 6–311G(2d,f) basis for the oxygen atoms and the small-core pseudopotential ECP10MDF basis set by Stoll *et. al* [47] for titanium. The most diffuse two s-, one p- and d-GTOs on Ti have been omitted, since otherwise the basis set would become linearly dependent in the HF calculations. In order to be able to obtain convergence at the HF stage, the exponents of the remaining most diffuse s- and p- GTOs (0.073438 and 0.071772 $Bohr^{-2}$, respectively) have been both upscaled to the value of 0.15 $Bohr^{-2}$. This [ECP-4s4p2d2f] basis set we denote as "basis set A". Diffuse orbitals for the titanium atoms have been partly reintroduced at the LMP2 stage through the dual basis set technique (see section 2). Since too diffuse orbitals (with exponents less than 0.06 $Bohr^{-2}$) are not treatable within the CRYSTAL code even for the Fock matrix generation only, we added one s-, p- and d- GTOs, all having an exponent of 0.06 $Bohr^{-2}$, resulting in the [ECP-5s5p3d2f] basis, denoted in the following

Table 1. Timings (in minutes) for different stages of LMP2 single point non-dedicated non-parallel calculations on a Xeon node with clock rate 3000 MHz for the rutile, anatase and brookite crystals.

System	Rutile		Anatase		Brookite
N. of symmetry operators	16		16		8
N. of corr. electrons in the cell	48		48		196
Elapsed time					
Basis set	A	B	A	B	A
HF	113		98		876
WF generation	240		299		3889
PAO generation	5	8	5	8	205
Multipole approximated ERIs	32	67	40	62	124
Density-fitted ERIs	244	635	208	542	1569
Solving LMP2 Equations	62	161	57	150	621
Total LMP2	343	871	310	762	2519
Total	696	1224	707	1159	7284

as "Basis set B". Additionally for the sake of comparison, we have performed DFT calculations employing the LDA, PBE and B3LYP functionals. Even with the basis set "A" none of the DFT calculations would converge. Therefore, for the DFT calculations we used a different basis set ("Basis set C"): small-core Hay-Wadt pseudopotential-411(311d) for Ti and 8–411(11d) for oxygen [48]. Since our previous tests have shown that the LMP2 energy and especially the singles contribution might depend quite heavily on the quality of the Fock matrix, tight overlap and pseudo-overlap tolerances of 10^{-7}, 10^{-7}, 10^{-7}, 10^{-15} and 10^{-50} have been used for the HF calculations.

In the LMP2 part of the calculations the valence electrons and, additionally, titanium 3*s* and 3*p* electrons have been correlated. For the valence WFs, which are centered on the oxygen atoms, four-atom domains have been chosen, comprising the oxygen and its three nearest-neighbor titanium atoms. The core orbitals are very localized and the corresponding domains include only the Ti atoms. This choice for the domains refers to a Boughton-Pulay criterion of 0.99. For the basis "A" the ERIs for WF pairs with the interorbital distance up to 7 Å have been generated explicitly via the DF approximation. Due to the presence of rather diffuse AOs in the basis, the default distance of 6 Å, beyond which the multipole approximation of the ERIs is invoked, is no longer adequate. In the basis "B", where even more diffuse functions are added, this distance has to be increased further to 8 Å. The cut-off distance for the WF pairs included in the LMP2 equations was set to 10 Å. The contribution to the LMP2 energy from the pairs beyond 10 Å was added via Lennard-Jones extrapolation (*vide supra*). For rutile and anatase the DRD-fitting has been used and for brookite the Direct-space fit.

Table 1 compiles the timings of the main stages of the single point LMP2 calculations. For rutile and anatase, which have unit cells with the same number of atoms and symmetry elements, the computational cost is comparable. The

Table 2. The difference between the total energies (per two TiO_2 units) of anatase and brookite crystals and the energy of rutile.

		Relative energy with respect to rutile, kJ/mol	
		Anatase	Brookite
HF	A	-5.8	6.7
HF+LMP2	A	1.7	-6.6
HF+LMP2	B	1.2	-
LDA	C	-0.3	-4.8
PBE	C	-12.7	-10.3
B3LYP	C	-16.7	-9.0

LMP2 part of the calculation for these two systems is approximately 3 times more expensive than the HF part, but not much more than the localization/symmetrization procedure for WFs. The use of the basis set "B" noticeably raises the cost of the LMP2 calculation: by more than a factor of 2, although the size of the basis "B" is just 10% larger than "A". The reason for this lies not in the number of functions in the basis "B" but rather in the inclusion of very diffuse AOs, which makes prescreening less effective and the direct-space ranges much more extended.

Linear scaling implies a factor of eight for the computational time when going from rutile or anatase to brookite. This factor is clearly reproduced in the HF and the main LMP2 parts of the calculations. The exception is the localization/symmetrization procedure, which in the present form scales more steeply and for systems with large unit cells can become the bottleneck. The PAO generation also reveals quite a steep scaling due to the reciprocal space technique, which does not scale linearly. On the other hand, even for the case of brookite this stage is still quite fast compared to other parts of the calculation. The multipole evaluation of the ERIs scales linearly and is generally inexpensive in all the cases, despite the fact that the symmetry treatment is not much involved at that stage. The most expensive step of the LMP2 calculations is the evaluation and transformation of 3-index ERIs for the density fitting. Although it scales also linearly and is generally not too expensive, some further optimizations are possible. Solving the LMP2 equations is computationally not very expensive, at least for the systems studied here, but has in its present form rather high memory requirements. In case of brookite with the basis "A" 12G of memory were required, and with the basis set "B" it was no longer possible to run the calculation on a node, having only 16G of memory. Therefore, for extended domains, basis sets, or unit cell sizes the memory requirements might become prohibitively high. However, this is a mere technical problem, which can be solved by implementing a paging scheme for the amplitudes.

The results of the calculations for these polymorphs are presented in table 2. Since no geometry optimization have been performed, only the general trends can be discussed. Our DFT results position anatase energetically lower than rutile in line with other DFT investigations of these systems [40,41]. This picture, however, contradicts most experimental data, which suggests the opposite order for the relative stability of the polymorphs. Furthermore, the high dependence of the results on the functional questions the reliability of DFT for this particular problem.

The HF calculation also places rutile above anatase. But interestingly the LMP2 method reverts this order, predicting anatase being slightly above rutile, which is in qualitative agreement with the experiment. A more quantitatively reliable LMP2 study demands a further investigation of the problem. The LMP2 energy of brookite is not only below anatase, which is consistent with the experiment, but also below rutile. The only existing experiments on the brookite-rutile transformation, detect a very small energetic difference between these two at the temperature of 971K [39]. Whether the discrepancy between the LMP2 prediction and the experimental result is caused by deficiency of the theoretical treatment or the experiment remains under question.

4. Conclusions

In this paper we described the general structure of the CRYSCOR code and explained briefly all essential steps of a periodic LMP2 calculation. As an application of the CRYSCOR code we considered the problem of the relative stability of three TiO_2 polymorphs.

The calculations demonstrate the efficiency of the LMP2 method in the periodic context and the good performance of its implementation. For systems with a few dozens of correlated electrons per unit cell and basis sets of triple-zeta quality, an average single point LMP2 calculation takes a few hours on a single 3GHz-processor. The scaling of the computational costs is close to linear so that relatively large systems with up to several hundred correlated electrons can be treated within a few days. Parallelization of the code could considerably speed up the calculations. The size of systems treatable by periodic LMP2 method is presently restricted by the localization procedure and a memory bottleneck in the LMP2 equation solver. Work addressing these issues is in progress. Usage of diffuse basis sets in the LMP2 calculations is not as harmful as for the periodic HF method. At the same time, diffuse functions noticeably increase the computational cost, while the importance of including them in the basis for relatively closely packed systems is moot question.

The study of the TiO_2 polymorphs reveals that the LMP2 method is a convenient and useful approach for studying systems where dynamic electronic correlation is of importance. In many such cases it provides a realistic picture of the physics whereas the DFT method fails. Our calculations on the TiO_2 polymorphs show that the post-HF inclusion of correlation via the LMP2 method leads to a slightly higher stability of the rutile polymorph than anatase, which is in agreement with existing experimental data, but not reproduced in the DFT calculations. Brookite is predicted to be the most stable structure, however, in order to draw a definite conclusion further studies (such as geometry optimizations, *etc.*) at the LMP2 level should be performed.

Acknowledgement

Financial support from the Deutsche Forschungsgemeinschaft DFG (priority program SPP1145) is gratefully acknowledged.

References

1. P. J. Knowles, M. Schütz, and H.-J. Werner, Ab initio methods for electron correlation in molecules. In Modern Methods and Algorithms of Quantum Chemistry. J. Grotendorst (Ed.), number 1 in NIC Series, page 69. John von Neumann Institute for Computing (NIC) (2000).
2. H.-J. Werner, P. J. Knowles, and F. R. Manby, J. Chem. Phys. **118** (2003) 8149.
3. M. Schütz and F. R. Manby, Phys. Chem. Chem. Phys. **5** (2003) 3349.
4. C. Pisani, L. Maschio, S. Casassa, M. Halo, M. Schütz, and D. Usvyat, J. Comp. Chem. (2008) DOI 10.1002/jcc.20975.
5. P. Y. Ayala, K. N. Kudin, and G. E. Scuseria, J. Chem. Phys. **115** (2001) 9698.
6. A. F. Izmaylov and G. E. Scuseria, Phys. Chem. Chem. Phys. **10** (2008) 3421.
7. G. Stollhoff and P. Fulde, Z. Phys. B **26** (1977) 251.
8. N. Flocke and R. J. Bartlett, J. Chem. Phys. **118** (2003) 5326.
9. H. Stoll, Phys. Rev. B **46** (1992) 6700.
10. B. Paulus, Phys. Rep. **428** (2006) 1.
11. C. Pisani, M. Busso, G. Capecchi, S. Casassa, R. Dovesi, L. Maschio, C. Zicovich-Wilson, and M. Schütz, J. Chem. Phys. **122** (2005) 094113.
12. L. Maschio, D. Usvyat, F. Manby, S. Casassa, C. Pisani, and M. Schütz, Phys. Rev. B **76** (2007) 075101.
13. D. Usvyat, L. Maschio, F. Manby, S. Casassa, M. Schütz, and C. Pisani, Phys. Rev. B **76** (2007) 075102.
14. F.R. Manby, D. Alfe, and M.J. Gillan, Phys. Chem. Chem. Phys., **8** (2006) 5178.
15. M. Marsman, A. Grüneis, J. Paier, and G. Kresse, J. Chem. Phys. **130** (2009) 184103.
16. R. Dovesi, V. R. Saunders, C. Roetti, R. Orlando, C. Zicovich-Wilson, F. Pascale, B. Civalleri, K. Doll, N. M. Harrison, I. J. Bush, P. D'Arco, and M. Llunell, CRYSTAL06 Users Manual (2006).
17. C. Pisani, R. Dovesi, and C. Roetti, Hartree-fock ab initio treatment of crystalline solids. In Lecture Notes in Chemistry Series, volume 48. Springer Verlag, Berlin (1988).
18. C. M. Zicovich-Wilson, R. Dovesi, and V. R. Saunders, J. Chem. Phys. **115** (2001) 9708.
19. S. Casassa, C. Zicovich-Wilson, and C. Pisani, Theor. Chem. Acc. **116** (2006) 726.
20. V. P. Smirnov and D. E. Usvyat, Phys. Rev. B **64** (2001) 245108.
21. C. M. Zicovich-Wilson, J. Phys: Conf. Series **117** (2008) 012030.
22. M. Schütz, G. Hetzer, and H.-J. Werner, J. Chem. Phys. **111** (1999) 5691.
23. R. A. Evarestov and I. I. Tupitsyn, Phys Solid State **44** (2002) 1656.
24. D. Usvyat and M. Schütz, Theor. Chem. Acc. **114** (2005) 276.
25. D. M. Schrader and S. Prager, J. Chem. Phys. **37** (1962) 3927.
26. B. I. Dunlap, J. W. D. Connolly, and J. R. Sabin, J. Chem. Phys. **71** (1979) 3396.
27. D. Kats, D. Usvyat, and M. Schütz, Phys. Chem. Chem. Phys. **10** (2008) 3430.
28. S. Varga, M. Milko, and J. Noga, J. Chem. Phys. **124** (2006) 034106.

29. S. Varga, Int. J. Quantum Chem. **108** (2008) 1518.
30. L. Maschio and D. Usvyat, Phys. Rev. B **78** (2008) 073102.
31. J. W. Mintmire and B. I. Dunlap, Phys. Rev. A, **25** (1982) 88.
32. F. R. Manby, P. J. Knowles, and A. W. Lloyd, J. Chem. Phys. **115** (2001) 9144.
33. D. Usvyat and M. Schütz, J. Phys: Conf. Series **117** (2008) 012027.
34. C. Pisani, S. Casassa, and L. Maschio, Z. Phys. Chem. **220** (2006) 913.
35. D. J. Thouless, Nucl. Phys. **21** (1960) 225.
36. K. Wolinski and P. Pulay, J. Chem. Phys. **118** (2003) 9497.
37. M. Halo, S. Casassa, L. Maschio, and C. Pisani, Chem. Phys. Lett. **467** (2009) 294.
38. A. Erba, S. Casassa, R. Dovesi, L. Maschio, and C. Pisani, J. Chem. Phys. **130** (2009) 074505.
39. M. R. Ranade *et al*, Proc. Natl. Acad. Sci. **99** (2002) 6476.
40. J. Muscat, V. Swamy, and N. M. Harrison, Phys. Rev. B **65** (2002) 224112.
41. F. Labat, P. Baranek, C. Domain, and C. Minot, J. Chem. Phys. **126** (2007) 154703.
42. A. Fahmi, C. Minot, B. Silvi, and M. Causa, Phys. Rev. B **47** (1993) 11717.
43. K. Rosciszewski, K. Doll, B. Paulus, P. Fulde, and H. Stoll, Phys. Rev. B **57** (1998) 14667.
44. A. Beltran, L. Gracia, and J. Andres, J. Phys. Chem. B **110** (2006) 23417.
45. J. K. Burdett, T. Hughbanks, G. J. Miller, J. W. Richardson, and J. V. Smith, J. Am. Chem. Soc. **109** (1987) 3639.
46. E. P. Meagher and G. A. Lager, Can. Mineral. **17** (1979) 77.
47. M. Dolg, U. Wedig, H. Stoll, and H. Preuss, J. Chem. Phys. **86** (1987) 866.
48. See http://www.crystal.unito.it.

Orbital-dependent Representation of Correlation Energy Functional

By H. Jiang and E. Engel*

Center for Scientific Computing, J. W. Goethe-Universität Frankfurt, Max-von-Laue-Str. 1, 60438 Frankfurt am Main, Germany

(Received April 7, 2009; accepted July 20, 2009)

Density Functional Theory / Correlation / Optimized Potential Method

Exchange-correlation energy functionals depending on the Kohn-Sham orbitals and eigenvalues resolve some of the well-known deficiencies of the local density and generalized gradient approximations. Such functionals can be derived in first-principles fashion by use of standard many-body techniques, if the Kohn-Sham Hamiltonian is utilized as non-interacting reference Hamiltonian. In this way one can establish an exact relation for the exchange-correlation functional, which opens several routes for the derivation of approximate functionals. Straightforward expansion in powers of the electron-electron coupling constant gives, to first order, the exact exchange of density functional theory and, to second order, a correlation functional which has the same structure as the second order Møller-Plesset term. This simplest first-principles correlation functional reproduces both the shell structure in the exact correlation potential and dispersion forces. On the other hand, it overestimates all correlation effects and is variationally instable for systems with a very small HOMO-LUMO gap. Both deficiences can be resolved by partial resummation of the Kohn-Sham perturbation expansion. In this contribution two such resummations are discussed, a minimum form designed to address the variational instability at no computational cost (based on the Epstein-Nesbet diagrams) and a more systematic variant (based on the ring diagrams).

1. Introduction

The concept of implicit density functionals for the exchange-correlation (xc) energy E_{xc} of density functional theory (DFT) establishes a bridge between standard density functional methods, based on the local density approximation (LDA) or generalized gradient approximation (GGA), and correlated *ab-initio* methods (for an overview of implicit functionals see [1,2]). It shares the common DFT background with the LDA and GGA approach. In particular, the basic

* Corresponding author. E-mail: engel@th.physik.uni-frankfurt.de

© by Oldenbourg Wissenschaftsverlag, München

equations to be solved are the single-particle Kohn-Sham (KS) equations ($\hbar = e = 1$),

$$\left\{-\frac{\nabla^2}{2m} + v_s(\boldsymbol{r})\right\} \phi_k(\boldsymbol{r}) = \varepsilon_k \phi_k(\boldsymbol{r}) , \tag{1}$$

with their multiplicative effective potential v_s,

$$v_s(\boldsymbol{r}) = v_{ext}(\boldsymbol{r}) + v_H(\boldsymbol{r}) + v_{xc}(\boldsymbol{r}) \tag{2}$$

$$v_H(\boldsymbol{r}) = \int d^3r' \frac{n(\boldsymbol{r}')}{|\boldsymbol{r} - \boldsymbol{r}'|} \tag{3}$$

$$v_{xc}(\boldsymbol{r}) = \frac{\delta E_{xc}}{\delta n(\boldsymbol{r})} . \tag{4}$$

Here $n(\boldsymbol{r})$ denotes the (ground state) density of the system, which can be expressed in terms of the N lowest (i.e. the occupied) KS orbitals,

$$n(\boldsymbol{r}) = \sum_k \Theta_k |\phi_k(\boldsymbol{r})|^2 \tag{5}$$

$$\Theta_k = \begin{cases} 1 & \text{if } \varepsilon_k \leq \varepsilon_F \\ 0 & \text{otherwise} \end{cases} , \tag{6}$$

with ε_F denoting the eigenvalue of the highest occupied orbital.

On the other hand, in the case of implicit functionals the actual representation of xc-effects via E_{xc} and v_{xc} does not solely rely on n. Rather, implicit xc-functionals are allowed to depend explicitly on the KS orbitals and eigenvalues, which are themselves unique functionals of the density (as v_s is uniquely determined by n). If such functionals are derived systematically by means of a KS-based many-body approach [3,4], the resulting expressions fully reflect the structure, complexity and informational content of the energy components met in correlated *ab-initio* methods.

The best example for this statement and, at the same time, the most important component of any orbital-dependent E_{xc} is the exact exchange of DFT,

$$E_x = -\frac{1}{2} \sum_{kl} \Theta_k \Theta_l (kl||lk) , \tag{7}$$

here expressed in terms of the KS-type Slater integral

$$(ij||kl) = \int d^3r \int d^3r' \frac{\phi_i^\dagger(\boldsymbol{r})\phi_k(\boldsymbol{r})\phi_j^\dagger(\boldsymbol{r}')\phi_l(\boldsymbol{r}')}{|\boldsymbol{r} - \boldsymbol{r}'|} . \tag{8}$$

E_x automatically emerges as the lowest order contribution to the full E_{xc} in a perturbation expansion with respect to the coupling-constant e^2, if the KS Hamiltonian is used as non-interacting reference Hamiltonian [3,5].

The potential corresponding to expressions like E_x has to be calculated via the optimized potential method (OPM, also referred to as optimized effective potential, OEP) [6]. For a functional depending on both ϕ_k and ε_k the associated $v_{\rm xc}$ is determined by the integral equation [7],

$$\int d^3r'\,\chi_{\rm s}(\boldsymbol{r},\boldsymbol{r}')\,v_{\rm xc}(\boldsymbol{r}') = \Lambda_{\rm xc}(\boldsymbol{r})\,, \tag{9}$$

where $\chi_{\rm s}$ is the static response function of the KS system,

$$\chi_{\rm s}(\boldsymbol{r},\boldsymbol{r}') = -\sum_k \Theta_k\phi_k^\dagger(\boldsymbol{r})G_k(\boldsymbol{r},\boldsymbol{r}')\phi_k(\boldsymbol{r}') + c.c. \tag{10}$$

$$G_k(\boldsymbol{r},\boldsymbol{r}') = \sum_{l\neq k}\frac{\phi_l(\boldsymbol{r})\,\phi_l^\dagger(\boldsymbol{r}')}{\varepsilon_l-\varepsilon_k}\,. \tag{11}$$

and the inhomogeneity is given by

$$\begin{aligned}\Lambda_{\rm xc}(\boldsymbol{r}) = &-\sum_k\int d^3r'\left[\phi_k^\dagger(\boldsymbol{r})G_k(\boldsymbol{r},\boldsymbol{r}')\frac{\delta E_{\rm xc}}{\delta\phi_k^\dagger(\boldsymbol{r}')}+c.c.\right]\\ &+\sum_k|\phi_k(\boldsymbol{r})|^2\frac{\partial E_{\rm xc}}{\partial\varepsilon_k}\,.\end{aligned} \tag{12}$$

For any functional at hand the derivative $\delta E_{\rm xc}/\delta\phi_k^\dagger$ can be evaluated analytically. Equation (9) (or a procedure which is equivalent to it) is then solved by means of a suitable basis set expansion of all ingredients.

Use of the exact exchange (EXX) resolves some well-known deficiencies of the LDA and GGA. In particular, one finds that the exact $v_{\rm x}$ decays like $-1/|\boldsymbol{r}|$ in the asymptotic regime of finite systems. This is an immediate consequence of the complete cancellation of the self-interaction contained in $v_{\rm H}$. As a result, the EXX spectrum of neutral atoms includes a Rydberg series of unoccupied states (unlike the LDA, GGA or Hartree-Fock (HF) spectrum), which, in turn, ensures the existence of atomic negative ions [8] (the differences between EXX and HF spectra originate from the multiplicative nature of $v_{\rm x}^{\rm EXX}$ - see [1] for a detailed comparison of EXX and HF). The exact elimination of the self-interaction is also the origin of a second important improvement observed: When applied to semiconductors, the EXX method yields band gaps in much better agreement with experimental data than the corresponding LDA and GGA values [9–11] By now, the EXX scheme has become a standard approach in DFT [12–22].

Given the improvements resulting from the EXX approach, the obvious next question concerns the correlation functional $E_{\rm c}$ to be combined with the exact $E_{\rm x}$. As is well known, the quality of LDA and, to some extent, also of GGA results is due to a cancellation of errors between exchange and correlation. This cancellation is no longer possible as soon as LDA or GGA correlation is added to the exact $E_{\rm x}$, so that many atomic and molecular properties are not described particularly well. As an example, Table 1 lists the mean absolute errors in the spectroscopic constants of the 31 first and second row diatomic molecules in the G2 test set [23], resulting from the combination of the exact $E_{\rm x}$ with the PBE-GGA for correlation. One observes a clear deterioration of the spectroscopic constants, when going from PBE to exact exchange.

Table 1. Bond lengths R_e, dissociation energies D_e (including the zero-point energies) and harmonic frequencies ω_e: Mean absolute deviation of LDA [24], PBE [25], PBE0 [26] EXX+PBE (combination of exact exchange with PBE correlation) results from experimental data for the first and second row diatomic molecules contained in the G2 test set [23].

method	R_e [Bohr]	D_e [eV]	ω_e [cm^{-1}]
LDA	0.027	1.027	48
PBE	0.026	0.329	42
PBE0	0.020	0.195	41
EXX+PBE	0.088	1.187	213

Three strategies have been followed to deal with this problem[1]. First of all, it turns out that only partial inclusion of the exact exchange in the form of a hybrid functional [26,29,30] gives much better results. The optimum fraction of exact exchange for standard molecular properties is found to be $1/4$ [26], which leads to the PBE0 hybrid. An indication of its quality is given in Table 1.

In a second approach models are employed to construct orbital-dependent correlation functionals for use with the full exact exchange [31–34]. These functionals have the advantage that they only depend on the occupied KS states, as correlation is expressed in terms of either the exact exchange energy density or E_x itself (plus some conventional density-dependence and, in some cases, the kinetic energy density). Thermochemical data obtained with these functionals are generally superior to those resulting from standard hybrid functionals.

Finally, KS-based many-body theory has been utilized to derive first-principles approximations for E_c [4,7,35–54]. In the following we will review the most basic functionals of this type and give some prototype results obtained with them.

2. Second order correlation

The simplest implicit density functional for E_c is obtained by a straightforward perturbation expansion of the exact E_{xc} in powers of e^2, using the KS Hamiltonian as non-interacting reference Hamiltonian. This expansion proceeds in analogy to the Møller-Plesset (MP) expansion on the basis of the HF Hamiltonian. The lowest (second) order term consists of two contributions [4],

$$E_c^{(2)} = E_c^{MP2} + E_c^{\Delta HF} . \tag{13}$$

E_c^{MP2} has the same form as the standard MP2 expression,

$$E_c^{MP2} = \frac{1}{4} \sum_{ijab} \Theta_i \Theta_j \overline{\Theta}_a \overline{\Theta}_b \frac{|(ab||ij) - (ab||ji)|^2}{\varepsilon_i + \varepsilon_j - \varepsilon_a - \varepsilon_b} , \tag{14}$$

[1] A fourth strategy is to give up use of the exact exchange in favor of a semi-local orbital-dependent representation of E_x in the form of the Meta-GGA [27, 28].

where $\overline{\Theta}_k \equiv 1 - \Theta_k$. $E_c^{\Delta HF}$ accounts for the fact that the present perturbation expansion is based on the KS, rather than the HF Hamiltonian,

$$E_c^{\Delta HF} = \sum_{ia} \Theta_i \overline{\Theta}_a \frac{|\langle i|v_x^{HF} - v_x|a\rangle|^2}{\varepsilon_i - \varepsilon_a} \tag{15}$$

$$\langle i|v_x^{HF}|a\rangle = -\sum_j \Theta_j (ij\|ja) \,. \tag{16}$$

$E_c^{\Delta HF}$ is well-defined as soon as v_x has been determined by solution of the OPM equation (9) for E_x. However, $E_c^{\Delta HF}$ is quantitatively much smaller than E_c^{MP2} and even vanishes for spin-saturated two-electron systems [35]. This contribution is therefore often neglected.

In contrast to the exact E_x, E_c^{MP2} depends on both the occupied and the unoccupied KS states, which is a common feature of all first-principles implicit correlation functionals (any treatment of correlation based on some noninteracting reference Hamiltonian necessarily involves unoccupied states - their presence is particularly required in the case of nearly degenerate situations, as one e.g. meets in the Be iso-electronic series). As a result applications are much more demanding than conventional DFT calculations. For this reason these functionals are often used in a perturbative fashion, i.e. their energy contribution is evaluated on top of an EXX or LDA/GGA calculation.

It can be easily demonstrated [7] that $E_c^{(2)}$ reproduces the $1/R^6$-behavior of the dispersion force. In fact, its contribution to the energy surface of two well-separated, finite systems A and B is given by

$$E_{c,int}^{(2)} = -\frac{C_6}{R^6} \tag{17}$$

$$C_6 = \int_0^\infty \frac{du}{2\pi} \sum_{ijkl=1}^{3} \left(\delta_{ij} - 3\frac{R^i R^j}{R^2}\right)\left(\delta_{kl} - 3\frac{R^k R^l}{R^2}\right) \times \alpha_{A,ik}(iu)\alpha_{B,jl}(iu) \,. \tag{18}$$

Here $\boldsymbol{R}$ is the Cartesian vector connecting A with B and $\alpha_A(iu)$ denotes the atomic or molecular polarizability tensor (at imaginary frequency),

$$\alpha_{ik}(\omega) = \int d^3r \int d^3r' \, r_i \, r'_k \, \chi_s^R(\boldsymbol{r}, \boldsymbol{r}', \omega) \,, \tag{19}$$

whose basic ingredient is the retarded KS response function,

$$\chi_s^R(\boldsymbol{r}, \boldsymbol{r}', \omega) = \sum_{i,k} [\Theta_i - \Theta_k] \frac{\phi_i^\dagger(\boldsymbol{r})\phi_k(\boldsymbol{r})\phi_k^\dagger(\boldsymbol{r}')\phi_i(\boldsymbol{r}')}{\omega - \varepsilon_k + \varepsilon_i + i\eta} \,. \tag{20}$$

The C_6 obtained from atomic KS polarizabilities via Eq.(18) are reasonably accurate for light atoms, but overestimate the exact C_6 (calculated from the full polarizabilities) substantially for heavier elements [55]. The same overestimation is found for the complete energy surface of noble-gas dimers [56] and for atomic

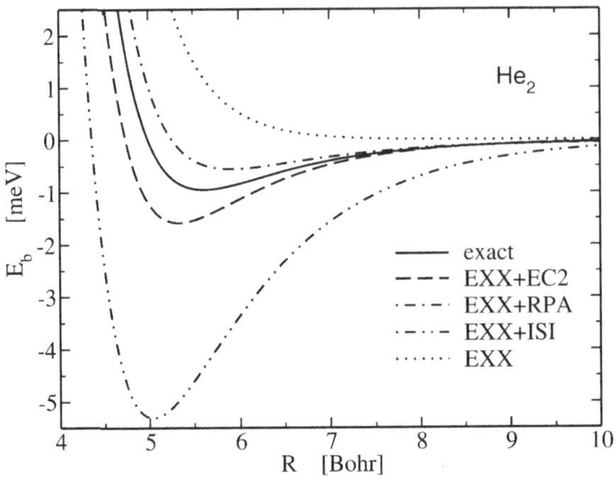

Fig. 1. Energy surface E_b of He_2: Exact E_x in combination with $E_c^{(2)}$ (EXX+EC2) [56], RPA and ISI versus EXX and exact data [57].

and molecular correlation energies [52,53]. As an illustration, Figure 1 shows the energy surface of the helium dimer.

Due to its perturbative nature, $E_c^{(2)}$ has particular difficulties whenever the HOMO-LUMO gap is small. This problem shows up in a variety of different forms. A first example is N_2. For this system the EXX HOMO-LUMO gap shrinks to zero with increasing internuclear separation R, which is a consequence of the $-1/r$-behavior of v_x for large r. As a result, $E_c^{(2)}$ increases with R, since the Slater integrals in the numerator of (14) do not vanish as fast as the denominator. This behavior sets in already quite early, so that the energy surface does not even have a local minimum [53] (in contrast to the conventional MP2 energy surface, for which the absence of Rydberg-type states in the HF spectrum prevents a vanishing HOMO-LUMO gap). The opposite R-dependence is found for Be_2, for which $E_c^{(2)}$ predicts a collapse of the dimer. This fundamental deficiency of $E_c^{(2)}$ has prevented large-scale tests of the functional. One of the few molecules for which a reasonably accurate results are obtained is H_2 (see Table 2).

While a perturbative use of orbital-dependent functionals is sufficient for many applications, the correlation potential v_c corresponding to $E_c^{(2)}$ is of obvious interest. It turned out [36] that the evaluation of v_c via Eq.(9) is not as straightforward as that of v_x. The presence of unoccupied, positive energy states in $E_c^{(2)}$ requires particular care [63]: Equation (9) does not allow for a solution with the correct boundary conditions, if the positive energy spectrum is treated as a continuum, i.e. via the dual space to the actual Hilbert space of discrete KS states. As soon as the positive energy spectrum is discretized by use of a basis set expansion [46] or by suitable boundary conditions [51] a solution can be obtained.

A result of the discretization by boundary conditions [51] is shown in Figure 2, in comparison with the exact v_c. $v_c^{\Delta HF} = \delta E_c^{\Delta HF}/\delta n$ has been neglected,

Table 2. Spectroscopic constants of prototype dimers: Combination of exact exchange (EXX) with different correlation functionals (EC2, Eq.(13) [4], RPA [37,38,58], RPA+ [59], HHEN [52] and ISI [60]) in comparison with PBE-GGA [25], PBE0-hybrid [26], and reference [61,62] results (EC2, RPA(+), HHEN and ISI have been evaluated with EXX orbitals).

	method	R_e [Bohr]	D_e [eV]	ω_e [cm^{-1}]
H_2	exact	1.401	4.747	4401
	EXX	1.386	3.637	4575
	EXX+EC2	1.413	5.000	4328
	EXX+ISI	1.393	4.636	4500
	EXX+RPA	1.398	4.688	4442
	EXX+RPA+	1.402	4.722	4417
	EXX+HHEN	1.394	4.665	4481
	PBE	1.417	4.538	4304
	PBE0	1.406	4.529	4407
Be_2	expt.	4.649	0.087	276
	EXX	unbound		
	EXX+EC2	collapsed		
	EXX+ISI	4.165	0.490	424
	EXX+RPA	4.582	0.018	281
	EXX+RPA+	4.618	–0.005	268
	EXX+HHEN	4.950	0.586	199
	PBE	4.592	0.426	352
	PBE0	4.728	0.275	303
N_2	expt.	2.074	9.905	2359
	EXX	2.011	4.973	2736
	EXX+EC2	unbound		
	EXX+ISI	2.235	12.225	1401
	EXX+RPA	2.077	9.603	2354
	EXX+RPA+	2.079	9.597	2345
	EXX+HHEN	2.185	12.466	1673
	PBE	2.082	10.583	2344
	PBE0	2.057	9.790	2474

only $v_c^{\text{MP2}} = \delta E_c^{\text{MP2}}/\delta n$ is included. In order to allow for an unambigous comparison, v_c^{MP2} is constructed from the ϕ_k and ε_k corresponding to the exact KS potentials. v_c^{MP2} reproduces the shell structure of the exact v_c, including the fact that v_c is asymptotically attractive on the left-hand side of the periodic table, but repulsive on the right-hand side [51]. This is an impressive demonstration of the first-principles nature of $E_c^{(2)}$, in view of the complete failure of the LDA and GGA [51,64].

On the other hand, the amplitudes of all shell features are by far overestimated by v_c^{MP2}. Even worse, one finds that $E_c^{(2)}$ is variationally instable, as soon

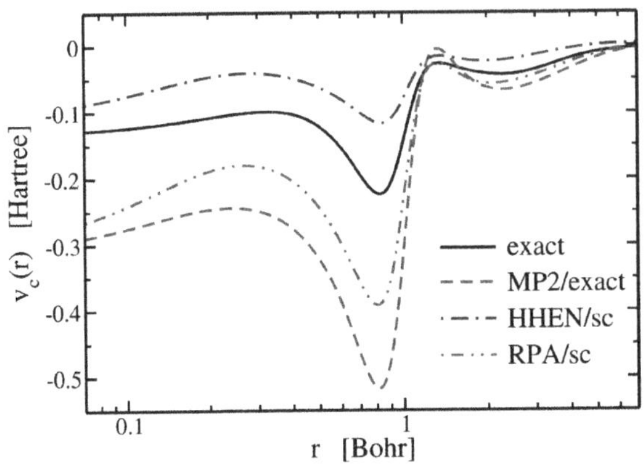

Fig. 2. (Color online) Correlation potential of Be atom: v_c^{MP2} obtained from exact KS states [51] (dash) versus selfconsistent HHEN [52] (dash-dot) and RPA [43] (dash-dot-dot) results as well as exact v_c [64] (solid).

as it is applied selfconsistently to systems with a very small HOMO-LUMO gap [49,51]. One is again faced with the fact that E_c^{MP2} is not bounded from below: If the HOMO-LUMO gap shrinks to zero, v_c^{MP2} increases so much that it can dominate v_s and thus seriously modify the resulting states. This process is self-amplifying, as the KS equations implement the minimization of the total energy functional, which is dominated by E_c^{MP2} as soon the gap approaches zero. Surprisingly enough, this effect did only show up for a few systems so far, with the beryllium atom being the prime example. In order to deal with this problem, an inclusion of higher order correlation is unavoidable.

3. Higher order correlation

The inclusion of higher order terms can follow different routes. Given the computational demands of MP2-type calculations, any attempt to limit the additional computational cost resulting from higher order terms is worthwhile. The most efficient way to address this problem is a representation of E_c–$E_c^{(2)}$ by an explicit density functional. In the interaction strength interpolation (ISI) [60] a model system is used to derive a gradient expansion for E_c which is valid in the limit of very strong interaction. Interpolation between $E_c^{(2)}$ and this strongly interacting limit via the adiabatic connection then yields an expression for E_c–$E_c^{(2)}$.

Alternatively, one can rely on the resummation of selected classes of diagrammatical contributions to the perturbation series. There is one such class which does not introduce any additional computational cost, the Epstein-Nesbet (EN) diagrams [65]. If one restricts the resummation to the hole-hole contributions (HHEN), the resulting functional reads [52]

$$E_{\rm c}^{\rm HHEN} = \frac{1}{4}\sum_{ijab} \frac{\Theta_i\Theta_j\overline{\Theta}_a\overline{\Theta}_b\,|(ij||ab)-(ij||ba)|^2}{\varepsilon_i+\varepsilon_j-\varepsilon_a-\varepsilon_b-(ij||ij)+(ij||ji)}\,. \qquad (21)$$

This expression only depends on matrix elements which are already required for the evaluation of $E_{\rm H}$, $E_{\rm x}$ or $E_{\rm c}^{(2)}$. Moreover, as $(ij\|ij) \geq (ij\|ji)$, the HHEN shift in the denominator of Eq.(21) enlarges the absolute value of the denominator, thus reducing the magnitude of the individual pair-correlation energies. $E_{\rm c}^{\rm HHEN}$ is the minimum extension of $E_{\rm c}^{(2)}$ by which the variational instability of $E_{\rm c}^{(2)}$ can be addressed.

However, the expression (21) suffers from its lacking invariance under unitary transformations in degenerate subspaces. For applications it is therefore necessary to redefine $E_{\rm c}^{\rm HHEN}$ as

$$E_{\rm c}^{\rm HHEN} = \frac{1}{4}\sum_{ijab} \frac{\Theta_i\Theta_j\overline{\Theta}_a\overline{\Theta}_b\,|(ij||ab)-(ij||ba)|^2}{\varepsilon_i+\varepsilon_j-\varepsilon_a-\varepsilon_b-\overline{(ij||ij)}+\overline{(ij||ji)}}\,, \qquad (22)$$

where $\overline{(ij\|ij)}$ denotes averaging of $(ij\|ij)$ over degenerate subspaces.

The second type of resummation which has been pursued is the random phase approximation (RPA) [37–39,41,54], which is a screened form of the direct contribution to $E_{\rm c}^{(2)}$,

$$E_{\rm c}^{\rm RPA} = \frac{1}{2}\int_0^\infty \frac{d\omega}{\pi}\,{\rm Tr}\left\{\ln\left[1-w\chi_{\rm s}^{\rm R}(i\omega)\right]+w\chi_{\rm s}^{\rm R}(i\omega)\right\} \qquad (23)$$

(w abbreviates the Coulomb interaction and the trace indicates integration over all spatial coordinates). Unlike $E_{\rm c}^{(2)}$, the expression (23) can also be applied to systems without any HOMO-LUMO gap. Due to its relation to the second order direct term, the RPA approaches the form (17) in the case of two finite systems at large separation. The RPA is somewhat less efficient than the HHEN approach, due to the ω-integration in (23). However, after suitable transformation [54] this integration can be performed with only a few mesh points, even if several shells are involved. Moreover, recently an analytic treatment of the ω-integration has been put forward [66].

On the other hand, the RPA misses important short-range correlations provided by the second order exchange (SOX) contribution to $E_{\rm c}^{(2)}$ [38,54,59]. Inclusion of the SOX term is therefore often mandatory. In order to be compatible with the RPA, this inclusion has to be based on the same screened interaction as used in the direct term establishing the RPA. However, the evaluation of the screened SOX energy is computationally very demanding, so that an LDA representation of this energy has been suggested [59]. Including all other higher order correlation corrections in the same fashion, one obtains

$$E_{\rm c}^{\rm RPA+} = E_{\rm c}^{\rm RPA} + E_{\rm c}^{\rm LDA} - E_{\rm c}^{\rm LDA-RPA}\,, \qquad (24)$$

where $E_{\rm c}^{\rm LDA-RPA}$ denotes the LDA for $E_{\rm c}^{\rm RPA}$.

Prototype results obtained with the combination of the exact $E_{\rm x}$ and the functionals (22)–(24) are included in Figures 1 and 2 as well as Table 2. Figure 2 demonstrates the variational stability of both the HHEN and the RPA functional for beryllium (see Ref.[51] for a comparison with some intermediate v_c^{MP2} obtained during the KS iteration before the HOMO-LUMO gap closes). Consistent with this observation, the beryllium and nitrogen dimers are obtained as bound. Quite generally, the overestimation of correlation is significantly reduced by these functionals. This is true both on the local level, i.e. for v_c, as well as for correlation, ionization and dissociation energies [52–54,67]. Both the HHEN and the RPA+ give rather accurate atomic correlation energies, while the pure RPA substantially overestimates these absolute energies (this effect is no longer as obvious in energy differences). While the accuracy of atomic data obtained with the HHEN and the RPA+ functional is comparable, the RPA+ provides much more accurate energy surfaces. In addition, the RPA(+) has the important advantage that it produces the correct dissociation limit in the case of dimers [38,40], as it captures the strong static left-right correlation which is relevant when the bond breaks. The RPA also gives a reasonably accurate energy surface for He_2 (see Figure 1 and Ref.[67]), as this functional reproduces the dispersion force (17). For the same reason the RPA yields an improved description of noble gas solids [44].

In contrast to the partial resummations of the perturbation series the ISI does not yield a consistent improvement. While atomic correlation energies are clearly improved and dimers like N_2 and Be_2 are predicted to be bound, the corresponding spectroscopic constants are not competitive. In fact, the ISI energy surface of He_2 is much further away from the exact result than the surface obtained with $E_c^{(2)}$ (Figure 1).

4. Concluding remarks

Many-body perturbation theory based on the KS Hamiltonian opens a route for the construction of fully nonlocal, orbital-dependent correlation functionals. Already the simplest functional of this type, obtained by lowest order perturbation theory, demonstrates the scientific potential of this approach: It correctly accounts for the dispersion interaction and leads to realistic atomic correlation potentials, reproducing both the correct asymptotic behavior and the correct shell-structure. However, it is obvious that this functional can only serve as a starting point. Its variational instability and its drastic overestimation of all correlation effects shows the need for partial resummation of the perturbation series. The variational instability can be addressed by an extremely simple extension, the resummation of the hole-hole Epstein-Nesbet diagrams, which does not introduce any additional computational cost. A more consistent improvement is obtained by the inclusion of screening in the form of the RPA, in particular, if the screened second order exchange term is included at the same time. The results available so far suggest that this functional reaches at least the same level of accuracy as GGA and hybrid functionals.

Acknowledgement

We would like to thank M. Hellgren for providing us with the RPA data of Ref.[43]. Financial support by the Deutsche Forschungsgemeinschaft within the priority program 1145 (grant EN 265/4–1,2) is gratefully acknowledged. The calculations for this work have been performed on the computer cluster of the Center for Scientific Computing of J.W.Goethe-Universität Frankfurt.

References

1. E. Engel, in A Primer in Density Functional Theory, Vol. 620 of Lect. Notes Phys., C. Fiolhais, F. Nogeira, and M. Marques (Eds.), Springer, Berlin (2003), p. 56.
2. S. Kümmel and L. Kronik, Rev. Mod. Phys. **80** (2008) 3.
3. L. J. Sham, Phys. Rev. B **32** (1985) 3876.
4. A. Görling and M. Levy, Phys. Rev. A **50** (1994) 196.
5. V. Sahni, J. Gruenebaum, and J. P. Perdew, Phys. Rev. B **26** (1982) 4371.
6. J. D. Talman and W. F. Shadwick, Phys. Rev. A **14** (1976) 36.
7. E. Engel, A. Facco Bonetti, S. Keller, I. Andrejkovics, and R. M. Dreizler, Phys. Rev. A **58** (1998) 964.
8. Y. Li, J. B. Krieger, and G. J. Iafrate, Chem. Phys. Lett. **191** (1992) 38.
9. T. Kotani, Phys. Rev. B **50** (1994) 14816.
10. T. Kotani, Phys. Rev. Lett. **74** (1995) 2989.
11. M. Städele, J. A. Majewski, P. Vogl, and A. Görling, Phys. Rev. Lett. **79** (1997) 2089.
12. S. Ivanov, S. Hirata, and R. J. Bartlett, Phys. Rev. Lett. **83** (1999) 5455.
13. A. Görling, Phys. Rev. Lett. **83** (1999) 5459.
14. W. G. Aulbur, M. Städele, and A. Görling, Phys. Rev. B **62** (2000) 7121.
15. E. Engel, A. Höck, R. N. Schmid, R. M. Dreizler, and N. Chetty, Phys. Rev. B **64** (2001) 125111.
16. W. Yang and Q. Wu, Phys. Rev. Lett. **89**, (2002) 143002.
17. S. Kümmel and J. P. Perdew, Phys. Rev. Lett. **90** (2003) 043004.
18. R. J. Magyar, A. Fleszar, and E. K. U. Gross, Phys. Rev. B **69** (2004).
19. A. Qteish, A. I. Al-Sharif, M. Fuchs, M. Scheffler, S. Boeck, and J. Neugebauer, Comput. Phys. Commun. **169** (2005) 28.
20. S. Sharma, J. K. Dewhurst, and C. Ambrosch-Draxl, Phys. Rev. Lett. **95** (2005) 136402.
21. S. Rohra, E. Engel, and A. Görling, Phys. Rev. B **74** (2006) 045119.
22. A. Heßelmann, A. W. Götz, F. D. Sala, and A. Görling, J. Chem. Phys. **127** (2007) 054102.
23. B. G. Johnson, P. M. Gill, and J. A. Pople, J. Chem. Phys. **98** (1993) 5612.
24. S. H. Vosko, L. Wilk, and M. Nusair, Can. J. Phys. **58** (1980) 1200.
25. J. P. Perdew, K. Burke, and M. Ernzerhof, Phys. Rev. Lett. **77** (1996) 3865.
26. J. P. Perdew, M. Ernzerhof, and K. Burke, J. Chem. Phys. **105** (1996) 9982.
27. J. P. Perdew, S. Kurth, A. Zupan, and P. Blaha, Phys. Rev. Lett. **82** (1999) 2544.
28. J. Tao, J. P. Perdew, V. N. Staroverov, and G. E. Scuseria, Phys. Rev. Lett. **91** (2003) 146401.
29. A. Becke, J. Chem. Phys. **98** (1993) 1372.
30. P. J. Stephens, F. J. Devlin, C. F. Chabalowski, and M. J. Frisch, J. Phys. Chem. **98** (1994) 11623.
31. A. Becke, J. Chem. Phys. **122** (2005) 064101.
32. P. Mori-Sanchez, A. J. Cohen, and W. Yang, J. Chem. Phys. **124** (2006) 091102.
33. A. J. Cohen, P. Mori-Sanchez, and W. Yang, J. Chem. Phys. **126** (2007) 191109.

34. J. P. Perdew, V. N. Staroverov, J. Tao, and G. E. Scuseria, Phys. Rev. A **78** (2008) 052513.
35. E. Engel and R. M. Dreizler, J. Comput. Chem. **20** (1999) 31.
36. A. Facco Bonetti, E. Engel, R. N. Schmid, and R. M. Dreizler, Phys. Rev. Lett. **86** (2001) 2241.
37. T. Kotani, J. Phys.: Condens. Matter **10** (1998) 9241.
38. F. Furche, Phys. Rev. B **64** (2001) 195120.
39. M. Fuchs and X. Gonze, Phys. Rev. B **65** (2002) 235109.
40. M. Fuchs, Y.-M. Niquet, X. Gonze, and K. Burke, J. Chem. Phys. **122** (2005) 094116.
41. N. E. Dahlen and U. von Barth, J. Chem. Phys. **120** (2004) 6826.
42. N. E. Dahlen and U. von Barth, Phys. Rev. B **69** (2004) 195102.
43. M. Hellgren and U. von Barth, Phys. Rev. B **76** (2007) 075107.
44. J. Harl and G. Kresse, Phys. Rev. B **77** (2008) 045136.
45. S. Ivanov, S. Hirata, and R. J. Bartlett, J. Chem. Phys. **116** (2002) 1269.
46. I. Grabowski, S. Hirata, S. Ivanov, and R. J. Bartlett, J. Chem. Phys. **116** (2002) 4415.
47. R. J. Bartlett, I. Grabowski, S. Hirata, and S. Ivanov, J. Chem. Phys. **122** (2005) 034104.
48. R. J. Bartlett, V. F. Lotrich, and I. V. Schweigert, J. Chem. Phys. **123** (2005) 062205.
49. P. Mori-Sanchez, Q. Wu, and W. Yang, J. Chem. Phys. **123** (2005) 062204.
50. J. G. Ángyán, I. C. Gerber, A. Savin, and J. Toulouse, Phys. Rev. A **72** (2005) 012510.
51. H. Jiang and E. Engel, J. Chem. Phys. **123** (2005) 224102.
52. H. Jiang and E. Engel, J. Chem. Phys. **125** (2006) 184108.
53. E. Engel and H. Jiang, Int. J. Quantum Chem. **106** (2006) 3242.
54. H. Jiang and E. Engel, J. Chem. Phys. **127** (2007) 184108.
55. M. Lein, J. F. Dobson, and E. K. U. Gross, J. Comput. Chem. **20** (1999) 12.
56. E. Engel, A. Höck, and R. M. Dreizler, Phys. Rev. A **61** (2000) 032502.
57. R. A. Aziz and M. J. Slaman, J. Chem. Phys. **94** (1991) 8047.
58. E. Engel and A. Facco Bonetti, Int. J. Mod. Phys. B **15** (2001) 1703.
59. Z. Yan, J. P. Perdew, and S. Kurth, Phys. Rev. B **61** (2000) 16430.
60. M. Seidl, J. P. Perdew, and S. Kurth, Phys. Rev. Lett. **84** (2000) 5070.
61. W. Kolos and L. Wolniewicz, J. Chem. Phys. **41** (1964) 3663.
62. K. P. Huber and G. L. Herzberg, Molecular Spectra and Molecular Structure. IV. Constants of Diatomic Molecules. Van Nostrand Reinhold, New York (1979).
63. E. Engel, H. Jiang, and A. F. Bonetti, Phys. Rev. A **72** (2005) 052503.
64. C. J. Umrigar and X. Gonze, Phys. Rev. A **50** (1994) 3827.
65. A. Szabo and N. S. Ostlund, Modern Quantum Chemistry. Dover, New York (1989).
66. F. Furche, J. Chem. Phys. **129** (2008) 114105.
67. F. Furche and T. V. Voorhis, J. Chem. Phys. **122** (2005) 164106.

Discontinuities of the Chemical Potential in Reduced Density Matrix Functional Theory

By N. N. Lathiotakis[1], S. Sharma[2,6,7,*], N. Helbig[4,6], J. K. Dewhurst[2,3,6], M. A. L. Marques[5,6], F. Eich[2,3,6], T. Baldsiefen[2,6], A. Zacarias[2,6], and E. K. U. Gross[2,6,7,**]

[1] Theoretical and Physical Chemistry Institute, National Hellenic Research Foundation, Vass. Constantinou 48, 11635 Athens, Greece
[2] Institut für Theoretische Physik, Freie Universität Berlin, Arnimallee 14, 14195 Berlin, Germany
[3] Fritz Haber Institute of the Max Planck Society, Faradayweg 4–6, 14195 Berlin, Germany
[4] Nano-Bio Spectroscopy group, Dpto. Física de Materiales, Universidad del País Vasco, Centro de Física de Materiales CSIC-UPV/EHU-MPC and DIPC, Av. Tolosa 72, E-20018 San Sebastián, Spain
[5] LPMCN, Université Claude Bernard Lyon I and CNRS, F-69622 Villeurbanne, France.
[6] European Theoretical Spectroscopy Facility
[7] Max-Planck-Institut für Mikrostrukturphysik, Weinberg 2, 06120 Halle, Germany.

(Received July 5, 2009; accepted October 19, 2009)

Discontinuities / Chemical Potential / Reduced Density Matrix Functional Theory

Using the discontinuity of the chemical potential as a function of excess charge, the fundamental gaps for finite systems and the band gaps of extended solids are determined within reduced density matrix functional theory. We also present the necessary and sufficient conditions for the one-body reduced density matrix of a system with fractional charge to be ensemble N-representable. The performance of most modern day reduced density matrix functionals is assessed for the gaps and the correlation energy of finite systems. Our results show that for finite systems the PNOF, BBC3, and power functionals yield very accurate correlation energies while for a correct description of the fundamental gap the removal of self-interaction terms is essential. For extended solids we find that the power functional captures the correct band gap behavior for conventional semiconductors as well as strongly correlated Mott insulators, where a gap is obtained in absence of any magnetic ordering.

---

* Corresponding author. E-mail: nsas@zedat.fu-berlin.de

** Corresponding author. E-mail: hardy@mpi-halle.de

Z. Phys. Chem. **224** (2010) 467–480
© by Oldenbourg Wissenschaftsverlag, München

1. Introduction

Reduced density matrix functional theory (RDMFT) has recently appeared as an alternative approach to handle complex systems. It has shown great potential for improving upon density functional theory (DFT) results for finite [1] and extended [2] systems. As the basic variable [3,4] RDMFT uses the one-body reduced density matrix (1-RDM)

$$\gamma(\mathbf{x},\mathbf{x}') = N\int d\mathbf{x}_2...d\mathbf{x}_N\Psi^*(\mathbf{x}',\mathbf{x}_2,...\mathbf{x}_N)\Psi(\mathbf{x},\mathbf{x}_2,...\mathbf{x}_N), \tag{1}$$

where Ψ denotes the many-body wave function and $\mathbf{x} = (\mathbf{r},\sigma)$. Diagonalization of this matrix produces a set of spinor valued natural orbitals [3], ϕ_j, and occupation numbers, n_j, leading to the spectral representation

$$\gamma(\mathbf{x},\mathbf{x}') = \sum_{j=1}^{\infty} n_j\phi_j(\mathbf{x})\phi_j^*(\mathbf{x}'), \tag{2}$$

where the necessary and sufficient conditions for ensemble N-representability [5] require $0 \leq n_j \leq 1$ for all j, and $\Sigma_{j=1}^{\infty} n_j = N$.

In terms of γ, the total ground state energy [4] of the interacting system is (atomic units are used throughout)

$$\begin{aligned} E[\gamma] = &-\frac{1}{2}\sum_{\sigma}\int \lim_{\mathbf{r}\to\mathbf{r}'} \nabla_{\mathbf{r}}^2\gamma(\mathbf{r},\sigma,\mathbf{r}',\sigma)\,d^3r' + \sum_{\sigma}\int \rho(\mathbf{r},\sigma)V_{\text{ext}}(\mathbf{r})\,d^3r \\ &+\frac{1}{2}\sum_{\sigma\sigma'}\int \frac{\rho(\mathbf{r},\sigma)\rho(\mathbf{r}',\sigma')}{|\mathbf{r}-\mathbf{r}'|}\,d^3r\,d^3r' + E_{\text{xc}}[\gamma], \end{aligned} \tag{3}$$

where $\rho(\mathbf{r},\sigma) = \gamma(\mathbf{x},\mathbf{x})$, V_{ext} is a given external potential, and E_{xc} we call the exchange-correlation (xc) energy functional. In principle, Gilbert's [4] generalization of the Hohenberg-Kohn theorem to the 1-RDM guarantees the existence of a functional $E[\gamma]$ whose minimum yields the exact γ and the exact ground-state energy of the systems characterized by the external potential $V_{\text{ext}}(\mathbf{r})$. In practice, however, the correlation energy is an unknown functional of the 1-RDM and needs to be approximated. Regarding these approximations, RDMFT has a twofold advantage over conventional DFT; one, the kinetic and exchange energies of the interacting system are explicit functionals of γ and two, the extra degrees of freedom, i.e. the off-diagonal elements of $\gamma(\mathbf{x},\mathbf{x}')$, make the theory amenable to a wider range of functional forms for the xc energy.

If the Hamiltonian of the system is diagonal in spin, the 1-RDM decouples in separate matrices for each spin channel. In order to describe systems with a specific spin S_z one can further restrict the natural orbitals to be spin independent $\phi_j(\mathbf{x}) = \varphi_j(\mathbf{r})$, and only allow the occupation numbers to aquire a spin-dependence [6]. Most of the commonly used approximations can then be cast in the form

$$E_{xc}[\gamma] = -\frac{1}{2}\sum_{\sigma}\sum_{j,k=1}^{\infty} f(n_{j\sigma},n_{k\sigma})\iint d^3r d^3r' \frac{\varphi_j(\mathbf{r})\varphi_j^*(\mathbf{r}')\varphi_k(\mathbf{r}')\varphi_k^*(\mathbf{r})}{|\mathbf{r}-\mathbf{r}'|} \tag{4}$$

with various choices for the function f. In this work we employ the Müller [7], GU [8], BBC1–3 [9], PNOF [10, 11], CGA [12], and the power [2] functionals. Other functionals of the above type are given in Refs. [13–18]. Eq. (3) can also be modified by including a density-dependent parameter which is determined from the homogeneous electron gas resulting in the so-called BBC++ functional [19].

Unlike DFT, RDMFT does not result in a set of single-particle Kohn-Sham like equations and the total energy in Eq. (3) has to be minimized directly with respect to the 1-RDM, or, alternatively, with respect to the occupation numbers and natural orbitals. In order to ensure the conservation of particle number as well as the orthonormality of the orbitals one introduces Lagrange multipliers and minimizes [20,21]

$$F[\gamma] = E[\gamma] - \sum_{\sigma}\mu_{\sigma}\left(\sum_{j=1}^{\infty} n_{j\sigma} - N_{\sigma}\right) - \sum_{j,k=1}^{\infty}\epsilon_{jk}\left(\int d^3r\,\varphi_j(\mathbf{r})\varphi_k^*(\mathbf{r}) - \delta_{jk}\right), \quad (5)$$

where $\varepsilon_{jk} = \varepsilon_{kj}^*$. One can show that the Lagrange multipliers μ_σ are identical to the (spin-dependent) chemical potentials [22, 23], i.e.

$$\mu_\sigma = \frac{\partial E}{\partial N_\sigma}. \quad (6)$$

Following the same path as in DFT [24–26] it can be proved that for the exact energy functional the chemical potential shows a discontinuity at integer particle numbers. The size of the discontinuity is identical to the fundamental gap, i.e. the difference between the ionization potential and the electron affinity, and is given as [23]

$$\Delta = \min_{\sigma}\left(\lim_{w_\sigma\to 0}\mu_\sigma(N_\sigma + w_\sigma, N_{\sigma'})\right) - \max_{\sigma}\left(\lim_{w_\sigma\to 0}\mu_\sigma(N_\sigma + w_\sigma, N_{\sigma'})\right), \quad (7)$$

where $\sigma' \neq \sigma$. In practice the band gap is determined by calculating $\mu(w)$ for $-1 \leq w \leq 1$ and constructing tangents (for details see Section 3 and Refs. [22,23,2]). This procedure involves the minimization of the total energy with respect to the 1-RDM of a system with fractional number of particles, γ^{N+w}. To perform this minimization, the necessary and sufficient conditions for ensemble N-representability of γ^{N+w} are needed. We present these conditions in Section 2. Results for finite and periodic systems are given in Sections 3.1 and 3.2, respectively, and we summarize our findings in Section 4.

2. Ensemble N-representability conditions

In this section we show that the necessary and sufficient ensemble N-representability conditions for a 1-RDM, $\gamma^{N+w}(\mathbf{r},\mathbf{r}')$, with fractional particle number are

$$\sum_i n_i = N + w \qquad \text{and} \qquad 0 \leq n_i \leq 1. \quad (8)$$

In order to show this, we first prove that the conditions in Eq. (8) are necessary. The 1-RDM for a fractional number of particles is defined as:

$$\gamma^{N+w}(\mathbf{r},\mathbf{r}') = w\,\gamma^{N+1}(\mathbf{r},\mathbf{r}') + (1-w)\,\gamma^{N}(\mathbf{r},\mathbf{r}'), \tag{9}$$

where γ^N and γ^{N+1} are ensemble N-representable N and N+1 particle 1-RDM, respectively. Using the triangular inequality (which is valid for positive definite matrices like γ), from Eq. (9) we get

$$\begin{aligned} &\{n_i^{N+w}\}_{\max} \le w\,\{n_i^{N+1}\}_{\max} + (1-w)\,\{n_i^{N}\}_{\max} \\ &\Rightarrow \{n_i^{N+w}\}_{\max} \le w + (1-w) = 1 \end{aligned} \tag{10}$$

and

$$\begin{aligned} &\{n_i^{N+w}\}_{\min} \ge w\,\{n_i^{N+1}\}_{\min} + (1-w)\,\{n_i^{N}\}_{\min} \\ &\Rightarrow \{n_i^{N+w}\}_{\min} \ge 0. \end{aligned} \tag{11}$$

Eqs. (10) and (11) clearly show that eigenvalues of any ensemble N-representable 1-RDM for a fractional number of particles, satisfy the conditions in Eq. (8).

As a second step, we demonstrate that the conditions in Eq. (8) are sufficient for ensemble N-representability of γ^{N+w}. To do this we first show that the set G of all the vectors $\mathbf{x} = (x_1, x_2 \ldots)$ such that $0 \le x_i \le 1$ and $\Sigma_i\, x_i = N+w$ is convex, with a set of extreme elements, G_e, that have N components to be 1, one component w and the rest 0. That this set is convex is obvious: let $\mathbf{x}$, $\mathbf{y} \in G$ then $\mathbf{z} = \alpha\,\mathbf{x} + (1-\alpha)\,\mathbf{y}$ for all $0 \le \alpha \le 1$, must also belong to G.

Elements of type $\mathbf{z} \in G_e$ are extreme elements of this convex set G which can be proved as follows, write element $\mathbf{z} \in G_e$ as

$$\mathbf{z} = \alpha\,\mathbf{x} + (1-\alpha)\,\mathbf{y} \qquad \forall\, 0 \le \alpha \le 1, \tag{12}$$

where $\mathbf{x}$, $\mathbf{y} \in G$. Element-wise, with elements of $\mathbf{z}$ arranged in descending order, one obtains the following set of equations

$$\alpha\, x_i + (1-\alpha)\, y_i = \begin{cases} 1, & 1 \le i \le N \\ w, & i = N+1 \\ 0, & i > N+1 \end{cases} \tag{13}$$

The first and last relations in Eq. (13) imply

$$\begin{aligned} &x_i = y_i = 1, \quad 1 \le i \le N \\ &x_i = y_i = 0, \quad i > N+1. \end{aligned} \tag{14}$$

and since $\mathbf{x}$,$\mathbf{y} \in G$ their components have the property $\Sigma_i\, x_i = \Sigma_i\, y_i = N+w$, which in turn means $x_i = y_i = w$ when $i = N+1$ and so $\mathbf{z} = \mathbf{x} = \mathbf{y}$ and hence $\mathbf{z}$ is extreme by definition.

We also need to show that no other element of G is extreme. This is done by showing any element $\mathbf{x} \in G \backslash G_e$ can be written as $\mathbf{x} = \alpha\, \mathbf{y}+(1-\alpha)\, \mathbf{z}$ for all $0 \le \alpha \le 1$. Let $\mathbf{x}$ have M non-zero components such that $M > N+1$. We can do a slight rearrangement and write elements of $\mathbf{x}$ such that $x_1 > x_2 > x_3 > ... > x_M$, this ensures $0 < x_M \le x_{M-1} < 1$. One can now choose $\varepsilon > 0$ such that $\varepsilon = \min(x_M, x_{M-1}, 1-x_M, 1-x_{M-1})$ and with this construct two vectors $\mathbf{y} \neq \mathbf{z} \in G$ with

$$\begin{aligned} \mathbf{y} &= (x_1, x_2, \ldots, x_{M-1} + \epsilon, x_M - \epsilon) \\ \mathbf{z} &= (x_1, x_2, \ldots, x_{M-1} - \epsilon, x_M + \epsilon) \end{aligned} \tag{15}$$

such that $\mathbf{x} = 1/2(\mathbf{y}+\mathbf{z})$. This demonstrates that any element of $G \backslash G_e$ cannot be extreme.

With this in hand any matrix $A(\mathbf{r},\mathbf{r}')$ satisfying conditions in Eq. (8) can be written as

$$A(\mathbf{r},\mathbf{r}') = \sum_i \{\mathbf{x}\}_i \phi_i(\mathbf{r})\phi_i^*(\mathbf{r}') = \sum_i \left\{\sum_j \alpha_j z^j\right\}_i \phi_i(\mathbf{r})\phi_i^*(\mathbf{r}') \tag{16}$$

where $\mathbf{x} \in G$ and $\mathbf{z} \in G_e$. Rearranging the terms, Eq. (16) reads

$$\begin{aligned} A(\mathbf{r},\mathbf{r}') &= \alpha_1 \sum_i z_i^1 \phi_i(\mathbf{r})\phi_i^*(\mathbf{r}') + \alpha_2 \sum_i z_i^2 \phi_i(\mathbf{r})\phi_i^*(\mathbf{r}') + \ldots \\ &= \alpha_1 \left[\sum_i^N \phi_i(\mathbf{r})\phi_i^*(\mathbf{r}') + w\phi_{N+1}(\mathbf{r})\phi_{N+1}^*(\mathbf{r}')\right] + \ldots \\ &= \alpha_1 \left[(1-w)\gamma_1^N + w\gamma_1^{N+1}\right] + \alpha_2 \left[(1-w)\gamma_2^N + w\gamma_2^{N+1}\right] + \ldots \\ &= (1-w)\left[\sum_j \alpha_j \gamma_j^N\right] + w\left[\sum_j \alpha_j \gamma_j^{N+1}\right] \end{aligned} \tag{17}$$

The terms in the square brackets in Eq. (17) are N and $N+1$ particle 1-RDM [5] thus making the left hand side of Eq. (17) ensemble N-representable.

3. The fundamental gap

3.1 Finite systems

In this section, we study the behavior of 1-RDM functionals for fractional total number of electrons for finite systems. As explained, the discontinuity of the chemical potential at integer particle number is a direct measure of the fundamental gap for the exact theory [22,23]. It is of great importance to examine if present day functionals reproduce this behavior.

The LiH molecule will serve as a test case for closed-shell systems. In Fig. 1 we show the chemical potential μ as a function of the fractional total number of particles M around $M = 4$, the particle number of the neutral system. As we see, there are functionals (GU, PNOF, power) that reproduce a step-like behavior in close resemblance to the discontinuity of the exact theory. Functionals that do not involve the removal of self-interaction (SI) terms, i.e. Müller, the BBC's as well as BBC++, do not produce any discontinuity. We also observe that for the GU, and PNOF functionals the size of the discontinuity is in agreement with the fundamental gap of LiH (see Table 1). For the power functional, a discontinuity appears for $\alpha > 0.6$ but it is very small at first and increases with increasing α. At the same time the cancellation of SI terms between the Hartree and the xc energies improves with increasing α thus confirming the importance of their removal for a correct description of the fundamental gap. One has to choose values close to Hartree-Fock ($\alpha = 1$) in order for the discontinuity to be comparable to the experimental gap. Unfortunately, the SI is a serious problem in the application of the GU and PNOF functionals for the band gap of periodic systems, since in the case of extended electronic states (Bloch-states) the SI terms are zero. A way-around [22] is to assume that the natural orbitals are localized Wannier-like states in which case the SI is non-zero. However, this imposes an extra restriction on the optimal 1-RDM, and we are essentially using a different functional. As was shown in Ref. [2] this limitation does not apply to the power functional.

The results shown in Fig. 1 were obtained by distributing the additional charge equally over the two spin channels, i.e. the system remained closed-shell. One can, however, restrict this charge to one spin channel at a time. This has more physical meaning, since, in the limit of a whole extra electron, the spin of the ground state of the system is altered. Adding an extra electron to the LiH molecule, for example, leads to a spin-doublet state [6,23]. If that process is done incrementally, the extra charge should go into the same spin in order to reach the spin doublet. In that way, there are two chemical potentials, $\mu^{\uparrow}$, $\mu^{\downarrow}$, one for each spin, and they can be functions of both $M^{\uparrow}$ and $M^{\downarrow}$ [6,23]. Keeping either $M^{\uparrow}$ or $M^{\downarrow}$ constant and equal to the neutral system value, $M^{\uparrow} = N^{\uparrow}$ or $M^{\downarrow} = N^{\downarrow}$, we obtain 4 different chemical potentials, namely, $\mu^{\uparrow}(M^{\uparrow},N^{\downarrow})$, $\mu^{\uparrow}(N^{\uparrow},M^{\downarrow})$, $\mu^{\downarrow}(N^{\uparrow},M^{\downarrow})$, and $\mu^{\downarrow}(M^{\uparrow},N^{\downarrow})$.

In Fig. 2 we show $\mu^{\uparrow}(M^{\uparrow},N^{\downarrow})$ and $\mu^{\downarrow}(M^{\uparrow},N^{\downarrow})$ for the open-shell treatment of the LiH molecule. As we see, only $\mu^{\uparrow}(M^{\uparrow},N^{\downarrow})$ shows a discontinuous behavior while $\mu^{\downarrow}(M^{\uparrow},N^{\downarrow})$ is perfectly continuous. We did not include the chemical potentials for a fractional number of particles in the down channel since those are identical to their spin-up counterparts for closed-shell neutral systems. Comparing the shape and size of discontinuities for open- and closed-shell treatments, Figs. 1 and 2, we note that i) they are of similar size, both being good measures of the fundamental gap, ii) the open-shell treatment resembles the true discontinuity better, being almost horizontal far from the step region.

If the neutral system is open-shell we have two functions that are discontinuous, $\mu^{\uparrow}(M^{\uparrow},N^{\downarrow})$ and $\mu^{\downarrow}(N^{\uparrow},M^{\downarrow})$ which are no longer equal. Although this might seem to complicate things the identification of the gap follows a physical principle: we have to pick up the smallest possible discontinuity [23] and that value is

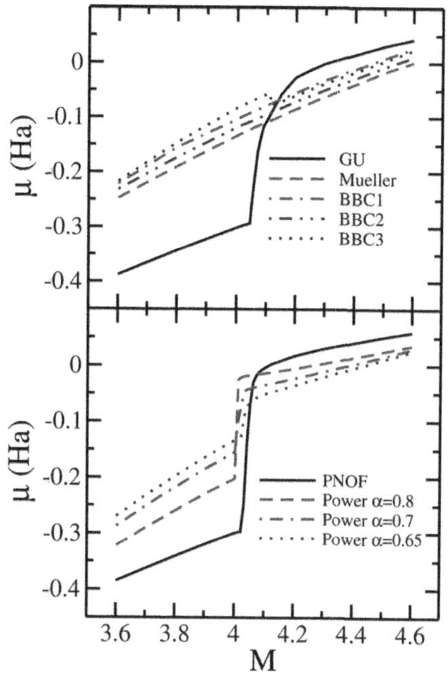

Fig. 1. The function $\mu(M)$ for the LiH molecule using the closed-shell treatment for the Goedecker-Umrigar, the Müller, BBC1, BBC2, BBC3, PNOF, and power functionals. The GU and the PNOF involve a complete removal of the SI terms and reproduce a pronounced step in resemblance of the discontinuity of the exact theory. For the power functional the degree of SI removal depends on the value of α which is directly reflected in the size of the step for different α.

the measure of the fundamental gap. The physical meaning behind this procedure is that we have to select the horizontal branches of the chemical potentials that correspond to the energetically favorable spin state when electrons are added to/removed from the system. (In the following we take the majority spin to be spin up.)

In Fig. 3, we show $\mu^{\downarrow}(N^{\uparrow}, M^{\downarrow})$ and $\mu^{\uparrow}(M^{\uparrow}, N^{\downarrow})$ for the Li atom doublet. In this case we have to select $\mu^{\uparrow}(M^{\uparrow}, N^{\downarrow})$ for negative w and $\mu^{\downarrow}(N^{\uparrow}, M^{\downarrow})$ for positive w. This choice reflects the fact that it is energetically favorable to remove an electron from the majority spin channel while the addition of an electron favours the minority spin. In Fig. 4, we include the same plots for the Na and F atoms. The case of Na is quite similar to Li while for F, we get the gap from the function $\mu^{\downarrow}(N^{\uparrow}, M^{\downarrow})$ only. This means that, for F, it is favorable to add an electron to and remove an electron from the minority spin yielding the positive ion triplet state.

In Table 1, we give the results obtained by backwards extrapolation of the chemical potential functions to extract numerical values for the gaps. Although

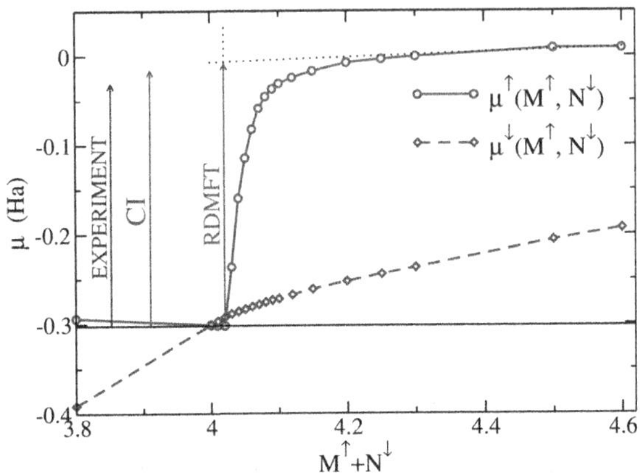

Fig. 2. Chemical potentials as functions of fractional electron numbers M for the LiH molecule using the GU functional in an open-shell treatment. $N^{\downarrow}$ is kept constant and equal to the integer, neutral-system value, $N^{\downarrow} = 2$. For comparison, the experimental and CI values of the fundamental gap are included.

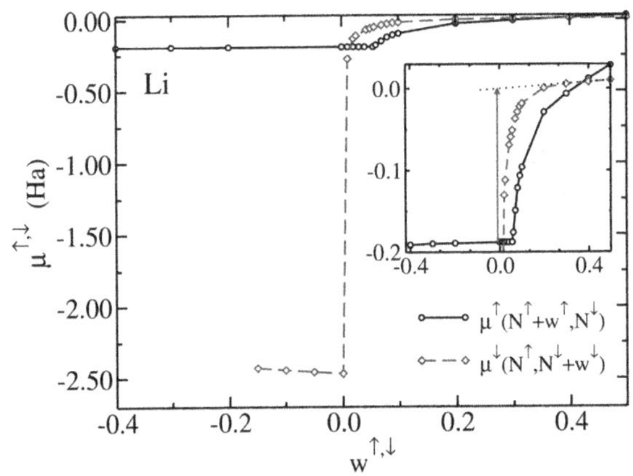

Fig. 3. The chemical potentials μ^{σ} of an electron fraction w^{σ} added (subtracted) to the neutral Li atom. In the inset, we show the region where the value of the fundamental gap is extracted from as the difference of the upper level of $\mu^{\downarrow}(N^{\uparrow}, M^{\downarrow})$ and the lower level of $\mu^{\uparrow}(M^{\uparrow}, N^{\downarrow})$.

this procedure has a certain degree of ambiguity, one can see that the extracted values agree very well with experiment and other theoretical calculations. For finite systems, one can also calculate the gap as the difference of the ionization potential and the electron affinity that are obtained from the total energy calculations for the N, N+1, and N–1 particle systems. The values for the gap obtained in this way are also given in Table 1 for comparison. One should keep in mind

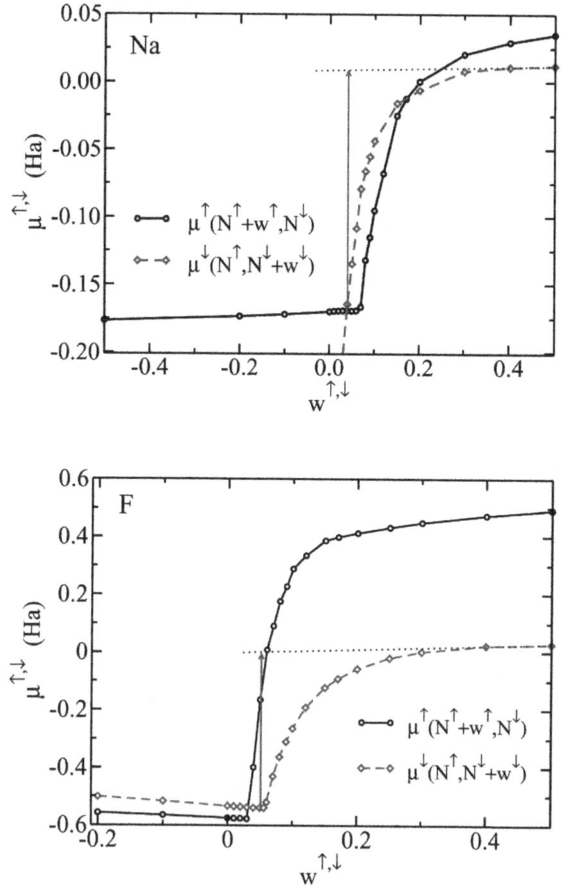

Fig. 4. Chemical potentials, μ^{σ}, as a function of an electron fraction w^{σ} added (subtracted) to the neutral system for the Na and F atoms.

that, for solid state systems, this direct procedure is not applicable because the addition or the removal of a single electron in an infinite solid is meaningless.

In order to design new functionals and/or improve upon the existing ones it is of great importance to know the limitations of the available approximate functionals [1,15,33–35]. As far as the calculation of the fundamental gaps is concerned we have seen that functionals which have their SI terms removed are far superior to functionals that contain self-interaction. However, another very important quantity to assess the quality of a functional is the correlation energy (the exchange part of E_{xc} is an explicit functional of γ). Therefore, we have tested the performance of several approximations for a set of 148 finite systems, known as the G2/97 molecule set [36,37]. The set contains 119 neutral systems and 29 radicals, 35 nonhydrogen systems, 22 hydrocarbons, 47 substituted hydro-

Table 1. The prediction for the fundamental gap for several atoms and small molecules using the size of the step of $\mu(M)$, and a direct calculation through the difference of the ionization potential (IP) from the electron affinity (EA) for the GU functional compared with experimental and other theoretical values. For the IP and EA we calculated the positive and negative ions with GU functional. [a] QCI from Ref. [27]; [b] from Ref. [28]; [c] Ionization potential from [27], electron affinity from [29]; [d] Closed-shell treatment; [e] Open-shell treatment; [f] CISD using the same basis set as in RDMFT; [g] Ionization potential from [30], electron affinity from [31].

System	RDMFT $\mu(M)$ step	RDMFT IP-EA	Other theoretical	Experiment
Li	0.18	0.202	0.175[a]	0.175[b]
Na	0.18	0.198	0.169[c]	0.169[b]
F	0.54	0.549		0.514[b]
LiH	0.27[d],0.29[e]	0.271	0.286[f]	0.271[g]

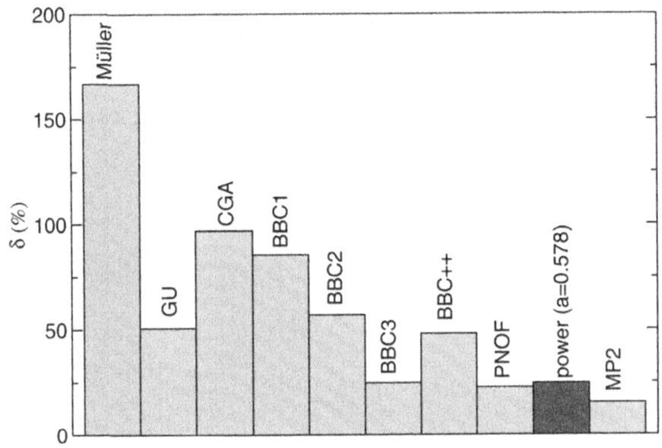

Fig. 5. Percentage deviation of the correlation energy, obtained using various 1-RDM functionals, from the exact CCSD(T) results.

carbons, and 15 inorganic hydrides. In Fig. 5, we show the relative error, δ in the correlation energy given by

$$\delta = 100\sqrt{\sum_{G2\text{ set}}\left[(E_c - E_c^{(\text{ref})})/E_c^{(\text{ref})}\right]^2/N_{\text{mol}}}\,, \tag{18}$$

where E_c is the correlation energy obtained with the method under consideration, $E_c^{(\text{ref})}$ the reference correlation energy which is obtained with the CCSD(T) method [32], and N_{mol} the number of systems included in the calculation. It is apparent from Fig. 5 that the BBC3, the PNOF, and the power functional, with $\alpha = 0.578$, yield very accurate correlation energies of finite systems (with PNOF functional being slightly better) [1]. However, for this value of α the gap is actually still zero while the GU functional which gave very good results for the fundamental gap performs significantly worse than other functionals for the correlation energy. It seems only the PNOF is of very high quality in both cases.

3.2 Periodic systems

A dramatic failure of most density functional approximations like the local density (LDA) or generalized gradient approximations (GGA) is the incorrect prediction of a metallic state for systems that are well known experimentally as highly correlated Mott insulators. Transition metal oxides (TMO) are prime examples of this. For some of them, like NiO and MnO, spin polarized calculations predict a very small band gap as a result of the anti-ferromagnetic ordering. In a spin unpolarized calculation, however, these systems are predicted to be metallic. On the other hand, these oxides are found experimentally to be insulators even above the Néel temperature [38,39]. In other words, long range magnetic order is not the reason for their insulating behavior and a real challenge for any *ab-initio* theory is to predict the insulating behavior for TMO in the absence of any spin ordering.

At this point it is also important to mention that periodic systems are fundamentally different from finite systems in that the addition of a finite amount of charge per unit cell to extended systems corresponds to an infinite number of electrons added to the solid as a whole (In order to maintain a finite energy density a compensating background charge is needed in addition). Thus the chemical potential of an insulating solid has a discontinuity at charge neutrality, with a slope, not necessarily zero, on either side corresponding to the continuum of states [2].

The calculations for solids are performed using the state-of-the-art full potential linearized augmented plane wave (FP-LAPW) method [40] implemented in the Elk code [41]. In Fig. 6, we show the dependence of the chemical potential on the excess charge per unit cell for 3 different solids; a prototype metal (Cu), semi-conductor (Si) and Mott insulator (NiO). For Cu one observes a nearly linear behaviour with no sign of discontinuity which is indicative of a metallic behaviour. On the other hand, for Si and NiO the behavior is qualitatively different with the curvature of the chemical potential changing sign with addition of excess charge. We interpret this as a sign of smoothed discontinuity [2], and an estimation of the fundamental gap for these systems can be made by constructing the two tangents [2,22,23] as shown in Fig. 6. In Fig. 7 are shown the percentage deviation of the calculated gap from the experimental data. Concentrating first on band insulators the agreement of the calculated gap with experiments is excellent, with an average deviation of the calculated gap being 5.5%. It must be stressed that this good agreement of band gaps is found for a wide range of semiconductors/insulators, with gap ranging from 1eV (Ge) to 14.2 eV (LiF) and the character of the material changing from predominantly covalent to ionic. In the case of TMO, the origin of the gap is profoundly different from conventional semiconductors/insulators. Here the gap is opened by strong Mott-Hubbard correlations. Gap formation by strong correlations is largely unaffected by the magnetic state, and so to focus exclusively on Mott-Hubbard physics we calculate these materials in the non-magnetic state. The power functional cor-

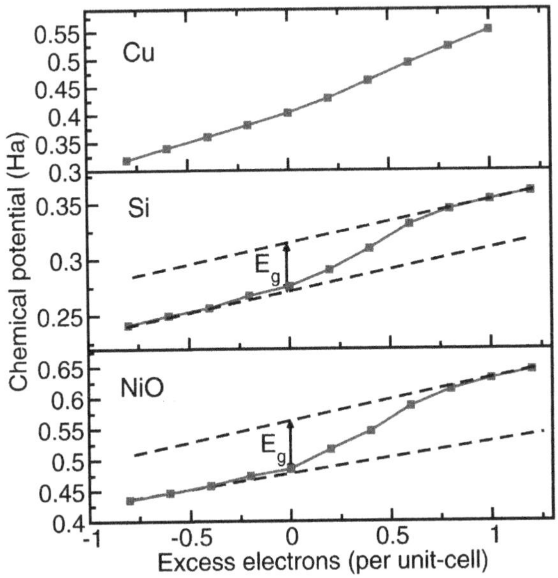

Fig. 6. Chemical potential (in Ha) versus the excess charge per unit-cell for (a) Cu, (b) Si and (2) NiO. The results are obtained using the power functional with $\alpha = 0.65$.

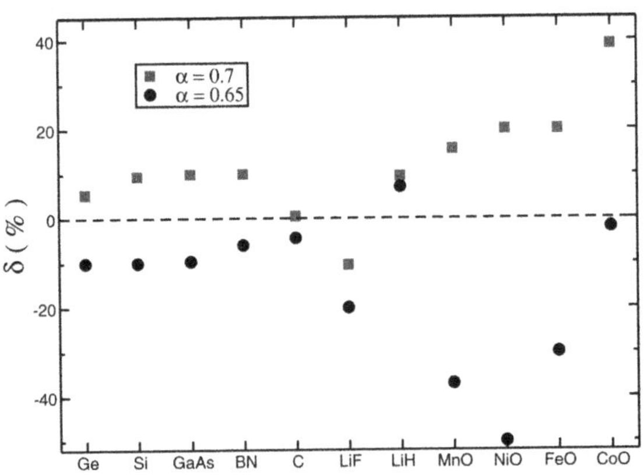

Fig. 7. Percentage deviation (δ) of calculated band gap from experiment for the power functional. Results for two different values of the exponent parameter α are shown.

rectly obtains an insulating behaviour for TMO in absence of any spin-ordering. On scanning the value of α from 0.5 to 0.9 we find that values between 0.65 and 0.7 are optimal for a wide set of materials [2]. However, for TMO the value of gap is much more sensitive to the choice of α compared to the band insulators.

At this point it is also worth mentioning that most of the band insulators and all the TMO have zero gap if the Müller functional is used.

4. Summary

We have used several recently proposed RDMFT functionals to assess their quality for finite and periodic systems. Our results show that the PNOF, BBC3, and power functionals perform very well for the correlation energy of finite systems. By using the discontinuity of the chemical potential, we have also determined the fundamental gaps of finite systems and found the removal of the SI terms to be of great importance. Most interestingly, the power functional also describes strong Mott-Hubbard correlations very well, as calculations of the non-magnetic TMO show. No other *ab-initio* approach has ever been able to achieve this. However, the optimal value of α is different for all three cases. The correlation energy of small atoms and molecules is best described with $\alpha = 0.578$ which, however, yields a zero gap. In extended solids a value of $\alpha = 0.6$–0.7 shows the best performance. The design of an *ab-initio* determination of α remains an important project for the future.

The encouraging results of the present work open up many future possibilities, such as the study of High T_C superconductors in their under-doped Mott insulating phase. It should also, hopefully, stimulate efforts to further improve approximate functionals and to develop the theory formally. The latter could include an extension to magnetic or temperature dependent RDMFT as well as a method to produce quasi particle spectra.

References

1. N. N. Lathiotakis, S. Sharma and E. Gross, Phys. Rev. A **79** (2009) 040501.
2. S. Sharma, J. K. Dewhurst and E. K. U. Gross, Phys. Rev. B **78** (2008) 201103.
3. P. O. Lödwin, Phys. Rev. **97** (1955) 1974.
4. T. L. Gilbert, Phys. Rev. B **12** (1975) 2111.
5. A. Coleman, Rev. Mod. Phys. **35** (1963) 668.
6. N. N. Lathiotakis, N. Helbig, and E. K. U. Gross, Phys. Rev. A **72** (2005) 030501.
7. A. M. K. Müller, Phys. Lett. **105A** (1984) 446.
8. S. Goedecker and C. J. Umrigar, Phys. Rev. Lett. **81** (1998) 866.
9. O. Gritsenko, K. Pernal, and E. J. Baerends, J. Chem. Phys. **122** (2005) 204102.
10. M. Piris, Int. J. Quant. Chem **106** (2006) 1093.
11. P. Leiva and M. Piris, J. Chem. Phys. **123** (2005) 214102.
12. G. Csányi, S. Goedecker and T. A. Arias, Phys. Rev. A **65** (2002) 032510.
13. M. A. L. Marques and N. N. Lathiotakis, Phys. Rev. A **77** (2008) 032509.
14. D. R. Rohr, K. Pernal, O. V. Gritsenko, and E. J. Baerends, J. Chem. Phys. **129** (2008) 164105.
15. N. N. Lathiotakis and M. A. L. Marques, J. Chem. Phys. **128** (2008) 184103.
16. J. Cioslowski and K. Pernal, Phys. Rev. B **71** (2005) 113103.
17. J. Cioslowski, K. Pernal, and M. Buchowiecki, J. Chem. Phys. **119** (2003) 6443.
18. K. Pernal and J. Cioslowski, J. Chem. Phys. **120** (2004) 5987.

19. N. N. Lathiotakis, N. Helbig, A. Zacarias and E. K. U. Gross, J. Chem. Phys. **130** (2009) 064109.
20. A. J. Cohen and E. J. Baerends, Chem. Phys. Lett. **364** (2002) 409.
21. S. Goedecker and C. J. Umrigar, Phys. Rev. A **55** (1997) 1765.
22. N. Helbig, N. N. Lathiotakis, M. Albrecht and E. K. U. Gross, Europhys. Lett. **77** (2007) 67003.
23. N. Helbig, N. N. Lathiotakis and E. K. U. Gross, Phys. Rev. A **79** (2009) 022504.
24. L. J. Sham and M. Schlüter, Phys. Rev. Lett. **51** (1983) 1888.
25. L. J. Sham and M. Schlüter, Phys. Rev. B **32** (1985) 3883.
26. W. Kohn, Phys. Rev. B **33** (1986) 4331.
27. J. A. Montgomery, Jr., J. W. Ochterski and G. A. Petersson, J. Chem. Phys. **101** (1994) 5900.
28. A. A. Radzig and B. M. Smirnov, Reference Data on Atoms and Molecules. Springer Verlag, Berlin (1985).
29. J. J. D. Groote and M. Masili, J. Chem. Phys. **120** (2004) 2767.
30. H. R. Ihle and C. H. Wu, J. Chem. Phys. **63** (1975) 1605.
31. S. B. Sharp and G. I. Gellene, J. Chem. Phys. **113** (2000) 6122.
32. J. A. Pople, M. Head-Gordon and K. Raghavachari, J. Chem. Phys. **87** (1987) 5968.
33. P. Leiva and M. Piris, Int. J. Quant. Chem. **107** (2007) 1.
34. P. Leiva and M. Piris, J. Theo. Comp. Chem. **4** (2005) 1165.
35. K. Pernal and J. Cioslowski, Chem. Phys. Lett. **412** (2005) 71.
36. L. A. Curtiss, K. Raghavachari, P. C. Redfern and J. A. Pople, J. Chem. Phys. **106** (1997) 1063.
37. L. A. Curtiss, P. C. Redfern, K. Raghavachari and J. A. Pople, J. Chem. Phys. **109** (1998) 42.
38. O. Tjernberg, S. Söderholm, G. Chiaia, R. Girard, U. O. Karlsson, H. Nylén and I. Lindau, Phys. Rev. B **54** (1996) 10245.
39. W. Jauch and M. Reehuis, Phys. Rev. B **70** (2004) 195121.
40. D. J. Singh, Planewaves Pseudopotentials and the LAPW Method. Kluwer Academic Publishers, Boston (1994).
41. http://elk.sourceforge.net (2004).

Coupling of Short-range Density-functional with Long-range Post-Hartree-Fock Methods

By Erich Goll*, Hans-Joachim Werner, and Hermann Stoll

Institut für Theoretische Chemie, Pfaffenwaldring 55, 70569 Stuttgart, Germany

(Received July 17, 2009; accepted October 16, 2009)

Short-range Density-functional / Long-range Post-Hartree-Fock

A review is given on recent work with range-separated hybrid density-functional approaches, using accurate wavefunction-based *ab initio* methods for the treatment of the long-range inter-electronic interaction. It is shown that in many cases results can be improved with respect to the borderline methods, since the hybrid approach is less basis-set dependent than in the pure *ab initio* case, and deficiencies of pure DFT for weak interactions are efficiently removed. However, there is still need for improvement of currently available short-range density functionals.

1. Introduction

A hybrid density-functional (DFT) - wavefunction-based *ab initio* method, based on a separation of the inter-electronic interaction into a short-range (*sr*) part (to be treated by DFT) and a long-range (*lr*) part (to be treated at the *ab initio* level) was suggested as early as 25 years ago [1]. However, it took 10 years until a short-range density functional (in the local-density approximation (LDA)) became available [2,3] and was implemented in atomic [2] and molecular [4] hybrid calculations. Since then, the range-separated hybrid approach has become quite popular, and many implementations have been made up to now, using *sr*-LDA [5,6], *sr*-GGA (generalized-gradient approximation) [7–10], or *sr*-meta-GGA [11] functionals on the DFT side, and *lr*-HF (Hartree-Fock) [12,13], *lr*-MCSCF (multi-configuration self-consistent field) [14], *lr*-RPA (random-phase approximation) [15,16], *lr*-MP2 (Møller-Plesset second-order perturbation theory) [9,17], *lr*-CI (configuration interaction) [18], or *lr*-CC (coupled cluster) [9,19,20] methods on the *ab initio* side. Also, a range separation of the mixing parameters in standard hybrid DFT has been suggested [21].

* Corresponding author. E-mail: goll@theochem.uni-stuttgart.de

Z. Phys. Chem. **224** (2010) 481–491
© by Oldenbourg Wissenschaftsverlag, München

The main motivation for the introduction of the hybrid scheme was (and is) the possibility to improve DFT by removing its well-known deficiencies for the treatment of, e.g., weak (van-der-Waals type) interactions or charge-transfer excitations (in time-dependent DFT). On the other side, it has been early recognized that there may be also advantages with respect to standard wavefunction-based *ab initio* methods which are plagued by the need for large one-particle basis sets and long configuration expansions caused by the short-range singularity in the inter-electronic interaction. As always, these advantages are not without drawbacks. The improvement of DFT by long-range *ab initio* corrections has to be paid for by adopting the notoriously high computational effort of *ab initio* methods, and the weaker basis-set dependence of short-range corrected *ab initio* methods may be connected with a deterioration of the high accuracy of the *ab initio* results. For this reason, most workers treat the hybrid scheme as a rung in Jacob's ladder of DFT, and keep the level of (and effort for) the *ab initio* corrections as low as possible, i.e., use *lr*-HF or *lr*-RPA.

In our own work, we focussed on high-level long-range *ab initio* approaches like *lr*-CC, instead, for two reasons. First, minimizing errors on the long-range side allows for an unbiased judgment of the reliability of available short-range functionals. Secondly, we keep the optimistic view that the hybrid approach may be useful also when treated as a DFT correction to high-level *ab initio* methods. The range of applicability of the latter methods has lately been much extended due to methodological progresses (local correlation methods [22–24], in particular, which lead to linearly scaling computational effort with molecular size [25,26]). We want to answer questions like: How much can we gain by mixing-in *sr*-DFT to such approaches? How efficient is the reduction of basis-set dependence? Can we develop short-range functionals which are sufficiently accurate for reaching the reliability of standard *ab initio* methods? In the present article, we try to (partially) answer these questions by reviewing material accumulated in papers [9,11,19,20,27,28] triggered by the SPP1145.

2. Methods

2.1 Splitting of inter-electronic interaction

The hybrid scheme has been described in detail in Refs. [9,11,19,20,27,28], e.g., and only a short résumé is given here. The scheme relies on a splitting of the inter-electronic interaction potential V_{ee} into a *sr* and a *lr* part, using the standard error function erf and its complement erfc = 1–erf:

$$V_{\mathrm{ee}} = V_{\mathrm{ee}}^{sr} + V_{\mathrm{ee}}^{lr} = \sum_{i<j} \frac{\mathrm{erfc}(\mu r_{ij})}{r_{ij}} + \sum_{i<j} \frac{\mathrm{erf}(\mu r_{ij})}{r_{ij}}, \tag{1}$$

where $\mu = 0$ and $\mu \rightarrow \infty$ lead to a pure DFT and a pure *ab initio* method, respectively. The appropriate short-range density functional can be constructed using Levy's constrained-search formalism [29]. We start writing the ground state energy as

$$E_0 = \min_{\rho}\left(E_0^{sr}[\rho] + \min_{\Psi\to\rho}\langle\Psi|T + V_{\text{ne}} + V_{\text{ee}}^{lr}|\Psi\rangle\right), \tag{2}$$

$$E_0^{sr}[\rho] = \min_{\Psi\to\rho}\langle\Psi|T + V_{\text{ne}} + V_{\text{ee}}|\Psi\rangle - \min_{\Psi\to\rho}\langle\Psi|T + V_{\text{ne}} + V_{\text{ee}}^{lr}|\Psi\rangle. \tag{3}$$

$E_0^{sr}[\rho]$, being a (universal) short-range energy functional, can be split into a Hartree term

$$J^{sr}[\rho] = \frac{1}{2}\int \rho(r_1)\frac{\text{erfc}(\mu r_{12})}{r_{12}}\rho(r_2)\,dr_1 dr_2 \tag{4}$$

and an exchange-correlation term E_{xc}^{sr} which is defined as everything remaining, so that E_0 can be rewritten as

$$E_0 = \min_{\Psi}\left(J^{sr}[\rho_\Psi] + E_{\text{xc}}^{sr}[\rho_\Psi] + \langle\Psi|T + V_{\text{ne}} + V_{\text{ee}}^{lr}|\Psi\rangle\right), \tag{5}$$

where ρ_Ψ is the one-particle density associated with Ψ. Then the wavefunction Ψ^{lr} minimizing equation (5) must be an eigenfunction of the Hamiltonian

$$\hat{H}_{sr\text{-KS}}^{lr} = V_{\text{H}}^{sr}[\rho] + V_{\text{xc}}^{sr}[\rho] + T + V_{\text{ne}} + V_{\text{ee}}^{lr}, \tag{6}$$

where $V_{\text{H}}^{sr} = \dfrac{\delta J^{sr}}{\delta\rho}$ (*sr* Hartree potential) and $V_{\text{xc}}^{sr} = \dfrac{\delta E_{\text{xc}}^{sr}}{\delta\rho}$ (*sr* exchange-correlation potential) are density- and μ-dependent one-electron potentials supplementing the long-range Hamiltonian $T + V_{\text{ne}} + V_{\text{ee}}^{lr}$. Standard *ab initio* methods of quantum chemistry can be applied to this problem with three modifications: a) all two-electron integrals must be calculated with the *lr* inter-electronic interaction instead of the full interaction; b) the additional one-electron potentials $V_{\text{H}}^{sr}[\rho]$ and $V_{\text{xc}}^{sr}[\rho]$ have to be included when calculating Ψ^{lr}; c) the ground-state energy E_0 must be calculated from equation (5) rather than from the expectation value of $\hat{H}_{sr\text{-KS}}^{lr}$. The evaluation of the two-electron integrals over Gaussian-type orbitals (GTO) is easily possible by means of a slight modification in standard quantum-chemical codes, see e.g. Ref. [4].

2.2 Short-range functionals

Short-range μ-dependent functionals using the local density approximation have been available for quite some time [3,5,6]. The construction of these functionals is based on calculations where the energy per particle of the homogeneous electron gas is determined for full and long-range interaction. According to equation (3) the difference between those two defines the energy per particle of the short-range functional.

In Refs. [9,19] we describe the construction of *sr*-GGA functionals. In order to obtain non-empirical functionals we just used a minimum set of parameters adjusted to physical constraints. For the *sr*-GGA we chose the functional form so as to optimally represent the exchange and the correlation energy of the helium atom.

For the construction of a *sr*-meta-GGA functional we used as an additional physical condition the exchange and the exchange-correlation energy of the hydrogen atom [11].

3. Results

Within this section we shortly recapitulate the results of our previous papers [9,11,19,20,27,28]. Computational details can be found there. If not mentioned otherwise, we refer to calculations with triple-zeta quality basis sets and a fixed value of 0.5 $bohr^{-1}$ for the coupling parameter μ in the *sr*-DFT/*lr-ab initio* approaches (abbreviated as DFT/*ab initio* in the following). This value of μ has been chosen as a compromise between computational efficiency (low sensitivity with respect to basis-set size and to triples within the CC treatment, see below) and accuracy [19]. In many cases, accuracy can be further improved by choosing higher μ values, but the accuracy for $\mu \rightarrow \infty$ is eventually limited by the basis-set quality and *ab-initio* correlation treatment applied.

3.1 Thermochemistry

As a suitable test set for the application of the hybrid methods in thermochemistry, the G2 set of molecules has been used, see Ref. [11]. As is well known, pure LDA shows quite a poor performance, greatly overestimating atomization energies. GGA functionals like PBE [30] give much better results which can still be improved by using meta-GGA ones like TPSS [31].

Admixture of *lr*-HF to the *sr* density functionals leads to a significant amelioration with respect to LDA and HF: in fact, LDA/HF even outperforms standard PBE (cf. Figure 1). However, in contrast to standard DFT, mean absolute deviations (MAD) for PBE/HF and TPSS/HF turn out to be twice as large as for LDA/HF. In order to achieve high accuracy results, it is - analogously to pure *ab initio* methods - inevitable to employ post-HF methods on the long-range side, such as *lr*-CC with single and double excitations and perturbative triples (*lr*-CCSD(T)). This reconstitutes the correct order of functionals in compliance with Jacob's ladder and leads to increasingly higher accuracy for the PBE/CCSD(T) and TPSS/CCSD(T) results as compared to the LDA/CCSD(T) ones.

Substituting *lr*-CCSD(T) by *lr*-CCSD does not affect the quality of the results: the MAD changes by 0.2 kcal/mol only. This shows that *lr* correlation - though playing a crucial role in the hybrid methods - needs not be included at the highest possible level. On the contrary, we find that the triples can be safely neglected for the computation of atomization energies, at least at such low values of μ as we typically use.

We find that LDA/CCSD results are of similar quality as PBE and CCSD results (see Figure 1), and PBE/CCSD results exceed in accuracy CCSD(T) (with triple-zeta basis set) and TPSS results, whereas TPSS/CCSD reaches the

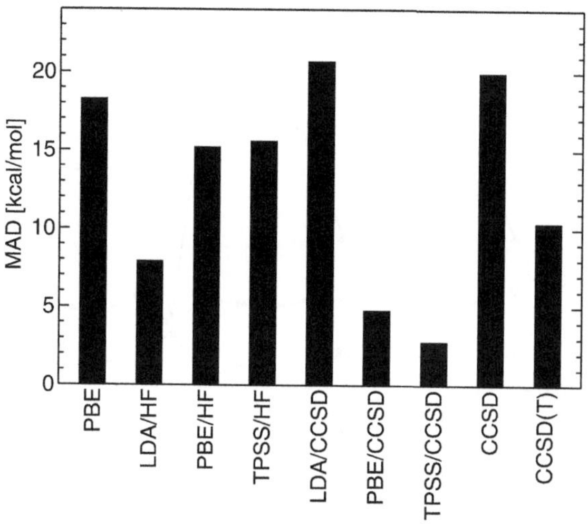

Fig. 1. Mean absolute deviations (MAD), with respect to experiment, of the atomization energies [kcal/mol] of the G2 set of molecules, in PBE, LDA/HF, PBE/HF, TPSS/HF, LDA/CCSD, PBE/CCSD, TPSS/CCSD, CCSD, and CCSD(T) calculations with aug-cc-pVTZ basis.

quality of global hybrid methods like B3LYP (MADs are 3.8 and 2.9 kcal/mol for B3LYP and TPSS/CCSD, respectively). However, one should keep in mind that B3LYP had empirically been adjusted to the original G2 set and thus has an advantage here.

The basis-set dependence of the hybrid methods has been studied in Ref. [11], by considering changes arising when enhancing the basis-set quality from triple-zeta to quadruple-zeta level. We find that admixture of DFT distinctly reduces the basis-set dependence of the *ab initio* methods. The basis-set dependence of DFT/CCSD and DFT/CCSD(T) is nearly identical with that of pure DFT. Actually, TPSS/CCSD with triple-zeta basis sets yields atomization energies superior to those of CCSD(T) with quadruple-zeta basis sets. This underlines the advantages of an implicit DFT treatment of *sr* correlation effects which neither requires extended basis sets nor high-order substitutions in the wavefunction.

For the ionization potentials, electron affinities, and proton affinities, the hybrid methods are rather similar to pure DFT and are surpassed in accuracy by pure CCSD(T). Thus, until the development of still better *sr* density functionals CCSD(T) should be used for an accurate determination of these properties [11].

3.2 Molecular properties

Standard density functionals may encounter difficulties with properties of transition metal compounds, cf. Ref. [27] and references therein. For example, DFT overestimates the dipole moments of group 11 hydrides and halides by roughly 1 D. Pure HF shows a similarly low quality.

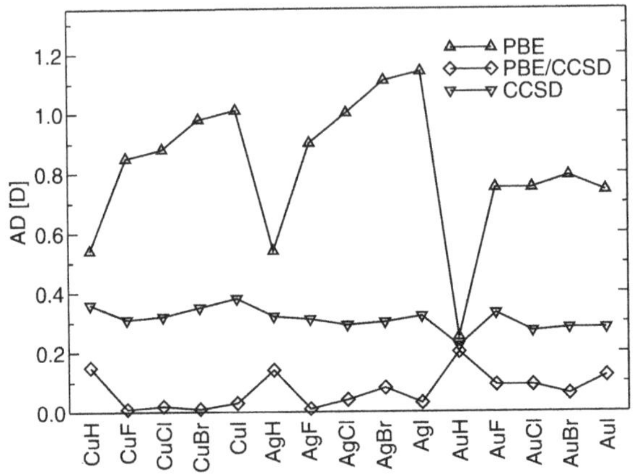

Fig. 2. Absolute deviations, with respect to basis-set extrapolated CCSD(T) values, of the dipole moments of group 11 hydrides and halides, in PBE, PBE/CCSD, and CCSD calculations with (aug)-cc-pVTZ basis.

Mixing of DFT and HF reduces the mean absolute deviation (MAD) and the standard deviation of the error ($\sigma(\Delta)$) with respect to both methods (to ~ 0.5 D for MAD). The importance of long-range correlation turns out when substituting long-range HF by long-range CCSD(T), which once again boosts the accuracy of the results, to ~ 0.1 D.

Though long-range correlation does have a significant influence on the results, use of perturbative triples in the mixed coupled cluster method does not: MAD and $\sigma(\Delta)$ are virtually the same, the difference between standard CCSD and CCSD(T) MADs on the other hand is over 0.2 D.

For the influence of the short-range density functional, we find that the differences between LDA/*ab initio* and PBE/*ab initio* MAD lie below 0.05 D, in contrast to pure DFT methods, where the MAD differences lie around 0.15 D.

PBE/CCSD clearly outperforms the two borderline methods PBE and CCSD, see Figure 2. In order to exceed results from DFT/CCSD calculations with triple-zeta basis sets, CCSD(T) with quadruple-zeta basis is needed, i.e., the *sr*-DFT functional accounts for basis-set effects beyond triple-zeta as well as for high-order substitutions.

Standard DFT also has problems in determining the polarizability of H_2 chains, see e.g. Ref. [32]. We performed calculations analogous to those of Ref. [32] for chains of n (n = 1–10) linearly arranged hydrogen molecules [33]. From Figure 3 it is clearly seen that the PBE results can be systematically improved towards the CCSD(T) limit (μ = 100) by mixing-in CCSD(T). For μ = 2 this limit is almost reached. Pure LDA expectedly yields polarizabilities even worse than pure PBE ones. But at μ = 0.25, LDA/CCSD(T) and PBE/CCSD(T)

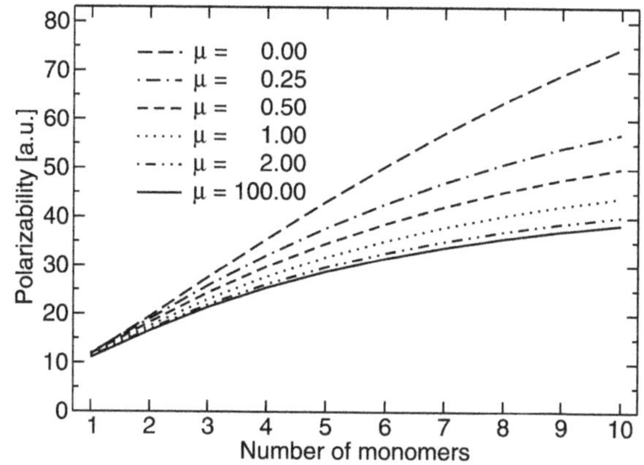

Fig. 3. Polarizability per molecule [a.u.] for hydrogen chains of different length (with intermolecular distances of 2.5 bohr), from PBE/CCSD(T) calculations at different values of μ with cc-pVTZ basis.

results already approach each other, and for $\mu = 0.5$ they even become virtually identical.

It is known that for this problem HF delivers results of comparable quality as sophisticated, wavefunction-based *ab initio* methods. Thus, it does not come as a surprise that substituting short-range DFT exchange by short-range exact exchange gives reasonable results independently of μ. For the same reason it is not astonishing that not only DFT/CCSD and DFT/CCSD(T) results come very close (as they regularly do), but also DFT/HF and DFT/CCSD(T) ones.

3.3 Noncovalent interactions

Noncovalent interactions are usually not completely described within the density functional theory framework. While charge transfer interactions and interactions between multipoles (Keesom interaction) or between multipoles and induced multipoles (Debye interaction) are still partially covered, density functional theory with local or semilocal density functionals (exactly like Hartree-Fock theory) is not suitable to describe interaction between induced multipoles (London interaction) at all.

A sound description of noncovalent interactions is mandatory, for example, for the quantum chemical computation of biological systems or catalytically active substances. Let us now review our results from papers on rare gas dimers [9], alkali-metal rare-gas dimers [19], and on different selected molecules [20,28] which are included in the test sets of Zhao and Truhlar [34] and of Hobza and coworkers [35].

For the computation of the rare gas dimers we applied a different μ than our standard value of 0.5 bohr^{-1}. Based on the idea that intraatomic effects should be mainly treated by DFT and interatomic effects mainly by *ab initio*, we chose

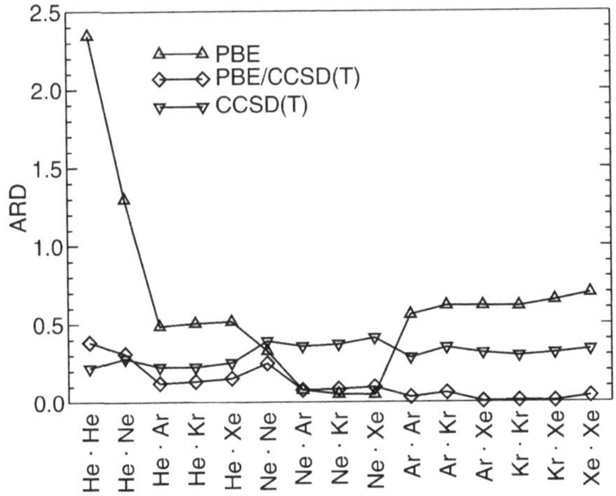

Fig. 4. Absolute relative deviations, with respect to experiment, of the dissociation energies of Rg · Rg' (Rg, Rg' = He, ..., Xe), in PBE, PBE/CCSD(T), and CCSD(T) calculations with aug-cc-pVTZ basis.

$\mu = 2/R_e$ (which equals $\mu = 1/R_{vdW}$ for homonuclear dimers). This leads to μ values between 0.24 and 0.36 bohr^{-1} which seems to be a reasonable choice for these simple model systems, as we see from results with the PBE/CCSD(T) method. With triple-zeta basis sets it yields much smaller MADs for bond lengths, dissociation energies, and harmonic wavenumbers than its borderline methods PBE and CCSD(T) [9] (the absolute relative deviations of the dissociation energies for PBE, PBE/CCSD(T), and CCSD(T) are shown in Figure 4). This is even true when using quadruple-zeta basis sets, except for the harmonic wavenumbers, where the standard CCSD(T) values become better than the PBE/CCSD(T) ones. The basis-set dependencies of the mixed method and the pure DFT methods are in fact very similar, i.e., both are close to zero, which means that CCSD(T) requires quintuple-zeta basis sets to match the quality of the triple-zeta PBE/CCSD(T) bond lengths and dissociation energies. Another interesting feature of PBE/CCSD(T) shows up here again: perturbative triples can be omitted without any loss of quality. Moreover, also the use *lr*-MP2 instead of *lr*-CC only slightly affects the quality of the results. (Note, however, that leaving out *lr* correlation altogether, in DFT/HF calculations, leads to repulsive potential curves [17]. Thus, as stated above, *lr* correlation is *not* negligible.).

For Ne · Ne we now check - turning back to our standard value of μ - whether the hybrid method describes rare gas dimers well also for nuclear distances far from equilibrium. Figure 5 displays the counterpoise corrected potential curve of the neon dimer between 3.5 and 6.4 bohr, calculated with different methods and basis sets. Parts of the potential curve have been amplified. Thereby one recognizes that PBE/CCSD(T) with aug-cc-pVTZ basis is able to reproduce CCSD(T) with aug-cc-pVQZ basis in the region of the (very flat) minimum, for

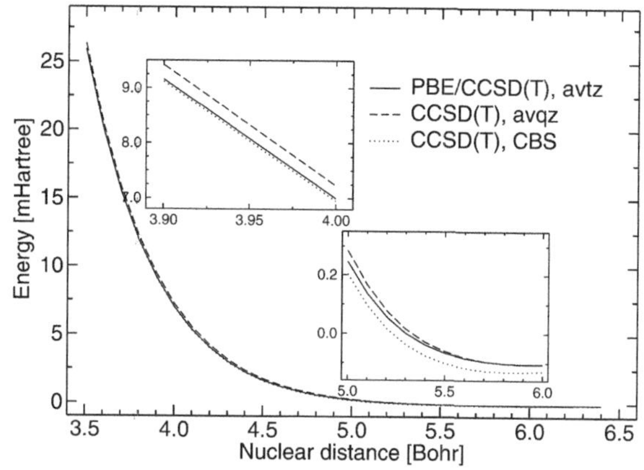

Fig. 5. Counterpoise corrected potential curve of the neon dimer, for PBE/CCSD(T) with aug-cc-pVTZ basis as well as for CCSD(T) with aug-cc-pVQZ basis and at the basis set limit (parts of the curve are magnified).

shorter bond lengths it even approaches the CCSD(T) basis set limit (CBS[45]).

The hybrid methods are not only useful for closed-shell, but likewise for open-shell systems, as we could show for the alkali-metal rare-gas dimers. For these, we made very similar observations as for the rare gas dimers [19]. The gain due to the use of hybrid methods is not as pronounced as for those, but still clearly visible.

In order to test the interaction of systems other than atoms we implemented local [20] and density fitting [28] methods. With these, we tested the performance of the hybrid methods in weak interaction between molecules like, for example, ethane, which is satisfying for these cases [20]. The results for stacking are of lower accuracy [28]; nevertheless, the hybrid methods still have smaller mean absolute deviations than either pure DFT or pure *ab initio* - except PBE/HF, which is worse than PBE [20,28].

We also find that interactions between multipoles and charge-transfer interactions are quite well-described by PBE/CCSD, and mostly are superior to both, PBE and CCSD (for the multipole interactions, PBE and PBE/CCSD errors are quite similar, because PBE shows an unexpectedly good performance here).

The hydrogen bonds are the most problematic case for the hybrid methods, and results do not improve but rather worsen here. This is a case where standard PBE is extremely accurate (i.e., even more accurate than CCSD(T) with triple-zeta basis set); the sr-PBE functional cannot maintain this accuracy for $\mu \neq 0$, yielding results comparable in accuracy to CCSD only. Moreover, our value of μ which is well-suited for weakly bound systems with large interaction distances, is probably not optimum for hydrogen bonds with shorter distances. Calculations with larger μ ($\mu = 1$ and $\mu = 2$) significantly improve the accuracy here [28].

4. Discussion and summary

Reviewing all the material mentioned above we find that for the μ values used in our calculations the hybrid methods have a basis-set dependence similar to the pure DFT method used. This distinctly reduces the basis-set dependence with respect to the pure *ab initio* methods except for the computation of the group 11 hydrides and halides, where standard DFT is rather basis-set dependent.

We also observe that the hybrid approach is less sensitive to methodological details than the corresponding borderline methods. This means, on the one hand, that the dependency on the functional used is smaller than in pure density functional theory, and, on the other hand, that the dependency on the correlation method used is smaller than in pure *ab initio* theory. The latter findings especially apply to the case of coupled-cluster methods, where in the hybrid approach the perturbative triples of CCSD(T) can be omitted without any loss of accuracy, which is in contrast to the importance of perturbative triples in pure coupled-cluster calculations. The dependency on the density functional is nearly zero for noncovalent interactions, in particular for London interactions; however, for hydrogen bonds and especially for atomization energies of molecules with strong chemical bonds, the choice of the functional affects the quality of the results. Development of improved *sr* density functionals is currently underway in our laboratory.

Regarding the mean absolute relative deviations (MARDs) we find that for the atomization energies of the G2 set of molecules the MARD reaches 1.0 % with TPSS/CCSD(T), using triple-zeta basis sets at $\mu = 0.5$. For the dissociation energies of noncovalent systems the MARD of PBE/CC roughly lies around 10 %, which actually is more systematic than, for example, the MARD of PBE, MP2, or CCSD.

References

1. H. Stoll and A. Savin, in Density Functional Methods in Physics. R. Dreizler and J. da Providencia (Eds.), Plenum, New York (1985) p. 177.
2. A. Savin and H.-J. Flad, Int. J. Quantum Chem. **56** (1995) 327.
3. A. Savin, in Recent Developments and Applications of Modern Density Functional Theory. J. M. Seminario (Ed.), Elsevier, Amsterdam (1996) p. 327.
4. T. Leininger, H. Stoll, H.-J. Werner, and A. Savin, Chem. Phys. Lett. **275** (1997) 151.
5. J. Toulouse, A. Savin, and H.-J. Flad, Int. J. Quantum Chem. **100** (2004) 1047.
6. S. Paziani, S. Moroni, P. Gori-Giorgi, and G. B. Bachelet, Phys. Rev. B **73** (2006) 155111.
7. H. Iikura, T. Tsuneda, T. Yanai, and K. Hirao, J. Chem. Phys. **115** (2001) 3540.
8. J. Toulouse, F. Colonna, and A. Savin, J. Chem. Phys. **122** (2005) 014110.
9. E. Goll, H.-J. Werner, and H. Stoll, Phys. Chem. Chem. Phys. **7** (2005) 3917.
10. O. A. Vydrov, J. Heyd, A. V. Krukau, and G. E. Scuseria, J. Chem. Phys. **125** (2006) 074106.
11. E. Goll, M. Ernst, F. Moegle-Hofacker, and H. Stoll, J. Chem. Phys. **130** (2009) 234112.

12. Y. Tawada, T. Tsuneda, S. Yanagisawa, T. Yanai, and K. Hirao, J. Chem. Phys. **120** (2004) 8425.
13. O. A. Vydrov and G. E. Scuseria, J. Chem. Phys. **125** (2006) 234109.
14. E. Fromager, J. Toulouse, and H. J. A. Jensen, J. Chem. Phys. **126** (2007) 074111.
15. J. Toulouse, I. C. Gerber, G. Jansen, A. Savin, and J. G. Ángyán, Phys. Rev. Lett. **102** (2009) 096404.
16. B. G. Janesko, T. M. Henderson, and G. E. Scuseria, J. Chem. Phys. **130** (2009) 081105.
17. J. G. Ángyán, I. C. Gerber, A. Savin, and J. Toulouse, Phys. Rev. A **72** (2005) 012510.
18. R. Pollet, A. Savin, T. Leininger, and H. Stoll, J. Chem. Phys. **116** (2002) 1250.
19. E. Goll, H.-J. Werner, H. Stoll, T. Leininger, P. Gori-Giorgi, and A. Savin, Chem. Phys. **329** (2006) 276.
20. E. Goll, H.-J. Werner, and H. Stoll, Chem. Phys. **346** (2008) 257.
21. T. Yanai, D. P. Tew, and N. C. Handy, Chem. Phys. Lett. **393** (2004) 51.
22. P. Pulay, Chem. Phys. Lett. **100** (1983) 151.
23. S. Saebø and P. Pulay, J. Chem. Phys. **86** (1987) 914.
24. S. Saebø and P. Pulay, J. Chem. Phys. **88** (1988) 1884.
25. M. Schütz, G. Hetzer, and H.-J. Werner, J. Chem. Phys. **111** (1999) 5691.
26. M. Schütz and H.-J. Werner, J. Chem. Phys. **114** (2001) 661.
27. E. Goll, H. Stoll, C. Thierfelder, and P. Schwerdtfeger, Phys. Rev. A **76** (2007) 032507.
28. E. Goll, T. Leininger, F. R. Manby, A. Mitrushchenkov, H.-J. Werner, and H. Stoll, Phys. Chem. Chem. Phys. **10** (2008) 3353.
29. M. Levy, Proc. Natl. Acad. Sci. USA **76** (1979) 6062.
30. J. P. Perdew, K. Burke, and M. Ernzerhof, Phys. Rev. Lett. **77** (1996) 3865.
31. J. Tao, J. P. Perdew, V. N. Staroverov, and G. E. Scuseria, Phys. Rev. Lett. **91** (2003) 146401.
32. M. van Faassen, P. L. de Boeij, R. van Leeuwen, J. A. Berger, and J. G. Snijders, Phys. Rev. Lett. **88** (2002) 186401.
33. E. Goll, Kopplung von Dichtefunktional- und ab-initio-Methoden. PhD thesis, Universität Stuttgart (2008).
34. Y. Zhao and D. G. Truhlar, J. Chem. Theor. Comp. **1** (2005) 415.
35. P. Jurecka, J. Spohner, J. Cerný, and P. Hobza, Phys. Chem. Chem. Phys. **8**, (2006) 1985.

Benchmark Studies for Explicitly Correlated Perturbation- and Coupled Cluster Theories

By Hans-Joachim Werner*, Gerald Knizia, Thomas B. Adler, and Oliver Marchetti

Institut für Theoretische Chemie, Universität Stuttgart, Pfaffenwaldring 55, 70569 Stuttgart, Germany

(Received August 3, 2009; accepted September 1, 2009)

Explicit Correlation / Local Correlation / Coupled-cluster Methods / F12-methods

The recently developed explicitly correlated MP2-F12 and CCSD(T)-F12x (x = a,b) methods are reviewed. The explicit correlation treatment leads to a dramatic improvement of the basis set convergence. Extensive benchmarks for reaction energies, atomization energies, electron affinities, ionization potentials, equilibrium structures, vibrational frequencies, and intermolecular interaction energies are presented which show that for many molecular properties the intrinsic accuracy of the CCSD(T) method is already reached with double-zeta (VDZ-F12) basis sets, while triple-zeta (VTZ-F12) basis sets yield results that are very close to the complete basis set limit. The steep scaling of the MP2-F12 method with molecular size can be reduced by local approximations. This has made it possible to carry out MP2-F12 calculations for molecules with up to 100 atoms. The errors caused by the local domain approximation are largely removed by the explicitly correlated terms, which account for the neglected configurations in an approximate way. Extensions to LCCSD(T)-F12 are discussed and preliminary results for LCCSD-F12 are presented.

1. Introduction

The coupled-cluster method with single and double excitations and a perturbative treatment of triple excitations [CCSD(T)] is nowadays considered as the golden standard of quantum chemistry. As long as a single reference treatment is sufficient, it yields highly accurate results for many properties, and chemical accuracy can be reached for energy differences such as reaction enthalpies [1,2]. However, the method suffers from two major problems: the very steep scaling of the computational cost with increasing molecular size, and the extremely slow conver-

* Corresponding author. E-mail: werner@theochem.uni-stuttgart.de

Z. Phys. Chem. **224** (2010) 493–511
© by Oldenbourg Wissenschaftsverlag, München

gence of the correlation energy with basis set size. The latter problem is severe even for very small molecules, since the computational cost increases with the fourth power of the basis set size N_{AO}, but the errors in the energy decrease only with N_{AO}^{-1}.

The slow convergence of the correlation energy with basis set size is due to the fact that the wavefunction cusp for $r_{12} \rightarrow 0$ cannot be described by an expansion in products of one-electron functions (orbitals). Already in 1929 Hylleraas realized for the He atom that this problem can be avoided by including terms into the wave function that depend explicitly on the interelectronic distances r_{12}. However, extensions to many-electron systems proved difficult since then very numerous many-electron integrals appear in the formalism. A way out of this problem was opened almost 25 years ago by the seminal work of Kutzelnigg [3]. He proposed to use resolutions of the identity (RIs) in order to factorize the many-electron integrals into sums of products of two-electron integrals, which are much easier to evaluate than the three-electron integrals.

A further breakthrough came when it was realized that a Slater-type function $F_{12} = \exp(-\gamma r_{12})$ leads to much better basis set convergence and numerical stability than the linear r_{12} factor. This was first proposed and demonstrated by Tenno [4]. Shortly afterwards several other MP2-F12 implementations followed [5–12]. Furthermore, new orbital [13] and RI [14,15] basis sets were optimized that are specially designed for F12 calculations. Many benchmark calculations with the MP2-F12 method showed that with triple-zeta basis sets quintuple zeta or even better accuracy can be obtained [8,10–12,16,17].

Most of the above mentioned methods were limited to MP2, which is not sufficiently accurate for reliable predictions. Extensions to coupled-cluster theory are therefore in the current focus. Formally complete treatments of CCSD-R12 or CCSD-F12 methods [18–25] are computationally two orders of magnitude more expensive than the standard CCSD method. In order to reduce the computational effort, Fliegl et al. [26,27] introduced simplified CC2(R12) and CCSD(T)(R12) methods, which have more recently been extended to CCSD(T)(F12) by Tew et al. [28,29]. In these treatments terms that are nonlinear in the amplitudes of the explicitly correlated configurations are neglected. Even simpler CCSD(T)-F12 approximations, denoted CCSD(T)-F12a and CCSD(T)-F12b, were proposed and implemented by our group [30,31]. As will be demonstrated in section 4 these methods yield excellent results, without much extra cost as compared to a standard CCSD(T) calculation with the same basis set. Furthermore, local approximations were introduced to extend the applicability of explicitly correlated methods to larger systems [10,32–34].

This paper summarizes our work performed within the priority program 1145 *Modern and universal first-principles methods for many electron systems in chemistry and physics* of the Deutsche Forschungsgemeinschaft. The methods were described in detail in Refs. [10–12,30–34]. Further benchmarks and applications were presented in Refs. [35–38].

In section 2 we will briefly outline the wavefunction ansätze and approximations of our MP2-F12 and CCSD(T)-F12x methods. The local approximations will be described in section 3. In section 4 we will present benchmarks for atomization energies, reaction energies, ionization potentials, electron affinities, intermolecular interactions, vibrational frequencies and equilibrium geometries which demonstrate that the intrinsic accuracy of the CCSD(T) method is already reached with double-zeta or triple-zeta basis sets.

2. Methods

The methods used in this paper have been implemented in the MOLPRO package of ab initio programs [39]. For closed-shell systems, MP2-F12 and CCSD(T)-F12 programs exist. For high-spin open-shell systems RMP2-F12 (spin restricted MP2), RCCSD(T)-F12 (partially spin adapted coupled cluster [40]) and UCCSD(T)-F12 (spin unrestricted coupled cluster) methods are available. All open-shell methods use RHF (spin-restricted Hartree-Fock) wavefunctions. In this section, we only outline the closed-shell methods [11,30,31]. The modifications needed in the open-shell case are described in Refs. [12,31].

In the following, the indices i,j,k,l denote correlated valence orbitals, o all occupied orbitals, a,b virtual orbitals, and α,β a complete orthonormal set of orbitals (the molecular orbitals are a subset). Summation over repeated dummy indices will be generally implied.

2.1 Explicitly correlated MP2 and CCSD wavefunctions

The MP2-F12 and CCSD-F12 wavefunction are defined as

$$\Psi_{\text{MP2-F12}} = (1 + \hat{T}_2)|\Psi_{\text{HF}}\rangle, \tag{1}$$

$$\Psi_{\text{CCSD-F12}} = \exp(\hat{T}_1 + \hat{T}_2)|\Psi_{\text{HF}}\rangle, \tag{2}$$

where $\hat{T}_1$ and $\hat{T}_2$ are one and two particle excitation operators:

$$\hat{T}_1 = t_a^i \hat{E}_{ai}, \tag{3}$$

$$\hat{T}_2 = \frac{1}{2}\left[T_{ab}^{ij}\hat{E}_{ai}\hat{E}_{bj} + \mathcal{T}_{\alpha\beta}^{ij}\hat{E}_{\alpha i}\hat{E}_{\beta j}\right], \tag{4}$$

with

$$\mathcal{T}_{\alpha\beta}^{ij} = \langle\alpha\beta|\hat{Q}_{12}F_{12}|kl\rangle T_{kl}^{ij}. \tag{5}$$

$\hat{E}_{\alpha i} = \eta_\alpha^\dagger \eta_i + \bar{\eta}_\alpha^\dagger \bar{\eta}_i$ are spin-summed one-electron excitation operators. The operator $\hat{T}_1$ and the first part of $\hat{T}_2$ are the same as in the standard CCSD. The second term in eq. (4) adds the explicitly correlated contributions to the MP2 or CCSD wavefunctions. Formally, these can be viewed as double excitations into a complete orthonormal virtual orbital basis. However, the amplitudes $T_{\alpha\beta}^{ij}$ are not independently varied, but contracted as defined in eq. (5). They depend on a small set of amplitudes T_{kl}^{ij}, which can either be optimized or determined from the wavefunction cusp conditions (see below).

F_{12} is the r_{12}-dependent correlation factor, where γ is a length-scale parameter which typically is chosen to be 1.0 a_0^{-1}. $\hat{Q}_{12}$ is a strong orthogonality projector which keeps the explicitly correlated terms orthogonal to the Hartree-Fock (HF) reference function and the conventional expansion of single and double excitations:

$$F_{12} = -\frac{1}{\gamma}e^{-\gamma r_{12}}, \tag{6}$$

$$\hat{Q}_{12} = (1-\hat{o}_1)(1-\hat{o}_2) - \hat{v}_1\hat{v}_2, \tag{7}$$

$$\hat{o} = |o\rangle\langle o|, \qquad \hat{v} = |a\rangle\langle a|. \tag{8}$$

The one-electron operators $\hat{o}$ and $\hat{v}$ project onto the occupied and virtual orbital spaces, respectively. Their subscripts denote the electron coordinate on which they act. Due to the projector the amplitudes $T_{\alpha\beta}^{ij}$ are zero if either α or β or both correspond to an occupied orbital, or if both α and β are virtual orbitals.

If this ansatz is used to derive the expressions for the MP2-F12 and CCSD-F12 amplitude equations and energy expressions, the one-electron projectors in eq. (7) lead to 3- and 4-electron integrals. As first proposed by Kutzelnigg [3], these can be avoided by inserting (approximate) resolutions of the identity (RIs) $\hat{u}$ into the projector $\hat{Q}_{12}$:

$$\hat{Q}_{12} \approx 1 - \hat{o}_1\hat{u}_2 - \hat{u}_1\hat{o}_2 + \hat{o}_1\hat{o}_2 - \hat{v}_1\hat{v}_2. \tag{9}$$

In practice, the RIs are approximated using the union of the orbital basis and an auxiliary basis set (CABS approach [41]). When this form of the projector is used, all many electron integrals factorize into sums of products of two-electron integrals (see, e.g., Ref. [16] for a review and Ref. [11] for details).

In principle, the range of orbitals *k,l* in eq. (5) is arbitrary and defines a wavefunction ansatz. In our methods, *k,l* run over all valence orbitals (in the open-shell cases with RHF reference functions, *k,l* run over all valence spatial orbitals that are either occupied with alpha or beta spin). With this ansatz, the optimized wavefunction is invariant with respect to unitary transformations among the closed-shell or open-shell valence orbitals (e.g., orbital localization). However, it has been found that this ansatz can lead to problems due to geminal basis superposition effects (GBSSE) [7]. Furthermore, when the orbitals *k,l* are far away from *i,j*, near singularities of the zeroth-order Hamiltonian may arise, which can lead to serious numerical problems in larger molecules. These problems can be avoided by using the "diagonal" (D) ansatz, in which *kl* is restricted to *ij* or *ji*:

$$T_{kl}^{ij} \stackrel{\mathrm{D}}{=} \delta_{ik}\delta_{jl}T_{ij} + \delta_{jk}\delta_{il}T_{ji}. \tag{10}$$

In most cases, this restricted ansatz gives even more accurate results than the full ansatz [11,12]. However, it is not orbital invariant and only size consistent if localized orbitals are used. Ten-no proposed to use fixed coefficients $T_{ij}^S = 1/2$ and $T_{ij}^T = 1/4$ for all singlet and triplet pairs, respectively [4,42]. These coefficients follow from the wavefunction cusp conditions [43,44]. In the current formalism, which does not use singlet and triplet coupling of the external electrons, this means

$$T_{kl}^{ij} \overset{\text{FIX}}{=} \frac{3}{8}\delta_{ik}\delta_{jl} + \frac{1}{8}\delta_{jk}\delta_{il}. \tag{11}$$

The fixed amplitude ansatz avoids GBSSE as well as singularities, and is orbital invariant and size consistent. Furthermore, it has the advantage that no amplitude equations have to be solved for the coefficients T_{kl}^{ij}. This ansatz is therefore used in the CCSD(T)-F12x methods described in the next section.

2.2 The CCSD(T)-F12x approximations

The exact CCSD-F12 equations [18–25] involve numerous multiple RI approximations, and the computational effort scales cubically with the size of the auxiliary basis set. Fortunately, the dependence on the auxiliary basis set can be entirely removed from the CCSD-F12 equations without much loss of accuracy. This is achieved in the CCSD-F12x methods [30,31] using the following approximations: (i) The fixed amplitude ansatz in eq. (11) is used. Alternatively, the amplitudes T_{kl}^{ij} can be determined in the preceding MP2-F12 calculation and then kept fixed in the CCSD-F12. (ii) All non-linear terms in the amplitude equations that depend on the amplitudes T_{kl}^{ij} are neglected. (iii) In the remaining terms, all contributions of the complementary auxiliary orbitals are neglected, except in the terms containing the Fock operator. The latter are the same as in the MP2-F12 residual and are treated exactly as in MP2-F12 (but using CCSD-F12 amplitudes). These approximations lead to great simplifications, but still include the most important couplings between the conventional amplitudes T_{ab}^{ij} and the explicitly correlated contributions. In the first step, a MP2-F12 calculation is performed. The computational effort for the subsequent CCSD-F12x iterations is virtually the same as for standard CCSD. A (T) perturbative energy correction for triple excitations can be added in the standard way, using the CCSD-F12x amplitudes. This correction is not explicitly correlated, and therefore the F12 treatment does not reduce the basis set incompleteness error of the triples (there is a small indirect effect by the singles and doubles amplitudes, which are affected by the coupling terms in the CCSD-F12x equations). A proper treatment of explicitly correlated (T) corrections, as recently proposed by Köhn [45], is complicated but may be considered at a later time. For the time being, we optionally correct for the basis set error of the triples approximately by scaling the triples energy as [31,38]

$$\Delta E_{(\mathrm{T}^*)} \quad = \quad \Delta E_{\mathrm{T}} \cdot \frac{E_{corr}^{\mathrm{MP2-F12}}}{E_{corr}^{\mathrm{MP2}}}. \tag{12}$$

This assumes that the relative basis set error is similar for the MP2 and (T) correlation energy contributions. The methods with scaled triples contribution are denoted by a star behind the T, i.e., CCSD(T*)-F12x.

The CCSD-F12a and CCSD-F12b methods differ only in the approximation used to compute the correlation energy. The CCSD-F12b method includes an additional coupling term in the energy expression, which usually slightly reduces the correlation energy (for details see Refs. [30, 31]). Extensive benchmarks have shown that CCSD-F12a somewhat overestimates the exact CCSD-F12 correlation energy, and this leads to a favorable and systematic error compensation when small (double or triple zeta) basis sets are used. The CCSD-F12b energy expression is closer to exact CCSD-F12 but slightly underestimates it. CCSD-F12b is recommended for large basis sets (quadruple-zeta or larger).

2.3 Perturbative CABS singles correction

In MP2-F12 and CCSD-F12 calculations the basis set errors of the Hartree-Fock energy are often much larger than the errors of the correlation energy. The error of the HF contribution can be reduced by including single excitations into the complementary auxiliary (CA) orbital space and by computing the second-order energy contribution perturbatively[12,30]. For closed-shell cases the CABS singles amplitudes are obtained by solving

$$0 = f^i_{\alpha'} + f_{\alpha'\beta'}t^i_{\beta'} - f_{ik}t^k_{\alpha'}, \tag{13}$$

where $f_{\alpha'\beta'}$ are the matrix elements of the closed-shell Fock operator and the indices α', β' run over the virtual and CA orbitals. This equation can easily be solved by block diagonalizing the Fock matrix in the occupied and virtual+CA spaces. The energy correction is then computed as

$$\Delta E_s = 2t^i_{\alpha'}f^i_{\alpha'}. \tag{14}$$

A spin-free extension for high-spin open-shell cases has been discussed in Ref. [12].

We consider the CABS singles correction as a correction to the HF energy and not as a correlation effect. The Fock matrix elements $f_{\alpha'\beta'}$ and $f_{\alpha' i}$ are needed in MP2-F12 calculations anyway and therefore the additional effort for this correction is negligible. It should be noted that this energy correction is entirely distinct from the CCSD-F12 calculations, and the singles amplitudes $t^i_{\alpha'}$ used here are not the same as those in the CCSD-F12.

3. Explicitly correlated local methods

Local correlation methods [46–48] make use of the short range character of dynamic electron correlation. In order to exploit this fact, a local orbital basis has to be used. The correlated (valence) Hartree-Fock orbitals (LMOs) are localized by a standard localization method. In the current work, we use Pipek-Mezey localization [49], but Boys localization [50] or natural localized orbitals (NLMOs) can also be employed in our program [51]. The choice of the localization method has usually only a small effect on the results. Since the orthogonal virtual orbitals cannot be well localized, projected atomic orbitals (PAOs) are used to span the virtual space [46–48]. In the following, PAOs will be denoted by a tilde over the orbital index.

3.1 LMP2-F12

In the LMP2-F12 method the excitation operator $\hat{T}_2$ is defined as

$$\hat{T}_2 = \frac{1}{2}\left[\sum_{ij\in P_d}\sum_{\tilde{a},\tilde{b}\in[ij]} T^{ij}_{\tilde{a}\tilde{b}}\hat{E}_{\tilde{a}i}\hat{E}_{\tilde{b}j} + \sum_{ij\in P_c}\sum_{\alpha,\beta} \mathcal{T}^{ij}_{\alpha\beta}\hat{E}_{\alpha i}\hat{E}_{\beta j}\right], \quad (15)$$

with

$$\mathcal{T}^{ij}_{\alpha\beta} = \sum_{kl}\langle\alpha\beta|\hat{Q}^{ij}_{12}\hat{F}_{12}|kl\rangle T^{ij}_{kl}. \quad (16)$$

As compared to the standard MP2-F12 ansatz, the following approximations are introduced:

(i) Orbital pairs ij where the LMOs i and j are very distant are neglected (*pair approximation*). This is indicated by the lists P_d and P_c. P_d is the list of all pairs that are included in the conventional LMP2 (strong, close, weak and distant pairs [48,52]). The list of explicitly correlated pairs P_c may optionally be smaller and exclude the weak and distant pairs.

(ii) The conventional double excitations are restricted to pair domains $[ij]$, which are spanned by subsets of PAOs that are spatially close to the LMOs from which the electrons are excited (*domain approximation*). The pair domain $[ij]$ is defined as the union of orbital domains $[i]$ and $[j]$. The orbital domain $[i]$ includes all PAOs located at a subset of atoms that is associated to the LMO i. Here we assume that the functions in each domain are orthonormalized and denoted by indices $\tilde{a}$, $\tilde{b}$.

(iii) Since the excitations are restricted to domains, a pair-specific projector $\hat{Q}^{ij}_{12}$[32] is used in eq. (16):

$$\hat{Q}^{ij}_{12} = (1-\hat{o}_1)(1-\hat{o}_2) - \sum_{\tilde{c},\tilde{d}\in[ij]} |\tilde{c}\tilde{d}\rangle\langle\tilde{c}\tilde{d}|. \quad (17)$$

The only difference to the ordinary projector in standard MP2-F12 or CCSD-F12 theory is that the summation in the last term is restricted to the pair domain $\tilde{c},\tilde{d} \in [ij]$. This is not an additional approximation but is implied by the local ansatz in the conventional part of the wave function.

Naturally, the restriction of the excitations to domains leads to a reduction of the computed correlation energy. This is denoted as *domain error* and amounts in conventional local correlation calculations to 1–2% of the correlation energy obtained with the same basis set in a non-local calculation. Even though this relative error is rather small, its absolute value can be significant and cause non-negligible errors when energy differences are computed. Due to the restriction of the summation in the last term of eq. (17), the domain error is to a large extent removed by the explicitly correlated terms. This is because the double excitations into virtual orbitals *outside* the domain [*ij*] are not entirely excluded as in standard local correlation methods, but for a pair *ij* implicitly approximated by

$$\Delta\hat{\mathcal{T}}_2^{ij} = \sum_{\tilde{a},\tilde{b}\notin[ij]} \bar{F}^{ij}_{\tilde{a}\tilde{b}} \hat{E}_{\tilde{a}i}\hat{E}_{\tilde{b}j}, \quad (18)$$

where $\bar{F}^{ij}_{\tilde{a}\tilde{b}} = \Sigma_{kl}\, T^{ij}_{kl} \langle kl \mid \hat{F}_{12} | \tilde{a}\tilde{b} \rangle$. This can be viewed as an externally contracted excitation operator, i.e., instead of fully optimized amplitudes T^{ij}_{ab} the fixed matrix elements $\bar{F}^{ij}_{\tilde{a}\tilde{b}}$ are used.

In order to make it possible to achieve low-order scaling in the local methods we use the diagonal (D) or fixed amplitude (FIX) ansatz combined with approximation 3*A(NOX). For details about this approximation see Refs. [11,33]. In order to reduce the scaling, local RI and local DF approximations can be introduced [33,34]. Local RI approximation imply that the CABS approach cannot be used. However, CABS can be simulated by using the union of the AO and RI basis sets to represent the resolution of the identities.

3.2 LCCSD-F12x

In the LCCSD-F12x method the excitation operators $\hat{T}_1$ and $\hat{T}_2$ are defined as

$$\hat{T}_1 = \sum_i \sum_{\tilde{a}\in[i]} t^i_{\tilde{a}} \hat{E}_{\tilde{a}i}, \quad (19)$$

$$\hat{T}_2 = \frac{1}{2} \sum_{ij\in P_s} \left[\sum_{\tilde{a},\tilde{b}\in[ij]} T^{ij}_{\tilde{a}\tilde{b}} \hat{E}_{\tilde{a}i}\hat{E}_{\tilde{b}j} + \sum_{\alpha\beta} \mathcal{T}^{ij}_{\alpha\beta} \hat{E}_{\alpha i}\hat{E}_{\beta j} \right]. \quad (20)$$

Only strong pairs, as defined by a list P_s, are included in $\hat{T}_2$. The remaining pairs are treated by LMP2-F12, as outlined in the previous section. The use of the pair specific projector automatically leads to linear scaling of the contributions of integrals over four PAOs [34]. The number of other integrals can be reduced by using local RI and local density fitting approximations, similar to LMP2-F12. Along with integral screening techniques this makes it possible to achieve a nearly linear overall scaling of the computational cost with molecular size. A detailed description of these approximations is beyond the scope of the current article and the reader is referred to the original publications [33,34].

4. Benchmarks

In several previous papers [11,12,31,35,36,38] we have presented benchmarks for atomization energies, reaction energies, ionization potentials, electron affinities, equilibrium distances, vibrational frequencies, and intermolecular interaction energies. All these calculations were performed with the augmented correlation consistent orbital basis sets (aug-cc-pV*n*Z) of Kendall et al. [53]. For the second-row atoms, we used the aug-cc-pV(*n*+*d*)Z basis sets of Dunning et al. [54]. For simplicity, the aug-cc-pV*n*Z and aug-cc-pV(*n*+*d*)Z basis sets will in the following be denoted AV*n*Z. As RI basis, we used the V*n*Z/JKFIT basis sets of Weigend [55], which turned out to work well for this purpose.

Recently, Peterson et al. optimized new atomic orbital (AO) and RI basis sets that are specifically designed for F12 calculations. In the current paper, we will repeat the benchmarks presented in Ref. [31] using these new basis sets. Except for the electron affinities, we will use the V*n*Z-F12 AO basis sets [13] along with the associated RI basis sets [14]. For the electron affinities, it turned out that the V*n*Z-F12 orbital basis sets are not well suited, probably due to the lack of sufficiently diffuse functions. We therefore computed the electron affinities using the aug-cc-pV*n*Z sets, along with the associated optimized RI basis sets that were published very recently [15].

The new OPTRI basis sets are slightly smaller than the JKFIT bases and are specially constructed for the CABS approach (they cannot be used without CABS). In particular, they contain less s and p functions, but higher angular momentum functions than in the JKFIT basis sets. Their main advantage is that they minimize the linear dependencies between the AO and the RI basis sets, which leads to improved numerical stability. For the density fitting, which is used to compute all integrals in the MP2-F12 part, the MP2FIT basis sets of Weigend et al. [56] are used. These were also employed in the previous calculations.

Peterson et al. recommended to use for the length scale parameter γ in the F12 correlation factor values of 0.9, 1.0, and 1.1 a_0^{-1} for the VDZ-F12, VTZ-F12, and VQZ-F12 basis sets, respectively. For the sake of simplicity, we have chosen $\gamma = 1.0\ a_0^{-1}$ for all basis sets; this value was also used in all previous benchmarks. It is possible that slightly better results could be obtained with the recommended values for the VDZ-F12 and VQZ-F12 basis sets, but the dependence on the value of this parameter is rather weak.

The molecular systems used in the next three subsections were the same as in Ref. [31]. For atomization energies, electron affinities, and ionization potentials the original G2 set [57] was used, with the exception that molecules containing alkali and alkali-earth elements were omitted, because no JKFIT basis sets were available. For the benchmarks of reaction energies, a total of 104 reactions were studied, among which 54 involved only closed-shell molecules and the remaining 50 also open-shell molecules. A list of these reactions and molecules can be found in Ref. [31].

The error statistics presented in the following sections refer to the same reference values as in [31]. Wherever possible, these were obtained by extrapolating the correlation energies of the AV5Z and AV6Z sets (CBS[56]). For some of the larger molecules, only the AVQZ and AV5Z results could be used for the

Table 1. Error statistics for the Hartree-Fock contribution to various properties with and without the CABS singles contribution. The OPTRI basis sets corresponding to the individual AO basis sets are used.

	Hartree-Fock			HF+Singles		
Orbital Basis	MAD	RMS	MAX	MAD	RMS	MAX
Ionization Potentials (meV)						
VDZ-F12	21.8	33.4	113.8	4.5	5.4	11.3
VTZ-F12	2.2	3.7	14.9	1.2	1.4	2.7
VQZ-F12	0.4	0.5	2.2	0.2	0.3	0.5
Electron Affinities (meV)						
AVDZ	42.2	76.2	325.5	6.7	10.5	42.4
AVTZ	8.1	11.8	35.1	2.4	3.2	6.7
AVQZ	2.4	3.6	12.8	1.2	1.6	3.7
Atomization Energies (kJ/mol)						
VDZ-F12	9.95	12.17	45.68	1.30	1.53	4.63
VTZ-F12	1.61	1.90	6.75	0.19	0.23	0.52
VQZ-F12	0.22	0.25	0.78	0.04	0.06	0.32
Reaction energies (kJ/mol)						
VDZ-F12	7.35	12.38	68.45	1.01	1.76	9.13
VTZ-F12	1.00	1.63	9.52	0.17	0.30	1.58
VQZ-F12[a]	0.15	0.21	0.98	0.07	0.12	0.66

a) Reactions without AV6Z reference values were excluded.

extrapolation (CBS[45]). In the tables we will present mean absolute deviations (MAD), root mean square deviations (RMS), and the absolute maximum errors (MAX).

4.1 Benchmarks for the CABS singles correction

In Table 1 the improvement of the Hartree-Fock energy differences by using the CABS singles correction is demonstrated. It is of interest to check how the new OPTRI basis sets perform in this respect, because they contain much fewer *s* and *p* functions than the JKFIT basis sets. In particular, functions with larger exponents are missing. As a result, the absolute values for the perturbative corrections are much smaller than for the JKFIT basis sets. Nevertheless, as seen in Table 1, the performance for energy differences is very good.

A more detailed comparison than presented here shows that the accuracy of the V*n*Z-F12/OPTRI results is for all properties slightly better than of the AV*n*Z/JKFIT or AV*n*Z/OPTRI ones. A comparison of the AV*n*Z/JKFIT or AV*n*Z/OPTRI (not shown) yields nearly the same accuracy for both RI basis sets.

Table 2. Error statistics using RMP2-F12 for various properties. The fixed amplitude ansatz and the OPTRI basis sets corresponding to the individual AO basis sets were used.

	RMP2			RMP2-F12		
Orbital Basis	MAD	RMS	MAX	MAD	RMS	MAX
Ionization Potentials (meV)						
VDZ-F12	234.8	252.3	484.3	30.3	34.4	78.9
VTZ-F12	111.1	120.8	205.2	5.6	6.3	14.8
VQZ-F12	55.9	61.3	105.4	1.1	1.4	3.4
Electron Affinities (meV)						
AVDZ	209.8	223.7	375.4	42.2	45.9	83.6
AVTZ	94.2	104.8	160.4	9.6	12.2	29.2
AVQZ	47.3	52.6	79.8	3.3	5.4	15.6
Atomization Energies (kJ/mol)						
VDZ-F12	60.66	66.37	130.01	2.44	3.09	7.97
VTZ-F12	27.03	29.68	61.29	0.45	0.59	1.69
VQZ-F12	12.73	14.02	29.07	0.21	0.27	0.84
Reaction energies (kJ/mol)						
VDZ-F12	23.38	32.53	113.79	1.80	2.98	16.75
VTZ-F12	9.44	13.22	44.14	0.44	0.65	2.79
VQZ-F12[a]	5.15	6.68	19.86	0.34	0.53	2.17

a) Reactions without CBS[56] reference values were excluded.

4.2 Benchmarks with RMP2-F12

Table 2 shows a comparison of standard RMP2 and RMP2-F12 results. The energy differences are computed from the total energies, including the Hartree-Fock contribution with CABS singles correction. The fixed amplitude ansatz has been used. It should be noted that these results cannot be compared directly with those in [12], since meanwhile the wavefunction ansatz has been changed: in our original work on RMP2-F12, the summation over k,l in eq. (5) included only the spin orbitals that are occupied in the RHF wavefunction. In our current methods, the summation includes both spin orbitals that can be formed from any occupied spatial orbital. It was found that this significantly improves the accuracy [31]. Furthermore, it is more consistent when the fixed amplitude ansatz is used. Table 2 demonstrates for RMP2-F12 the dramatic improvement by the F12 treatment. For all properties, the double-zeta F12 results are better than the standard quadruple zeta ones.

4.3 Benchmarks with CCSD(T)-F12x

Table 3 shows the results for the RCCSD(T^*)-F12a and RCCSD(T^*)-F12b methods. Again the accuracy is excellent, even though the errors are somewhat larger than in the RMP2-F12 case. This is mainly due to the fact that the basis set error of the triples contribution is not removed by the F12 treatment. In particular for

Table 3. Error statistics using RCCSD(T*)-F12x for various properties. The fixed amplitude ansatz and the OPTRI basis sets corresponding to the individual AO basis sets were used. The triples energies were scaled using eq. (12).

	RCCSD(T*)-F12a			RCCSD(T*)-F12b		
Orbital Basis	MAD	RMS	MAX	MAD	RMS	MAX
Ionization Potentials (meV)						
VDZ-F12	57.0	70.5	163.7	77.0	88.4	196.0
VTZ-F12	7.3	10.7	31.8	19.2	23.2	51.7
VQZ-F12	7.5	8.3	17.0	4.4	5.4	12.1
Electron Affinities (meV)						
AVDZ	46.3	54.2	120.1	73.4	79.4	153.3
AVTZ	5.7	7.2	15.7	18.9	21.6	38.2
AVQZ	8.8	10.2	17.4	5.2	7.0	19.5
Atomization Energies (kJ/mol)						
VDZ-F12	3.70	4.42	10.19	6.99	8.28	20.84
VTZ-F12	1.50	1.72	3.44	2.03	2.39	5.52
VQZ-F12	1.91	2.12	4.27	0.51	0.61	1.44
Reaction energies (kJ/mol)						
VDZ-F12	2.93	3.86	11.16	3.37	4.44	13.38
VTZ-F12	0.95	1.25	3.73	1.24	1.75	6.03
VQZ-F12^a	0.69	0.92	3.00	0.33	0.44	1.53

a) Reactions without CBS[56] reference values were excluded.

the CCSD(T)-F12b method, significant improvements of all results are achieved by scaling the triples contribution, as explained in section 2.2. For CCSD(T)-F12a, which somewhat overestimates the correlation energy, the improvements by the scaling are less systematic, since there is an error compensation between the overshooting of the CCSD-F12a correlation energy and the underestimation of the triples energy. Overall, for the double and triple zeta basis sets the F12a method yields more accurate results, due to the above-mentioned error compensation. For larger basis sets the F12b approximation is preferable, since it converges more smoothly to the basis set limit.

Table 4 shows computed CCSD(T)-F12a and CCSD(T)-F12b equilibrium distances and harmonic vibrational frequencies for the same 13 diatomic molecules studied in Ref. [31]. The accuracy is outstanding, in particular for CCSD(T)-F12b. Already with the VDZ-F12 basis set the RMS error of the vibrational frequencies is less than 3 cm^{-1}. This is about the same accuracy as achieved with CCSD(T) and the AV5Z basis set [31]. Also the RMS error for the equilibrium distances obtained with CCSD(T)-F12b/VDZ-F12 is smaller than the one for CCSD(T)/AVQZ. Somewhat surprisingly, CCSD(T)-F12b yields for these properties with all basis sets more accurate results than CCSD(T)-F12a. Furthermore, in contrast to most other properties, the scaling of the triples energy does not improve the frequencies and distances. The reason for this behavior is not yet understood in detail.

Table 4. Error statistics for harmonic vibrational frequencies and equilibrium distances for 13 diatomic molecules using CCSD(T)-F12x.

	CCSD(T)-F12a			CCSD(T)-F12b		
Orbital Basis	MAD	RMS	MAX	MAD	RMS	MAX
Harmonic Vibrational Frequencies (cm^{-1})						
VDZ-F12	5.1	5.6	8.5	2.2	2.9	5.7
VTZ-F12	3.1	3.4	6.3	1.7	2.1	4.9
VQZ-F12	0.8	1.1	2.1	0.6	0.7	1.3
Equilibrium Distances (pm)						
VDZ-F12	0.14	0.17	0.40	0.09	0.11	0.27
VTZ-F12	0.07	0.09	0.23	0.05	0.07	0.18
VQZ-F12	0.02	0.02	0.06	0.01	0.01	0.04

The RMS deviations of the CCSD(T)/CBS values from the experimental equilibrium distances and harmonic frequencies amount to 0.17 pm and 10.3 cm^{-1}, respectively. Thus, the basis set incompleteness errors are already for the VDZ-F12 basis set smaller than the intrinsic errors of the CCSD(T) method. To further improve the results, it would be necessary to include higher excitations, core correlation, and relativistic effects.

A similar accuracy was also achieved in Ref. [36] for nine polyatomic molecules with up to 6 atoms. Due to the efficiency of the CCSD(T)-F12x methods, it was possible to automatically generate many-dimensional potential energy surfaces around the equilibrium geometries and to compute anharmonic frequencies using vibrational configuration interaction methods. Using CCSD(T)-F12a/AVTZ, the mean absolute deviation of all computed frequencies from experimental values was only 4 cm^{-1}.

4.4 Intermolecular interaction energies

Intermolecular interaction energies are of great interest due to their importance in many biological systems. For example, hydrogen bonding and π–stacking interactions between nucleic acid base pairs are crucial for the folding of nucleic acids such as DNA and RNA. A particular challenge is the accurate prediction of dispersion energies, which dominate π–stacking interactions. The calculation of dispersion interactions is difficult because they are pure electron correlation effects, which cannot be described by simple methods such as Hartree-Fock or standard density functional theory.

We have applied the explicitly correlated MP2-F12 and CCSD(T)-F12a methods to calculate intermolecular interaction energies for the S22 benchmark set of Jurečka, Šponer, Černý, and Hobza [58]. This includes clusters with strong electrostatic interactions or hydrogen bonds, as well as predominantly dispersion bonded and mixed clusters.

The interaction energies range from less than 1 kcal/mol to over 20 kcal/mol. In some clusters with strong dispersion interactions, the correlation contribution

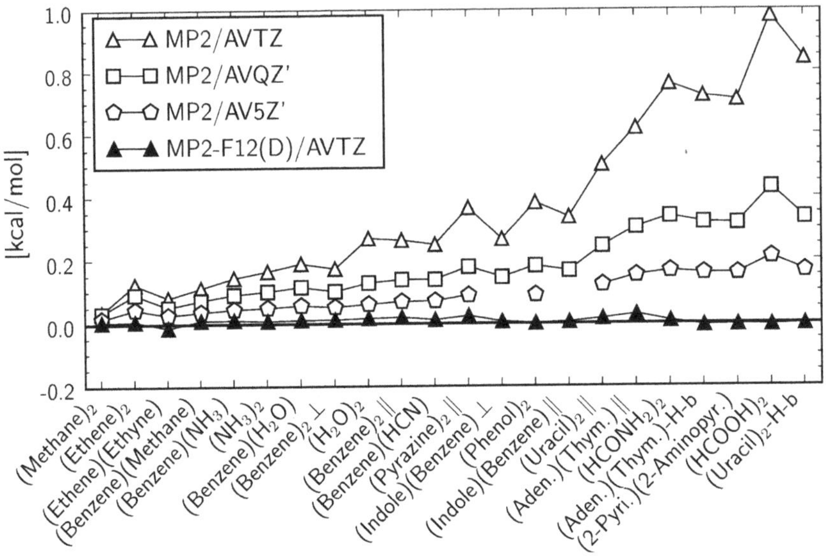

Fig. 1. Errors of the total MP2/AV*n*Z' and MP2-F12/AVTZ interaction energies for the S22 set relative to the MP2/CBS[45] estimates of the basis set limits.

to the interaction energy is larger than the total interaction energy and may exceed 25 kcal/mol. In all calculations, the counterpoise correction [59] was applied to reduce basis set superposition errors (BSSE).

The deviations of the MP2 and MP2-F12 interaction energies from MP2/CBS[45] reference values are shown in Figure 1 (from Ref. [35]). In the standard MP2 calculations the diffuse functions on the hydrogen atoms were omitted. This reduces the BSSE and has little effect on the results. The resulting mixed basis sets are denoted AV*n*Z'. In the MP2-F12 calculations the diagonal approximation was used. Figure 1 shows that the conventional MP2 results converge very slowly with increasing basis set size. In contrast, the MP2-F12 method yields amazingly accurate results already with small basis sets. Using the triple zeta basis set, the MP2-F12 interaction energies are much more accurate than the MP2/AV5Z' results.

Figure 2 shows a similar comparison for CCSD(T^*)-F12a [38]. In this case the CCSD(T) reference calculations were only possible for a reduced set of 11 dimers. The results are similar to MP2-F12: Using the AVTZ basis, CCSD(T^*)-F12a yields highly accurace results. Even with the AVDZ basis set the errors are smaller than 0.2 kcal/mol for all dimers.

4.5 Benchmarks with LMP2-F12 and LCCSD-F12

In this section we will summarize our results obtained with local MP2-F12 and CCSD-F12 methods. In all cases, the domains were determined using the method

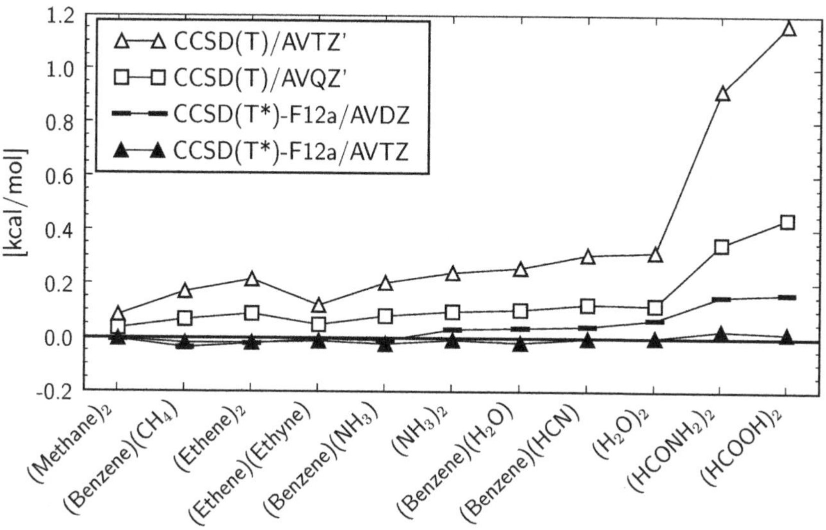

Fig. 2. Errors of CCSD(T) and CCSD(T*)-F12 interaction energies relative to CCSD(T)/CBS[34] results for different basis sets.

of Boughton and Pulay [60], using a completeness criterion of 0.985 (triple zeta basis sets) or 0.98 (double zeta basis sets). Details of this method, as implemented in MOLPRO, can be found in Refs. [48,61]. Unless otherwise noted, in all F12 calculations the diagonal ansatz with approximation 3*A has been used, and the VTZ/JKFIT basis sets were employed for the RI.

Table 5 shows the computed correlation energies for some large molecules (cf. Ref. [33]). The F12 energy contributions as well as the total correlation energies are compared. It can be seen that the F12 energy contributions in the MP2-F12 and LMP2-F12 calculations differ by about 100 mH. This is due to the fact that in the LMP2-F12 method this contribution corrects for the domain errors in the LMP2 wavefunction. However, the total correlation energies differ in all cases by less than 10 mH. This demonstrates that the explicitly correlated terms remove the domain errors very efficiently. This is also true for energy differences. Examples for reactions involving large molecules can be found in Ref. [33]

Table 6 presents a comparison of CCSD and LCCSD correlation energies with and without explicit correlation [34]. All pairs are fully included in the LCCSD and LCCSD-F12 calculations, i.e., the difference between the local and non-local results solely reflects the domain error. The conventional CCSD and LCCSD energies differ by several mH, but this error is reduced to about 0.1 mH in the explicitly correlated methods. This shows that the domain error is also removed in LCCSD-F12. As compared to the estimated complete basis set limits, the CCSD-F12 and LCCSD-F12 correlation energies are somewhat too large.

Table 5. Comparison of MP2-F12 and LMP2-F12 correlation energies ($-E_{corr}$ in mH) for some large molecules. The AVTZ orbital basis was used and all pairs apart from very distant pairs were treated in F12. No local DF or local RI approximations were employed.

Molecule	MP2-F12		LMP2-F12	
	E_{F12}	E_{tot}^{corr}	E_{F12}	E_{tot}^{corr}
Progesterone ($C_{21}H_{30}O_2$)	397.7	4466.9	477.9	4460.5
Cholesterol ($C_{27}H_{46}O$)	470.6	5457.1	576.3	5448.7
Borrelidin ($C_{28}H_{43}NO_6$)	654.0	7100.4	763.1	7091.8

Table 6. LCCSD-F12a correlation energies (-E_{corr} in mH) obtained with the VTZ-F12 basis set in comparison to the extrapolated CCSD/CBS[45] values. The fixed amplitude Ansatz was used.

Molecule	CCSD	LCCSD	CCSD-F12	LCCSD-F12	CBS
C_2H_4O	611.8	609.2	658.0	658.1	650.4
CH_3CHO	606.1	603.8	651.6	651.7	643.9
C_2H_5OH	648.2	644.3	697.3	697.3	689.1
NH_2CONH_2	851.9	847.6	918.0	917.9	908.2
$HCOOCH_3$	844.7	841.1	911.7	911.7	901.7

This is mainly due to the 3*A approximation used in these calculations. Previous studies have shown that this has little effect on computed energy differences. This is confirmed by the results presented below.

Finally, in Table 7 we present the error statistics for the 54 closed-shell reactions, computed with various local and non-local methods. The fixed amplitude ansatz with approximation 3*A has been used in all these explicitly correlated calculations, while the results in Tables 2 and 3 were obtained with the theoretically more accurate approximation 3C. Nevertheless, the errors of the 3*A results are equally small. The local and non-local variants yield virtually the same accuracy, both for MP2-F12 and CCSD-F12x, i.e., the domain errors are almost fully removed. This is a very promising result for future applications of the LCCSD(T)-F12x methods for large molecules. The implementation of a fully local linear scaling LCCSD(T)-F12x implementation is currently in progress.

5. Conclusions

We have presented benchmark results using new AO and RI basis sets for explicitly correlated methods. Overall, the results are of the same quality as our previous results obtained with the standard AV*n*Z basis sets and JKFIT basis sets for the RI. Whatever basis set is used and property is considered, the explicitly correlated results are uniformly much more accurate than those obtained with the conventional MP2 and CCSD(T) methods and the same basis set. While the additional effort for MP2-F12 relative to MP2 is still significant (typically a

Table 7. Error statistics for reaction energies (in kJ/mol) computed using the LMP2-F12 and LCCSD-F12 methods. The fixed amplitude ansatz with approximation 3*A and the VnZ-F12/OPTRI auxiliary basis sets were used in the F12 calculations.

Method	VDZ-F12			VTZ-F12		
	MAD	RMS	MAX	MAD	RMS	MAX
MP2	10.1	10.8	68.9	4.3	4.7	31.9
LMP2	14.5	13.8	78.3	6.1	5.9	35.1
MP2-F12	1.2	1.1	6.6	0.5	0.5	2.8
LMP2-F12	1.3	1.2	6.5	0.6	0.7	3.3
CCSD	10.5	10.0	57.5	4.6	4.4	27.8
LCCSD	14.3	13.6	66.5	6.1	5.4	30.2
CCSD-F12	1.8	1.5	6.2	0.7	0.6	2.6
LCCSD-F12	1.8	1.5	6.1	0.7	0.6	2.4

factor of 10), the additional effort becomes negligible for CCSD(T)-F12x calculations on larger molecules. This is because the cost of the MP2-F12 calculation becomes negligible relative to the triples calculation, which scales much more steeply with molecular size. Overall, for the largest systems presented in this work, the CCSD(T)-F12x calculations are only a few percent more expensive than standard CCSD(T) calculations with the same basis. Of course, such comparisons depend on the efficiency of the CCSD(T) itself. In this respect we may note that our CCSD(T) program is considered to be one of the fastest currently available. For example, the complete CCSD(T)-F12/AVTZ calculation for the T-structure of the benzene dimer (including counterpoise correction) took 118 h on a single Xeon 3.0 GHZ processor (MacPro). Of this time, 5.9 h was spent for the MP2-F12 parts, i.e., the additional effort for the F12 correction is only 5% of the corresponding CCSD(T)/AVTZ calculation [35].

All canonical MP2-F12 and CCSD(T)-F12 calculations presented in this paper were performed with default values as implemented in version 2009.1 of MOLPRO and require no special input. Thus, they can be performed in a black box manner, similar to standard MP2 and CCSD(T) calculations. In conclusion, these methods are ready for public use and are highly recommended for future applications.

The local variants are very promising to extend the applicability of the CCSD(T)-F12 method to larger molecules. It has been demonstrated that the domain error is almost entirely removed by the explicitly correlated terms. Preliminary investigations have shown for LMP2 how the F12 energy correction depends on pair approximations. Future work is necessary to determine what is the best choice of pair approximations in LCCSD(T)-F12 calculations in order to obtain a best compromise between cost and accuracy.

Acknowledgement

This article summarizes work that was funded in the priority program 1145 of the Deutsche Forschungsgemeinschaft. Further support by the Fonds der Chemischen Industrie is gratefully acknowledged. T.B.A. would also like to thank the Studienstiftung des deutschen Volkes.

References

1. A. Karton, E. Rabinovich, J. M. L. Martin, and B. Ruscic, J. Chem. Phys. **125** (2006) 144108.
2. A. Tajti, P. G. Szalay, A. G. Császár, M. Kállay, J. Gauss, E. F. Valeev, B. A. Flowers, J. Vázquez, and J. F. Stanton, J. Chem. Phys. **121** (2004) 11599–11613.
3. W. Kutzelnigg, Theor. Chim. Acta **68** (1985) 445–469.
4. S. Ten-no, Chem. Phys. Lett. **398** (2004) 56–61.
5. A. J. May and F. R. Manby, J. Chem. Phys. **121** (2004) 4479–4485.
6. D. P. Tew and W. Klopper, J. Chem. Phys. **123** (2005) 074101.
7. D. P. Tew and W. Klopper, J. Chem. Phys. **125** (2006) 094302.
8. E. F. Valeev, J. Chem. Phys. **125** (2006) 244106.
9. J. Noga, S. Kedžuch, and J. Šimunek, J. Chem. Phys. **127** (2007) 034106.
10. F. R. Manby, H.-J. Werner, T. B. Adler, and A. J. May, J. Chem. Phys. **124** (2006) 094103.
11. H.-J. Werner, T. B. Adler, and F. R. Manby, J. Chem. Phys. **126** (2007) 164102.
12. G. Knizia and H.-J. Werner, J. Chem. Phys. **128** (2008) 154103.
13. K. A. Peterson, T. B. Adler, and H.-J. Werner, J. Chem. Phys. **128** (2008) 084102.
14. K. E. Yousaf and K. A. Peterson, J. Chem. Phys. **129** (2008) 184108.
15. K. E. Yousaf and K. A. Peterson, Chem. Phys. Lett. **476** (2009) 303.
16. A. J. May, E. Valeev, R. Polly, and F. R. Manby, Phys. Chem. Chem. Phys. **7** (2005) 2710.
17. W. Klopper, F. R. Manby, S. Ten-no, and E. F. Valeev, Int. Rev. Phys. Chem. **25** (2006) 427.
18. J. Noga, W. Kutzelnigg, and W. Klopper, Chem. Phys. Lett. **199** (1992) 497.
19. J. Noga and W. Kutzelnigg, J. Chem. Phys. **101** (1994) 7738–7762.
20. J. Noga and P. Valiron, Chem. Phys. Lett. **324** (2000) 166.
21. J. Noga, S. Kedžuch, J. Šimunek, and S. Ten-no, J. Chem. Phys. **128** (2008) 174103.
22. T. Shiozaki, M. Kamiya, S. Hirata, and E. F. Valeev, Phys. Chem. Chem. Phys. **10** (2008) 3358.
23. T. Shiozaki, M. Kamiya, S. Hirata, and E. F. Valeev, J. Chem. Phys. **129** (2008) 071101.
24. A. Köhn, G. W. Richings, and D. P. Tew, J Chem Phys **129** (2008) 201103.
25. T. Shiozaki, M. Kamiya, S. Hirata, and E. F Valeev, J. Chem. Phys. **130** (2009) 054101.
26. H. Fliegl, W. Klopper, and C. Hättig, J. Chem. Phys. **122** (2005) 084107.
27. H. Fliegl, C. Hättig, and W. Klopper, Int. J. Quantum Chem. **106** (2006) 2306.
28. D. P. Tew, W. Klopper, C. Neiss, and C. Hättig, Phys. Chem. Chem. Phys. **9** (2007) 1921–1930.
29. D. P. Tew, W. Klopper, and C. Hättig, Chem. Phys. Lett. **452** (2008) 326–332.
30. T. B. Adler, G. Knizia, and H.-J. Werner, J. Chem. Phys. **127** (2007) 221106.
31. G. Knizia, T. B. Adler, and H.-J. Werner, J. Chem. Phys. **130** (2009) 054104.
32. H.-J. Werner, J. Chem. Phys. **129** (2008) 101103.
33. T. B. Adler, H.-J. Werner, and F. R. Manby, J. Chem. Phys. **130** (2009) 054106.
34. T. B. Adler and H.-J. Werner, J. Chem. Phys. **130** (2009) 241101.
35. O. Marchetti and H.-J. Werner, Phys. Chem. Chem. Phys. **10** (2008) 3400.
36. G. Rauhut, G. Knizia, and H.-J. Werner, J. Chem. Phys. **130** (2009) 054105.
37. P. Botschwina, R. Oswald, G. Knizia, and H.-J. Werner, Z. Phys. Chem. **223** (2009) 447.

38. O. Marchetti and H.-J. Werner, J. Phys. Chem. A **113** (2009) 115800.
39. H.-J. Werner, P. J. Knowles, R. Lindh, F. R. Manby, and M. Schütz et al., MOLPRO, version 2009.1, a package of ab initio programs, 2009, see http://www.molpro.net.
40. P. J. Knowles, C. Hampel, and H.-J. Werner, J. Chem. Phys. **99** (1993) 5219–5227.
41. E. F. Valeev, Chem. Phys. Lett. **395** (2004) 190–195.
42. S. Ten-no, J. Chem. Phys. **121** (2004) 117–129.
43. T. Kato, Comm. Pure Appl. Math. **10** (1957) 151–177.
44. R. T. Pack and W. Byers Brown, J. Chem. Phys. **45** (1966) 556–559.
45. A. Köhn, J. Chem. Phys. **130** (2009) 131101.
46. P. Pulay, Chem. Phys. Lett. **100** (1983) 151–154.
47. S. Saebø and P. Pulay, Annu. Rev. Phys. Chem. **44** (1993) 213–236.
48. C. Hampel and H.-J. Werner, J. Chem. Phys. **104** (1996) 6286–6297.
49. J. Pipek and P. G. Mezey, J. Chem. Phys. **90** (1989) 4916–4926.
50. S. F. Boys, *Localized Orbitals and Localized Adjustment Functions.* In: *Quantum Theory of Atoms, Molecules, and the Solid State.* P. O. Löwdin (Ed.) p. 253–262. Academic Press, New York (1966).
51. R. Mata and H.-J. Werner, Mol. Phys. **105** (2007) 2753–2761.
52. M. Schütz and H.-J. Werner, J. Chem. Phys. **114** (2001) 661–681.
53. R. A. Kendall, T. H. Dunning, and R. J. Harrison, J. Chem. Phys. **96** (1992) 6796.
54. Jr. T.H. Dunning, K. A. Peterson, and A.K. Wilson, J. Chem. Phys. **114** (2001) 9244–9253.
55. F. Weigend, Phys. Chem. Chem. Phys. **4** (2002) 4285.
56. F. Weigend, A. Köhn, and C. Hättig, J. Chem. Phys. **116** (2002) 3175.
57. L. A. Curtiss, K. Raghavachari, G. W. Trucks, and J. A. Pople, J. Chem. Phys. **94** (1991) 7221–7230.
58. P. Jurečka, J. Šponer, J. Černý, and P. Hobza, Phys. Chem. Chem. Phys. **8** (2006) 1985.
59. F. Bernardi and S. F. Boys, Mol. Phys. **19** (1970) 553.
60. J. W. Boughton and P. Pulay, J. Comput. Chem. **14** (1993) 736–740.
61. H.-J. Werner and K. Pflüger, Ann. Reports in Comput. Chem. **2** (2006) 53.

Fully Automated Implementation of the Incremental Scheme for Correlation Energies

By Joachim Friedrich*, Michael Hanrath, and Michael Dolg

Institut für Theoretische Chemie, Greinstr. 4, 50939 Köln, Germany

(Received July 1, 2009; accepted October 12, 2009)

Incremental Scheme / Bethe-Goldstone Expansion / Local Correlation Methods

The applicability and performance of a fully automated implementation of the incremental scheme for the evaluation of correlation contributions in wavefunction-based quantum chemical calculations on large molecules is briefly reviewed and a typical example from organic chemistry is discussed in more detail. The accuracy of relative energies is analyzed for various water hexamers at the CCSD(T)/aug-cc-pVDZ level. Furthermore the accuracy of the potential energy surface of octane is analyzed at the CCSD level of theory. Finally, CCSD(T)/aug-cc-pVTZ calculations with 1656 basis functions in C_1 symmetry are reported for $(H_2O)_{18}$, where a full calculation is infeasible within reasonable time.

1. Introduction

It is well known in the quantum chemistry community that electron correlation has to be included in accurate quantitative electronic structure calculations. An economic and thus widely used way to do this is density functional theory (DFT). One of the major drawbacks of DFT approaches is that they are not systematically improvable. Therefore the alternative strategy to include electron correlation by means of wavefunction-based post-Hartree-Fock (HF) methods like many-body perturbation theory (MBPT), configuration interaction (CI) or the coupled cluster (CC) ansatz is still attractive. These methods are systematically improvable, however their steep scaling with respect to the one-particle basis set often limits their applicability. The goal of local correlation methods is to combine the systematic improvability of the post-HF methods with high efficiency. This has been explored by several groups since more than 2 decades now [1–39]. One of these approaches is the incremental scheme of Stoll [4–6]. It is based

* Corresponding author. E-mail: joachim_friedrich@gmx.de

Z. Phys. Chem. **224** (2010) 513–525
© by Oldenbourg Wissenschaftsverlag, München

on the Bethe-Goldstone expansion as introduced to quantum chemistry by Nesbet [40–42]. The basic idea of the incremental scheme is to partition the total system into comparatively small localized orbital domains and to write the correlation energy as a sum of contributions of single domains, as well as correction terms arising from pairs, triples, ... of domains, i.e.

$$E_{corr} = \sum_{i} \Delta\varepsilon_i + \sum_{i>j} \Delta\varepsilon_{ij} + \sum_{i>j>k} \Delta\varepsilon_{ijk} + \ldots \tag{1}$$

$$\Delta\varepsilon_i = \varepsilon_i \qquad \Delta\varepsilon_{ij} = \varepsilon_{ij} - \Delta\varepsilon_i - \Delta\varepsilon_j \qquad \text{etc.}$$

Here ε_i stands for the correlation energy of the domain i, ε_{ij} for the one of the joint domains i and j, etc. Since for systems with well localized orbitals the correlation energy increments $\Delta\varepsilon_{ij\ldots}$ quickly become smaller with increasing order the summation eqn. 1 can be truncated once the desired accuracy is reached. For the convenient treatment of higher order terms eqn. 1 is expressed in a notation based on simple set theory [43]:

$$E_{\mathrm{corr}} = \sum_{\substack{\mathbb{X} \\ \mathbb{X}\in\mathcal{P}(\mathbb{D})\wedge|\mathbb{X}|\leq O}} \Delta\varepsilon_{\mathbb{X}} \tag{2}$$

$\mathbb{D}$ is the set of domains, $\mathcal{P}(\mathbb{D})$ stands for the power set of the set of the domains and O denotes the order of the expansion. The summation index $\mathbb{X}$ is a set of domain indices. Since the power set of $\mathbb{D}$ is equivalent to the index permutations from the summations in eqn. 1 we can use the $\mathcal{P}(\mathbb{D})$ to label the individual increments. The orders of eqn. 1 are given by the cardinality of the sets $\mathbb{X}$. When $\varepsilon_{\mathbb{X}}$ represents the correlation energy of the unified subsystems of $\mathbb{X}$ the general correlation energy increment $\Delta\varepsilon_{\mathbb{X}}$ is given as:

$$\Delta\varepsilon_{\mathbb{X}} = \varepsilon_{\mathbb{X}} - \sum_{\substack{\mathbb{Y} \\ \mathbb{Y}\in\mathcal{P}(\mathbb{X})\wedge|\mathbb{Y}|<|\mathbb{X}|}} \Delta\varepsilon_{\mathbb{Y}} \tag{3}$$

here the summation index $\mathbb{Y}$ runs over the power set of $\mathbb{X}$.

Since its initial applications by Stoll in 1992 [4–6] the incremental scheme was successfully applied to various periodic systems [44] such as polymers [45,46], van der Waals crystals [47], ionic crystals [48,49], semiconductors [50] and metals [51,52]. Since the incremental scheme is not limited to ground state energies it has also been applied to band structures [53]. Besides the various applications to periodic systems the incremental scheme has also been applied to molecular systems [43,54,55]. Since in cases of large molecules with low or no point group symmetry a much higher number of increments than in a simple periodic system has to be evaluated, we designed and implemented the first fully automated version of the incremental scheme [43]. Avoiding the extensive hand work needed in earlier applications of the incremental scheme we were able to go to higher orders and investigate systematically the performance of incremental energy expansions for a large variety of molecular systems [43,56,57]. Furthermore we extended the approach to open-shell systems within the framework of the restricted open-shell coupled cluster approach with single and double excitations (RCCSD) [58], implemented incremental MP2 and CCSD(T) energy expansions, CCSD dipole moment and quadrupole moment expansions [59] and devised an efficient treatment of core-valence correlation [60]. A significant improvement of the performance was recently achieved by introducing domain-specific basis sets [61]. Finally, we investigated to which extent point group symmetry can be exploited [62]. The present contribution provides a brief overview over these developments and demonstrates the accuracy and efficiency of our implementation of the incremental scheme for a variety of test systems.

2. Implementation

The current C++ based implementation of the incremental scheme is applicable in connection with unmodified standard quantum chemistry codes that offer size-extensive correlation schemes. Interfaces exist at present to DALTON, MOLPRO and TURBOMOLE. In the following we summarize some general aspects of our implementation.

2.1 Generation of the one-site domains

The first step in our fully automated incremental approach is the generation of the one-site domains. The set of occupied localized HF orbitals $\mathbb{O}$ is divided into disjoint subsets under the side condition that the orbitals in one domain should be close to each other in real space. This problem can be reformulated into a graph partitioning problem for an edge weighted graph where standard library routines can be applied. We define the connectivity of two localized molecular orbitals (LMO) ϕ_i,ϕ_j according to:

$$C_{ij} = \begin{cases} 0, & \text{if } D_{ij} > t_{con} \\ \frac{w}{D_{ij}}, & \text{if } D_{ij} \leq t_{con} \wedge \frac{w}{D_{ij}} < 10^8 \\ 10^8, & \text{if } D_{ij} \leq t_{con} \wedge \frac{w}{D_{ij}} \geq 10^8 \end{cases} \tag{4}$$

where w is a constant factor of 10^4, D_{ij} is the distance of the corresponding centers of charge and t_{con} is a input parameter to control the connectivity. With the connectivity matrix defined in eqn. 4 we use METIS graph-partitioning [63] and obtain a set of sufficiently local domains. The domain size input parameter *dsp* roughly corresponds to the number of orbitals per domain and allows to limit the size of the domains, so that the calculations needed for the highest order increments still remain feasible on the available hardware.

2.2 Distance screening

The scaling of eqn. 2 with respect to the number of calculations $\mathcal{N}_{calc}$ is given by the binomial coefficient:

$$\mathcal{N}_{calc} = \sum_{\lambda=1}^{O} \binom{|\mathbb{D}|}{\lambda} \tag{5}$$

This behavior makes the application of eqn. 2 for higher orders very expensive. However, upon analyzing the contributions of the individual energy increments one finds for their absolute values that they decay with increasing order and, for a given order, also with increasing distances between the underlying domains. Therefore, the incremental summation can usually be terminated at low order, i.e. 3rd or 4th order, and for a given order all contributions arising from domains which are too far apart can be neglected. In our implementation we use order-dependent distance thresholds $t_{dist} = f_1/O$ or $t_{dist} = f_2/(O-1)^2$ and exclude those increments for which all distances between the centers of charge of two groups of underlying domains is larger than t_{dist}. For properly chosen values of the parameter f_μ the contribution of the neglected increments is small, i.e. below 1 kcal/mol in our typical applications.

From our experience the values of f_μ are transferable for different basis sets, different methods and for similar systems. In order to adapt the parameter f_μ for a so far not investigated system one can perform a MP2 calculation in a small basis to check the accuracy of this threshold.

2.3 Domain-specific basis sets

For larger systems it is usually not necessary to use the complete virtual space to correlate the electrons in a specific localized region. Furthermore it would be desirable to use further approximations for the environment of a domain, to reduce the number of the two electron integrals. In our approach this is achieved by the introduction of a domain-specific basis set, e.g. for a given domain we use the orbitals of the domain to define a region where the large original AO basis set is applied and in the environment we reduce the AO basis set. The region is defined by the distance threshold t_{main} and the centers of charge of the orbitals in the domain. Around every center of charge in the domain we construct a sphere with the radius t_{main}. The unification of the area in the spheres defines the active part of the molecule. The complement is the environment, where a smaller AO basis set is used. In order to obtain a set of localized orthogonal orbitals in the new basis we perform a HF calculation with a subsequent Boys localization. Finally we identify the orbitals of the domain according to their centers of charge [61]. Template sets of localized orbitals can be used to remove the ambiguities in the localization procedure and to obtain for different basis sets localized orbital sets as similar as possible, e.g. within the domain-specific basis set approach [64].

2.4 Parallel implementation

The incremental scheme is inherently parallel by construction, since the calculations of the domain correlation energies are independent. The price one has to pay for this lack of communication is a different complexity and runtime for increments of different order. This can be easily seen from the fact that second order increments require the unification of the corresponding one-site domains. So even for ideal one-site domains of equal size the two-site domains have a doubled size. Therefore, the parallel efficiency as defined by Janowski et al. [65] or the speedup as defined by Harding et al. [66] or Olson [67] suffers if N_{calc} approaches the number of CPUs (N_{CPU}). However, in most practical cases it is $N_{calc} \gg N_{CPU}$ and the parallel efficiency is high [68]. Our implementation is parallelized in a master/slave structure using the socket++ library. The master process keeps track of the administrative tasks and the slaves perform the correlation calculations for the domains. In the current implementation slave-processes can be started at any time of the calculation. If the master has no more jobs to provide, the slaves terminate automatically. This avoids idle processes in calculations with a huge number of slaves. At the end of the calculation the master calculates the correlation energy according to eqn. 2 and terminates. Due to the fact that only little communication is required and the calculations in the subspaces require significantly less RAM and disk space, one can run large correlation calculations on a cluster of cheap standard personal computers.

By using a list of relevant increments it is possible to restart calculations after complete system failures or to automatically reevaluate faulty or missing increments after failures of individual slaves.

3. Selected applications

3.1 Computational details

The wavefunction-based correlation energies were obtained with the MOLPRO quantum chemistry code [69,70]. Localized orbitals were obtained using the Boys-localization [71]. If nothing else is stated the geometries were optimized at the RI-BP86/SV(P) level of theory using the TURBOMOLE quantum chemistry program package [72–77]. Stationary points were characterized by analyzing the Hessian matrix [78].

3.2 Large molecules

In order to demonstrate the performance of the incremental scheme for typical organic molecules we included $C_{11}H_{19}O_3N$ depicted in Fig. 1 as a test case. The convergence of the incremental MP2, CCSD and CCSD(T) correlation energies for the cc-pVDZ and cc-pVTZ basis sets are shown in Table 1. For the cc-pVDZ basis the MP2, CCSD and CCSD(T) energies converge rapidly to the value of

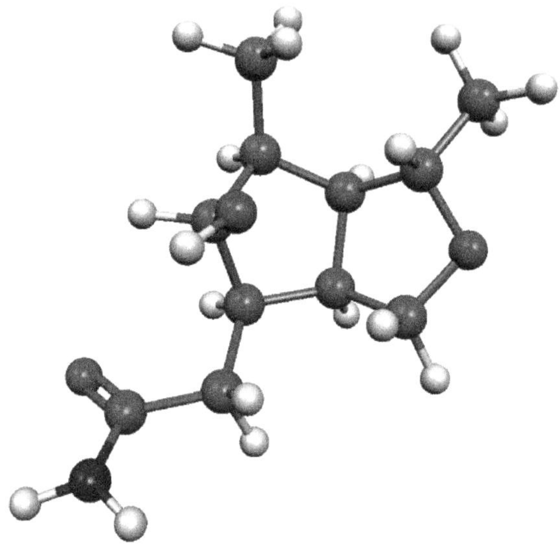

Fig. 1. RI-BP86/SV(P) optimized structure of 2-[(3S,3aS,4S,5S,6R,6aS)-5-hydroxy-3,4-dimethyl-3,3a,4,5,6,6a-hexahydro-1H-cyclopenta[c]furan-6-yl]acetamide.

the canonical reference calculation. Since the errors at a given order are of the same order of magnitude for all methods, we conclude that the incremental expansions of the MP2, CCSD and CCSD(T) energies converge equally well. Using a distance screening of $t_{dist} = 30/(O-1)^2$ the error with respect to the reference result is already below 1 kcal/mol at third order and drops below 0.3 kcal/mol at fourth order. Comparing the errors in the cc-pVDZ and the cc-pVTZ basis sets for MP2 and CCSD at a given order, we find that they are on the same order of magnitude. Therefore we conclude that increasing the basis set does not significantly change the convergence behavior of the incremental series. Since the CCSD/cc-pVTZ calculation was already expensive, we could not perform the perturbative treatment of the triple excitations on our current hardware. However, as judged from the errors of MP2 and CCSD in the cc-pVTZ basis in combination with the similar errors at the cc-pVDZ level, we expect the error of the CCSD(T) calculation to be also of the order of 1 kcal/mol.

3.3 Intermolecular Interactions

3.3.1 Relative Energies of $(H_2O)_6$ Clusters

Usually chemists are interested in the relative energy of two points on the potential energy surface. Therefore we test our approach for the relative energies of six small water clusters. The dsp parameter was set to 4 orbitals. In combination with the frozen-core approximation this leads to single water molecules as do-

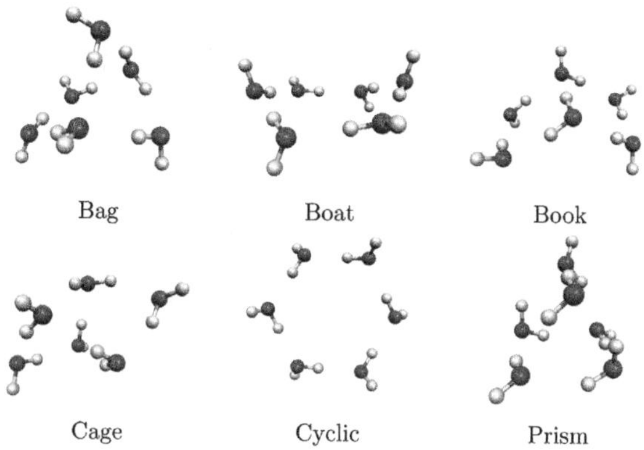

Fig. 2. Different structures of the water hexamer taken from Day et al. [79].

Table 1. Convergence of the incremental MP2, CCSD and CCSD(T) correlation energies for $C_{11}H_{19}O_3N$ (Fig. 1).

O_λ	MP2		CCSD		CCSD(T)	
	$\Sigma\Delta\varepsilon(\lambda)$ au	error kcal/mol	$\Sigma\Delta\varepsilon(\lambda)$ au	error kcal/mol	$\Sigma\Delta\varepsilon(\lambda)$ au	error kcal/mol
cc-pVDZ[a]						
1	-1.69113	354.80	-1.91631	290.24	-1.95182	318.57
2	-0.55714	5.19	-0.47945	-10.62	-0.52030	-7.93
3	-0.00956	-0.81	0.01591	-0.64	0.01115	-0.93
4	0.00096	-0.21	0.00067	-0.22	0.00103	-0.29
cc-pVDZ[b]						
1	-1.69113	354.80	-1.91631	290.24	-1.95181	318.57
2	-0.55714	5.19	-0.47945	-10.62	-0.52030	-7.93
3	-0.00937	-0.69	0.01588	-0.66	0.01123	-0.88
cc-pVTZ[c]						
1	-2.09831	428.59	-2.30124	352.25	-2.36192	-
2	-0.67391	5.71	-0.57949	-11.38	-0.64073	-
3	-0.01065	-0.98	0.01657	-0.98	0.01050	-

$\Sigma\Delta\varepsilon(\lambda)$:= sum of the λ-th order contributions
frozen core = 15 orbitals, dsp = 4, t_{con} = 3 bohr, t_{main} = 3 bohr
fit basis for the environment: H = STO-3G; C,N,O = 6–31G (section 2.3)
[a]correlated orbitals = 43, virtual orbitals = 247, f_2 = 30
[b]correlated orbitals = 43, virtual orbitals = 247, f_2 = 20
[c]correlated orbitals = 43, virtual orbitals = 658, f_2 = 20

mains. The results of the CCSD(T)/aug-cc-pVDZ calculations at the geometries of Day et al. [79] are given in Table 2. The energies are relative to the prism isomer, which has the lowest absolute energy. The errors are given with respect

Table 2. Incremental CCSD(T)/aug-cc-pVDZ energies of water hexamers in Fig. 2 relative to the prism isomer at different orders of the expansion (E_{rel}). Errors are given with respect to the exact CCSD(T)/aug-cc-pVDZ relative energies ($error_{rel}$) and with respect to the corresponding absolute energy ($error_{abs}$). All values are given in kcal/mol.

order	E_{rel}	$error_{rel}$	$error_{abs}$	E_{rel}	$error_{rel}$	$error_{abs}$
		Boat			Book	
1	-0.33	-3.59	18.61	-0.63	-1.75	20.45
2	3.73	0.47	-0.18	1.41	0.30	-0.35
3	3.15	-0.10	0.01	1.10	-0.02	0.09
can.	3.25	0.00	0.00	1.11	0.00	0.00
		Cage			Cyclic	
1	0.13	-0.07	22.12	-0.87	-2.97	19.23
2	0.24	0.04	-0.61	2.58	0.49	-0.16
3	0.25	0.05	0.17	1.99	-0.11	0.01
can.	0.20	0.00	0.00	2.10	0.00	0.00
		Bag			Prism	
1	-0.11	-1.89	20.30	0.00	0.00	22.20
2	1.99	0.20	-0.45	0.00	0.00	-0.65
3	1.81	0.02	0.13	0.00	0.00	0.11
can.	1.79	0.00	0.00	0.00	0.00	0.00

frozen core = 6 orbitals, dsp = 4, t_{con} = 3 bohr, t_{main} = 3 bohr, correlated orbitals = 24, f_2 = no truncation
fit basis for the environment: H = STO-3G; O = 6–31G (section 2.3)

to the full CCSD(T)/aug-cc-pVDZ reference calculation in the canonical basis. For these small water clusters the relative energy is well represented at the second order of the incremental series, with a maximum error of 0.49 kcal/mol. At the third order the maximum error is -0.11 kcal/mol. Therefore we conclude that the incremental scheme can provide accurate relative energies for water clusters. We did not consider corrections to the basis set superposition error (BSSE), since we compare our numbers with the theoretical values at the same level. From the small errors in the absolute energies (Tab. 2) one can see that the incremental scheme is a good approximation to the CCSD(T) energy and therefore we end up with the same BSSE as in the standard calculation.

3.3.2 Correlation Energy of $(H_2O)_{18}$

In order to demonstrate the applicability of our approach to larger systems with more than 100 correlated electrons and more than 1000 basis functions, we calculated the $(H_2O)_{18}$ cluster shown in Fig. 3. The rapid convergence of the incremental series is obvious from Table 3. In this case the MP2 calculation in the aug-cc-pVDZ basis is already expensive (2 days, 61 GB RAM, 305 GB hard disk). The incremental CCSD(T) calculation however took on 51 nodes about 12 hours using at most 2.4 GB RAM and 13.6 GB disk space. Furthermore we were able to calculate the CCSD(T) energy of the $(H_2O)_{18}$ cluster in the aug-cc-pVTZ basis

Table 3. Convergence of incremental MP2, CCSD and CCSD(T)/aug-cc-pVXZ (X = D,T) correlation energies for $(H_2O)_{18}$ (Fig. 3).

	MP2		CCSD		CCSD(T)	
O_λ	$\Sigma\Delta\varepsilon(\lambda)$	$E_{corr}(\lambda)$	$\Sigma\Delta\varepsilon(\lambda)$	$E_{corr}(\lambda)$	$\Sigma\Delta\varepsilon(\lambda)$	$E_{corr}(\lambda)$
	au	au	au	au	au	au
aug-cc-pVDZ[a]						
1	-3.87367	-3.87367	-4.00955	-4.00955	-4.09665	-4.09665
2	-0.18915	-4.06282	-0.17502	-4.18457	-0.20202	-4.29867
3	0.00653	-4.05629	0.00690	-4.17767	0.00678	-4.29189
exact MP2		-4.05695				
aug-cc-pVTZ[b]						
1	-4.75157	-4.75157	-4.83790	-4.83790	-4.98294	-4.98294
2	-0.20214	-4.95371	-0.18322	-5.02112	-0.21276	-5.19571
3	0.00518	-4.94853	0.00569	-5.01544	0.00527	-5.19044

$\Sigma\Delta\varepsilon(\lambda)$:= sum of the λ-th order contributions
$E_{corr}(\lambda)$:= correlation energy at λ-th order
frozen core=18 orbitals, dsp=4, t_{con}=3 bohr, t_{main}=3 bohr, correlated orbitals=72, f_2=25 bohr
fit basis for the environment: H=STO-3G; O=6-31G (section 2.3)
[a] 738 basis functions
[b] 1656 basis functions

with 1656 contracted Gaussians. This is equivalent to a CCSD(T) calculation with 72 occupied and 1566 virtual orbitals. The maximum disk space was less than 160 GB and the RAM space 7.6 GB. With the applied thresholds for the distance between the domains and the domain-specific basis set, the error of the MP2 calculation is about 0.4 kcal/mol at the aug-cc-pVDZ level. Since the errors of the CCSD and the CCSD(T) methods are usually similar to the MP2 value, we expect them to be of the order of 1 kcal/mol. Since the third-order corrections in the aug-cc-pVTZ basis are similar to the corresponding values in the aug-cc-pVDZ basis, we expect the error of the aug-cc-pVTZ calculations to be similar, too.

3.4 Potential Energy Surfaces

Potential energy surfaces (PES) are a challenging problem for local correlation methods [25,57,80,81]. Even though the CCSD method itself does not dissociate chemical bonds very accurately, the stretching of bonds is a critical and important test for a local correlation method. For instance in the beginning of a geometry optimization or in a transition state one or more bonds might be elongated. In order to be useful, the accuracy of a local approximation to the exact CCSD energy should also be comparable at these points of the potential energy surface. Therefore we studied the performance of the incremental scheme for the dissociation of the central C-C bond of octane at the CCSD level (Fig. 4). If we consider the PES at the second-order level of the incremental series, we find a curve

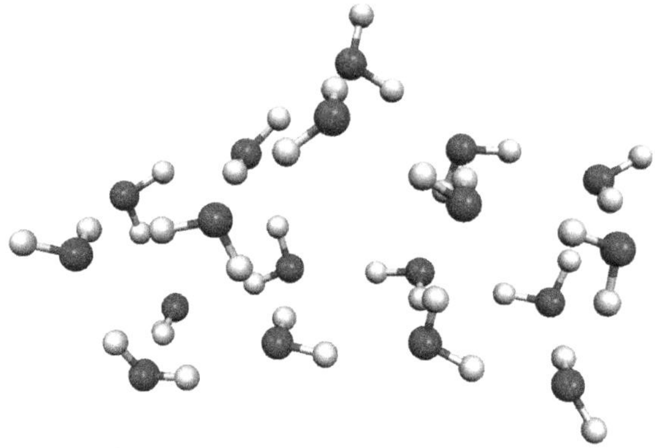

Fig. 3. Structure of $(H_2O)_{18}$ taken from Day et al. [79].

which is smooth to the eye at this scale. Zooming into the PES at the minimum we find that the incremental PES are shifted from the exact PES. At the fourth-order level the PES is smooth to the eye and on top of the full CCSD PES. A more critical criterion to test the accuracy of the incremental PES is the error with respect to the canonical CCSD PES. Without distance screening we find an error of a few milli-atomic units (mau) at second order, which is not constant with respect to the distance and thus affects the minimum geometry. Upon increasing the order the error falls below 0.1 mau at third order and is almost negligible at fourth order. If a distance screening is applied we observe some discontinuities of the PES, however these jumps are already small at third and fourth order for the standard thresholds [57].

4. Conclusion

In this contribution we demonstrated that by using our fully automated and highly parallelized implementation the incremental scheme can be applied to larger molecules to evaluate MP2, CCSD and CCSD(T) energies within chemical accuracy of 1 kcal/mol. Relative energies as well as potential energy surfaces can be obtained with sufficiently high accuracy. We also showed that one can apply the incremental scheme to increase the applicability of a standard CCSD(T) code, e.g. we were able to calculate the CCSD(T) correlation energy of $(H_2O)_{18}$ in a aug-cc-pVTZ basis set with 1656 contracted Gaussian basis functions in C_1 symmetry. Our current efforts concentrate on the incremental evaluation of molecular properties and energy gradients as well as on an extension to wavefunctions depending explicitly on the interelectronic distance.

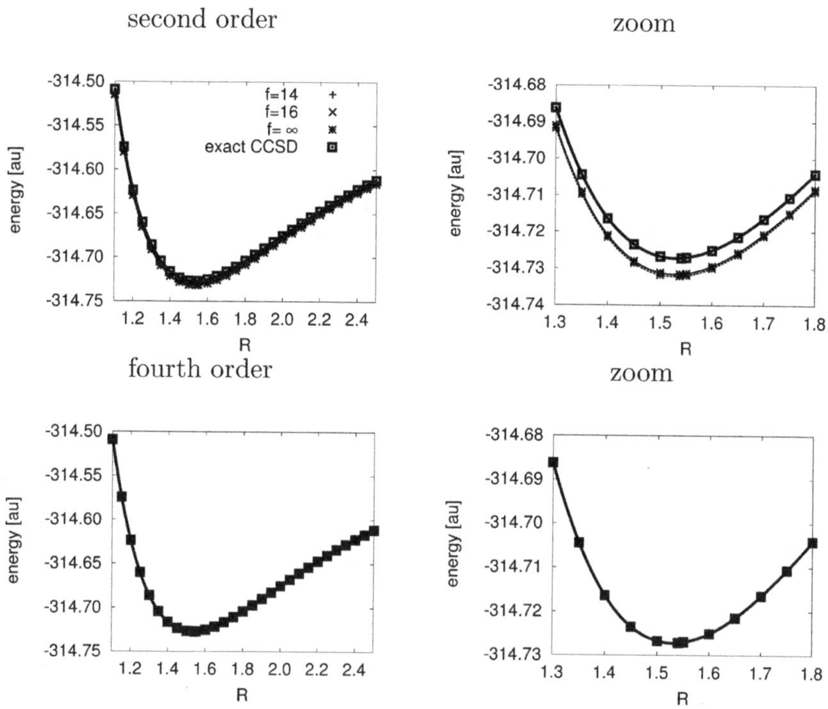

Fig. 4. CCSD/6–31G** single point calculations at the RI-BP86/SV(P) geometries. The geometries were obtained with a relaxed scan along the central C-C bond of octane. A distance screening of f/O was applied. The domain-specific basis set was not used in this example.

Acknowledgement

This work was supported by the German Research Foundation (DFG) in the framework of the Priority Program (Schwerpunktprogramm) 1145.

References

1. W. Förner, J. Ladik, P. Otto, J. Cizek, Chem. Phys. **97** (1985) 251.
2. P. Pulay, S. Saebø, Theor. Chim. Acta. **69** (1986) 357.
3. S. Saebø, P. Pulay, J. Chem. Phys. **86** (2),(1987) 914.
4. H. Stoll, Chem. Phys. Lett. **191** (1992) 548.
5. H. Stoll, Phys. Rev. B: Condens. Matter Mater. Phys. **46** (1992) 6700.
6. H. Stoll, J. Chem. Phys. **97** (1992) 8449.
7. S. Saebø, P. Pulay, Annu. Rev. Phys. Chem. **44** (1993) 213.
8. J. W. Boughton, P. Pulay, J. Comput. Chem. **14** (1993) 736.
9. C. Hampel, H.-J. Werner, J. Chem. Phys. **104** (1996) 6286.
10. M. Schütz, G. Hetzer, H. J. Werner, J. Chem. Phys. **111**(13)(1999), 5691.
11. P. E. Maslen, M. Head-Gordon, Chem. Phys. Lett. **283** (1998) 102.
12. P. E. Maslen, M. Head-Gordon, J. Chem. Phys. **109** (1998) 7093.

13. P. E. Maslen, M. S. Lee, M. Head-Gordon, Chem. Phys. Lett. **319** (2000) 205.
14. M. S. Lee, P. E. Maslen, M. Head-Gordon, J. Chem. Phys. **112** (2000) 3592.
15. M. Schütz, J. Chem. Phys. **113**(22),(2000) 9986.
16. M. Schütz, H.-J. Werner, J. Chem. Phys. **114** (2001) 661.
17. D. Walter, A. B. Szilva, K. Niedfeld, E. A. Carter, J. Chem. Phys. **117** (2002) 1982.
18. W. Li, S. Li, J. Chem. Phys. **121** (2004) 6649.
19. N. Flocke, R. J. Bartlett, J. Chem. Phys. **121** (2004) 10935.
20. D. G. Fedorov, K. Kitaura, J. Chem. Phys. **121** (2004) 2483.
21. D. G. Fedorov, K. Kitaura, J. Chem. Phys. **123** (2005) 134103.
22. V. Deev, M. A. Collins, J. Chem. Phys. **122** (2005) 154102.
23. J. E. Subotnik, M. Head-Gordon, J. Chem. Phys. **123** (2005) 64108.
24. S. Li, J. Shen, W. Li, Y. Jiang, J. Chem. Phys. **125** (2006) 074109.
25. J. E. Subotnik, A. Sodt, M. Head-Gordon, J. Chem. Phys. **125** (2006) 074116.
26. A. Auer, M. Nooijen, J. Chem. Phys. **125** (2006) 024104.
27. V. Weijo, P. Manninen, P. Jørgenson, O. Christiansen, J. Olsen, J. Chem. Phys. **127** (2007) 074106.
28. M. Kobayashi, Y. Imamura, H. Nakai, J. Chem. Phys. **127** (2007) 074103.
29. B. Doser, D. S. Lambrecht, C. Ochsenfeld, Phys. Chem. Chem. Phys. **10** (2008) 3335.
30. T. F. Hughes, N. Flocke, R. J. Bartlett, J. Phys. Chem. A **112** (2008) 5994.
31. M. Kobayashi, H. Nakai, J. Chem. Phys. **129** (2008) 044103.
32. T. S. Chwee, A. B. Szilva, R. Lindh, E. A. Carter, J. Chem. Phys. **128** (2008) 224106.
33. E. E. Dahlke, H. R. Leverentz, D. G. Truhlar, J. Chem. Theory Comput. **4** (2008) 33.
34. H.-J. Werner, J. Chem. Phys. **129** (2008) 101103.
35. T. B. Adler, H.-J. Werner, F. R. Manby, J. Chem. Phys. **130** (2009) 054106.
36. M. Kobayashi, H. Nakai, Int. J. Quantum Chem. **109** (2009) 2227.
37. W. Li, P. Piecuch, J. R. Gour, Theory and Applications of Computational Chemistry - 2008, AIP Conference Proceedings **1102** (2009) 68.
38. M. S. Gordon, J. M. Mullin, S. R. Pruitt, L. B. Roskop, L. V. Slipchenko, J. A. Boatz, J. Phys. Chem. B **113** (2009) 9646.
39. W. Li, P. Piecuch, J. R. Gour, S. Li, J. Chem. Phys. **131** (2009) 114109.
40. R. K. Nesbet, Phys. Rev. **155**, (1967)51.
41. R. K. Nesbet, Phys. Rev. **175** (1968) 2.
42. R. K. Nesbet, Adv. Chem. Phys. **14** (1969) 1.
43. J. Friedrich, M. Hanrath, M. Dolg, J. Chem. Phys. **126** (2007) 154110.
44. H. Stoll, B. Paulus, P. Fulde, J. Chem. Phys. **123** (2005) 144108.
45. M. Yu, S. Kalvoda, M. Dolg, Chem. Phys. Lett. **224** (1997) 121.
46. A. Abdurahman, A. Shukla, M. Dolg, J. Chem. Phys. **112** (2000) 4801.
47. K. Rosciszewski, B. Paulus, P. Fulde, H. Stoll, Phys. Rev. B **60** (1999) 7905.
48. K. Doll, M. Dolg, P. Fulde, H. Stoll, Phys. Rev. B **52** (1995) 4842.
49. K. Doll, M. Dolg, P. Fulde, H. Stoll, Phys. Rev. B **55** (1997) 10282.
50. M. Albrecht, B. Paulus, H. Stoll, Phys. Rev. B **56** (1997) 7339.
51. E. Voloshina, B. Paulus, Phys. Rev. B **75** (2007) 245117.
52. N. Gaston, B. Paulus, U. Wedig, M. Jansen, Phys. Rev. Lett. **100** (2008) 226404.
53. V. Bezugly, U. Birkenheuer, Chem. Phys. Lett. **399** (2004) 57.
54. S. Kalvoda, B. Paulus, M. Dolg, H. Stoll, H.-J. Werner, Phys. Chem. Chem. Phys. **3** (2001) 514.
55. B. Paulus, Int. J. Quantum Chem. **100** (2004) 1026.
56. J. Friedrich, M. Hanrath, M. Dolg, J. Phys. Chem. A **111** (2007) 9830.
57. J. Friedrich, M. Hanrath, M. Dolg, Chem. Phys. **338** (2007) 33.
58. J. Friedrich, M. Hanrath, M. Dolg, J. Phys. Chem. A **112** (2008) 8762.

59. J. Friedrich, S. Coriani, T. Helgaker, M. Dolg, J. Chem. Phys. **131** (2009) 154102.
60. J. Friedrich, K. Walczak, M. Dolg, Chem. Phys. **356** (2009) 47.
61. J. Friedrich, M. Dolg, J. Chem. Phys. **129** (2008) 244105.
62. J. Friedrich, M. Hanrath, M. Dolg, Chem. Phys. **346** (2008) 266.
63. G. Karypis, V. Kumar, SIAM J. Sci. Comput. **20** (1),(1998) 359.
64. J. Friedrich, to be published.
65. T. Janowski, A. R. Ford, P. Pulay, J. Chem. Theory Comput. **3** (2007) 1368.
66. M. E. Harding, T. Metzroth, J. Gauss, A. A. Auer, J. Chem. Theory Comput. **4** (2008) 64.
67. R. M. Olson, J. L. Bentz, K. R. A, M. W. Schmidt, M. S. Gordon, J. Chem. Theory Comput. **3** (2007) 1312.
68. J. Friedrich, M. Dolg, J. Chem. Theory Comput. **5** (2009) 287.
69. H.-J. Werner, P. J. Knowles, R. Lindh, F. R. Manby, M. Schütz, et al., MOLPRO, a package of ab initio programs designed by H.-J. Werner and P. J. Knowles version 2006. tech. rep., University of Birmingham (2006).
70. C. Hampel, K. Peterson, H.-J. Werner, Chem. Phys. Lett. **190** (1992) 1.
71. J. M. Foster, S. F. Boys, Rev. Mod. Phys. **32** (1960) 300.
72. A. D. Becke, Phys. Rev. A **38** (6),(1988 3098).
73. J. P. Perdew, Phys. Rev. B **33** (12),(1986) 8822.
74. R. Ahlrichs, M. Bär, H.-P. Baron, R. Bauernschmitt, S. Böcker, M. Ehrig, K. Eichkorn, S. Elliott, F. Furche, F. Haase, M. Häser, H. Horn, C. Huber, U. Huniar, C. Kölmel, M. Kollwitz, C. Ochsenfeld, H. Öhm, A. Schäfer, U. Schneider, O. Treutler, M. von Arnim, F. Weigend, P. Weis, H. Weiss, Turbomole 5.10, Institut für Physikalische Chemie, Universität Karlsruhe, Kaiserstr. 12, D-76128 Karlsruhe (2008).
75. O. Treutler, R. Ahlrichs, J. Chem. Phys. **102** (1995) 346.
76. K. Eichkorn, O. Treutler, H. Oehm, M. Haeser, R. Ahlrichs, Chem. Phys. Lett. **240** (1995) 283.
77. K. Eichkorn, F. Weigend, O. Treutler, R. Ahlrichs, Theor. Chem. Acc. **97** (1997) 119.
78. P. Deglmann, K. May, F. Furche, R. Ahlrichs, Chem. Phys. Lett. **384** (2004) 103.
79. P. Day, R. Pachter, M. S. Gordon, G. N. Merrill, J. Chem. Phys. **112** (2000) 2063.
80. N. J. Russ, T. D. Crawford, J. Chem. Phys. **121** (2004) 691.
81. R. A. Mata, H.-J. Werner, J. Chem. Phys. **125** (2006) 184110.

Tensor Product Multiscale Many-Particle Spaces with Finite-Order Weights for the Electronic Schrödinger Equation

By Michael Griebel and Jan Hamaekers*

Institute for Numerical Simulation, University of Bonn, Wegelerstraße 6, 53115 Bonn, Germany

(Received July 10, 2009; accepted September 16, 2009)

Schrödinger Equation / Numerical Approximation / Sparse Grid Method / Antisymmetric Sparse Grids / Finite-Order Weights

In this article we combine the favorable properties of efficient Gaussian type orbitals basis sets, which are applied with good success in conventional electronic structure methods, and tensor product multiscale bases, which provide guaranteed convergence rates and allow for adaptive resolution. To this end, we develop and study a new approach for the treatment of the electronic Schrödinger equation based on a modified adaptive sparse grid technique and a certain particle-wise decomposition with respect to one-particle functions obtained by a nonlinear rank-1 approximation. Here, we employ a multiscale Gaussian frame for the sparse grid spaces and we use Gaussian type orbitals to represent the rank-1 approximation. With this approach we are able to treat small atoms and molecules with up to six electrons.

1. Introduction

For many problems in quantum chemistry it is sufficient to consider nuclei as classical point-like particles and to only treat the electrons as quantum-mechanical particles. Then, the Born-Oppenheimer approximation of the time-dependent Schrödinger equation leads to a stationary electronic Schrödinger equation

$$H\Psi = E\Psi$$

with the spin independent molecular Hamiltonian

$$H = -\frac{1}{2}\sum_{i=1}^{N}\Delta_i - \sum_{k=1}^{N_{\text{nuc}}}\sum_{i=1}^{N}\frac{Z_k}{|\mathbf{x}_i - \mathbf{R}_k|} + \sum_{i<j}^{N}\frac{1}{|\mathbf{x}_i - \mathbf{x}_j|} + \sum_{k<l}^{N_{\text{nuc}}}\frac{Z_k Z_l}{|\mathbf{R}_k - \mathbf{R}_l|}$$

* Corresponding author. E-mail: hamaekers@ins.uni-bonn.de

Z. Phys. Chem. **224** (2010) 527–543
© by Oldenbourg Wissenschaftsverlag, München

written in atomic units. Here, N_{nuc} denotes the number of clamped nuclei $(\mathbf{R}_k, Z_k)$ and N is the number of electrons with coordinates $(\mathbf{x}_i)$. The solution of this N-electron problem lives in a $3N$-dimensional space and respects the antisymmetry condition imposed by Pauli's principle.

The high dimensionality of this eigenvalue problem turns out to be the main bottleneck for a direct numerical treatment, since any discretization on e.g. uniform grids with $O(m)$ points in each direction would involve $M = O(m^{3N})$ degrees of freedoms. Furthermore, only a convergence estimate of type

$$\|\Psi - \Psi_M\|_{\mathcal{H}^s} \leq c(N) M^{-r/(3N)} \|\Psi\|_{\mathcal{H}^{r+s}}$$

can be achieved, where $\|\cdot\|_{\mathcal{H}^s}$ is the usual Sobolev norm in $\mathcal{H}^s$, $r+s$ denotes the isotropic smoothness of Ψ and c is a constant which may depend on N but not on M. Here, we encounter the curse of dimensionality, i.e. the rate of convergence deteriorates exponentially with the number N of electrons. Therefore, further model approximations were developed and implemented in quantum chemistry which lead to reduced complexity. Examples are Hartree-Fock or multi-configuration methods (HF/MCHF) and a variety of additive methods (configuration interaction), exponential methods (coupled cluster), perturbative methods (Möller-Plesset theory), quantum Monte Carlo methods (VMC/DMC) and density functional theory methods (DFT). These first principles electronic structure methods are used with great success in practice. However, it is not yet completely clear how to systematically improve their approximation error in a mathematically rigorous way.

On the other hand, sparse grid techniques are frequently used to treat high dimensional problems. Sparse grids spaces are multiscale approximation spaces based on tensor products which promise to circumvent the above-mentioned curse of dimensionality of a conventional discretization using full grids, at least to some extent. We envision, up to logarithmic terms, a convergence estimate of the type

$$\|\Psi - \Psi_M\|_{\mathcal{H}^s} \leq c(N) M^{-r/3} \|\Psi\|_{\mathcal{H}^{r,s}_{\mathrm{mix}}}, \tag{1}$$

where the rate of convergence does no longer exponentially deteriorate with the number N of particles. Note however, that a more restrictive smoothness requirement, namely the boundedness of a certain (r,s)-th mixed derivative is involved. In [1], Yserentant showed that the solution of the Schrödinger equation lies in certain spaces of mixed dominating smoothness, i.e. $\Psi \in \mathcal{H}^{1/2,1}_{\mathrm{mix}}$, when the physically relevant antisymmetry of the wave function is taken into account. These spaces are closely related to the smoothness assumption of the conventional sparse grid method.

A first application of the sparse grid combination technique to the Schrödinger equation was studied in [2]. Furthermore, sparse grid approaches using Fourier basis functions and Meyer basis functions were implemented and studied in [3] and [4], respectively. Here, it turned out that, in principle, the sparse grid approach indeed possesses favourable approximation rates and cost complexities for the solution of Schrödinger's equation. However, the computations are limited due to large constants involved in the approximation rates and cost complexities. Actually, these constants can be exponentially dependent on the number of particles. Also in the framework of the variational Monte Carlo (VMC) approach wavelets and sparse grids have been successfully applied for the representation of Jastrow factors [5,6].

In this article we propose a new approach for the electronic Schrödinger equation based on the combination of a nonlinear low-rank approximation method and sparse grid bases which provide guaranteed approximation rates. This combination is realized with help of a general particle-wise subspace splitting of the many-particle space which builds on so-called weighted spaces of finite order $q \ll N$. Then it is only the lower dimensionality of the q-th order terms, which exponentially enters the work count complexity for multilinear approximation problems [7–9]. In this way, we implement tensor product multiscale many-particle spaces with finite-order weights which are built from non-orthogonal Hartree-Fock orbitals and a variant of sparse grid spaces for many-particle functions. Here, to obtain favorably small constants, we present a multiscale frame of Gaussians and develop a heuristic h-adaptive refinement scheme for the resolution of cusps.

The remainder of this article is organized as follows: In Section 2 we recall the Pauli principle and many-particle Sobolev spaces of dominating mixed smoothness. Furthermore, we discuss the discretization with general sparse grid spaces and give estimates for approximation rates for electronic wave functions. In Section 3 we introduce weighted many-particle spaces based on a particle-wise subspace splitting and discuss our numerical approach for the solution of the electronic Schrödinger equation. Here, we introduce a multiscale Gaussian frame and describe our adaptive scheme. In Section 4 we present the results of our numerical experiments. Finally, some concluding remarks are given in Section 5.

2. Pauli principle, regularity of the wave function and general sparse grid spaces

In general, an electronic wave function depends on the positions $\mathbf{x}_i$ of the electrons and on their associated spin coordinates $s_i \in \{+\frac{1}{2}, -\frac{1}{2}\}$. It furthermore has to obey the antisymmetry conditions

$$\Psi(P\vec{\mathbf{x}}, P\vec{s}) = (-1)^{|P|}\Psi(\vec{\mathbf{x}}, \vec{s}), \quad \forall P \in \mathcal{S}_N,$$

which reflect the Pauli principle. Here, $\mathcal{S}_N$ denotes the symmetric group. Note that for a given spin distribution $\vec{s}$ the spatial part $\Psi_{\vec{s}}(\vec{\mathbf{x}}) := \Psi(\vec{\mathbf{x}}, \vec{s})$ obeys the partial antisymmetry condition

$$\Psi_{\vec{s}}(P\vec{\mathbf{x}}) = (-1)^{|P|}\Psi_{\vec{s}}(P\vec{\mathbf{x}}), \quad \forall P \in \mathcal{S}_{\vec{s}} := \{P \in \mathcal{S}_N \; : \; P\vec{s} = \vec{s}\}.$$

In particular, the minimal eigenvalue of all eigenvalue problems for the spatial components is equal to the minimal eigenvalue of the full eigenvalue problem [10,11].

Although there are 2^N possible different spin distributions $\vec{s}$, the bilinear form $\langle \Psi(P\,\cdot)|H|\,\Psi(P\,\cdot)\rangle$ is invariant under all permutations $P \in \mathcal{S}_N$ of the position coordinates $\vec{\mathbf{x}}$. Thus, it is sufficient to consider N+1 eigenvalue problems which are associated with N+1 different class representative spin vectors with a total spin projection of values $M_S = -\frac{N}{2}, \ldots, -1, 0, 1, \ldots, \frac{N}{2}$ for even N and a total spin projection of values $M_S = -\frac{N}{2}, \ldots, -\frac{1}{2}, \frac{1}{2}, \ldots, \frac{N}{2}$ for odd N. Here, we choose the N+1 different class representative spin vectors $\vec{s}^{(M_S)} \in \{+\frac{1}{2}, -\frac{1}{2}\}^N$, where

$$s_j^{(M_S)} := \begin{cases} +\frac{1}{2} & \text{for } j \leq \frac{N}{2} + M_S, \\ -\frac{1}{2} & \text{for } j > \frac{N}{2} + M_S. \end{cases}$$

In the following we denote the number of spin-up particles by $N_\uparrow$ and the number of spin-down particles by $N_\downarrow$. Then $M_S = \frac{1}{2}(N_\uparrow - N_\downarrow)$. The functions of the N-particle space $\mathcal{L}^2((\mathbb{R}^3)^N)$ which obey the partial antisymmetry condition for a given $\vec{s}^{(M_S)}$ form a linear subspace $\mathcal{L}^2_{(M_S)} \subset \mathcal{L}^2$. We define the projection into this subspace, i.e. the antisymmetrization operator $\mathfrak{A}^{(M_S)}$: $\mathcal{L}^2((\mathbb{R}^3)^N) \to \mathcal{L}^2_{(M_S)}((\mathbb{R}^3)^N)$ by

$$\mathfrak{A}^{(M_S)}\Psi(\vec{\mathbf{x}}) := \frac{1}{N_\uparrow! N_\downarrow!} \sum_{P \in \mathcal{S}_{\vec{s}^{(M_S)}}} (-1)^{|P|} \Psi(P\vec{\mathbf{x}}).$$

The spatial part with respect to M_S of an electronic wave function is at least in $\mathcal{H}^1 \cap \mathcal{L}^2_{(M_S)}$. Let us assume that the smallest energy with respect to M_S, i.e.

$$E^{(M_S)}_{\min} = \min_{\Psi \in \mathcal{H}^1 \cap \mathcal{L}^2_{(M_S)}, \|\Psi\|_{\mathcal{L}^2}=1} \langle \Psi, H\Psi \rangle_{\mathcal{L}^2},$$

exhibits multiplicity one. Furthermore, let $\{V_\kappa\}_{\kappa \in \mathbb{N}}$ be an arbitrary dense family of finite-dimensional subspaces in $\mathcal{H}^1 \cap \mathcal{L}^2_{(M_S)}$. Let finally E_κ and Ψ_κ denote Galerkin approximations associated with the lowest state in the subspace V_κ, i.e.

$$E_\kappa = \min_{\Psi \in V_\kappa, \|\Psi\|_{\mathcal{L}^2}=1} \langle \Psi, H\Psi \rangle_{\mathcal{L}^2}, \quad \Psi_\kappa = \operatorname{argmin}_{\Psi \in V_\kappa, \|\Psi\|_{\mathcal{L}^2}=1} \langle \Psi, H\Psi \rangle_{\mathcal{L}^2}.$$

Then, $E^{(M_S)}_{\min} \leq E_\kappa$ for all $\kappa \in \mathbb{N}$ and thereby a relation between an estimate for the accuracy of an eigenfunction and an estimate for the approximation error of the lowest eigenvalue can be deduced: There exist $C_1, C_2 > 0$ and $\tilde{\kappa} \in \mathbb{N}$ such that the relation

$$\begin{aligned} E^{(M_S)}_{\min} - E_\kappa &\leq C_1 \langle \Psi^{(M_S)}_{\min} - \Psi_\kappa, H(\Psi^{(M_S)}_{\min} - \Psi_\kappa) \rangle_{\mathcal{L}^2} \\ &\leq C_2 \|\Psi^{(M_S)}_{\min} - \Psi_\kappa\|^2_{\mathcal{H}^1} \end{aligned} \tag{2}$$

holds for all $\kappa \geq \tilde{\kappa}$; see [11–12]. An upper bound for the approximation error in the $\mathcal{H}^1$-norm, i.e. for the right hand side of (2), can then be estimated with the help of the regularity results of [1,11,13] similar to (1). This works since the spatial part $\Psi^{(M_S)} := \Psi_{\vec{s}^{(M_S)}}$ of a weak solution of the electronic Schrödinger equation is the Sobolev space $\mathcal{H}^{1/2,1}_{\text{mix}}$ of dominating mixed smoothness and it is even in $\mathcal{H}^{1,1}_{\text{mix}}$ for an electronic wave function of totally parallel spin; see [1] for details.

Such types of spaces are defined for $-\infty < t,\ r < \infty$, $N \in \mathbb{N}$ by[1]

$$\mathcal{H}^{t,r}_{\text{mix}}((\mathbb{R}^3)^N) := \Big\{ f \in \mathcal{S}'((\mathbb{R}^3)^N) :$$

$$\|f\|_{\mathcal{H}^{t,r}_{\text{mix}}} := \int_{(\mathbb{R}^3)^N} |\omega^t_{\text{mix}}(\vec{\mathbf{k}})\omega^r_{\text{iso}}(\vec{\mathbf{k}})\hat{f}(\vec{\mathbf{k}})|^2\, d\vec{\mathbf{k}} < \infty \Big\}$$

with $\omega_{\text{iso}}(\vec{\mathbf{k}}) := \sqrt{1+\Sigma_{p=1}^N|\mathbf{k}_p|_\infty^2}$ and $\omega_{\text{mix}}(\vec{\mathbf{k}}) := \sqrt{\Pi_{p=1}^N(1+|\mathbf{k}_p|_\infty^2)}$. Here, $\mathcal{S}'$ denotes the space of all tempered distributions and $\hat{f}$ denotes the Fourier transform of f. Note that the regularity of a function is directly related to the decay properties of its Fourier transform. In particular, the standard isotropic Sobolev spaces as well as the standard Sobolev spaces of dominating mixed smoothness [14], both generalized to the N-particle case [3,4], are included in the definition of $\mathcal{H}^{t,r}_{\text{mix}}$. They can be written as

$$\mathcal{H}^r((\mathbb{R}^3)^N) = \mathcal{H}^{0,r}_{\text{mix}}((\mathbb{R}^3)^N) \quad \text{and} \quad \mathcal{H}^t_{\text{mix}}((\mathbb{R}^3)^N) = \mathcal{H}^{t,0}_{\text{mix}}((\mathbb{R}^3)^N),$$

respectively.

In the following, we consider $f \in \mathcal{H}^{t,r}_{\text{mix}}$ with certain decay properties for $|\vec{\mathbf{x}}| \to \infty$. Note that the decay of a function f for $|\vec{\mathbf{x}}|_2 \to \infty$ in the spatial space is directly related to the regularity of the Fourier transform $\hat{f}$ in Fourier space. In this way, for $t+r \geq 0$, $t \geq 0$, $\hat{t}+\hat{r} \geq 0$, $\hat{t} \geq 0$, we define

$$\mathcal{H}^{t,r;\hat{t},\hat{r}}_{\text{mix}} := \Big\{ f \in \mathcal{L}^2((\mathbb{R}^3)^N) \,:\, \|f\|_{\mathcal{H}^{t,r;\hat{t},\hat{r}}_{\text{mix}}} := \|f\|_{\mathcal{H}^{t,r}_{\text{mix}}} + \|\hat{f}\|_{\mathcal{H}^{\hat{t},\hat{r}}_{\text{mix}}} < \infty \Big\}.$$

In [15] we study its discretization by general sparse grid spaces and the resulting linear approximation properties. To this end, let $\{\phi_{l,\mathbf{j}}\}_{(l,\mathbf{j})\in\mathbb{N}\times\mathbb{Z}^3}$ be a set of multiscale basis functions which span the one-particle space $\mathcal{L}^2(\mathbb{R}^3)$, where the indices l and $\mathbf{j}$ are associated with the scale and the location, respectively. Then, the set of multiscale tensor product basis functions

$$\mathcal{B} := \left\{ \phi_{\vec{l},\vec{\mathbf{j}}} := \bigotimes_{p=1}^{N} \phi_{l_p,\mathbf{j}_p} \right\}_{(\vec{l},\vec{\mathbf{j}})\in\mathbb{N}^N\times(\mathbb{Z}^3)^N}$$

spans the N-particle space $\mathcal{L}^2((\mathbb{R}^3)^N)$. A general finite-dimensional sparse grid space is given by

$$V^{SG}_{\Lambda} := \operatorname{span}\Big(\mathcal{B}_\Lambda := \big\{\phi_{\vec{l},\vec{\mathbf{j}}} \in \mathcal{B} \,:\, (\vec{l},\vec{\mathbf{j}}) \in \Lambda\big\}\Big) \tag{3}$$

for a finite subset of indices $\Lambda \subset \mathbb{N}^N \times (\mathbb{Z}^3)^N$. The specific choice of the basis functions $\{\phi_{l,\mathbf{j}}\}$ and of the set Λ leads to different variants of sparse grids, e.g. regular sparse grids, energy-norm based sparse grids and dimension adaptive sparse grids [16].

---

1 In particular, the Sobolev space $\mathcal{H}^{t,r}_{\text{mix}}((\mathbb{R}^3)^N)$ is a Hilbert space together with the Hermitian inner product $\langle f,g\rangle_{\mathcal{H}^{t,r}_{\text{mix}}} := \langle \omega^t_{\text{mix}}\,\omega^r_{\text{iso}}\,\hat{f}\ ,\ \omega^t_{\text{mix}}\,\omega^r_{\text{iso}}\,\hat{g}\rangle_{\mathcal{L}^2}$.

For example, in [15] we construct sparse grid basis sets by a Littlewood-Paley like multiscale decomposition of general hyperbolic cross spaces. In certain cases this leads to Meyer wavelets [17]. For the details of this construction of general sparse grid basis functions we refer to [15,18].

Now let us go back to the Schrödinger equation ($\mathcal{H}^{1/2,1}_{\text{mix}}$-regularity) and let us additionally assume that the Fourier transform $\hat{\Psi}^{(M_S)}$ is in $\mathcal{H}^{\hat{t},0}_{\text{mix}}$, $\hat{t} > 0$. Let further V^{SG}_Λ be an appropriate general sparse grid space with $M = |\Lambda|$ degrees of freedom.[2] Then, the estimate

$$\inf_{\tilde{\Psi} \in V^{SG}_\Lambda} \|\Psi^{(M_S)} - \tilde{\Psi}\|_{\mathcal{H}^1} \lesssim \left(\frac{M}{\log_2(M)^{2(N-1)}} \right)^{-\frac{1}{6(1+\tau)}} \|\Psi^{(M_S)}\|_{\mathcal{H}^{\frac{1}{2},1;\hat{t},0}_{\text{mix}}} \tag{4}$$

holds, where $\tau = 3/(2\hat{t})$. For a proof see [15]. We have $\tau \to 0$ if the Fourier transform $\hat{\Psi}^{(M_S)}$ is smooth, i.e. if $\Psi^{(M_S)}$ decays sufficiently fast.[3] Moreover, up to logarithmic terms, the convergence rate is independent of the number of electrons N and is almost the same as in the two-electron case. Thus, we obtain a rate of order $-\frac{1}{6(1+\tau)}$ and hence, due to (2), a rate of order $-\frac{2}{6(1+\tau)}$ for the minimal eigenvalue. Note however that the term $\|\Psi^{(M_S)}\|_{\mathcal{H}^{\frac{1}{2},1;\hat{t},0}_{\text{mix}}}$ and the constants involved in the approximation and complexity order estimates usually depend on the number of particles.

Due to Pauli's principle it is sufficient to consider the partially antisymmetric subspace $\mathfrak{A}^{(M_S)}(V^{SG}_\Lambda) \subset V^{SG}_\Lambda$. This subspace can be spanned by basis functions in the form of the product of two Slater determinants, i.e.

$$\mathfrak{A}^{(M_S)} \bigotimes_{p=1}^{N} \phi_{l_p, \mathbf{j}_p} = \frac{1}{N_\uparrow!} \bigwedge_{p_\uparrow=1}^{N_\uparrow} \phi_{l_{p_\uparrow}, \mathbf{j}_{p_\uparrow}} \otimes \frac{1}{N_\downarrow!} \bigwedge_{p_\downarrow=N_\uparrow+1}^{N} \phi_{l_{p_\downarrow}, \mathbf{j}_{p_\downarrow}}.$$

In particular, the order for the number of degrees of freedom related to $\mathfrak{A}^{(M_S)}(V^{SG}_\Lambda)$ stays the same as in the case of V^{SG}_Λ. However, the involved constant is now reduced by the factor $1/(N_\uparrow! N_\downarrow!)$, i.e. $|\mathfrak{A}^{(M_S)}(V^{SG}_\Lambda)| \leq \frac{1}{N_\uparrow! N_\downarrow!} |V^{SG}_\Lambda|$. Furthermore, the order of the achieved accuracy does not change when we switch to the partially antisymmetric case. For details see [3,4,15].

2 Here, V^{SG}_Λ corresponds to the space $V^{0;0}_{L;\tau L}$, $\tau = \frac{3}{2\hat{t}}$, introduced in [15]. There $V^{0;0}_{L;J}$ is spanned by $\{\phi_{\vec{l},\vec{\mathbf{j}}} : \vec{l} \in I_L, \vec{\mathbf{j}} \in \mathcal{J}_J(\vec{l})\}$, where $I_L := \{\vec{l} \in \mathbb{N}^N : |\vec{l}|_1 \leq L + N - 1\}$, $\mathcal{J}_J(\vec{l}) := \bigcup_{\vec{\iota} \in I_J} Q_{\vec{\iota}+\vec{l}+\vec{1}}$, $Q_{\vec{\alpha}} := Q_{\alpha_1} \times \ldots \times Q_{\alpha_N}$ and $Q_\alpha := \{\mathbf{k} \in \mathbb{Z}^3 : |\mathbf{k}|_\infty \leq 2^\alpha\}$. In particular with respect to the level parameter L it holds for the number of degrees of freedom $M = |V^{0;0}_{L;\tau L}| = O(2^{3L(1+\tau)}(\tau L^2)^{N-1})$.

3 It can be shown that eigenfunctions associated with eigenvalues in the discrete spectrum (bound-states) decay exponentially; compare [11,19].

3. Adaptive approximation spaces of finite order

General sparse grid spaces can also be employed for many-particle spaces with so-called finite-order weights [7,9]. The resulting dimension-wise decompositions and discretization schemes then allow to get rid of the remaining exponential dependence on the number of particles with respect to the logarithmic terms in (4), since for finite-order weights of order q the problem of the approximation of an N-particle function reduces to the problem of the approximation of q-particle functions. Note that our construction scheme includes the common CI spaces as a special case, but allows for a more flexible choice of finite-dimensional approximation spaces.

3.1 Particle-wise decomposition

In the following we focus on a particle-wise decomposition of the partially antisymmetric many-particle space $\mathcal{L}^2_{(M_S)}$. Here, for a shorter notation we set $\mathcal{N}:= \{1, \ldots, N\}$, $\mathcal{N}_\uparrow := \{0, \ldots, N_\uparrow\}$ and $\mathcal{N}_\downarrow := \{N_\uparrow+1, \ldots, N\}$. Let $\{g_p\}_{p\in\mathcal{N}}$ be $\mathcal{L}^2$-normalized one-particle functions, where $\{g_p\}_{p\in\mathcal{N}_\uparrow}$ as well as $\{g_p\}_{p\in\mathcal{N}_\downarrow}$ are linear independent. Then we have the direct sum decompositions

$$\mathcal{L}^2(\mathbb{R}^3) = U_\uparrow \oplus W_\uparrow \text{ and } \mathcal{L}^2(\mathbb{R}^3) = U_\downarrow \oplus W_\downarrow, \tag{5}$$

where $U_\uparrow := \operatorname{span}(\{g_p\}_{p\in\mathcal{N}_\uparrow})$ and $U_\downarrow := \operatorname{span}(\{g_p\}_{p\in\mathcal{N}_\downarrow})$. With this splitting of the one-particle space, the subspace splitting

$$\mathcal{L}^2_{(M_S)} = \mathfrak{A}^{(M_S)}\left(\bigotimes_{p\in\mathcal{N}_\uparrow}(U_\uparrow \oplus W_\uparrow) \otimes \bigotimes_{p\in\mathcal{N}_\downarrow}(U_\downarrow \oplus W_\downarrow)\right) = \bigoplus_{u\subset\mathcal{N}} \mathcal{W}_u^{(M_S)} \tag{6}$$

follows for the N-particle space, where we set

$$\mathcal{W}_u^{(M_S)} := \mathfrak{A}^{(M_S)} \bigotimes_{p=1}^{N} \mathcal{W}_{u,(p)}^{(M_S)}, \quad \mathcal{W}_{u,(p)}^{(M_S)} := \begin{cases} \operatorname{span}\{g_p\} & \text{for } p \in \mathcal{N}\setminus u, \\ W_\uparrow & \text{for } p \in u \cap \mathcal{N}_\uparrow, \\ W_\downarrow & \text{for } p \in u \cap \mathcal{N}_\downarrow. \end{cases}$$

Correspondingly, any function $f \in \mathcal{L}^2_{(M_S)}$ can be decomposed as

$$f = \sum_{u\subset\mathcal{N}}^{N} \mathfrak{A}^{(M_S)}(F_u) = \sum_{u\subset\mathcal{N}}^{N} \mathfrak{A}^{(M_S)}\left(\vec{\mathbf{y}} \mapsto f_u(\vec{\mathbf{y}}_u) \prod_{p\in\mathcal{N}\setminus u} g_p(\mathbf{y}_p)\right) \tag{7}$$

with the help of appropriate linear projections such that $\mathfrak{A}^{(M_S)}F_u \in \mathcal{W}_u^{(M_S)}$ and $f_u \in \otimes_{p\in u}\mathcal{W}_{u,(p)}^{(M_S)}$. Moreover, in the case of orthogonal direct sums $V_\uparrow \oplus W_\uparrow$, $V_\downarrow \oplus W_\downarrow$ and orthonormal $\{g_p\}_{p\in\mathcal{N}_\uparrow}$, $\{g_p\}_{p\in\mathcal{N}_\downarrow}$, the orthogonality relation $\langle F_u, F_{u'}\rangle = 0$ holds for all $u \neq u'$. Note that a similar type of a particle-wise decomposition was introduced by Sinanoğlu in quantum chemistry for the analysis of many-electron wave functions [20]. There, orthogonal Hartree-Fock orbitals were suggested as one-particle functions $\{g_p\}_{p\in\mathcal{N}}$. Note furthermore the close relation of this particle-wise decomposition to the classical ANOVA decomposition [8].

Now, we consider decompositions

$$\mathcal{L}^2(\mathbb{R}^3) = U_\uparrow + W \text{ and } \mathcal{L}^2(\mathbb{R}^3) = U_\downarrow + W, \tag{8}$$

which involve non-direct sums instead of the direct sums in (5). Then, similar to (6) we obtain

$$\mathcal{L}^2_{(M_S)} = \mathfrak{A}^{(M_S)}\left(\bigotimes_{p\in\mathcal{N}_\uparrow}(U_\uparrow + W) \otimes \bigotimes_{p\in\mathcal{N}_\downarrow}(U_\downarrow + W)\right) = \sum_{u\subset\mathcal{N}} \mathcal{W}_u^{(M_S)} \tag{9}$$

for the N-particle space.[4] Here, in contrast to (6) and (7), a corresponding decomposition of a function in $\mathcal{L}^2_{(M_S)}$ is no longer unique. However, appropriate weighted norms can still be introduced with the help of the infimum over all decompositions of a function; see also [21]. This is discussed in the following.

3.2 Weighted many-particle approximation spaces

In the setting of (8) and (9) for $t+r \geq 0$, $t \geq 0$, $\{g_p\}_{p\in\mathcal{N}}$ and non-negative $\{\gamma_u\}_{u\subset\mathcal{N}}$, weighted partially antisymmetric many-particle spaces can be introduced by

$$\mathcal{H}^{t,r;\hat{t},\hat{r};(M_S)}_{\text{mix},\{g_p\},\{\gamma_u\}}((\mathbb{R}^3)^N) := \left\{ f \in \mathcal{L}^2_{(M_S)} : \|f\|_{\mathcal{H}^{t,r;\hat{t},\hat{r}}_{\text{mix},\{g_p\},\{\gamma_u\}}} < \infty \right\}$$

with the norm

$$\|f\|^2_{\mathcal{H}^{t,r;\hat{t},\hat{r};(M_S)}_{\text{mix},\{g_p\},\{\gamma_u\}}} := \inf\Bigg\{ \sum_{u\subset\mathcal{N}} \gamma_u^{-1} \|\tilde{f}_u\|^2_{\mathcal{H}^{t,r;\hat{t},\hat{r}}_{\text{mix}}} :$$

$$\{\tilde{f}_u \in W^{|u|}\}_{u\subset\mathcal{N}}, f = \sum_{u\subset\mathcal{N}} \mathfrak{A}^{(M_S)}(\vec{\mathbf{y}} \mapsto \tilde{f}_u(\vec{\mathbf{y}}_u) \prod_{p\in\mathcal{N}\setminus u} g_p(\mathbf{y}_p))\Bigg\}.$$

Note that $\{\gamma_u\}_{u\subset\mathcal{N}}$ is called a set of weights of finite order q if $\gamma_u = 0$ for all γ_u with $|u| > q$. In that case, the problem of the approximation of an N-particle function reduces to the problem of the approximation of $\binom{N}{q}$ q-particle functions. In the framework of the decomposition (9) we use finite-order weights $\{\gamma_u\}_{u\subset\mathcal{N}}$ to switch certain subspaces $\mathcal{W}_u^{(M_S)}$ on or off, i.e. we set

$$\mathcal{V}^{(M_S)}_{\{\gamma_u\}_u} := \sum_{u\subset\mathcal{N},\gamma_u>0} \mathcal{W}_u^{(M_S)}. \tag{10}$$

Clearly, $\mathcal{V}^{(M_S)}_{\{\gamma_u\}_u} \subset \mathcal{L}^2_{(M_S)}$ since $\{u \subset \mathcal{N} : \gamma_u > 0\}$ is a subset of $\{u \subset \mathcal{N}\}$. The restriction of the decomposition (9) via the finite-order weights $\{\gamma_u\}_{u\subset\mathcal{N}}$ resembles a first step towards approximation, but the involved subspaces $\mathcal{W}_u^{(M_S)}$ are still infinite-dimensional and need further discretization. Thus, the idea is to construct a finite-dimensional subspace of $\mathcal{V}^{(M_S)}_{\{\gamma_u\}_u}$ by choosing a specific finite-dimensional subspace of $\mathcal{W}_u^{(M_S)}$ for each u with $\gamma_u > 0$ separately. Altogether, this induces a discretization of $\mathcal{V}^{(M_S)}_{\{\gamma_u\}_u}$.

4 Note that we have $\mathcal{W}^{(M_S)}_{u,(p)} = \begin{cases} \text{span}\{g_p\} & \text{for } p \in \mathcal{N}\setminus u, \\ W & \text{for } p \in u. \end{cases}$

To this end, let $\{\phi_\nu\}_{\nu\in\mathbb{N}}$ be an appropriate frame of $W = \mathcal{L}^2(\mathbb{R}^3)$. Note that for reasons of simplicity we switch here from the notation with two indices $(l, \mathbf{j})$ associated to level and location as in Section 3 to a notation with just one unspecified index ν. Then, for each $u = \{p_1 < \ldots < p_{|u|}\} \subset \mathcal{N}$ with $\gamma_u > 0$ we introduce a finite-dimensional subspace $\mathcal{W}_{u,\mathcal{B}_u}^{(M_S)}$ of $\mathcal{W}_u^{(M_S)}$ by the span of the finite set $\mathcal{B}_u$ of partially antisymmetric N-particle functions. Here, $\mathcal{B}_u$ is chosen as appropriate finite subset of

$$\mathcal{B}_u^{(M_S)} := \left\{ \mathfrak{A}^{(M_S)}\left(\vec{\mathbf{y}} \mapsto \prod_{q=1}^{|u|} \phi_{\nu_q}(\mathbf{y}_{p_q}) \prod_{p\in\mathcal{N}\setminus u} g_p(\mathbf{y}_p)\right)\right\}_{(\nu_1,\ldots,\nu_{|u|})\in\mathbb{N}_u^{(M_S)}}, \tag{11}$$

where

$$\mathbb{N}_u^{(M_S)} := \left\{(\nu_1,\ldots,\nu_{|u|}) \in \mathbb{N}^{|u|} \ : \ \nu_1 < \ldots < \nu_{n_\uparrow}, \nu_{n_\uparrow+1} < \cdots < \nu_{|u|}\right\}$$

with $n_\uparrow := |u \cap \mathcal{N}_\uparrow|$. Note that for a shorter notation we indicate subspaces directly by their spanning system instead of a set of indices as in (3). In this way, a finite-dimensional subspace of $\mathcal{V}_{\{\gamma_u\}_u}^{(M_S)}$ can be defined by

$$V_{\{\gamma_u\}_u,\{\mathcal{B}_u\}_u}^{SG} := \sum_{u\subset\mathcal{N},\gamma_u>0} \mathcal{W}_{u,\mathcal{B}_u}^{(M_S)}. \tag{12}$$

Note that the specific choice of the one-particle functions $\{g_p\}_{p\in\mathcal{N}}$, the multiscale frame $\{\phi_\nu\}_{\nu\in\mathbb{N}}$, the weights $\{\gamma_u\}_{u\subset\mathcal{N}}$ and the family of finite systems $\{\mathcal{B}_u\}_{u\subset\mathcal{N},\gamma_u>0}$ is still open.

For the $\{g_p\}_{p\in\mathcal{N}}$ we employ a set of non-orthogonal Hartree-Fock orbitals, which are given in terms of atomic orbitals. In particular, these orbitals are written in the form of a finite expansion of isotropic and modulated Gaussians; see [15,22]. The specific choices of $\{\phi_\nu\}_{\nu\in\mathbb{N}}$, $\{\gamma_u\}_{u\subset\mathcal{N}}$ and $\{\mathcal{B}_u\}_{u\subset\mathcal{N},\gamma_u>0}$ are discussed in the following.

3.3 Multiscale Gaussian frame

For the one-particle frame $\{\phi_\nu\}_{\nu\in\mathbb{N}}$ we employ a wavelet-like frame based on Gaussians, which exhibits exponential decay in real space as well as in Fourier space and in particular allows for local adaptivity and the computation of all inner products by analytic formulae.[5] We introduce this wavelet-like frame in the following.

By dilation and translation, we define the functions

$$\varphi_{\sigma,c,l,\mathbf{j}}(\mathbf{x}) := \left(c2^l\right)^{\frac{3}{2}} \varphi_\sigma(c2^l\mathbf{x} - \mathbf{j})$$

for $c > 0$, $l \in \mathbb{N}_0$ and $\mathbf{j} \in \mathbb{Z}^3$, where we employ for the generating function $\varphi_\sigma(\mathbf{x}) := (\sigma\sqrt{\pi})^{-\frac{3}{2}} e^{-\frac{1}{2\sigma^2}|\mathbf{x}|_2^2}$ a normalized isotropic Gaussian. Moreover, we introduce a wavelet-like function in terms of generating functions of two scales by

$$\psi_\sigma(\mathbf{x}) := C_{\psi_\sigma}\left(\varphi_{\frac{\sigma}{2}}(\mathbf{x}) - 2^{-\frac{3}{2}}\varphi_\sigma(\mathbf{x})\right),$$

5 Note that from the Balian-Low theorem there follows that no orthonormal frame with exponential decay in both real space and in Fourier space exists.

with the normalization constant $C_{\psi_\sigma} = (1-\frac{16}{25}\gamma\sqrt{5}+\gamma^2)^{-\frac{1}{2}}$. In particular, we define the functions

$$\psi^{[\mathbf{z}]}_{\sigma,c,l,\mathbf{j}}(\mathbf{x}) := \left(c2^l\right)^{\frac{3}{2}} \psi_\sigma(c2^l\mathbf{x} - \mathbf{j} - \tfrac{1}{2}\mathbf{z})$$

by dilation and translation for $c > 0$, $l \in \mathbb{N}_0$, $\mathbf{j} \in \mathbb{Z}^3$ and $\mathbf{z} \in \mathcal{Z} := \{0,1\}^3 \backslash \mathbf{0}$. Note that $\psi^{[\mathbf{z}]}_{\sigma,c,l,\mathbf{j}}$ is normalized and its zeroth and first moments vanish. Finally, we introduce a wavelet-like frame by

$$\mathfrak{b}_{\sigma,c} := \left\{\varphi_{\sigma,c,0,\mathbf{j}} \;:\; \mathbf{j} \in \mathbb{Z}^3\right\} \cup \bigcup_{\mathbf{z}\in\mathcal{Z}} \left\{\psi^{[\mathbf{z}]}_{\sigma,c,l,\mathbf{j}} \;:\; l \in \mathbb{N}_0, \mathbf{j} \in \mathbb{Z}^3\right\}. \tag{13}$$

Here, the scaling-like functions $\varphi_{\sigma,c,0,\mathbf{j}}$ are enumerated by $\mathbf{j}$ where the wavelet-like functions $\psi^{[\mathbf{z}]}_{\sigma,c,l,\mathbf{j}}$ are enumerated by $(l, \mathbf{j}, \mathbf{z})$ due to its construction. Now, to invoke (11) we need an enumeration of the overall set $\mathfrak{b}_{\sigma,c}$. We assume ν to be the enumeration index, i.e. $\{\phi_\nu\}_{\nu\in\mathbb{N}} = \mathfrak{b}_{\sigma,c}$.

Note further that a frame like $\mathfrak{b}_{\sigma,c}$ can particularly be seen as an approximation to a Meyer wavelet basis [23]. In [4] we studied the application of an orthonormal tensor product basis built from Meyer wavelets and Meyer scaling functions in practice. It turned out that, although the wavelet system allows us to resolve the cusps locally, the sub-exponential decay of the basis functions in real space and the expensive numerical integration of the one-particle operator integrals and in particular of the two-particle operator integrals leads to impractically huge constants. Now with the frame $\mathfrak{b}_{\sigma,c}$ we are able to substantially reduce the involved constants as we will see later.

3.4 Adaptive scheme

We choose the weights $\{\gamma_u\}_{u\subset\mathcal{N}}$ in an a priori way based on quantum chemistry knowledge. To be precise, we employ a specific set of finite-order weights of order three so that we obtain a sum of subspaces in the form

$$\mathcal{V}^{(M_S)}_{three} := \mathcal{V}^{zero} + \mathcal{V}^{one} + \mathcal{V}^{two\uparrow\downarrow} + \mathcal{V}^{two\uparrow\uparrow} + \mathcal{V}^{three} \subset \sum_{u\subset\mathcal{N},|u|\le 3} \mathcal{W}^{(M_S)}_u, \tag{14}$$

where

$$\mathcal{V}^{zero} := \mathcal{W}^{(M_S)}_{\emptyset},\; \mathcal{V}^{one} := \sum_{p_1\in\mathcal{N}} \mathcal{W}^{(M_S)}_{\{p_1\}},\; \mathcal{V}^{two\uparrow\downarrow} := \sum_{p_1\in\mathcal{N}_\uparrow}\sum_{p_2\in\mathcal{N}_\downarrow} \mathcal{W}^{(M_S)}_{\{p_1,p_2\}},$$

$$\mathcal{V}^{two\uparrow\uparrow} := \sum_{p_1,p_2\in\mathcal{N}_\uparrow} \mathcal{W}^{(M_S)}_{\{p_1,p_2\}} + \sum_{p_1,p_2\in\mathcal{N}_\downarrow} \mathcal{W}^{(M_S)}_{\{p_1,p_2\}},$$

$$\mathcal{V}^{three} := \sum_{p_1,p_2\in\mathcal{N}_\uparrow}\sum_{p_3\in\mathcal{N}_\downarrow} \mathcal{W}^{(M_S)}_{\{p_1,p_2,p_3\}} + \sum_{p_1\in\mathcal{N}_\uparrow}\sum_{p_2,p_3\in\mathcal{N}_\downarrow} \mathcal{W}^{(M_S)}_{\{p_1,p_2,p_3\}},$$

compare (10).

The sets $\{\mathcal{B}_u\}_{u\subset\mathcal{N},\gamma_u>0}$ are then constructed by an heuristic h-adaptive scheme in an a posteriori fashion. We denote the resulting sequence of finite-dimensional spaces by

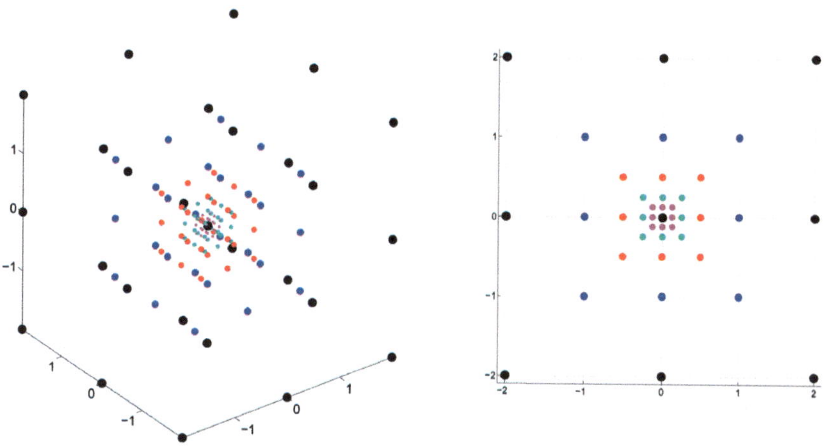

Fig. 1. Localization peaks of basis functions in $\mathfrak{b}^{(5)}_{1,1/2}$. On the right hand side we depict a slice plane, i.e. the (x_1,x_2)-plane, of the view on the left hand side.

$$\left\{V_\kappa^{SG} := V^{SG}_{\{\gamma_u\}_u,\{\mathcal{B}_u^{[\kappa]}\}_u} \subset \mathcal{V}^{(M_S)}_{\{\gamma_u\}_u}\right\}_{\kappa\in\mathbb{N}_0}.$$

To this end, we first choose the initial spanning system $\{\mathcal{B}_u^{[0]}\}_{u,\gamma_u>0}$. The initial approximation space V_0^{SG} is built with the help of one-particle subspaces (with just a few degrees of freedom) which provide an accurate representation of atomic orbitals. We define those one-particle subspaces by the span of the initial system

$$\mathfrak{b}^{(L)}_{\sigma,c} := \left\{\varphi_{\sigma,c,0,\mathbf{j}} \,:\, |\mathbf{j}|_\infty \le 1\right\} \cup \bigcup_{\mathbf{z}\in\mathcal{Z}} \left\{\psi^{[\mathbf{z}]}_{\sigma,c,l,\mathbf{j}} \,:\, l \le L-2, \mathbf{j} = -\mathbf{z}\right\} \subset \mathfrak{b}_{\sigma,c}, \tag{15}$$

see Figure 1. Now, let us consider an atomic system with N electrons and a nucleus of atomic charge Z_1 centered at $\mathbf{R}_1$, where we assume $\mathbf{R}_1 = \mathbf{0}$ for reasons of simplicity. Let the parameters σ, c and L be fixed. Then, in the case of spanning systems associated with finite-order weights $\gamma_u > 0$ with $|u| = 1$ (i.e. related to $\mathcal{V}^{one}$), we set

$$\mathcal{B}^{[0]}_{\{p_1\}} = \mathcal{B}^{(M_S)}_{\{p_1\}} \cap \left\{\mathfrak{A}^{(N,M_S)}\Big(\vec{\mathbf{y}} \mapsto w(\mathbf{y}_{p_1}) \prod_{q\in\mathcal{N}\setminus\{p_1\}} v_q(\mathbf{y}_q)\Big) \,:\, \phi \in \mathfrak{b}^{(L)}_{\sigma,c}\right\}$$

for $p_1 \in \mathcal{N}$. For spanning systems according to finite-order weights $\gamma_u > 0$, $|u| \ge 2$ the idea is to employ certain tensor product functions with localization peaks at or close to the electronic cups. In this way, for example for the case of $\mathcal{V}^{two\uparrow\downarrow}$, we set

$$\mathcal{B}^{[0]}_{\{p_1,p_2\}} = \mathcal{B}^{(M_S)}_{\{p_1,p_2\}} \cap$$
$$\left\{\mathfrak{A}^{(N,M_S)}\Big(\vec{\mathbf{y}} \mapsto \phi(\mathbf{y}_{p_1})\phi(\mathbf{y}_{p_2}) \prod_{q\in\mathcal{N}\setminus\{p_1,p_2\}} \phi_q(\mathbf{y}_q)\Big) \,:\, \phi \in \mathfrak{b}^{(L)}_{\sigma,c}\right\},$$

for $p_1 \in \mathcal{N}_\uparrow$, $p_2 \in \mathcal{N}_\downarrow$; see also Figure 2 (left). The sets $\mathcal{B}_{\{p_1,p_2\}}$ and $\mathcal{B}_{\{p_1,p_2,p_3\}}$ related to $\mathcal{V}^{two\uparrow\uparrow}$ and $\mathcal{V}^{three}$, as well as the initial sets in the case of two atoms,

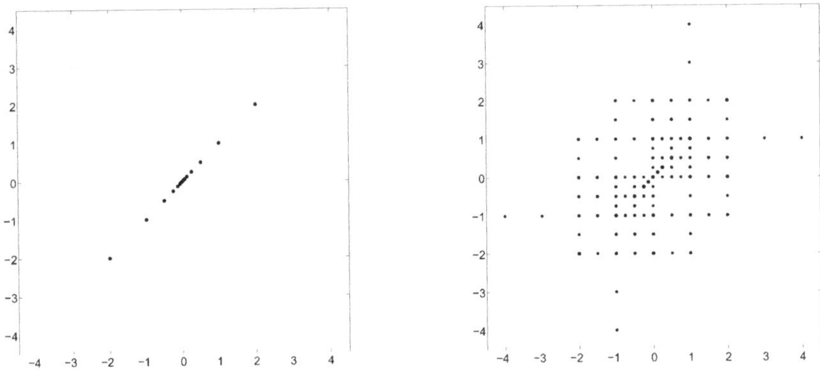

Fig. 2. Localization peaks of frame functions in the case of He, where we depict the $(x_{1,(1)}, x_{2,(1)})$-plane. Left: Initial set $\mathcal{B}^{[0]}_{\{1,2\}}$ for $L = 7$ and $c = \frac{1}{2}$. Right: Basis set $\mathcal{B}^{[9]}_{\{1,2\}}$ constructed by our adaptive algorithm.

are constructed in a similar way. Next, we invoke a coarsening step on the a priori chosen set $\{\mathcal{B}_u^{[0]}\}_{u,\gamma_u>0}$. Finally, the sequence of spanning systems $\{\mathcal{B}_u^{[\kappa]} \subset \mathcal{B}_u^{(M_S)}\}_{\kappa\in\mathbb{N}}$ is build by a simple h-adaptive refinement scheme in an a posteriori fashion. To this end, the solution of the general linear eigenvalue problem and a refinement and expansion of the spanning system is performed iteratively; for details see [15]. The resulting sequence of approximation spaces

$$\left\{ V_\kappa^{SG} = V^{SG}_{\{\gamma_u^{[\kappa]}\}_u, \{\mathcal{B}_u^{[\kappa]}\}_u} \right\}_{\kappa>0}$$

hopefully gives an efficient representation of the involved cusps. Figure 2 (right) shows the localization peaks of the basis set $\mathcal{B}^{[9]}_{\{1,2\}}$ constructed by our algorithm in the case of the He atom. Note that the span of $\mathcal{B}^{[9]}_{\{1,2\}}$ is a subspace of $\mathcal{V}^{two\uparrow\downarrow}$.

For technical details of our adaptive scheme to build the sequence of approximation spaces $\{V_\kappa^{SG}\}$ see [15]. For a further reading on adaptive wavelet techniques see [24].

4. Numerical methods and experiments

For each finite-dimensional subspace V_κ^{SG} the Galerkin discretization results in a generalized linear eigenvalue problem, i.e. $Av = EBv$. We compute the entries of the corresponding stiffness A and the mass matrices B with the help of the so-called Löwdin rules [25] for Slater determinants. Here, an N-particle integral over a product of Slater determinants is reduced to the computation of determinants of matrices with entries put together from values of certain one- and two-particle integrals. Furthermore, we perform the assembly of the system matrices A and B in parallel in a straightforward way. For the parallel solution of the eigenvalue problem $Av = EBv$ we invoke the scalable library for eigenvalue problem computations (SLEPc) [26]. This software package provides a wrapper to the software library BLOPEX, which is an implementation of the parallelized locally optimal block preconditioned conjugate gradient method (LOBPCG) [27]. For the molecules considered in this article, we have to treat full matrices. Thus, we invoke a Cholesky decomposition as implemented in the software package PLAPACK [28] to perform preconditioning in an efficient way.

4.1 Applications

In the following we apply our novel method to several atomic and diatomic systems with up to six electrons. Here, we aim at the determination of the total energies of the considered molecular systems up to the so-called chemical accuracy, i.e. 1 kcal/mol $\approx 1.595_{-3}$ hartree. Note that the main task is to efficiently describe electron correlation [29]. We give the results of our numerical approach in Table 1 and Figure 3. They suggest that our new method is indeed convergent and that the measured rates are in the expected range. In particular, the results demonstrate that our new method allows us to efficiently describe electron-nuclei and electron-electron cusps. Nevertheless, we see a dependence of the involved constants on the molecular system size.

For the two- and three-electron systems chemical accuracy is easily reached. For four-electron systems we obtain an approximation error of a size smaller than ten milli-hartree. In particular, we obtain in the case of two-electron, three-electron and four-electron systems proportions E_c of the correlation energy larger than 98 %, 96 % and 88 %, respectively. In the case of the five- and six-electron molecular systems, our results seem to be still in the pre-asymptotic range. Note here that the involved number of degrees of freedom is restricted by physical memory limitations. Nevertheless, for the five- and six-electron systems proportions E_c of the correlation energies are achieved in the range of 71 % to 87 %.

Note furthermore that for the studied systems Li, Be, LiH, B, BeH and BH the size of the approximation error of our new method is less than the error obtained by VMC methods which employ a single-determinant Jastrow-Slater trial wave function. The better accuracy could be expected since the present approach improves the reference single-determinant in an exact way like CI and CC methods, instead of employing an inexact multiplicative Jastrow factor ansatz. Moreover, our results are in the range of those computed by diffusion Monte Carlo (DMC) methods which are based on a single-determinant Jastrow-Slater trial wave function; compare [30–32].

5. Concluding remarks

In this article we introduced and studied new tensor product multiscale many-particle spaces with finite-order weights and applied them in the numerical treatment of the electronic Schrödinger equation. These spaces are constructed from a particle-wise subspace splitting of the N-particle space. In particular, this con-

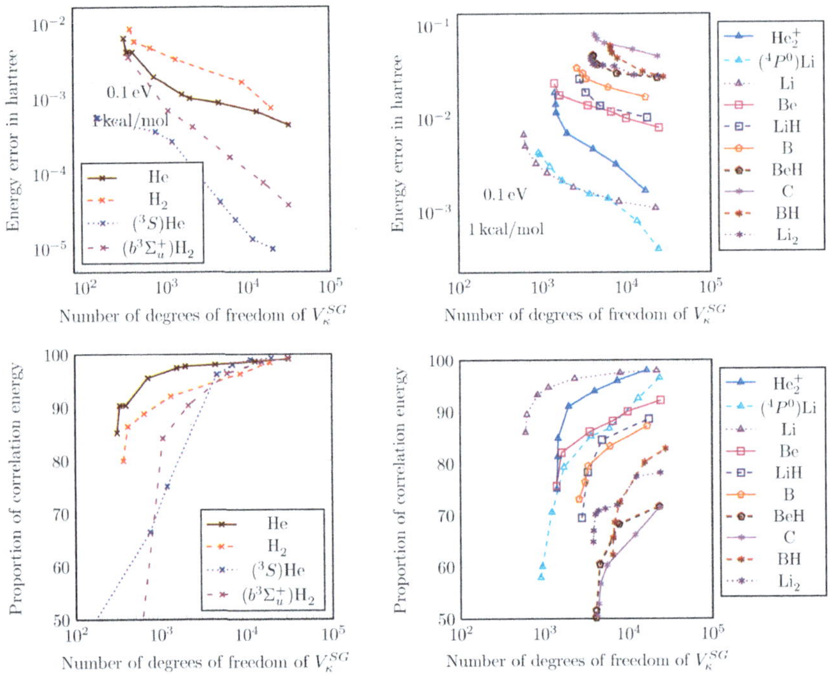

Fig. 3. Error of the approximately computed energy of the ground-state of several atomic and diatomic systems with up to six electrons.

struction provides a systematic improvement of a nonlinear rank-1 approximation by its combination with a tensor product multiscale approximation scheme and allows for convergence with guaranteed approximation rates. We demonstrated these rates numerically for atoms and diatomic molecules with up to six electrons. Note that to our knowledge this is the first time that systems with more than two electrons were successfully treated by direct numerical approximation by means of an application of tensor product multiscale bases in the framework of ab initio methods (except for HF and DFT methods).

Both the multiscale frame and the adaptive scheme can probably still be improved. Multi-wavelet like frames based on Hermite-Gaussian functions together with an *h*- and *p*-adaptive refinement strategy may lead to improved approximation properties. Furthermore, from a theoretical point of view, such an adaptive best *M*-term approximation requires a new, not yet existing mixed Besov regularity theory for the electronic Schrödinger equation. Here, exponential Jastrow factors in the two-electron case have been studied in [33] to deduce some preliminary assertions on possible convergence rates. Their results yield a convergence rate of order $-1/2$ for the related two-particle correlation functions, compared to a rate of order $-1/4$ when applying a linear approximation scheme. Moreover, our scheme might be extended to varying weights and thereby subspaces in an adaptive way (particle-wise adaptivity) similar to dimension-wise

Table 1. Numerical results for the total energy $\tilde{E}_{tot}$ and the Hartree-Fock energy $\tilde{E}^{HF}_{tot}$. With respect to the applied parameters see Table 2. The exact values for the total energies E_{tot} and the Hartree-Fock limits E^{HF}_{tot} are taken from literature; see [15] and the references therein. We give the proportion of correlation energy $E_c := [(\tilde{E}_{tot}-\tilde{E}^{HF}_{tot})/(E_{tot}-E^{HF}_{tot})]$ in percentage. E^{VMC}_{tot} are the results computed by variational quantum Monte Carlo methods which are based on a single-determinant Jastrow-Slater trial wave function. These are taken from [30–32].

	$\tilde{E}_{tot}$	E_{tot}	$\tilde{E}^{HF}_{tot}$	E^{HF}_{tot}	E_c	E^{VMC}_{tot}
He	-2.90330	-2.90372	-2.86166	-2.86168	99.00	
H_2	-1.17376	-1.17447	-1.12854	-1.13366	98.26	
(^3S)He	-2.17522	-2.17523	-2.17424	-2.17420	99.10	
$(b^3\Sigma_u^+)H_2$	-0.89704	-0.89708	-0.89057	-0.89288	99.14	
He_2^+	-4.99305	-4.99464	-4.91647	-4.92285	97.78	
$(^4P^0)$Li	-5.36764	-5.36801	-5.35830	-5.35830	96.18	
Li	-7.47702	-7.47806	-7.43271	-7.43273	97.58	-7.47683
Be	-14.65978	-14.66736	-14.57296	-14.57302	90.94	-14.63110
LiH	-8.06084	-8.07055	-7.97054	-7.98735	88.05	-8.04593
B	-24.63768	-24.65391	-24.52921	-24.52906	87.00	-24.60562
BeH	-15.22016	-15.24680	-15.04328	-15.15318	71.55	-15.21210
C	-37.80020	-37.84500	-37.65886	-37.68862	71.35	-37.81471
BH	-25.26089	-25.28790	-25.08382	-25.13195	82.68	-25.21220
Li_2	-14.96810	-14.99540	-14.86166	-14.87152	77.97	-14.98255

Table 2. Parameters for the frame $\mathfrak{b}_{\sigma,c}$ and its subset $\mathfrak{b}^{(L)}_{\sigma,c}$; compare (13) and (15). For all systems we set $\sigma = 1$. For all diatomic systems we set $c = 2/R$. The respective bond distances R are given in bohr.

	He	(^3S)He	$(^4P^0)$Li	Li	Be	B	C
c	$\frac{1}{2}$	$\frac{1}{4}$	$\frac{1}{4}$	1	$\frac{1}{4}$	1	1
L	7	7	7	6	8	7	7

	H_2	$(b^3\Sigma_u^+)H_2$	He_2^+	LiH	BeH	BH	Li_2
R	1.4	2.0	2.042	3.015	2.537	2.329	5.051
L	5	5	6	6	7	7	7

adaptive approaches for high-dimensional quadrature [34,35]. Also, an adaptive scheme which applies multiscale frames including ridgelet-like two-particle functions [36], the application of two-particle functions for the electron-electron cusps similar to the R12/F12 methods [37] and the extended geminal model [38] could be promising for the future.

Acknowledgement

This work was partially supported by the Deutsche Forschungsgemeinschaft in the priority program 1145 "Modern and universal first-principles methods for many-electron systems in chemistry and physics". We are grateful to Harry Yserentant (TU Berlin) for fruitful discussions and suggestions.

References

1. H. Yserentant, Numerische Mathematik **98** (2004) 731.
2. J. Garcke and M. Griebel, Journal of Computational Physics **165** (2000) 694.
3. M. Griebel and J. Hamaekers, Mathematical Modelling and Numerical Analysis **41** (2007) 215.
4. M. Griebel and J. Hamaekers, *A wavelet based sparse grid method for the electronic Schrödinger equation.* In Proceedings of the International Congress of Mathematicians. M. Sanz-Solé, J. Soria, J. Varona, and J. Verdera (Eds.), volume III, pages 1473–1506, Madrid, Spain (2006) European Mathematical Society.
5. H. Luo, D. Kolb, H. Flad, and W. Hackbusch, Physical Review B **75** (2007) 125111.
6. H. Luo, D. Kolb, H. Flad, W. Hackbusch, and T. Koprucki, Journal of Chemical Physics **117** (2002) 3625.
7. I. Sloan, X. Wang, and H. Woźniakowski, Journal of Complexity **20** (2004) 46.
8. M. Griebel, *Sparse grids and related approximation schemes for higher dimensional problems.* In Foundations of Computational Mathematics (FoCM05), Santander. L. Pardo, A. Pinkus, E. Suli, and M. Todd (Eds.), pages 106–161, Cambridge University Press (2006).
9. G. Wasilkowski and H. Woźniakowski, Foundations of Computational Mathematics **9** (2009) 105.
10. H. Yserentant, On the electronic Schrödinger equation. Report 191, SFB 328, Universität Tübingen (2003).
11. H. Yserentant, *Regularity properties of wavefunctions and the complexity of the quantum-mechanical N-body problem.* Preprint, (2007).
12. R. Schneider, *Analysis of the projected coupled cluster method in electronic structure calculation.* ePrint 2008–007, Institut für Mathematik, Technische Universität Berlin, Germany (2008).
13. H. Yserentant, Numerische Mathematik **105** (2007) 659.
14. H. Schmeisser and H. Triebel, Fourier Analysis and Functions Spaces, John Wiley (1987).
15. J. Hamaekers, Tensor Product Multiscale Many-Particle Spaces with Finite-Order Weights for the Electronic Schrödinger Equation. Dissertation, Institut für Numerische Simulation, Universität Bonn (2009).
16. H. Bungartz and M. Griebel, Acta Numerica **13** (2004) 1.
17. Y. Meyer, Wavelets and operators. Volume 37 of Cambridge Studies in Advanced Mathematics, Cambridge University Press (1992).
18. H. Yserentant, Numerische Mathematik **101** (2005) 381.
19. S. Agmon, Lectures on Exponential Decay of Solutions of Second-order Elliptic Equations: Bounds on Eigenfunctions of N-body Schrödinger Operators. Volume 29 of Mathematical Notes, Princeton University Press (1982).
20. O. Sinanoğlu, Reviews of Modern Physics **35** (1963) 517.
21. M. Griebel and P. Oswald, Numerische Mathematik **66** (1994) 449.
22. D. Allison, N. Handy, and S. Boys, Molecular Physics **26** (1973) 715.
23. G. Kyriazis and P. Petrushev, Transactions of the American Mathematical Society **354** (2002) 749.
24. A. Cohen, W. Dahmen, and R. DeVore, *Adaptive wavelet techniques in numerical simulation.* In Encyclopedia of Computational Mechanics. R. De Borste, T. Hughes, and E. Stein (Eds.), volume 1, pages 157–197, John Wiley & Sons (2004).
25. P.-O. Löwdin, Physical Review **97** (1955) 1474.

26. V. Hernandez, J. Roman, and V. Vidal, ACM Transactions on Mathematical Software **31** (2005) 351.
27. A. Knyazev, SIAM Journal on Scientific Computing **23** (2001) 517.
28. R. van de Geijn et al., Using PLAPACK: Parallel Linear Algebra Package. MIT Press (1997).
29. D. Tew, W. Klopper, and T. Helgaker, Journal of Computational Chemistry **28** (2007) 1307.
30. A. Lüchow and J. Anderson, Journal of Chemical Physics **105** (1996) 7573.
31. M. Brown, J. Trail, P. Rios, and R. Needs, Journal of Chemical Physics **126** (2007) 224110.
32. J. Toulouse and C. Umrigar, Journal of Chemical Physics **128** (2008) 174101.
33. H. Flad, W. Hackbusch, and R. Schneider, Mathematical Modelling and Numerical Analysis **41** (2007) 261.
34. T. Gerstner and M. Griebel, Computing **71** (2003) 65.
35. M. Holtz, Sparse Grid Quadrature in High Dimensions with Applications in Finance and Insurance. Dissertation, Institut für Numerische Simulation, Universität Bonn (2008).
36. D. Donoho, Proceedings of the National Academy of Sciences of the United States of America **96** (1999) 1828.
37. W. Klopper, F. Manby, S. Ten-No, and E. Valeev, International Reviews in Physical Chemistry **25** (2006) 427.
38. I. Røeggen, *Extended Geminal Models.* In Correlation and Localization. P. Surján (Ed.), volume 203 of Topics in Current Chemistry, pages 89–103, Springer (1999).

On Occupied-orbital Dependent Exchange-correlation Functionals: From Local Hybrids to Becke's B05 Model

By Martin Kaupp*, Alexei Arbuznikov**, and Hilke Bahmann

Institut für Physikalische und Theoretische Chemie, Universität Würzburg, Am Hubland, D-97074, Würzburg, Germany

(Received July 24, 2009; accepted September 24, 2009)

Local Hybrid Functionals / Localized-local Hybrid potentials / Nondynamical Correlation Model / Occupied-orbital-dependent Functional / Optimized Effective Potential.

Recent work in the field of occupied-orbital dependent (OOD) exchange-correlation functionals in density functional theory is reviewed. The main emphasis is put on the development of so-called local hybrid functionals, and on the nontrivial self-consistent implementation of complex OOD functionals. Local hybrids employ a so-called local mixing function (LMF) to govern their position-dependent exact-exchange admixture. Recently proposed LMFs have provided local hybrids of remarkable accuracy in the computation of thermochemical data and classical reaction barriers, and with good performance for some magnetic-resonance parameters. These local hybrids mix only local and exact exchange and exhibit very few semi-empirical parameters. Further refinement and the efficient implementation of local hybrids offers the prospect of a new level of accuracy in Kohn-Sham density functional calculations. Two levels of the self-consistent implementation of OOD functional are discussed: one may either stop after the derivation of the functional derivatives with respect to the orbitals (FDOs), leading to nonlocal potentials. This is discussed for local hybrids and for general OOD functionals up to and including the complicated B05 real-space model of nondynamical correlation. Alternatively, one may append an additional transformation to local and multiplicative potentials based on the optimized effective potential (OEP) approach or of approximations to the OEP. Numerical results for various properties are reviewed briefly, ranging from nonself-consistent energies via FDO-based calculations to OEP-transformed potentials.

---

* Corresponding author. E-mail: kaupp@mail.uni-wuerzburg.de

** Corresponding author. E-mail: arbouznikov@mail.uni-wuerzburg.de

Z. Phys. Chem. **224** (2010) 545–567
© by Oldenbourg Wissenschaftsverlag, München

1. Introduction

By including an explicit dependence on orbitals into the exchange-correlation functionals [1] of Kohn-Sham density functional theory [2,3], qualitatively new levels of accuracy in the description of various molecular properties may be achieved, that were unattainable within either the local spin-density approximation (LSDA) [4–7] or the generalized gradient approximation (GGA) [8–13]. This review deals with aspects of an important subclass of such orbital-dependent functionals, those that depend on the occupied orbitals only (occupied-orbital dependent – OOD – functionals). Among several other types of functionals, the OOD class includes the now almost ubiquitous (global) hybrid functionals [14–16], with the most popular example B3LYP [17,12,18], as well as certain meta-GGA functionals like TPSS [19] (together with its global hybrid extension, TPSSh [20]). Within a widely accepted classification of functionals according to a "Jacob's ladder to the heaven of chemical accuracy" [21,22], these OOD functionals occupy the "fourth rung" ("hyper-GGA") and the "third rung" ("meta-GGA"), respectively (see below).

In this review we will first reconsider briefly the classification of functionals (Section 2). We will then proceed to discuss some recent developments in the area of OOD functionals, with particular emphasis on our own contributions over the last five years. The aim is to extend the accuracy of Kohn-Sham density functional theory by functionals that exhibit an improved balance between the minimization of self-interaction errors and the description of nondynamical correlation. The main topics will be the elaboration of the relatively new class of local hybrid functionals (Section 3), and important aspects of the self-consistent implementation of OOD functionals up to and including the very complicated B05 real-space model of nondynamical correlation [23,24] (Section 4). A brief overview of computational results obtained with these novel approaches is given in Section 5, before we summarize and conclude in Section 6.

2. A classification of exchange-correlation functionals

Classification of the functionals in terms of the "Jacob's ladder" [21,22] reflects mainly the historical development by choosing LSDA as the lowest rung of the ladder, followed by GGA as second, meta-GGA as third, and hyper-GGA as fourth. The fifth rung is represented by general nonlocal functionals depending also on the unoccupied orbitals [25–29]. The conceptual value of the historical Jacob's ladder classification is undisputed. However, perusal of the abovementioned modern functionals suggests that not all of them are well characterized in this way. For example, Becke's B05 real-space model of nondynamical correlation (NDC) [23,24] (see end of Section 4.1) is regarded as a hyper-GGA but involves neither LSDA nor GGA in its exchange part. Next, the very first global hybrid functional (Becke's "half-and-half" model) [14] involved only LSDA and exact exchange. The popular LYP correlation functional [12] is usually considered of GGA type. However, it has nothing to do with the real-space cutoff procedure [8,10] of the gradient expansion for the exchange energy of the slowly-varying electron gas density which constitutes the essence of the GGA. Moreover, in its original form, the LYP functional includes not only the gradient, but also the Laplacian of the density, and could thus be classified formally as of meta-GGA type. The list of examples can be further continued.

From the point of view of implementation, a less specific classification, may be preferable. It consists of only three categories, albeit subclasses of functionals may of course be defined for finer distinctions.

(i) The lowest category is represented by explicit density functionals, E_{xc}^{ρ}, that include only ingredients depending on the electron density, ρ, or its derivatives, i.e., LSDA [4–7], GGA [8–13], and Laplacian-dependent meta-GGA [30]. Here self-consistent implementation is straightforward, as a local and multiplicative exchange-correlation potential, v_{xc}^{ρ}, may be evaluated explicitly, using the well-known relationship

$$v_{xc,\sigma}^{\rho}(\mathbf{r}) = \frac{\delta E_{xc}^{\rho}}{\delta\rho_\sigma(\mathbf{r})} = \left[\frac{\partial\varepsilon_{xc}^{\rho}}{\partial\rho_\sigma}\right](\mathbf{r}) - \left[\nabla\left(\frac{\partial\varepsilon_{xc}^{\rho}}{\partial\nabla\rho_\sigma}\right)\right](\mathbf{r}) + \left[\nabla^2\left(\frac{\partial\varepsilon_{xc}^{\rho}}{\partial\nabla^2\rho_\sigma}\right)\right](\mathbf{r}), \quad \sigma = \alpha, \beta \tag{1}$$

where

$$E_{xc}^{\rho} = \int \varepsilon_{xc}^{\rho}\left[\rho_\alpha(\mathbf{r}), \rho_\beta(\mathbf{r}); \nabla\rho_\alpha(\mathbf{r}), \nabla\rho_\beta(\mathbf{r}); \nabla^2\rho_\alpha(\mathbf{r}), \nabla^2\rho_\beta(\mathbf{r})\right]d\mathbf{r} \tag{2}$$

(ii) The second (higher) category is the OOD class, with which we are primarily concerned in this review (see above). It has become clear that an explicit dependence on the individual occupied orbitals (i.e., not just through the density and/or its gradients) may be needed to obtain improved accuracy. The kinetic energy density

$$\tau_\sigma(\mathbf{r}) = \frac{1}{2}\sum_{i_\sigma}^{occ}\left|\nabla\varphi_{i_\sigma}(\mathbf{r})\right|^2, \tag{3}$$

and the exact exchange energy density

$$\varepsilon_{x,\sigma}^{exact}(\mathbf{r}) = -\frac{1}{2}\sum_{i_\sigma, j_\sigma}^{occ}\varphi_{i_\sigma}^{*}(\mathbf{r})\varphi_{j_\sigma}(\mathbf{r})\int\frac{\varphi_{j_\sigma}^{*}(\mathbf{r}')\varphi_{i_\sigma}(\mathbf{r}')}{|\mathbf{r}-\mathbf{r}'|}d\mathbf{r}' \tag{4}$$

are the most important ingredients of this type in state-of-the-art functionals. In general, OOD functionals, E_{xc}^{OOD} ("hyper-GGAs" according to the classification of Perdew and Schmidt [21]), may be formally represented as

$$E_{\mathrm{xc}}^{\mathrm{OOD}} = \int \varepsilon_{\mathrm{xc}}^{\mathrm{OOD}} \Big[\rho_\alpha, \nabla\rho_\alpha, \nabla^2\rho_\alpha, \tau_\alpha, \varepsilon_{\mathrm{x},\alpha}^{\mathrm{exact}}; \\ \rho_\beta, \nabla\rho_\beta, \nabla^2\rho_\beta, \tau_\beta, \varepsilon_{\mathrm{x},\beta}^{\mathrm{exact}}\Big] \mathrm{d}\mathbf{r} \tag{5}$$

As mentioned above, two important subclasses of OOD functionals are τ-dependent meta-GGA (τ-MGGA) [19]

$$E_{\mathrm{xc}}^{\tau\text{-MGGA}} = \int \varepsilon_{\mathrm{xc}}^{\tau\text{-MGGA}} \left[\rho_\alpha(\mathbf{r}), \nabla\rho_\alpha(\mathbf{r}), \tau_\alpha(\mathbf{r}); \\ \rho_\beta(\mathbf{r}), \nabla\rho_\beta(\mathbf{r}), \tau_\beta(\mathbf{r})\right] \mathrm{d}\mathbf{r}, \tag{6}$$

and traditional (global) hybrid functionals [14–17,20]

$$E_{\mathrm{xc}}^{\mathrm{hybr}} = a_0\, E_{\mathrm{x}}^{\mathrm{exact}} + \overline{E}_{\mathrm{xc}}^{\mathrm{DFT}}\left[\rho_\sigma, \nabla\rho_\sigma, \ldots\right], \\ 0 \le a_0 \le 1, \quad \sigma = \alpha, \beta, \tag{7}$$

where

$$E_{\mathrm{x}}^{\mathrm{exact}} = \sum_{\sigma=\alpha,\beta} \int \varepsilon_{\mathrm{x},\sigma}^{\mathrm{exact}}(\mathbf{r})\, \mathrm{d}\mathbf{r}, \tag{8}$$

and $\overline{E}_{\mathrm{xc}}^{\mathrm{DFT}}[\rho, \nabla\rho, \ldots]$ is some "conventional" density functional contribution based on the LSDA, GGA, or meta-GGA. Other representatives are self-interaction corrected LSDA [31], the B05 nondynamical correlation functional [23,24], and the MCY hyper-GGAs [32].

For $E_{\mathrm{xc}}^{\mathrm{OOD}}$ we cannot apply Eq. (1) and thus have no explicit functional derivative with respect to the density. One can still evaluate functional derivatives with respect to the occupied orbitals (FDOs) followed by the substitution

$$\mathrm{v}_{\mathrm{xc}}^{\mathrm{OOD}}(\mathbf{r})\varphi_i(\mathbf{r}) \leftarrow \frac{1}{2}\frac{\delta E_{\mathrm{xc}}^{\mathrm{OOD}}}{\delta\varphi_i(\mathbf{r})} \equiv \hat{\mathrm{v}}_{\mathrm{xc}}^{\mathrm{OOD}}(\mathbf{r})\varphi_i(\mathbf{r}). \tag{9}$$

Operator $\hat{\mathrm{v}}_{\mathrm{xc}}^{\mathrm{OOD}}$ is nonmultiplicative and, in certain cases, nonlocal (see below). Substitution (9) leads us outside the genuine Kohn-Sham formalism, which requires a local and multiplicative potential. Nevertheless, this is the usual way in which, for example, global hybrid functionals or τ-dependent meta-GGA functionals have been implemented self-consistently in standard quantum-chemical codes [33–35] and have been used subsequently with success for property calculations (see, e.g., refs. 36–39, etc.). This approach is sometimes referred to as "generalized Kohn-Sham method" (GKS) [40,41]. A way to transform the nonmultiplicative operators $\hat{\mathrm{v}}_{\mathrm{xc}}^{\mathrm{OOD}}$ into local and multiplicative Kohn-Sham potentials is provided by the optimized effective potential (OEP) method [42], which is formally equivalent to the evaluation of the functional derivative with respect to the density [43] (see Section 4). Sections 3 and 4 will be concerned with more elaborate OOD functionals than those mentioned so far.

(iii) The third and highest category of functionals are general nonlocal functionals that involve information also about the unoccupied orbitals [25–29] (corresponding to the fifth rung of the Jacob's ladder classification, see above). These functionals are important from a fundamental point of view, as the insight provided from the presumably very accurate Kohn-Sham potentials obtained along this way may aid in the construction of approximate, computationally more efficient methods. From a practical point of view, however, the complexity of these methods is comparable to (or higher than) that of standard post-Hartree-Fock ab initio methodology. We will not consider them further here.

3. Local hybrid functionals

Global hybrid functionals (7) have been very successful in describing various properties. An explanation of this success lies in the mixing of exact exchange, which is free from self-interaction error (SIE), with local or semilocal DFT exchange, which partially simulates nondynamical correlation (NDC). The balance between minimization of SIE and inclusion of NDC is controlled by the constant, fractional EXX admixture a_0. It has become clear over the past decade that in many cases this may not be a sufficiently flexible construction [21,44,45]. It is usually not possible to find a unique value of a_0 that would provide optimum accuracy for a wide range of properties or for different types of molecular or solid-state systems. For example, with GGA approximations for E_{xc}^{DFT}, the best performance for thermochemical properties (in particular for atomization energies) requires rather modest EXX admixtures, a_0, of typically about 0.16 – 0.30 (cf. some theoretical justifications [46]), whereas larger values tend to be optimum for, e.g., reaction barriers [47] or for some linear response properties [36,37,48–51].

Two more flexible extensions of the hybrid-functional concept have therefore been suggested. One new family of functionals are the so-called range-separated hybrids, where the mixing coefficient is made to depend on the interelectronic distance. This *ansatz* has been discussed in detail elsewhere [52–58], and we will refrain from covering it here. The second natural extension, which we will be concerned with, are the so-called local hybrid functionals. In contrast to global hybrids (7), here the exact-exchange admixture is position dependent in real space and is governed by a so-called local mixing function (LMF), $g(\mathbf{r})$:

$$E_{xc}^{\text{loc-hyb}} = \sum_{\sigma=\alpha,\beta} \int \left\{ g_\sigma(\mathbf{r})\, \varepsilon_{x,\sigma}^{\text{exact}}(\mathbf{r}) + \left[1 - g_\sigma(\mathbf{r})\right] \times \varepsilon_{x,\sigma}^{\text{DFT}}(\mathbf{r}) \right\} d\mathbf{r} \; + \; E_c^{\text{DFT}}, \quad 0 \le g_\sigma(\mathbf{r}) \le 1. \tag{10}$$

The term "local hybrid functional" has been introduced by Jaramillo et al. [44], together with the first explicit construction, albeit the possibility of this generalization had been envisioned earlier [21,59]. In ref.44, the ratio of the von Weizsäcker kinetic energy density

$$\tau_{W,\sigma}(\mathbf{r}) = \left|\nabla\rho_\sigma(\mathbf{r})\right|^2 / 8\rho_\sigma(\mathbf{r}) \tag{11}$$

to the noninteracting local kinetic energy density, τ_σ(Eq. (3)),

$$g_\sigma^t(\mathbf{r}) = \tau_{W,\sigma}(\mathbf{r}) / \tau_\sigma(\mathbf{r}) \equiv t_\sigma(\mathbf{r}) \tag{12}$$

has been suggested as the LMF. It fulfills the condition $0 < g_\sigma < 1$ and thereby brings in 100% exact exchange in one-electron regions (this is thought to be favorable to reduce SIE) but no exact exchange in homogeneous regions. The corresponding functionals *Lh*-BLYP and *Lh*-PBEPKZB (i..e. $\varepsilon_X^{DFT} = \varepsilon_X^{B88}$ [11] or ε_X^{PBE} [13], and $\varepsilon_c^{DFT} = \varepsilon_c^{LYP}$ [12] or ε_c^{PKZB} [60]) provided reasonable results for the dissociation behavior of two-center, three-electron symmetric radicals, and for the reaction barriers in linear hydrogen abstraction, but overall very poor atomization energies (mean absolute errors for the G2–1 test set [61] were ca. 20 and 13 kcal/mol for *Lh*-BLYP and *Lh*-PBEPKZB, respectively [44]).

We have subsequently been successful in constructing more accurate local hybrids [62–64]. As we considered the originally proposed variable t (Eq. (12)) physically appealing and a good starting point for an LMF, we initially considered simple polynomials of this function,

$$g_\sigma^{\text{poly}}(\mathbf{r}) = \sum_{k=0}^{N} c_k\, t_\sigma^k(\mathbf{r}). \tag{13}$$

While constructions with GGA exchange gave nonintuitive and unsuccessful LMFs of this type [45,65], a simple LMF

$$g_\sigma(\mathbf{r}) \;=\; 0.48\, t_\sigma(\mathbf{r}), \quad \sigma = \alpha, \beta \tag{14}$$

provided the best performance, when we mixed the exact-exchange energy density only with its LSDA counterpart

$$\varepsilon_{X,\sigma}^{\text{LSDA}}(\mathbf{r}) = -\frac{3}{2}\left(\frac{3}{4\pi}\right)^{1/3} \rho_\sigma^{4/3}(\mathbf{r}) \,. \tag{15}$$

This mixture of the simplest conceivable ingredients for exchange gave strikingly accurate results for thermochemistry and reaction barriers (significantly better than B3LYP [17,12,18], see Section 5), in conjunction with LSDA correlation. This simple local hybrid features only one semi-empirical parameter. It will in the following be referred to as "*Lh*-SVWN, $g = 0.48t$" [62], where "S" stands for "Slater" exchange [4] and VWN for the Vosko-Wilk-Nusair local correlation functional [6].

A partial explanation of the success of local hybrids without GGA exchange may be related to the gauge problem: energy densities (unlike integrated energies) are not unique quantities. They may be calibrated arbitrarily by adding some gauge function which integrates to zero [66,67]. Any mismatch in the gauge origin of mixed energy densities creates an error that hampers refinement of LMFs and has been suggested to affect thermochemical data [68,69]. Here LSDA exchange may be advantageous [64], due to the absence of enhancement factors (in the homogeneous limit, the gauges coincide). The success of such local hybrids when combined with LSDA correlation may further benefit from efficient error cancellation between LSDA exchange and correlation in spatial regions where LSDA exchange dominates over exact exchange.

To achieve accurate results for atomization energies and barriers, the scaling factor 0.48 of this t-LMF was mandatory. That is, we had to sacrifice the correct

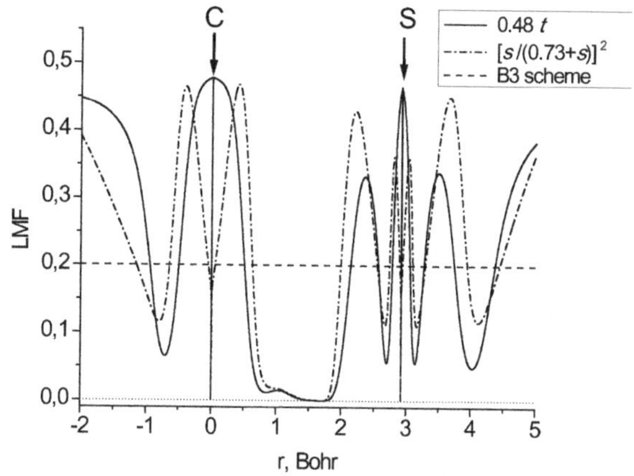

Fig. 1. Semiempirical local mixing functions (*t*- and *s*-LMFs).

asymptotic behavior ($g = 1$ in one-electron regions). This does not affect thermochemistry or barriers, but might influence properties that depend crucially on the long-range asymptotics. This was a prime motivation to consider alternative LMFs. In ref. [63] we reported on the so called "*s*-LMFs", which depend on the dimensionless density gradient,

$$s_\sigma(\mathbf{r}) = \frac{|\nabla \rho_\sigma(\mathbf{r})|}{2(3\pi^2)^{1/3}\rho_\sigma^{4/3}(\mathbf{r})}, \qquad \sigma = \alpha, \beta \tag{16}$$

As s varies from 0 to infinity, a mapping $[0;\infty] \rightarrow [0;1]$ was needed. Among the many possible mappings, several have been tried, and we consider here two of the more successful ones, with

$$g_\sigma(\mathbf{r}) = \left[s_\sigma(\mathbf{r})/(\lambda + s_\sigma(\mathbf{r}))\right]^2, \tag{17}$$

$$g_\sigma(\mathbf{r}) = \operatorname{erf}\left[\lambda\, s_\sigma(\mathbf{r})\right], \tag{18}$$

(λ is an adjustable parameter).

The scaled *t*-LMF and one selected *s*-LMF are displayed in Figure 1 as one-dimensional cuts for the carbon monosulfide molecule (along the C-S bond; as a trivial "special" case, the horizontal line corresponds to constant $a_0 = 0.20$ of the B3 scheme [17]). The correct asymptotic behavior of the *s*-LMF, as well as its undesirable negative cusp at the nuclei is apparent. The *t*-LMF has a more physical behavior near the nuclei but converges only to 0.48 asymptotically. Yet, its superior performance for thermochemistry and barriers (see section 5) may be conserved and the correct asymptotic behavior recovered by mixing it with a suitable *s*-LMF [64].

Further progress [64] in the improvement of thermochemical accuracy has been achieved with spin-polarized LMFs. Their development was motivated by conceptual similarities between local hybrids and explicit NDC functionals like the B05 model of Becke et al. [23,24]. Indeed, local hybrids may be written in a form close to such NDC functionals by starting from 100% exact exchange and viewing the "local hybrid" as a correction term that describes NDC. To mimic only some structural aspects of NDC functionals, we have employed spin polarization

$$\zeta(\mathbf{r}) = \left[\rho_\alpha(\mathbf{r}) - \rho_\beta(\mathbf{r})\right]/\rho(\mathbf{r}), \tag{19}$$

as an additional variable to introduce "cross terms" into the LMF and therefore into the NDC term. By introducing ζ, we have modified the so far most successful t- and s-LMFs [62,63], and obtained a t-ζ- and an s-ζ-LMF (argument $\mathbf{r}$ is omitted for brevity):

$$g_\alpha = (b + c\zeta)t_\alpha, \quad g_\beta = (b - c\zeta)t_\beta, \tag{20}$$

$$g_\alpha = \mathrm{erf}\left[(b + c\zeta)s_\alpha\right], \quad g_\beta = \mathrm{erf}\left[(b - c\zeta)s_\beta\right]. \tag{21}$$

These two new LMFs exhibit still only two adjustable parameters and provided another notable improvement in atomization energies over the underlying one-parameter LMFs, with only minor deterioration of reaction barriers (see Section 5).

The LMFs discussed so far exhibit an *ad hoc* character insofar as they are on one side based on physical reasoning and well-established inhomogeneity parameters like $t = \tau_W/\tau$ or the dimensionless density gradient s, but they lack any deeper rooting in first principles. Nevertheless, the numerical results (Section 5) suggest that local hybrids built upon these variables may incorporate a great deal of the "right physics". That is, the position-dependent mixing of LSDA and exact exchange by the LMFs formed from these variables do seem to provide an already reasonable spatial balance between minimal SIE and simulation of NDC. Yet, further insight into the formal basis of local hybrids is clearly desirable. Some important questions are, for instance: (i) even if we know that the LMF should go to one asymptotically at long range to ensure complete elimination of SIE in one-electron regions, at what rate should it approach 1.0? (ii) How should the "perfect" LMF behave near the nuclei? Since the initial basis for the first global hybrids [17,14] has been the adiabatic connection (AC) formalism [70], it seems natural to employ the AC for an "*ab initio*" construction of LMFs and local hybrids. This has been initiated by us recently [71]. Without going into the rather cumbersome details of the derivations of such "*ab initio* AC-LMFs", let us note that a local version of the coupling-constant integral was introduced, i.e. an expression for the exchange-correlation energy density in terms of the corresponding coupling-constant-resolved quantity:

$$\varepsilon_{\mathrm{xc}}(\mathbf{r}) = \int_0^1 \varepsilon_{\mathrm{xc},\lambda}(\mathbf{r})\, d\lambda. \tag{22}$$

It turned out, that the LMFs obtained in this procedure depend exclusively on the dynamical correlation functional one starts with [71]. Employing further information on the properties of the underlying energy densities with respect to uniform density scaling, we have constructed three LMFs. The first was based on LSDA correlation (within the PW91 parametrization) and is given by

$$g^{\text{AC-LSDA}} = 1 - 0.5\left(1 + \frac{r_s}{2e_c^{\text{LSDA}}}\frac{\partial e_c^{\text{LSDA}}}{\partial r_s}\right)^{-1}. \tag{23}$$

Two LMFs were constructed from the PW91 [7,72] and PBE [13] GGA correlation functional, respectively, resulting in the general expression

$$g^{\text{AC-GGA}} = 1 - 0.5\left(1 + \frac{r_s}{2e_c^{\text{GGA}}}\frac{\partial e_c^{\text{GGA}}}{\partial r_s} - \frac{\xi}{4e_c^{\text{GGA}}}\frac{\partial e_c^{\text{GGA}}}{\partial \xi}\right)^{-1}. \tag{24}$$

In Eqs. (23)–(24), r_s is the Seitz radius

$$r_s = [3/4\pi\rho]^{1/3}, \tag{25}$$

ζ is the spin polarization (Eq. (19)), and ξ is the inhomogeneity parameter employed in the PW91 and PBE correlation functionals:

$$\xi = \frac{|\nabla\rho|}{2\left[(1+\zeta)^{2/3} + (1-\zeta)^{2/3}\right](3/\pi)^{1/6}\rho^{7/6}}. \tag{26}$$

Energy densities $e_c^{\text{LSDA/GGA}}$ are those per electron, related to the familiar quantities $\varepsilon_c^{\text{LSDA/GGA}}$ as

$$e_c^{\text{LSDA/GGA}}(\mathbf{r}) = \varepsilon_c^{\text{LSDA/GGA}}(\mathbf{r})/\rho(\mathbf{r}). \tag{27}$$

According to the suggested model [71], and within the strict AC formalism, there are only a few degrees of freedom in the structure of a local hybrid (10): (i) the same level of DFT approximation (either LSDA or GGA in the given case) must be employed for both exchange and correlation parts, and also for generating the LMF; (ii) as one can see from Eqs. (23), (24) and mentioned above, the LMF is completely determined by the correlation functional. In contrast, no restrictions of this kind were employed for the semi-empirical local hybrids discussed above.

AC-LMFs constructed on the basis of the LSDA and GGA (PW91) correlation are shown in Figure 2 for carbon monosulfide (for comparison, we display also the successful semi-empirical *t*-LMF (14)). The AC-LSDA LMF is rather featureless and represents poorly even the atomic shell structure in the core and semi-core regions. More structure is found for the GGA-based LMF. Yet both AC-LMFs exhibit clearly an incorrect asymptotic behavior. From the practical point of view, the resulting local hybrids (both LSDA- and GGA-based) turn out [71] to be inferior compared to the best semi-empirical local hybrids (although a semi-empirical scaling may improve results; Section 5). The fundamental AC formalism seems thus of limited practical usefulness in constructing accurate local hybrids. This should not really surprise us: beyond a certain complexity

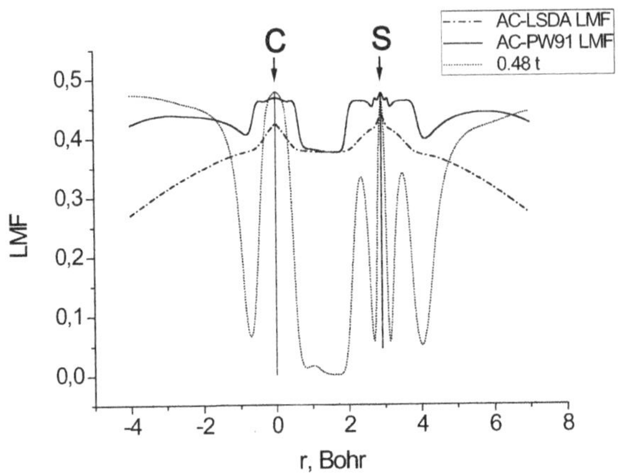

Fig. 2. Local mixing functions (LMFs) derived on the basis of the adiabatic connection (AC) formalism.

of an exchange-correlation functional, first-principles arguments alone may be insufficient for the detailed construction of an accurate DFT method [21]. Nevertheless, information extracted from the basic physical formalism provides useful insight and may be employed when constructing improved functionals. In particular, one can observe striking qualitative similarities between the AC-PW91 LMF and successful semiempirical *t*-LMF (see Figure 2; further points have been noted [71]). Moreover, the two dimensionless quantities $(r_s/2e_c^{LSDA/GGA})(\partial e_c^{LSDA/GGA}/\partial r_s)$ and $(-\xi/4e_c^{GGA})(\partial e_c^{GGA}/\partial \xi)$, which emerged in a natural way when deriving Eqs. (23)–(24), may turn out to be useful ingredients of future local hybrids.

We should finally mention local hybrids developed recently in other research groups. Perdew and co-workers [69] have constructed a sophisticated local hybrid based on the requirement to satisfy as many exact constraints as possible. This required five semi-empirical parameters and a sophisticated LMF that depends itself on the exact-exchange energy density (4). Performance for atomization energies and reaction barriers is far superior to the AC-LMFs discussed above [71] but overall somewhat inferior to the so far best semi-empirical local hybrids [45,62,64]. There have also been suggestions and tests of local hybrids based on density-matrix similarity metrics [73].

4. On the self-consistent implementation of OOD functionals: from localized global and local hybrid potentials to the first self-consistent implementation of the B05 functional

Evaluation of molecular properties other than total energy inevitably requires a self-consistent implementation of a functional. As has been discussed above, there are two levels of the self-consistent implementation of occupied-orbital dependent functionals: (i) one can stop at the stage of evaluating the FDOs and make replacement (9), or (ii) nonlocal/nonmultiplicative contributions to FDOs may be further "localized", i.e. transformed into local and multiplicative components of the exchange-correlation potential using some OEP-based techniques. In our studies we have employed both approaches since each of them has its own advantages and shortcomings. For instance, in the case of pure exact exchange, localized potentials are known to give a more realistic description of the virtual orbital subspace [74], which is crucial for the description of linear-response properties or evaluations of excited states. In the absence of current-dependent terms in the potential, magnetic property calculations may be carried out in an uncoupled approximation (see below). On the other hand, the exact (numerical) solution of the integral OEP equation [42] is typically not efficient from a computational point of view and numerically unstable for molecules. Therefore, a number of approximate OEP schemes have been developed [74–80]. For some properties (for instance, hyperfine coupling constants), the results turn out to be sensitive towards the details of the approximation employed. This leads to a number of as yet unsolved questions regarding the reliability of the OEP-based methods for certain response properties.

Nonlocal/nonmultiplicative FDOs are free from additional inaccuracies and artefacts introduced by an approximate OEP treatment. On the other hand, the nonlocal terms in the potential may give rise to linear response terms in case of magnetic perturbations. This leads to the need of a more involved iterative coupled-perturbed Kohn-Sham (CPKS) scheme [81]. This is computationally less efficient, and qualitative analyses of the response properties are made less transparent by the arising coupling terms. Moreover, one leaves the genuine Kohn-Sham framework, which is theoretically less satisfying. That is, there is a choice to be made between FDO and "FDO plus OEP" approaches.

4.1. Functional derivatives with respect to the orbitals ("nonlocal implementations")

We consider first FDOs derived from typical τ-dependent meta-GGA (τ-MGGA) functionals (6):

$$\frac{\delta E_{xc}^{\tau\text{-MGGA}}}{\delta\varphi_i} = 2\left[\frac{\partial\varepsilon_{xc}^{\tau\text{-MGGA}}}{\partial\rho} - \nabla\left(\frac{\partial\varepsilon_{xc}^{\tau\text{-MGGA}}}{\partial\nabla\rho}\right)\right]\varphi_i$$
$$-\nabla\left(\frac{\partial\varepsilon_{xc}^{\tau\text{-MGGA}}}{\partial\tau}\right)\cdot\nabla\varphi_i - \left(\frac{\partial\varepsilon_{xc}^{\tau\text{-MGGA}}}{\partial\tau}\right)\nabla^2\varphi_i. \tag{28}$$

As one can see, the right-hand side (RHS) of Eq. (28) contains differential operators that make it nonmultiplicative, but still (semi)local (i.e., at each point of space **r**, the FDO is determined by quantities evaluated at the same point **r** only).

A qualitatively different situation is observed for hybrid functionals. The well known FDOs derived from global hybrids (Eq. (7)),

$$\frac{\delta E_{xc}^{\text{hybr}}}{\delta\varphi_i} = 2\left(a_0\hat{v}_x^{\text{exact}}\varphi_i + \overline{v}_{xc}^{\text{DFT}}\varphi_i\right) \equiv 2\hat{\overline{v}}_{xc}^{\text{hybr}}\varphi_i, \tag{29}$$

contain an essentially nonlocal exact-exchange operator $\hat{v}_x^{\text{exact}}$, which is defined by its action on an arbitrary one-particle function ψ as

$$\left[\hat{v}_x^{\text{exact}}\psi\right](\mathbf{r}) = -\sum_j^{\text{occ}}\varphi_j(\mathbf{r})\int\frac{\varphi_j(\mathbf{r}')\psi(\mathbf{r}')}{|\mathbf{r}-\mathbf{r}'|}d\mathbf{r}', \tag{30}$$

and $\overline{v}_{xc}^{\text{DFT}}$ is given simply by Eq. (1) (provided that $\overline{E}_{xc}^{\text{DFT}}$ from Eq. (7) does not contain other essentially orbital-dependent contributions). A noticeably more sophisticated expression arises in the case of local hybrids [65,82]:

$$\frac{\delta E_{xc}^{\text{loc-hyb}}}{\delta\varphi_i} = \left[g\,\hat{v}_x^{\text{exact}}\varphi_i + \hat{v}_x^{\text{exact}}(g\varphi_i)\right]$$
$$+ 2\left[(1-g)v_x^{\text{DFT}} + v_c^{\text{DFT}}\right]\varphi_i + \frac{\partial g}{\partial\varphi_i}\left(\varepsilon_x^{\text{exact}} - \varepsilon_x^{\text{DFT}}\right)$$
$$-\nabla\left[\frac{\partial g}{\partial\nabla\varphi_i}\left(\varepsilon_x^{\text{exact}} - \varepsilon_x^{\text{DFT}}\right)\right] + 2\nabla g\cdot\frac{\partial\varepsilon_x^{\text{DFT}}}{\partial\nabla\rho}\varphi_i. \tag{31}$$

It is interesting to note that in the limit of $g(\mathbf{r}) = \text{const} = a_0$, Eq. (31) is greatly simplified and naturally transforms to Eq. (29) for global hybrids (as it should be).

Finally, for the general OOD functional (5), which may not be reducible to a local hybrid form (10), the FDO expressions become still more complicated. As we have shown in ref. 83, it is convenient to decompose the total expression for the FDO into a sum of terms in such a way, that each of them arises from variation of only one of the ingredients (i.e., separate contributions due to variation of ρ_σ, $|\nabla\rho_\sigma|$, $\nabla^2\rho_\sigma$, τ_σ, $\varepsilon_{x,\sigma}^{\text{exact}}$, $\sigma = \alpha, \beta$, and $\nabla\rho_\alpha\cdot\nabla\rho_\beta$). Without giving the specific cumbersome expressions for each of these terms, here we focus only on the $(\delta E_{xc}^{\text{OOD}}/\delta\varphi_i)_{\varepsilon_x^{\text{exact}}}$ term (spin labels are suppressed):

$$\left(\frac{\delta E_{\mathrm{xc}}^{\mathrm{OOD}}}{\delta \varphi_i}\right)_{\varepsilon_{\mathrm{x}}^{\mathrm{exact}}} = \left(\frac{\partial \varepsilon_{\mathrm{xc}}^{\mathrm{OOD}}}{\partial \varepsilon_{\mathrm{x}}^{\mathrm{exact}}}\right) \hat{\mathrm{v}}_{\mathrm{x}}^{\mathrm{exact}} \varphi_i + \hat{\mathrm{v}}_{\mathrm{x}}^{\mathrm{exact}} \left\{ \left(\frac{\partial \varepsilon_{\mathrm{xc}}^{\mathrm{OOD}}}{\partial \varepsilon_{\mathrm{x}}^{\mathrm{exact}}}\right) \varphi_i \right\}, \tag{32}$$

which introduces an essential nonlocality into an FDO. As one can see, the RHS of Eq.(32) resembles the first term on the RHS of Eq. (31), apart from the fact that the LMF g in the latter is replaced by a more general quantity $\partial \varepsilon_{\mathrm{xc}}^{\mathrm{OOD}} / \partial \varepsilon_{\mathrm{x},\sigma}^{\mathrm{exact}}$.

In [83] we have chosen Becke's real-space model for NDC [23,24] (also known as B05 functional, with promising performance in atomization energies [24] and reaction barriers [84]) as one of the currently most complicated OOD functionals to demonstrate the implementation by this general formalism. The rather cumbersome FDOs for this functional were implemented and tested [83]. Particular difficulties with the self-consistent implementation of the B05 functional are associated with the fact that the underlying NDC energy density involves an implicitly defined quantity, which is available only numerically, as a result of the iterative solution of a nonlinear equation (to generate a corresponding initial guess, we have constructed an efficient analytical ansatz [85]). More importantly, this energy density is only piece-wise continuous. This leads to discontinuities in the corresponding FDOs and, as a consequence, to some as yet unsolved technical problems. So far SCF convergence was only possible for closed-shell systems (detailed discussion and numerical results are given in ref. 83).

As mentioned above, implementation of OOD functionals at the FDO level, without further OEP transformation, requires iterative CPKS schemes for second-order magnetic properties, as is well known for global hybrids [81]. Given the more complicated form of local hybrids (31) or of the even more general OOD functional (32), one might expect complications to arise from the position-dependent admixture of different exchange-energy densities. We have recently derived generalized CPKS schemes for OOD functionals and implemented them for the specific example of electronic g-tensors with local hybrids [86]. It turned out that, fortunately, no additional coupling terms arise from derivatives of $E_{\mathrm{xc}}^{\mathrm{OOD}}$ with respect to variables other than the exact-exchange energy density itself (although those derivatives may depend on the latter), keeping the formalism at a manageable complexity. Section 5 provides a brief discussion of the numerical results.

4.2. OEP-based transformations ("localized implementations")

Let us now proceed to the second, possible but not mandatory, step in the self-consistent treatment of OOD functionals: the OEP-based transformation of the nonlocal, nonmultiplicative parts of the potential to a localized and multiplicative form. As noted above, full numerical solution of the integro-differential OEP equations for molecules experiences numerical difficulties, and therefore approximations are usually applied. These range from the traditional KLI approximation

[75] to basis-set expansion methods [77–80]. In our first self-consistent implementation of local hybrids, we employed the so-called localized Hartree-Fock (LHF) approach [74], which is equivalent to the common energy denominator approximation (CEDA) [76] (KLI may be viewed as a "diagonal approximation" to LHF/CEDA). Here the target potential, v_{xc}^{OOD}, is constructed from two contributions, the orbital-weighted FDO (or "average potential",v_{xc}^{av}) and the correction (or response [87]) term (v_{xc}^{corr}). Originally [74], the LHF approach has been applied to pure exact exchange only. We and others [49,50,88–90] used LHF subsequently to construct localized potentials from global hybrid functionals ("localized global hybrid potentials", LGH). As in ref. 74, a resolution of the identity (RI) was used for the evaluation of the average potential (which for $\hat{v}_{xc}^{OOD} = \hat{v}_{x}^{exact}$ is called Slater potential), to avoid computation of two-electron integrals other than the standard ones that are available in any quantum-chemical code. LGH potentials have been employed in calculations of nuclear shielding constants [49,89,90] and electronic g-tensors [50] (see also ref.88). We note that the solution of uncoupled perturbation equations was sufficient in these cases, due to the absence of coupling terms from any nonlocal exact-exchange terms. Some results are discussed in Section 5. FDOs (28) of a τ-dependent meta-GGA functional (PKZB) [60] have also been transformed into multiplicative potentials by a modified LHF approach [91]. Matters are somewhat simpler in this case than with exact exchange [74]: as nonlocal operators are absent in (28), no RI is required in the evaluation of the average potential. Instead, its matrix elements in the atomic orbital (AO) basis are evaluated using three-dimensional partial integration (see Eqs. (29)–(31) of Ref. 91 for details).

Applying a similar LHF-based procedure to the FDOs (31) of local hybrids gives "localized local hybrid" (LLH) potentials [65,82]. Without going into the partly tedious details, we note again some technical peculiarities associated with the evaluation of the average potential. In particular, two RIs had to be employed. As this was the first self-consistent implementation of local hybrids, there was a need to confirm the correctness of the derivations and of the implementation. We compared in particular total energies and atomization energies with those obtained in nonself-consistent calculations. It turned out that an uncontraction of the original orbital basis was necessary in the evaluation of the RIs to obtain consistently total energies that were a few mH below the nonself-consistent results, as they should be. Then atomization energies agreed well between self-consistent and nonself-consistent calculations [65]. Subsequently, the LLH potentials were used to compute nuclear shielding constants [82] in uncoupled perturbation calculations (cf. Section 5).

5. Examples of computational results

To illustrate the so far already remarkable performance of local hybrids, we summarize a few selected computational results. Atomization energies and reaction barriers were so far obtained mostly in nonself-consistent calculations, based on the orbitals obtained with global hybrids. The first self-consistent implementation was done within our MAG-ReSpect code [92], with the possibility to extract the KS orbitals either at the FDO level or after LHF-based OEP transformation. We have meanwhile also carried out an FDO-based implementation in a development version of the Turbomole package [93]. Independently, the group of Scuseria has done an FDO-based implementation in a development version of the Gaussian program [41]. We will in the following report largely results obtained with MAG-ReSpect, except for data with ζ-dependent spin-polarized LMFs, which were obtained with the more recent Turbomole implementation. Extended Gaussian basis sets of triple- (IGLO-III [94]) or quadruple-zeta quality (cc-pVQZ [95] and QZVP [96]) have been employed. To evaluate the exact-exchange energy density (4) and closely related Slater potential, we always used the corresponding uncontracted basis sets to reduce any errors arising from RIs.

The G2/97 [61] and G3/99 [97] atomization energy databases have been used in our thermochemical studies. To validate our local hybrids for activation barriers of chemical reactions, we have employed the HTBH38 [98] and NHTBH38 [99] test sets. The small AE6 and BH6 sets [100] have been used when optimizing the two-parameter ζ-dependent LMFs [64]. Differences in the computational results for the different implementations and computational protocols are not zero but sufficiently small to discuss the results from different publications on reasonably equal footing. Computations of nuclear shielding and electronic g-tensors used also test sets that had been employed previously to validate density functionals. See also the original papers cited for further computational details and references to the experimental data.

5.1. Thermochemical results (atomization energies)

Table 1 gives mean absolute errors (MAE) and mean signed errors (MSE) for atomization energies obtained with various local hybrid functionals (Section 3) compared to the B3LYP global hybrid [17,12,18]. In addition to the overall results for the full G3/99 set (223 molecules), we list results for the subsets G2–1 (55 molecules), G2/97 (extra 93 molecules), and G3/99 (75 molecules), to show how the performance of the various functionals changes once more molecules are included. It is remarkable, that the best thermochemical results can be obtained when only local LSDA exchange (without any GGA correction) is mixed with exact exchange, indeed with local correlation (VWN) added on top. Among the one-parameter local hybrids, the best results are provided by *Lh*-SVWN with the scaled *t*-LMF (14). While the scaling parameter 0.48 had been optimized on the small G2–1 subset, the functional shows a remarkable stability with respect to the extension of the test set: the MAE for the full G3 set (225 species) remains essentially the same as for the G2–1 set! This contrasts with the performance of the popular B3LYP functional, which deteriorates significantly when going to the larger sets. This type of deterioration seems to be

Table 1. Mean absolute errors (MAE) and mean signed errors (MSE) for atomization energies (in kcal mol^{1}) of the G3 test set and its subsets.

Functional	Local mixing function (LMF) [b]	G2–1 (55) [a] MAE	G2–1 (55) [a] MSE	G2 (93) [a] MAE	G2 (93) [a] MSE	G3 (75) [a] MAE	G3 (75) [a] MSE	Σ G3 (223) [a] MAE	Σ G3 (223) [a] MSE	Ref.
Lh-SVWN	$0.48\ t_\sigma$	3.73	0.98	4.35	1.53	3.31	-0.12	3.85	0.84	64
Lh-SVWN	erf $(0.22\ s_\sigma)$	4.91	1.66	5.51	2.12	5.22	0.00	5.26	1.30	64
Lh-SLYP	$[s_\sigma / (0.73 + s_\sigma)]^2$	2.60	-0.71	5.56	4.70	8.89	8.27	5.95	4.57	45
Lh-SVWN	$(0.446 \pm 0.0531\zeta)t_\sigma$ [c]	2.66	0.20	2.97	1.43	3.16	-0.58	2.96	0.45	64
Lh-SVWN	erf $[(0.197 \pm 0.0423\zeta)s_\sigma]$ [c]	3.30	0.60	3.21	1.74	2.97	-1.23	3.15	0.46	64
Lh-SVWN	AC-LSDA [d]	13.3	13.0	48.3	48.2	93.7	93.7	55.0	54.8	71
Lh-BPW91	AC-PW91 [e]	11.7	-11.3	20.3	-20.3	29.7	-29.7	21.3	-21.2	71
Lh-BPW91	AC-PW91, downshifted [f]	3.5	-1.3	4.9	-0.5	8.9	-5.2	5.9	-2.3	71
B3LYP	-	2.53	-0.35	4.61	-4.10	10.8	-10.7	6.17	-5.41	64

[a] The number of molecules in the test sets are given in parentheses. [b] Variables t_σ , s_σ and ζ are defined by Eqs. (12), (16), and (19), respectively; $\sigma = \alpha, \beta$. [c] Plus sign for $\sigma = \alpha$ and minus sign for $\sigma = \beta$. [d] See Eq. (23). [e] See Eq. (24). [f] LMF given by Eq. (24), downshifted by 0.34: $g^{\text{AC-GGA}} - 0.34$.

relatively general for many global hybrids and has been attributed partly to an incorrect description of medium-range interpair correlation effects [101]. The *Lh*-SVWN functional with the *s*-LMF $g = \text{erf}(0.22\ s)$ performs inferior to the best *t*-LMF. Nevertheless, the atomization energies are also extremely stable upon enlargening the test set. In contrast, one *Lh*-SLYP functional included, which employs LYP correlation in (10), also exhibits a deterioration for larger test sets akin to the B3LYP data. Why this is so, and to what extent this behavior depends on a balance between dynamical correlation and the simulation of non-dynamical correlation by position-dependent exchange mixing remains to be studied in more detail.

Further significant improvement in the atomization energies is obtained with the two-parameter local hybrids with *t*-ζ- and *s*-ζ-LMFs (Table 1). The reduction in the MAE is particularly notable for the *s*-LMF. Again the best results are obtained when only LSDA ingredients are involved. We have investigated partial GGA or meta-GGA (TPSS) corrections to exchange and correlation, but optimization for thermochemistry generally resulted in vanishing GGA or meta-GGA coefficients (data not shown [102]). We note in passing that the parameters of the ζ-dependent local hybrids have been optimized for the small AE6 and BE6 test sets (six atomization energies plus six reaction barriers taken into account with equal weights 100]).

As has been mentioned above, nonempirical local hybrids based on the adiabatic connection (AC) formalism yield significantly inferior thermochemical results. This is particularly so for *Lh*-SVWN with AC-LSDA (Table 1). The situation is somewhat improved when passing from the local hybrid with a featureless AC-LSDA LMF to the *Lh*-BPW91 with a more structured AC-PW91 LMF (cf. Figure 2). In this case the accuracy is comparable to (or better than) that of Becke's initial "half-and-half" functional [14] (see detailed discussion in ref. 45).

Table 2. Mean absolute errors (MAE) and mean signed errors (MSE) for classical reaction barriers (in kcal mol[1]) of the HTBH38 and NHTBH38 test sets[a].

Functional	Local mixing function (LMF)	HTBH38 MAE	HTBH38 MSE	NHTBH38 MAE	NHTBH38 MSE	Ref.
Lh-SVWN	0.48 t_σ	2.5	-2.1	2.5	-1.5	64
Lh-SVWN	erf (0.22 s_σ)	3.5	-3.4	4.3	-3.7	64
Lh-SLYP	$[s_\sigma / (0.73 + s_\sigma)]^2$	5.5	-5.5	5.3	-4.7	45
Lh-SVWN	$(0.446 \pm 0.0531\zeta)\, t_\sigma$	3.3	-2.9	2.6	-1.9	64
Lh-SVWN	erf $[(0.197 \pm 0.0423\zeta)\, s_\sigma]$	4.3	-4.3	4.4	-4.4	64
Lh-SVWN	AC-LSDA	5.6	-5.6	4.8	-3.1	71
Lh-BPW91	AC-PW91	2.0	1.3	2.3	1.6	71
Lh-BPW91	AC-PW91, downshifted	5.6	-5.6	6.2	-6.1	71
B3LYP	-	4.3	-4.3	5.0	-4.8	64

[a] See Footnotes to Table 1

In general, it seems that AC-based LMFs admix too much exact exchange, which is profitable for reaction barriers (see below), but unfavorable for thermochemistry. It is interesting to note that a simple *ad hoc* shifting down of the AC-PW91 LMF by 0.34 improves the thermochemistry dramatically. Of course this corresponds to a loss of the nonempirical character of the LMF (and barriers deteriorate, see below). AC-LMFs provide nevertheless an interesting point of reference for the further development of local hybrids.

5.2. Classical reaction barriers

Table 2 shows MAE and MSE for classical reaction barrier heights obtained with the same set of functionals. It is apparent that the thermochemically optimized one-parameter local hybrids *Lh*-SVWN with both *t*- and *s*-LMFs provide significantly higher accuracy than B3LYP. An only moderate deterioration of the barriers is observed for the two-parameter ζ-dependent local hybrids, which has to do with a slightly reduced overall exact-exchange admixture. That is, these simple functionals do not yet allow a completely independent optimization of thermochemical data and barriers without compromise. Yet, they are already a clear improvement over standard global hybrids like B3LYP for both thermochemistry and barriers. They provide thus an excellent starting point for further refinement. We note in passing, that the good performance of these local hybrids for barriers has been partially rationalized by the build-up of local maxima in the LMF in the bond region upon stretching bonds from their equilibrium distance [45].

The best reaction barriers are actually obtained with the *ab initio* local hybrid *Lh*-BPW91 using the AC-PW91 LMF. This reflects the relatively large exact-exchange admixture which, however, is detrimental for the thermochemical performance. Similar observations pertain to global hybrids with large EXX admixtures like the half-and-half functionals [45].

5.3. Magnetic resonance parameters

Here we focus mainly on nuclear shielding constants, as they exhibit a high sensitivity both to the functional and to its self-consistent implementation. In addition to the mean absolute and signed errors, Table 3 gives also the results of linear regression analyses for the nonhydrogen shieldings (32 values) of 22 main-group molecules. Starting with global hybrids, a significant dependence on the nonlocal FDO vs. LHF-based localized implementation is apparent. Nonlocal exact-exchange admixture gives rise to large coupling-terms in the CPKS treatment, which lead to a large overestimation of the (typically negative) paramagnetic shielding contributions for molecules with low-lying excited states. This becomes more and more pronounced with increased EXX admixture (compare B3PW91 with 20% exact exchange and EXX-PW91 with 50% exact exchange), consistent with earlier results [103,104]. No such dramatic dependence, and significantly improved agreement with experiment, is obtained in the LHF-based localized implementation (LGH potentials). This is particularly apparent for the B-EXX-PW91 functional with 50% exact exchange, which exhibits a particularly small standard deviation, better than the more common B3PW91. We note, however, that calculations with a localization by a basis-set expansion OEP method give different results, which do not seem to favor such large exact-exchange admixtures [105]. This may point to deficiencies in the virtual orbital space obtained in the LHF-based localization (albeit deficiencies in the basis-expansion OEP may also not be ruled out). This should be kept in mind when discussing the LLH-based results below.

So far, no FDO-based CPKS scheme for nuclear shieldings with local hybrids has been implemented (the need for gauge-including atomic orbitals for accurate nuclear shieldings complicates matters, and the implementation is ongoing work). Table 3 gives some results based on an LHF-based localization (LLH potentials, uncoupled calculations) for selected local hybrids. We deliberately compare for each LMF results with the thermochemically optimal single parameter ($0.48t$ and $[s/(0.73+s)]^2$, respectively), and data obtained after reoptimizing (within the LLH uncoupled scheme) the parameter for the nuclear-shielding test set at hand ($0.90t$ and $[s/(0.40+s)]^2$). Performance in the latter case is even better than with the best LGH-based treatment (BEXX-PW91 with 50% EXX). More interestingly, already the thermochemically optimized LMFs give rather accurate shieldings. We keep in mind, however, the caveat of possible shortcomings of the LHF-based localization for such sensitive response properties and postpone a final judgement until reliable basis-expansion OEP transformations have been tested.

Recently, we have evaluated the first generalized CPKS implementation of magnetic linear-response properties for an FDO-based implementation of local hybrids for the example of electronic g-tensors of test sets of main-group radicals and transition-metal complexes, respectively [86] (cf. Section 4). We have focused so far only on the thermochemically optimized one- or two-parameter local

Table 3. Computed isotropic nuclear shielding constants of 22 main-group molecules (in ppm) with global and local hybrid functionals at different levels of their implementation compared to experiment[a].

Functional	a_0 or $g(\mathbf{r})$ [b]	MAE (ppm)	MSE (ppm)	Linear regression analysis [c] Intercept (ppm)	Slope	Regression coefficient	Standard deviation (ppm)	Ref.
FDO (nonlocal) implementation of global hybrids								
B3PW91	0.2	50.5	-49.8	-60.7	1.176	0.9923	60.2	49
B-EXX-PW91	0.5	77.5	-73.8	-95.6	1.350	0.9753	126	49
OEP (localized) implementation: LGH potentials								
B3PW91 [d]	0.2	24.5	-23.2	-26.4	1.051	0.9983	24.9	49
B-EXX-PW91 [e]	0.5	13.9	-4.6	-5.4	1.013	0.9988	20.5	49
OEP (localized) implementation: LLH potentials								
Lh-SVWN [f]	0.48 t_σ	22.5	-19.2	-22.4	1.051	0.9982	25.6	82
Lh-SVWN [f]	0.90 t_σ	14.9	-1.0	-1.0	1.000	0.9988	20.5	82
Lh-SLYP [g]	$[s_\sigma / (0.73 + s_\sigma)]^2$	17.0	-11.2	-13.2	1.032	0.9983	24.5	82
Lh-SLYP [g]	$[s_\sigma / (0.40 + s_\sigma)]^2$	15.0	-1.3	-1.8	1.007	0.9985	22.9	82

[a] Data for 32 nonhydrogen nuclei are included in the test set.
[b] Cf. Eqs. (7) and (10). [c] σ (calc.) = A σ (expt.) + B, with σ (expt.) and B in ppm; the standard deviation is $[\sum_{j=1}^{n}(\sigma_j-\bar{\sigma})^2 / (n-1)]^{1/2}$. Perfect agreement with experiment corresponds to A = 1, B = 0, zero standard deviation, and regression coefficient equal to 1. [d] B3(L)-PW91 in the original work. [e] B-EXX(L)-PW91 in the original work. [f] *Lh*(L)SVWN in the original work. [g] *Lh*(L)-SLYP in the original work.

hybrids discussed above. The results were encouraging and sobering at the same time: the agreement with experiment for both main-group and transition-metal systems was slightly superior to that obtained with standard global hybrids like B3LYP or B3PW91. Similar results were obtained in preliminary screening studies for hyperfine couplings or nuclear quadrupole couplings for sensitive cases like copper halides (unpublished). This is good news insofar as the much better accuracy of these local hybrids for thermochemistry and barriers compared to standard functionals is accompanied by rather good performance for magnetic properties. Yet, there is obviously still substantial room for improved LMFs regarding such properties. This will require detailed examination of the properties of LMFs in different spatial regions.

6. Conclusions and outlook

Among the known exchange-correlation functionals in Kohn-Sham DFT, occupied-orbital dependent (OOD) functionals promise the currently most rewarding compromise between computational efficiency and accuracy in many cases. One new family of OOD functionals are local hybrids, which have been demonstrated to exhibit very promising results for a large range of properties, even when only LSDA and exact exchange are "hybridized", and even with currently very simple one- or two-parameter local mixing functions (LMFs). Local hybrids may in fact be the next generation of functionals for wider application. Our current efforts focus on their more efficient self-consistent implementation in fast computer programs, and on the further refinement of the LMFs to allow yet more universal and accurate functionals to be constructed. Notably, appreciable insight into the operating mechanisms of local hybrids has been obtained by (i) the *ab initio* construction of LMFs from a local version of the adiabatic connection and by (ii) graphical and analytical analyses of LMFs. In particular, when we hybridize LSDA and exact exchange only, all sophistication of a local hybrid rests in the LMF, which is a conveniently studied three-dimensional real-space function.

We have furthermore reviewed here the nontrivial ingredients needed for the self-consistent implementation of local hybrids, and of even more complicated OOD functionals. The mandatory first step is the computation of the functional derivatives of the exchange-correlation functional with respect to the Kohn-Sham orbitals. Then we have the choice between direct use of these "FDOs", which contain nonlocal and nonmultiplicative contributions, and a subsequent localization of these contributions by some variant of the optimized effective potential approach. We have noted the advantages and disadvantages of these two routes and gave computational examples for both. Clearly, the results of magnetic linear response properties depend significantly on the way of their implementation, a fact that needs to be carefully evaluated in each case. We have furthermore demonstrated the self-consistent implementation at the FDO level of one of the most complicated OOD functionals known, the B05 real-space model of nondynamical correlation. So far, convergence in actual calculations could be reached only for closed-shell systems, due to discontinuities in the FDOs of this peculiar functional. Here further progress may be expected from (i) suitable modifications of the weighting functions of the nondynamical correlation terms to get around the numerical difficulties, and (ii) transfer of some of the insight obtained from this complicated but physically appealing functional to less difficult approaches. An example of the latter procedure has led to the introduction of spin polarization as an ingredient of local hybrids with the currently best performance for thermochemistry.

Acknowledgement

The reviewed work of the authors has been funded by Deutsche Forschungsgemeinschaft within Priority Program 1145, "*Modern and universal first-principles methods for many-electron systems in chemistry and physics*" (projects KA1187/6–1 and KA1187/10–1). Further contributions are due to A. Rodenberg and B. Zarzycki. We also thank R. Ahlrichs (Karlsruhe) and the Turbomole developer consortium for access to the Turbomole source code.

References

1. S. Kümmel and L. Kronik, Rev. Mod. Phys. **80** (2008) 3.
2. R. G. Parr and W. Yang. *Density-Functional Theory of Atoms and Molecules.* Oxford University Press, New York (1989).
3. W. Koch and M. C. Holthausen. *A Chemist's Guide to Density Functional Theory,* 2nd Ed. Wiley, New York (2001).
4. P. A. M. Dirac, Proc. Cambridge Phil. Soc. **26** (1930) 376; J. C. Slater, Phys.Rev. **81** (1951) 385.
5. U. Von Barth and L. Hedin, J. Phys. C. **5** (1972) 1629.
6. S. J. Vosko, L. Wilk, and M. Nusair, Can. J. Phys. **58** (1980) 1200.
7. J. P. Perdew and Y. Wang, Phys. Rev. B. **45** (1992) 13244.
8. J. P. Perdew, Phys. Rev. Lett. **55** (1985) 1665.
9. J. P. Perdew and W. Yue, Phys. Rev. B. **33** (1986) 8800.
10. J. P. Perdew, Phys. Rev. B. **33** (1986) 8822.
11. A. D. Becke, Phys. Rev. A. **38** (1988) 3098.
12. C. Lee, W. Yang, and R. G. Parr, Phys. Rev. B **37** (1988) 785; B. Miehlich, A. Savin, H. Stoll, and H. Preuss, Chem. Phys. Lett. **157** (1989) 200.
13. J. P. Perdew, K. Burke, and M. Ernzerhof, Phys. Rev. Lett. **77** (1996) 3865.
14. A. D. Becke, J. Chem. Phys. **98** (1993) 1372.
15. A. D. Becke, J. Chem. Phys. **107** (1997) 8554.
16. C. Adamo and V. Barone, J. Chem. Phys. **110** (1999) 6158.
17. A. D. Becke, J. Chem. Phys. **98** (1993) 5648.
18. P. J. Stephens, F. J. Devlin, C. F. Chabalowski, and M. J. Frisch, J. Phys. Chem. **98** (1994) 11623.
19. J. Tao, J. P. Perdew, V. N. Staroverov, and G. E. Scuseria, Phys. Rev. Lett. **91** (2003) 146401.
20. V. N. Staroverov, G. E. Scuseria, J. Tao, and J. P. Perdew, J. Chem. Phys. **119** (2003) 12129.
21. J. P. Perdew and K. Schmidt, in: *Density Functional Theory and its Application to Materials,*V. Van Doren, C. Van Alsenoy, and P. Geerlings, Eds., AIP, Melville, New York (2001).
22. J. P. Perdew, A. Ruzsinszky, J. Tao, V. N. Staroverov, G. E. Scuseria, and G. I. Csonka, J. Chem. Phys. **123** (2005) 062201.
23. A. D. Becke, J. Chem. Phys. **119** (2003) 2972.
24. A. D. Becke, J. Chem. Phys. **122** (2005) 064101.
25. A. Görling and M. Levy, Phys. Rev. A. **50** (1994) 196.
26. E. Engel and R. M. Dreizler, J. Comput. Chem. **20** (1999) 31.
27. R. J. Bartlett, I. Grabowski, S. Hirata, and S. Ivanov, J. Chem. Phys. **122** (2005) 034104.
28. R. J. Bartlett, V. F. Lotrich, and I. V. Schweigert, J. Chem. Phys. **123** (2005) 062205.
29. S. Grimme, J. Chem. Phys. **124** (2006) 034108; T. Schwabe, and S. Grimme, Phys. Chem. Chem. Phys. **9** (2007) 3397.
30. M. Filatov and W. Thiel, Phys. Rev. A. **57** (1998) 189.
31. J. P. Perdew and A. Zunger, Phys. Rev. B. **23** (1981) 5048.
32. P. Mori-Sánchez, A. J. Cohen, and W. Yang, J. Chem. Phys. **124** (2006) 091102; Ibid. **127** (2007) 034101.
33. T. Van Voorhis and G. E. Scuseria, J. Chem. Phys. **109** (1998) 400.
34. R. Neumann, R. Nobes, and N. Handy, Mol. Phys. **87** (1996) 1.
35. C. Adamo, M. Ernzerhof, and G. E. Scuseria, J. Chem. Phys. **112** (2000) 2643.
36. M. Bühl, Chem. Phys. Lett. **267** (1997) 251; M. Bühl, O. L. Malkina, and V. G. Malkin, Helv. Chim. Acta. **79** (1996) 742.

37. M. Kaupp, R. Reviakine, O. L. Malkina, A. Arbuznikov, B. Schimmelpfennig, and V. G. Malkin, J. Comput. Chem. **23** (2002) 794.
38. F. Neese, J. Chem. Phys. **115** (2001) 11080.
39. C. Remenyi, R. Reviakine, A. V. Arbuznikov, J. Vaara, and M. Kaupp, J. Phys. Chem. A. **108** (2004) 5026.
40. A. Seidl, A. Görling, P. Vogl, J. A. Majewski, and M. Levy, Phys. Rev. B **53** (1996) 3764.
41. B. G. Janesko, A. V. Krukau, and G. E. Scuseria, J. Chem. Phys. **129** (2008) 124110.
42. R. T. Sharp and G. K. Horton, Phys. Rev. **90** (1953) 317; J. D. Talman and W. F. Shadwick, Phys. Rev. A **14** (1976) 36.
43. A. Görling, J. Chem. Phys. **123** (2005) 062203.
44. J. Jaramillo, G. E. Scuseria, and M. Ernzerhof, J. Chem. Phys. **118** (2003) 1068.
45. M. Kaupp, H. Bahmann, and A. V. Arbuznikov, J. Chem. Phys. **127** (2007) 194102.
46. J. P. Perdew, M. Ernzerhof, and K. Burke, J. Chem. Phys. **105** (1996) 9982.
47. J. L. Durant, Chem. Phys. Lett. **256** (1996) 595; H. Chermette, H. Razafinjanahary, and L. Carrion, J. Chem. Phys. **107** (1997) 10643.
48. M. Munzarová and M. Kaupp, J. Phys. Chem. A **103** (1999) 9966.
49. A. V. Arbuznikov and M. Kaupp, Chem. Phys. Lett. **386** (2004) 8.
50. A. V. Arbuznikov and M. Kaupp, Chem. Phys. Lett. **391** (2004) 16.
51. P. J. Wilson, R. D. Amos, and N. C. Handy, Chem. Phys. Lett. **312** (1999) 475; P. J. Wilson and D. J. Tozer, J. Chem. Phys. **116** (2002) 10139; B. Champagne, E. Botek, M. Nakano, T. Nitta, and K. Yamaguchi, J. Chem. Phys. **122** (2005) 114315; M. Miura, Y. Aoki, and B. Champagne, J. Chem. Phys. **127** (2007) 084103.
52. T. Leininger, H. Stoll, H.-J. Werner, and A. Savin, Chem. Phys. Lett. **275** (1997) 151.
53. H. Iikura, T. Tsuneda, T. Yanai, and K. Hirao, J. Chem. Phys. **115** (2001) 3540.
54. T. Yanai, D. P. Tew, and N. C. Handy, Chem. Phys. Lett. **393** (2004) 51.
55. I. C. Gerber and J. G. Ángyán, Chem. Phys. Lett. **415** (2005) 100.
56. M. J. G. Peach, T. Helgaker, P. Salek, T. W. Keal, O. B. Lutnæs, D. J. Tozer, and N. C. Handy, Phys. Chem. Chem. Phys. **8** (2006) 558.
57. O. A. Vydrov, J. Heyd, A. V. Krukau, and G. E. Scuseria, J.Chem. Phys. **125** (2006) 074106.
58. T. M. Henderson, B. G. Janesko, and G. E. Scuseria, J. Phys. Chem. A **112** (2008) 12530.
59. F. G. Cruz, K.-C. Lam, and K. Burke, J. Phys. Chem. A **102** (1998) 4911.
60. J. P. Perdew, S. Kurth, A. Zupan, and P. Blaha, Phys. Rev. Lett. **82** (1999) 2544.
61. L.A. Curtiss, K. Raghavachari, P. C. Redfern, and J. A. Pople, J. Chem. Phys. **106** (1997) 1063.
62. H. Bahmann, A. Rodenberg, A. V. Arbuznikov, and M. Kaupp, J. Chem. Phys. **126** (2007) 011103.
63. A. V. Arbuznikov and M. Kaupp, Chem. Phys. Lett. **440** (2007) 160.
64. A. V. Arbuznikov, H. Bahmann, and M. Kaupp, J. Phys. Chem. A, **113** (2009) 11898.
65. A. V. Arbuznikov, M. Kaupp, and H. Bahmann, J. Chem. Phys. **124** (2006) 204102.
66. K. Burke, F. G. Cruz, and K.-C. Lam, J. Chem. Phys. **109** (1998) 8161.
67. K. Burke, F. G. Cruz, and K.-C. Lam, Int. J. Quantum Chem. **70** (1998) 583.
68. J. Tao, V. N. Staroverov, G. E. Scuseria, and J. P. Perdew, Phys. Rev. A **77** (2008) 012509.
69. J. P. Perdew, V. N. Staroverov, J. Tao, and G. E. Scuseria, Phys. Rev. A **78** (2008) 052513.
70. D. C. Langreth and J. P. Perdew, Solid State Commun. **17** (1975) 1425; O. Gunnarsson and B. I. Lundqvist, Phys. Rev. B **13** (1976) 4274; M. Levy in *Density Func-*

tional Theory, edited by J. Keller and J. L. Gasquez (Springer, New York, 1983); J. Harris, Phys. Rev. A. **29** (1984) 1648; J. Harris and R. O. Jones, J. Phys. F: Metal Phys. **4** (1974) 1170.
71. A. V. Arbuznikov and M. Kaupp, J. Chem. Phys. **128** (2008) 214107.
72. J. P. Perdew, J. A. Chevary, S. H. Vosko, K. A. Jackson, M. R. Pederson, D. J. Singh, and C. Fiolhais, Phys. Rev. B **46** (1992) 6671.
73. B. G. Janesko and G. E. Scuseria, J. Chem. Phys. **127** (2007) 164117; Ibid. **128** (2008) 084111.
74. F. Della Sala and A. Görling, J. Chem. Phys. **115** (2001) 5718.
75. J. B. Krieger, Y. Li, and G. J. Iafrate, Phys. Rev. A **45** (1992) 101.
76. O. V. Gritsenko and E. J. Baerends, Phys. Rev. A **64** (2001) 042506.
77. A. Görling, Phys. Rev. Lett. **83** (1999) 5459.
78. W. Yang and Q. Wu, Phys. Rev. Lett. **89** (2002) 143002.
79. V. K. Staroverov, G. E. Scuseria, and E. R. Davidson, J. Chem. Phys. **124** (2006) 141103.
80. A. Heßelmann, A. W. Götz, F. Della Sala, and A. Görling, J. Chem. Phys. **127** (2007) 054102; and references therein.
81. W. Kutzelnigg, Israel J. Chem. **19** (1980) 193; M. Schindler and W. Kutzelnigg, J. Chem. Phys. **76** (1982) 1919
82. A. V. Arbuznikov and M. Kaupp, Chem. Phys. Lett. **442** (2007) 496.
83. A. V. Arbuznikov and M. Kaupp, J. Chem. Phys. **131** (2009) 084103.
84. R. M. Dickson and A. D. Becke, J. Chem. Phys. **123** (2005) 111101.
85. A. V. Arbuznikov and M. Kaupp, J. Mol. Struct: THEOCHEM **762** (2006) 151.
86. A. V. Arbuznikov, M. Kaupp, J. Chem. Theory Comput. **5** (2009) 2985.
87. O. V. Gritsenko, R. van Leeuwen, E. van Lenthe, and E. J. Baerends, Phys. Rev. A **51** (1995) 1944.
88. A. V. Arbuznikov and M. Kaupp, Int. J. Quantum Chem. **104** (2005) 261.
89. W. Hieringer, F. Della Sala, and A. Görling, Chem. Phys. Lett. **383** (2004) 115.
90. A. M. Teale and D. J. Tozer, Chem. Phys. Lett. **383** (2004) 109.
91. A. V. Arbuznikov and M. Kaupp, Chem. Phys. Lett. **381** (2003) 495.
92. V. G. Malkin, O. L.Malkina, R. Reviakine, A. V. Arbuznikov, M. Kaupp, B. Schimmelpfennig, I. Malkin, M. Repiský, S. Komorovský, P. Hrobarik, E. Malkin, T. Helgaker, and K. Ruud, MAG-ReSpect program, version 2.1 (2009).
93. Turbomole 5.10. See, e.g: R. Ahlrichs, M. Baer, M. Haeser, H. Horn, and C. Koelmel, Chem. Phys. Lett. **162** (1989) 165; O. Treutler and R. Ahlrichs, J. Chem. Phys. **102** (1995) 346.
94. W. Kutzelnigg, U. Fleischer, and M. Schindler, In: P. Diehl, E. Fluck, H. Günther, R. Kosfeld (Eds.), NMR – Basic Principles and Progress, Vol. 23, Springer, Heidelberg, 1990, p. 165–262; and references therein.
95. T. H. Dunning, Jr., J. Chem. Phys. **90** (1989) 1007; D. E. Woon and T. H. Dunning, Jr., J. Chem. Phys. **98** (1993) 1358.
96. F. Weigend and R. Ahlrichs, Phys. Chem. Chem. Phys. **7** (2005) 3297.
97. L.A. Curtiss, K. Raghavachari, P. C. Redfern, and J. A. Pople, J. Chem. Phys. **112** (2000) 7374.
98. Y. Zhao, B. J. Lynch and D. G. Truhlar, Phys. Chem. Chem. Phys. **7** (2005) 43.
99. Y. Zhao, N. González-García, and D. G. Truhlar, J. Phys. Chem. A **109** (2005) 2012.
100. B. J. Lynch and D. G. Truhlar, J. Phys. Chem. A, **107** (2003) 8996.
101. S. Grimme, Angew. Chem., Int. Ed. Engl. **45** (2006) 4460.
102. H. Bahmann, A. V. Arbuznikov, and M. Kaupp, unpublished results.
103. P. J. Wilson, R. D. Amos, and N. C. Handy, Mol. Phys. **97** (1999) 757.
104. P. J. Wilson and D. J. Tozer, Chem. Phys. Lett. **337** (2001) 341.
105. O. B. Lutnæs, A. M. Teale, T. Helgaker, and D. J. Tozer, J. Chem. Theory Comput. **2** (2006) 827.

Ab initio Electron Dynamics with the Multi-Configuration Time-Dependent Hartree-Fock Method

By Mathias Nest[1,*], Tillmann Klamroth[2], and Peter Saalfrank[2]

[1] Theoretische Chemie, TU München, Lichtenbergstr. 4, 80807 Garching, Germany
[2] Institut für Theoretische Chemie, Karl-Liebknecht-Str. 24–25, 14476 Potsdam, Germany

(Received June 19, 2009; accepted September 22, 2009)

MCTDHF / Correlated Many Electron Dynamics / Ultrafast Quantum Dynamics

This paper gives an overview of the theory and some selected applications of the Multi-Configuration Time-Dependent Hartree-Fock (MCTDHF) method. This method can be seen as an explicitly time-dependent version of the Complete Active Space Self-Consistent Field (CASSCF) method of standard quantum chemistry. Strengths and shortcomings are discussed, and some comparisons with other methods of correlated electron dynamics are given.

1. Introduction

Ab initio quantum chemistry methods have been used for decades to determine the geometric and electronic structure of molecules, as well as their properties with great success. However, the focus has almost always been on static and equilibrium properties. In fact, until recently there was no need to study the ultrafast motion of electrons, because the timescale of electron motion is shorter by orders of magnitude than what could be achieved by experiments. For example, the classical round trip time of the electron in a hydrogen atom is about 152 attosecond (1 as = 10^{-18} s). Since a few years, however, sub-fs laser pulses are available and it became possible to follow the motion of electrons in real time [1–4].

Also due to the experimental progress, there is the need for a theoretical description of these processes. Consequently, various methods of standard quantum chemistry have been extended to the solution of the time-dependent Schrö-

* Corresponding author. E-mail: mathias.nest@mytum.de

Z. Phys. Chem. **224** (2010) 569–581
© by Oldenbourg Wissenschaftsverlag, München

dinger equation in real time. Examples are explicitly time-dependent configuration interaction (TDCI) methods[5–7], time-dependent Hartree-Fock (TDHF), and its multi-configuration variant (MCTDHF)[8–17]. Also, density functional theory has been applied in the time-domain[18–20], as well as approaches with a restricted number of active electrons [21–23].

From this set of methods, we will focus on MCTDHF, provide an overview over its development, outline selected applications that illustrate its possibilities and limitations, and present comparisons with other methods. The MCTDHF method can be seen as an explicitly time-dependent CASSCF method and is systematically improvable towards the Full-CI (FCI) limit. It is thus able to treat, for example, non-dynamical correlation, long-range charge transfer, and Van-der-Waals interactions which can be problem cases for some of the other methods mentioned above.

The paper is organized as follows. Section 2 presents the theory of MCTDHF, with a focus on the equations of motion. Section 3 discusses several applications, which deal with correlated electronic motion in small molecules, a Na_8 cluster, and a one-dimensional model system mimicking a metal film. Section 4 summarizes this work. Atomic units are used throughout if not stated otherwise.

2. Theory

We begin by giving a short review of the Multi-Configuration Time-Dependent Hartree-Fock method in order to clarify concepts and outline the notation, which follows closely the standard notation of the MCTDH (Multi-Configuration Time-Dependent Hartree) method for nuclear quantum dynamics [24,25]. In MCTDHF, the correlated state of an N-electron system is expressed by a time-dependent superposition of determinants, which are likewise made of time-dependent spin orbitals $\varphi_j(\vec{x},t)$:

$$\Psi(\vec{x}_1,\ldots,\vec{x}_N,t) = \sum_J A_J(t)|\varphi_{j_1}(\vec{x}_1,t)\ldots\varphi_{j_N}(\vec{x}_N,t)| \quad (1)$$

$$= \sum_{j_1=1}^{n}\cdots\sum_{j_N=1}^{n} A_{j_1\ldots j_N}(t)\prod_{k=1}^{N}\varphi_{j_k}(\vec{x}_k,t) \quad (2)$$

The capital letter J is a composite index which enumerates the N spin orbitals appearing in the determinant on the right of Eq. (1), and $\vec{x}_i = (\vec{r}_i,s_i)$ is a composite variable for position and spin of electron i. The summation is restricted to ordered N-tuples J, because only they correspond to Slater determinants. If the determinants are expanded into Hartree products (second line, Eq. (2)), one sees that the other (non-ordered) coefficients are obtained by permutations of the indices, and corresponding changes of sign. For brevity of notation, the determinants will also be referred to by the symbol $|J(t)\rangle$. For a given number of spin orbitals, n, there are $\binom{n}{N}$ of them. We will employ standard CASSCF notation $(N,n/2)$ to refer to a system with N electrons in $n/2$ spatial orbitals. In the following, only singlet determinants with spin-restricted orbitals will be considered.

With the above ansatz, it is our goal to solve the time-dependent electronic Schrödinger equation in the fixed-nuclei approximation, i.e.,

$$i\frac{\partial\Psi(t)}{\partial t} = H(t)\Psi(t) \tag{3}$$

In contrast to nuclear quantum dynamics, the time evolution of electronic wave functions is governed by a small class of very similar Hamiltonians. More precisely, the Hamilton operators are always of the form

$$H(t) = \sum_{i=1}^{N}\left(\frac{\vec{p}_i^2}{2} + V(\vec{r}_i,t)\right) + \sum_{i>j}^{N}\frac{1}{r_{ij}} = \sum_{i=1}^{N} h_i + V_{ee} \tag{4}$$

for a general, molecular N-electron system, with r_{ij} denoting interelectronic distances, and V_{ee} being the electron-electron repulsion. The external potential $V(\vec{r},t)$ is produced by the (fixed) nuclei with and may also contain the coupling to an additional laser field in the dipole approximation, $\vec{\mu}\vec{F}(t)$, where $\vec{F}(t)$ is the field and $\vec{\mu}$ the dipole operator. In the case of a molecule with N_A nuclei with charges Z_A we have for the one-electron Hamiltonians

$$h_i = \frac{p_i^2}{2} - \sum_{A=1}^{N_A}\frac{Z_A}{r_{iA}} + \vec{r}_i\vec{F}(t) \quad , \tag{5}$$

where r_{iA} is the distance between electron i and nucleus A.

Because of the special ansatz for the wave functions, Eq.(1), the time evolution is not derived from the time-dependent Schrödinger equation, but from the Dirac-Frenkel variational principle. One finds for the coefficients

$$\dot{A}_J = -i\sum_{L}\langle J(t)|H|L(t)\rangle A_L \tag{6}$$

and for the orbitals

$$\dot{\varphi}_j = -i\sum_{kl}(1-P)_{jk}\left(h_{kl} + \sum_{m}\rho_{km}^{-1}\langle V_{ee}\rangle_{ml}\right)\varphi_l \quad . \tag{7}$$

The new quantities introduced in Eq. (7) are the projector on the space spanned by the single particle functions $P = \Sigma_{j=1}^{n}|\varphi_j\rangle\langle\varphi_j|$, the inverse of the single electron reduced density matrix ρ, and the mean field matrix $\langle V_{ee}\rangle$, where V_{ee} is the two-electron part of the Hamiltonian. The notation O_{kl} corresponds to a representation of an operator O in the basis of the spin orbitals. We integrate the equations of motion Eq. (6) and Eq. (7) with an adaptive stepsize Runge-Kutta integrator of order eight[32].

The spin orbitals are represented either, in the case of one-dimensional model systems[10,11], on a grid, or, in the case of real molecules [12,13,15–17], by a linear combination of atomic orbitals (LCAO-MO). In order to solve Eqs.(6) and

(7), it is necessary to represent the operators involved in the basis of choice. If a grid basis is used, it is easy to obtain this with standard methods[26], because all operators are either diagonal in position or momentum representation. An advantage of this choice of basis is, that the basis functions are orthogonal, and that a long grid allows also the treatment of ionization, i.e. unbound electron motion. Disadvantages of a grid basis are, that in 3-dimensional space extremely many points are required, and that the treatment of the Coulomb singularity can be very awkward[27–29]. Therefore, orbitals of realistic molecular systems are usually expanded in Gaussian Type Orbitals (GTOs). Since the latter are typically atom-centered, they cannot easily be used to describe unbound and large-amplitude electron dynamics.

When using K atomic orbitals (AOs), we have $2K$ spin orbitals. Denoting the AOs by greek letters $|\mu\rangle$, $|\nu\rangle$, …, the integrals for the operators $\vec{r}_i$, $\Sigma_i\vec{p}_i^2/2$, $1/r_{ij}$ as well as the overlap matrix elements $S_{\mu\nu} = \langle\mu|\nu\rangle$ were obtained from the GAMESS program package[30] in all examples below. Because it is cumbersome to work with non-orthogonal basis functions during a propagation, we transform to a new, orthogonal basis set according to the symmetric orthogonalization scheme[31].

Two properties of the equations of motion will be especially important for the applications in later sections. First, the equation for the orbitals is in general, namely when $n < 2N$, highly non-linear. The origin of this is the fact, that the projector P and the mean-fields $\langle V_{ee}\rangle$ contain the orbitals on which they act. With increasing n the operator $(1-P)$ tends to zero, and the linear, exact Schrödinger equation is recovered, corresponding to the Full-CI (FCI) limit. Thus, the MCTDHF method can be systematically improved from the MCTDHF(N,$N/2$) limit, which corresponds to TDHF, towards MCTDHF(N,K) which is the FCI solution within a given basis. Second, a matrix representation of the full N-electron Hamiltonian (see Eq. (4)) in the basis of determinants $|J(t)\rangle$ is explicitly time-dependent, through the time-dependence of the orbitals. This can be visualized, e.g., by time-dependent energy eigenvalues. Disturbing as this feature seems at first glance, it will turn out to be rather useful, too. Both effects will be referred to again, later.

3. Selected applications

3.1 Inverse photoemission

One of the first applications of MCTDHF was the simulation of an inverse photoemission experiment in a one-dimensional model system [10], mimicking a metal film. The idea of this kind of experiments is, that an electron is injected into a metal, and falls into a previously unoccupied level by electron-electron scattering. The radiation that is emitted during this process carries information about the unoccupied levels of the system. Similarly, such an experimental setup

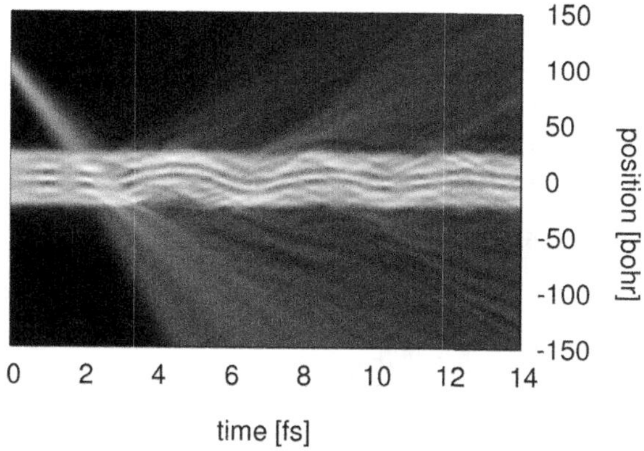

Fig. 1. Time dependent electron density for a typical electron scattering/transmission event. (The square root of the density is shown, for better visibility.) Visible are effects like excitation and dephasing of coherent oscillations, trapping in resonances, elastic and inelastic transmission, see text.

has been used to find one of the first signatures of plasmons in the kinetic energy distribution of transmitted electrons[33].

Instead of a real metal in three dimensions, we have chosen a thin metal film of thickness $d = 50$ a$_0$, modeled by a jellium potential. In the one-dimensional jellium model, the electron-electron repulsion is $V_{ee} = \Sigma_{i>j}^{N} 1/\sqrt{r_{ij}^2+a}$ and the external potential is $V(\vec{r}_i) = -\int[(\rho^+(R))/(\sqrt{(r_i-R)^2+a})]\, dR$. Here, $a = 1$ a$_0$ is a screening parameter and $\rho^+(R) = N/d$ is the background, jellium density which is constant within the film, and zero outside. The conduction band of the film is modeled by 5 electrons in their Hartree-Fock ground state, and a 6th electron is located outside the film, in an "orbital" of Gaussian shape and with momentum directed towards the surface. All calculations were done on the MCTDHF(6,5) level of theory. (Qualitative reliability of the results was checked by comparison with MCTDHF (6,4) results.) A grid representation with 2048 points in the interval [-600 a$_0$, +600 a$_0$] was used.

Fig. 1 shows the time-dependent electron density $\rho(r,t) = \int|\Psi(t)|^2\, ds_1\, dx_2 \ldots dx_N$ for a typical event. Light areas represent high density, dark areas low density. Several phenomena of correlated motion of electrons are visible: The incident electron comes from the upper left corner, and enters the metal film (the horizontal band) after about 2 fs. This leads to the excitation of coherent oscillations, which can also be seen to dephase on a time scale of about 10 fs. A part of the incoming electron density is trapped in resonances, and emitted only after some delay (see, in particular, the upper half of Fig. 1). The lower half shows transmitted electron density, with a separation between direct, elastic and delayed, inelastic transmission. Another process, which is not visible in this figure, but which can be extracted from the underlying wave function, is electron impact ionization. Altogether, these processes are examples for the correlated motion of electrons.

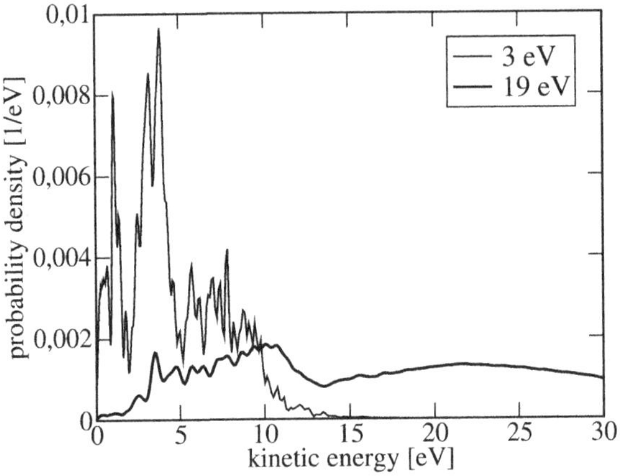

Fig. 2. Final kinetic energy distributions of transmitted electrons, for different initial kinetic energies of the incoming electron. Slow electrons show a much more structured spectrum, because they interact more strongly with metal electrons than fast electrons.

As an exemplary result from Ref. [10], we want discuss the kinetic energy distribution of the transmitted electrons, for two selected initial energies of the incoming electron, see Fig. 2. A clear trend is visible, from a highly structured distribution at a low impact energy, to a smooth distribution at a high energy. This is somewhat counterintuitive, because a high energy process should rather be more "violent" and disordered. However, there are two effects which explain the trend as shown in Fig. 2. First, the slower electrons spend more time inside the metal film, and have thus more time to interact with the metal electrons. Second, if the incident electron has a high momentum, then the orbital it occupies is already orthogonal to the metal electron orbitals. Otherwise, the orbitals have to orthogonalize, leading to an increased interaction.

3.2 Vertical Excitation Energies and Time-Dependent State Averaging

The calculation of excitation energies is usually done by solving the time-independent Schrödinger equation, i.e. by solving the eigenvalue problem of the electronic Hamiltonian. However, the time-dependent Schrödinger equation contains generally the same information, and it depends on the special situation at hand, which approach is more suitable. In the time-dependent approach, pioneered by Heller[26,34], a wave packet is created and propagated, which is composed of several eigenstates. The eigenvalue spectrum can then be obtained from the Fourier transform of various quantities. One of the most often used is the dipole moment

$$\langle \vec{\mu} \rangle (t) \quad = \quad \sum_{J,K} A_J^*(t) A_K(t) \langle J(t) | \vec{\mu} | K(t) \rangle \tag{8}$$

$$= \sum_{n,m} c_n^* c_m e^{i(E_n - E_m)t} \langle n|\vec{\mu}|m\rangle \quad . \tag{9}$$

Here, the first line gives the MCTDHF form, while (9) makes use of the eigenstate representation with c_m denoting the coefficient of state $|m\rangle$ in the wave packet. A Fourier transform of the time-dependent dipole expectation value gives peaks at all possible eigenstate energy differences E_n–E_m, with a height proportional to the amplitudes of the states and the transition dipole moment $\langle n|\vec{\mu}|m\rangle$. In order to obtain a spectrum which is not too crowded, it is therefore advisable to employ wave packets where one of the c_n (usually the one of the ground state) is close to 1.

We will apply this formalism to the calculation of the first excited state of LiH [13]. There are several ways to create the initial wave packet. One commonly applied method is to simulate an ultrashort, broadband laser pulse to excite the system. We have chosen a 1 fs pulse, with $\sin^2$ envelope, a laser frequency corresponding to a photon energy of 5.44 eV, and a peak fluence of 3.5×10^{12} W/cm^2. These parameters were not chosen to simulate an experiment, but rather to provide a numerical tool for the calculation of the excitation energies. All calculations in this section were carried out with the 6–31G* basis set.

What accuracy can we expect for the energies of excited states obtained from the procedure described above? Fig. 3 shows the error for the first excited state, E_1^{FCI}–E_1, relative to the (time-independent) Full-CI solution, for a variety of (stationary) CI and (time-dependent) MCTDHF-type methods. The configuration interaction variants CIS (CI Singles), CIS(D) (CI Singles with perturbative Doubles [35], and CISD (CI Singles and Doubles) have been chosen. It should be noted that CISD corresponds to FCI in the valence space of LiH, and gives therefore a much better result for LiH, than for other molecules. A comparison with the MCTDHF calculations with active spaces (4,3), (4,4), and (4,5) shows, that the error of the time-dependent approach is about a factor of 10 smaller than the errors of CIS and CIS(D). Generally, the results become better for larger active spaces. It should also be noted that a diagonalization of the Hamiltonian matrix, evaluated with the determinants at time $t = 0$ ($\langle J(0)|H|L(0)\rangle$), gives a first excited state which lies above the fourth true eigenenergy. The reason for this is, that the orbitals at that time are optimized to represent the ground state. Thus, the information about the electronic structure of the molecule is really generated only during the propagation, and through the non-linear evolution of the orbitals.

As a next step, the procedure was applied to methane [13], which has 10 electrons. It was treated on the MCTDHF (8,6) level of theory, with a neglect of core excitations. Again an error of the order of a few meV was found. Additionally, it could be shown that the transition dipole moments inferred from the oscillator strengths are often qualitatively correct.

The way MCTDHF generates information about the electronic structure calculation, i.e. via the time-dependence of both the coefficients and the orbitals,

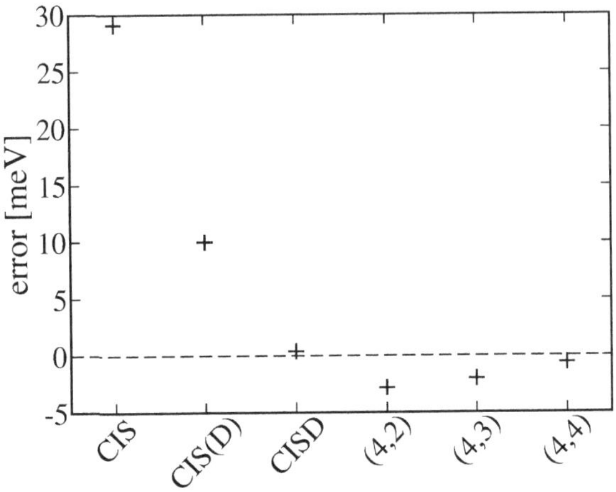

Fig. 3. Error of the energies of the lowest excited state of LiH, relative to a FCI/6–31G* calculation. CIS and CIS(D) perform less well than even the smallest MCTDHF method, while CISD, which is a Full-CI in the valence space, gives almost the correct excitation energy. The error of the MCTDHF (4,3), (4,4) and (4,5) calculations is generally a factor of 10 smaller than CIS(D).

leads to a time-dependent version of the state-averaging effect known from standard MCSCF calculations[37]. To illustrate this effect, we show the time evolution of the lowest two eigenvalues of $\langle J(t)|H|L(t)\rangle$ of a water molecule, see Fig. 4, after laser pulse excitation. A 6–31G** basis set and a (6,5) active space have been used. The orbitals adapt to the excited state of the system, while the laser pulse is 'on'. Thus the lowest state becomes less well represented, while simultaneously the representation of the first excited state improves. However, both values do not match the eigenenergies of the water molecule, which can be obtained by Fourier transforming the autocorrelation function (dashed lines in Fig. 4). Therefore, it is the interplay of the time-dependence of the coefficients and the orbitals which generate electronic structure information in MCTDHF. While this feature explains the efficiency of MCTDHF, it also leads to problems when the method is used to simulate control experiments, which rely on resonance conditions to be fulfilled. This problem has been studied further by Mundt and Tannor in [36], to which we refer the reader for a more detailed analysis.

3.3 Laser manipulation of LiH

One of the prospects since the availability of ultrashort laser pulses is, to manipulate the electronic state of molecules in a controlled fashion. However, the extremely short duration of the pulses also implies a very large spectral width, so that standard control techniques, which rely on resonance conditions, can not be applied. That this does not preclude any control at all, will now be shown for the example of a LiH molecule [15,16]. This molecule has been chosen, because

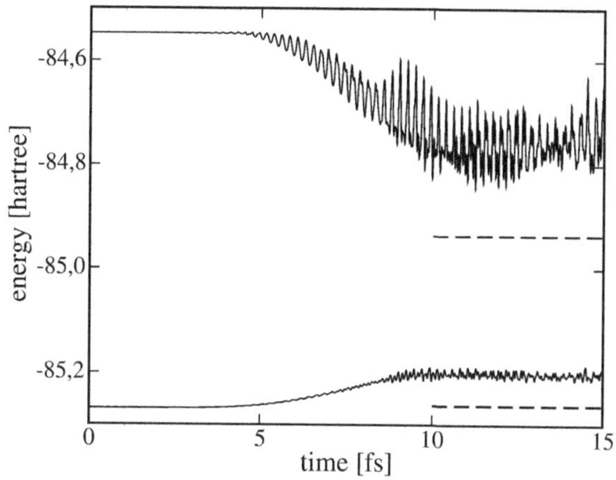

Fig. 4. The two lowest eigenvalues of the time-dependent Hamiltonian matrix of H_2O, after excitation by a 1 fs $\sin^2$ laser pulse with frequency 0.34 a.u. and a maximal field strength of 10.28 GV/m. The horizontal lines show the corresponding eigenenergies of the molecule obtained from a Fourier transform of the autocorrelation function.

its four electrons can be treated with high accuracy, and it is the smallest truly heteroatomic molecule for which the creation and control of an electronic wave packet can be attempted. One precondition for a quantitative treatment is a rather large basis set. For this, we chose a triple zeta basis set, with diffuse and polarization functions, 6–311++G(2df,2p), which contains 49 cartesian functions. These have been combined to ten molecular spin orbitals, giving rise to 210 time-dependent determinants. A controlled manipulation is possible with, and requires, a fully characterized pulse: The laser (= carrier) frequency ω, the width σ, fluence, polarization, and carrier-envelope phase ϕ can all be used to steer the electronic motion. The envelope function was chosen to be gaussian.

It is a general problem of electron dynamics, that it usually assumes the fixed nuclei approximation. This implies, that only ultrashort, few-cycle pulses with a duration σ shorter than the vibrational period of the molecule (= 24 fs for LiH in the electronic ground state) can be used to achieve a certain goal. These pulses are typically too short to fulfill the π-pulse condition for complete population inversion in a two-level system at reasonable fluences. Therefore it was one purpose of this study to show, to what extent control over the electronic motion is possible even with much shorter pulses.

As an exemplary result, we show in Fig. 5 the x and z expectation values, parameterized with time, of an electron wavepacket created with a laser pulse with two different polarizations, but the same carrier frequency $\omega = 0.13\ E_h/\hbar$, width $\sigma = 1$ fs, and maximal intensity of $2.2 \cdot 10^{13}$ W/cm^2. The molecule was oriented along the z-direction, with a Li-H bond length of 3.08 a_0. The resulting wave packets are quite different for the different polarizations. On the left, a polarization along z (the molecular axis) has been used, and a fairly simple

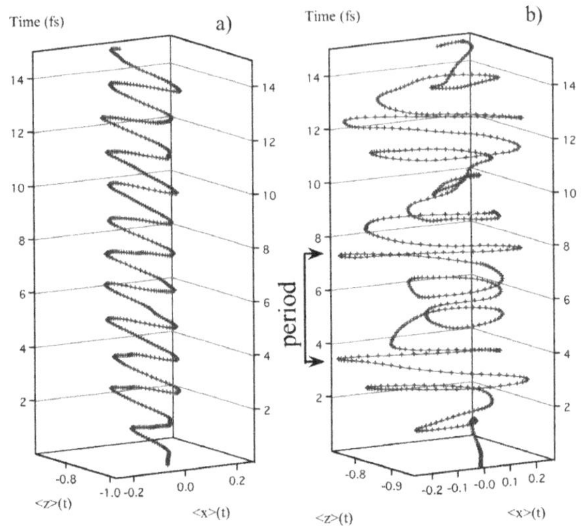

Fig. 5. Electronic wave packets in LiH. Depending on the polarization of the laser pulse, either a simple oscillation along the molecular axis is excited, or a rotational motion, with a 4 fs period. (Reproduced from Ref. [16], with permission from IOP Publishing Ltd.).

one-dimensional oscillation is observed. If the polarization is changed to the *xz* direction, then a wave packet is created by populating the two lowest-lying excited states of LiH, $^1\Sigma^+$ and $^1\Pi$, of which one also has a non-vanishing dipole moment along *x*. Both excited states are close in energy resulting in a comparatively slow circular motion with a period of 4 fs. Apart from the polarization, also the carrier-envelope phase changes the behaviour of the electronic wave packet, but without such a clear signal as the former. In Ref. [16] it was also shown, how this 'slow' 4 fs period can be tested by a pump-probe experiment.

3.4 Onset of thermalization in Na_8 clusters

Small sodium clusters provide an interesting testing ground for theoretical and experimental electron dynamics, because their electronic structure is in between that of molecules and extended metallic systems. Therefore, they allow to study the emergence of statistical phenomena with increasing system size, and the effect of finite size on many-electron phenomena. Here, we want to discuss the onset of thermalization between the eight valence electrons of a Na_8 cluster. Thermalization is usually thought of as typical for very large large systems, with particle numbers approaching Avogadro's number. But as long as the recurrence time is larger than the equilibration time, thermalization can also be observed in much smaller, finite size systems.

In our simulations [17], we excited the valence electrons of a Na_8 cluster at an RHF-optimized geometry, to a non-equilibrium state with intense laser pulses

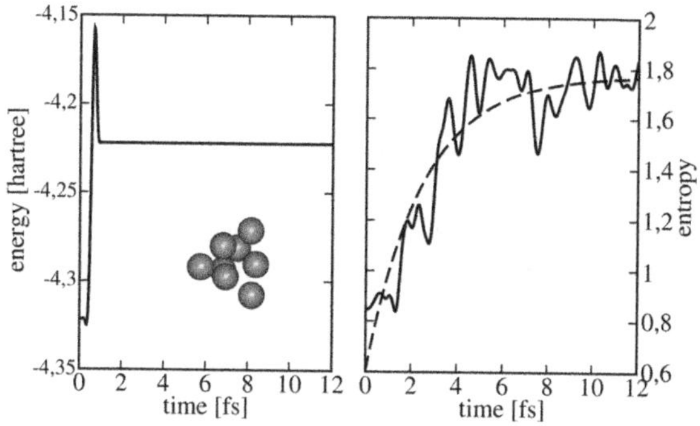

Fig. 6. Onset of ultrafast electron thermalization in a Na_8 cluster. The entropy rises on a fs timescale after the initial 1 fs laser pulse has stopped. The dashed line shows a mono-exponential fit.

of 1 fs duration and various intensities. Core electrons were replaced by an effective core potential. As an illustration, Fig. 6 shows the evolution of the energy and the entropy during and after excitation, on the level MCTDHF(8,6). On the left, the energy rises during the first fs, but then stays constant, as expected. On the right, we show the entropy, which rises on a timescale of several fs even after the pulse is over. This is a signature of energy redistribution through electron-electron collissions, which take place under energy conservation. The entropy has been calculated from the reduced density matrix according to (in units of k_B)

$$S = -\mathrm{Tr}\{\rho \ln \rho\} \tag{10}$$

with a normalization such that the entropy of the Hartree-Fock ground state is zero. The oscillations around the mono-exponential least square fit (dashed line) are a finite size effect, possibly increased by the nonlinear equations of motion of MCTDHF.

The notion of thermalization implies that it is natural to look for an electronic temperature before and after the process. For the example above, we find, by fitting a Fermi distribution to the populations of the Hartree-Fock orbitals, that the temperature rises by about 6400 K. This amount is not unusual for electronic systems: Temperatures of several thousand K have been found in surface science experiments [38] after laser excitation. The reason for reporting the temperature increase, rather than the absolute temperature here, is that the correlated ground state contains small populations in excited orbitals. In other words, the correlation between the electrons mimics a finite temperature. This phenomenon, based on the fact that real electrons have a strong interaction, in contrast to quasiparticles of statistical physics, is elaborated upon further in the article [17].

When relaxation processes like the one described above are to be simulated, it is important to take the various strengths and weaknesses of the several possible methods into account. For example, TD-CIS does not show any thermalization, because electron-electron scattering is described badly with only singly excited determinants. Another important point is the number of molecular orbitals, which are used to calculate the one electron density matrix: If this number is small, then the maximum entropy that can be represented by the density matrix is small, too. Therefore, TD-CISD calculations have usually shown a much higher degree of thermalization, then the various MCTDHF calculations.

According to Fermi-Liquid Theory [39], the lifetime of an excited level in a metal decreases as

$$\tau_n \propto \frac{1}{(\epsilon_n - \epsilon_F)^2} \quad , \tag{11}$$

where ε_n is a suitably chosen single particle energy, and ε_F the Fermi energy. In order to estimate the rate $1/\tau_n$, we used the monoexponential fits (see Fig. 6), and analysed the exponential prefactor as a function of laser fluence. Only TD-CISD (and a version of TD-CASCI), which can be regarded as the highest-quality method tried, was able to reproduce a superlinear increase. Again, this can be related to restrictions on the reduced electronic density matrices for the other methods.

4. Summary

MCTDHF is a powerful and accurate method for the treatment of correlated electron dynamics in real time, as long as systems with a moderate number of electrons are to be described. Moderate refers here to a comparison with time-independent electronic structure theory, which today deals routinely with a much larger number of electrons. As an example, the electronic wave function of methane has 30 spatial degrees of freedom, which is already fairly big for a quantum dynamical problem. The main computational effort goes into the four index transform of the two-electron integrals from atomic to molecular orbitals, which has to be done for each time-step. In contrast, in correleted time-independent calculations the transformation has to be done only once. On the other hand, MCTDHF allows for the non-perturbative treatment of electron dynamics driven by external perturbations of arbitrary temporal shape.

Another advantage of MCTDHF is, that it includes multiple excitations. This make this method especially suitable for the description of high energy processes. Here, for molecules, the challenge lies with limitations posed by the Gaussian Type Orbitals, which restrict the spatial amplitude of electronic motion. New approaches will have to be developed, if quantum chemical accuracy and ionization or electron scattering are to be included at the same time.

Acknowledgement

The authors gratefully acknowledge financial support of the Deutsche Forschungsgemeinschaft through the Schwerpunktprogramm SPP 1145.

References

1. M. Hentschel, R. Kienberger, C. Spielmann, G.A. Reider, N. Milosevic, T. Brabec, P. Corkum, U. Heinzmann, M. Drescher, F. Krausz, Nature **414** (2001) 501.
2. R. Kienberger, M. Hentschel, M. Uiberacker, C. Spielmann, M. Kitzler, A. Scrinzi, M. Wieland, T. Westerwalbesloh, U. Kleineberg, U. Heinzmann, M. Drescher, F. Krausz, Science **297** (2002) 1144.
3. P. Bucksbaum, Nature **421** (2003) 593.
4. A. Föhlisch, P. Feulner, F. Hennies, A. Fink, D. Menzel, D. Sanchez-Portal, P. Echenique, W. Wurth, Nature **436** (2005) 7049.
5. T. Klamroth, Phys. Rev. B **68** (2003) 245421.
6. P. Krause, T. Klamroth, P. Saalfrank, J. Chem. Phys. **123** (2005) 074105.
7. P. Krause, T. Klamroth, P. Saalfrank, J. Chem. Phys. **127** (2007) 034107.
8. T. Kato, H. Kono, Chem. Phys. Lett. **392** (2004) 533.
9. F. Zanghellini, M. Kitzler, C. Fabian, T. Brabec, A. Scrinzi, Laser Physics **13** (2003) 1064.
10. M. Nest, T. Klamroth, P. Saalfrank, J. Chem. Phys. **122** (2005) 124102.
11. M. Nest, T. Klamroth, Phys. Rev. A, **72** (2005) 012710.
12. M. Nest, J. Theor. Comp. Chem., **6** (2007) 563.
13. M. Nest, P. Ramanathan, P. Saalfrank, J. Chem. Phys. **126** (2007) 214106.
14. E. Persson, K. Schiessl, A. Scrinzi, J. Burgdörfer, Phys. Rev. A **74** (2006) 013818.
15. F. Remacle, M. Nest, R.D. Levine, Phys. Rev. Lett. **99** (2007) 183902.
16. M. Nest, F. Remacle, R.D. Levine, New J. Phys. **10** (2008) 025019.
17. T. Klamroth, M. Nest, Phys. Chem. Chem. Phys. **11** (2009) 349.
18. F. Calvayrac, P.-G. Reinhard, E. Suraud, C.A. Ullrich, Phys. Rep. **337** (2000) 493.
19. T. Kunert, R. Schmidt, Eur. Phys. J. D **25** (2003) 15.
20. Y. Kurzweil, R. Baer, Phys. Rev. B **73** (2006) 075413.
21. A. Takahashi, S. Mukamel, J. Chem. Phys. **100** (1994) 2366.
22. I. Barth, J. Manz, Angew. Chem. Intern. Ed. **45** (2006) 2962.
23. I. Barth, J. Manz, Y. Shigeta, K. Yagi, J. Am. Chem. Soc. **128** (2006) 7043.
24. M.H. Beck, A. Jäckle, G.A. Worth, H.-D. Meyer, Phys. Rep. **324** (2000) 1.
25. H.-D. Meyer, G. A. Worth, Theor. Chem. Acc. **109** (2003) 251.
26. N. Balakrishnan, C. Kalyanaraman, N. Sathyamurthy, Phys. Rep. **280** (1997) 79.
27. S.W. deLeeuw, J.W. Perram, E.R. Smith, Proc. R. Soc. London, Ser. A **373** (1980) 27.
28. J.P. Hansen, MD dynamics simulation of Coulomb sytems, in MD simulation of Statistical Mechanical Systems, North Holland, Amsterdam, (1986).
29. G.J. Martyna, M.E. Tuckerman, J. Chem. Phys. **110** (1999) 2810.
30. M.W. Schmidt, K.K. Baldridge, J.A. Boatz, S.T. Elbert, M.S. Gordon, J.H. Jensen, S. Koseki, N. Matsunaga, K.A. Nguyen, S.J. Su, T.L. Windus, M. Dupuis, and J.A. Montgomery, J. Comput. Chem. **14** (1993) 1347.
31. A. Szabo, N.S. Ostlund, Modern Quantum Chemistry, Dover, New York (1989).
32. E. Hairer, S.P. Norsett, and G. Wanner, Solving Ordinary Differential Equations I, 2.ed., Springer Verlag, (1993).
33. C.J. Powell, J.B. Swan, Phys. Rev. **115** (1959) 869.
34. E. J. Heller, Acc. Chem. Res. **14** (1981) 368.
35. M. Head-Gordon, R.J. Rico, M. Oumi, T.J. Lee, Chem. Phys. Lett. **219** (1994) 21.
36. M. Mundt, D.J. Tannor, submitted
37. F. Jensen, Introduction to Computational Chemistry, John Wiley and Sons, Chichester, 2007.
38. J. Hohlfeld, S.-S. Wellershoff, J. Güdde, U. Conrad, V. Jähnke, E. Matthias, Chem. Phys. **251** (2000) 237.
39. N.W. Ashcroft, N.D. Mermin, Solid State Physics, Brooks Cole (1976).

The Density Matrix Renormalization Group Algorithm in Quantum Chemistry

By Konrad Heinrich Marti and Markus Reiher*

ETH Zurich, Laboratorium für Physikalische Chemie, Wolfgang-Pauli-Str. 10, 8093 Zürich, Switzerland

(Received July 30, 2009; accepted September 7, 2009)

Quantum Chemistry / Electron Correlation / Density Matrix Renormalization Group / Transition Metal Chemistry

In this work, we derive the density matrix renormalization group (DMRG) algorithm in the language of configuration interaction. Furthermore, the development of DMRG in quantum chemistry is reviewed and DMRG-specific peculiarities are discussed. Finally, we discuss new results for a dinuclear μ-oxo bridged copper cluster, which is an important active-site structure in transition-metal chemistry, an area in which we pioneered the application of DMRG.

1. Introduction

In electronic structure theory, the exponential scaling in the dimension of the Hilbert space with system size poses the biggest challenge for the exact description of strongly correlated systems of electrons. The numerical renormalization group algorithm [1,2] is based on the definition of a subspace on which the original Hamiltonian is projected. The eigenstates of the subspace Hamiltonian which correspond to the largest eigenvalues form a basis for the reduced-dimensional description of the enlarged system. The success of the numerical renormalization group lies in the decimation procedure which restricts the size of the Hilbert space. However, the method was not able to describe the ground-state properties of some many-body Hamiltonians such as the Hubbard or Heisenberg models where a clear energy scale separation is not present.

To cure this deficiency, White [3,4] proposed a different basis-state selection procedure based on the reduced many-particle density matrix yielding the density matrix renormalization group (DMRG) algorithm. In 1999, White and Martin

* Corresponding author. E-mail: markus.reiher@phys.chem.ethz.ch

Z. Phys. Chem. **224** (2010) 583–599
© by Oldenbourg Wissenschaftsverlag, München

applied this DMRG algorithm for the first time to the full molecular spin-free non-relativistic Hamiltonian to solve the electronic Schrödinger equation for a quantum chemical problem [5]. Before that, several other groups applied DMRG to study molecular system with semi-empirical Hamiltonians such as the Pariser-Parr-Pople Hamiltonian [6–8]. In this review, however, we will only focus on the work considering the full ab initio Hamiltonian. White and Martin [5] benchmarked the DMRG algorithm against FCI calculations for water. Further developments in the quantum chemical DMRG community are summarized later in this work.

For an excellent review ranging from the early beginnings of DMRG in the condensed-matter community to the quantum chemical applications of DMRG, we refer to the work of Schollwöck [9] and for an overview of most recent trends in DMRG to the paper by Hallberg [10].

In this work, we shall derive the *quantum-chemical* DMRG algorithm within the standard language of quantum chemistry, namely from the point of view of a superposition of Slater deteminants (section 2). In section 3, work on DMRG in quantum chemistry is reviewed. Section 4 then continues with a discussion of peculiarities innate to the DMRG algorithm. The capabilities of DMRG for strongly correlated electronic structures are then discussed at the example of transition metal complexes in section 5.

2. Derivation of the DMRG Algorithm

The essential technique employed by all quantum chemical methods to calculate a many-electron wave function is the expansion into a suitable set of basis functions. Of course, two different basis sets are required. The quantum mechanical N-electron state is expanded into a set of *N-electron* basis functions (e.g., Slater determinants or configuration state functions). However, these N-electron basis functions need to be constructed themselves. They can be set up from a Hartree product of *one-electron* functions antisymmetrized with respect to pair permutations of any two electronic coordinates in order to fulfill the Pauli principle.

2.1 Expansion into Slater Determinants

In order to understand how DMRG constructs the electronic wave function $\Psi^{(N)}{}_{A,el}$ for N electrons in total electronic state A, we start from the full configuration-interaction (FCI) ansatz,

$$\Psi^{(N)}_{A,el} = \Phi^{(N)}_{A,\mathrm{FCI}} = \sum_{I}^{M} C_I^{(A)} \, \Phi_I^{(N)}(\psi_{I_1}, \psi_{I_2}, \ldots, \psi_{I_N}) \qquad (1)$$

(with $M = \infty$) which is an exact representation of an N-electron wave function if *all* Slater determinants $\Phi^{(N)}{}_I$ that can be constructed from a complete one-electron basis set $\{\psi_i\}$ - i.e., from the spin orbitals - are included in the superposition of configurations in Eq. (1). Note that then the one-electron basis set must be complete (implying an infinite number of elements), and hence the total number of configurations will be infinite.

In the basis of these N-electron functions $\Phi^{(N)}{}_I$, the electronic Hamiltonian $\hat{H}_{el}$ becomes a matrix $\mathbf{H} = \{H_{KL}\}$, which requires the evaluation of $3N$-dimensional integrals $H_{KL} = \langle \Phi^{(N)}{}_K \mid \hat{H}_{el} \mid \Phi^{(N)}{}_L \rangle$. Such a high-dimensional integration is most conveniently carried out in second quantization which is in this context a clever book-keeping scheme for the implementation of the Slater-Condon rules for expectation values of particle number conserving operators [11]. In addition, second quantization allows one to conveniently work in Fock space where a simple tensor product structure ansatz for the wave function is possible and the antisymmetry requirement is elegantely built into the algebra of the elementary creation and annihilation operators.

It is important to note here that for all such configuration-expansion methods - like FCI or the truncated CI methods - the N-electron basis set is explicitly constructed, and hence the Hamiltonian matrix elements can be explicitly calculated for these N-electron basis functions. However, it is only feasible if the one-electron basis set is finite so that the N-electron basis set is also finite though still extremely large. Since this large number of configurations can at most be handled up to about one billion [12], the standard approach in quantum chemistry is to reduce the number of configurations by imposing additional restrictions on the sum in Eq. (1), which can be a simple truncated CI expansion $\Psi^{(N)}{}_{A,el} \approx \Phi^{(N)}{}_{A,\mathrm{CI}}$ with M being finite, i.e., with a limited number of configurations M.

For large M, configurations can no longer be picked manually and an automatic scheme is to be defined though this may compromise the accuracy of the approximation. One scheme is the systematic substitution of one-electron functions in a reference N-electron basis function (e.g., substitution of molecular spin orbitals in the Hartree-Fock Slater determinant), which is usually denoted `excitation' (it must not be confused with true electronic excitation as these occur between states A expressed by different sets of CI expansion coefficients $\{C^{(A)}{}_I\}$, but expanded in the *same* set of N-electron functions).

Names for the CI substitution hierarchy are well-known: singles excitations, doubles excitations, triples excitations etc. as well as combinations of them. To improve on the expansion one may even start `excitations' from a couple of preselected configurations (multi-reference CI). However, the most efficient scheme to select configurations is the coupled-cluster ansatz, which features excitation operators for singles, doubles, etc. in an exponential whose series expansion contains product excitations. If applied to an actual vacuum the excitation rank gets truncated at the number of total electrons [13].

Of course, also the choice of the one-electron basis functions employed for the construction of configurations can determine the number M of relevant configurations. In principle, any type of *one-electron* functions can be chosen for the construction of the determinants. Even atom-centered Gaussian basis functions $\{\chi_i\}$, which are, in general, non-orthogonal, can be employed. For such non-orthogonal one-electron basis sets, the second-quantized electronic Hamiltonian involves then creation and annihilation operators that fulfill anticommutation relations involving elements of the (one-electron) overlap matrix [13].

2.2 Construction of the DMRG Wave Function

The DMRG wave function is constructed in a completely different way when compared to the CI excitation hierarchy as we shall see now. Since the limiting factor of all large CI or coupled-cluster calculations is the factorial growth of the number of N-electron basis functions with the excitation (substitution) grade and with the number of one-electron basis functions, we need to find a solution to the dilemma of including all possible configurations - because we cannot know in advance for any arbitrary molecular structure with a given number of electrons which Slater determinants will be most important - but without explicitly representing them. Hence, we must give up the explicit knowledge of the composition of our N-electron basis functions, but want to assure at the same time that *any* important configuration can be picked up. How can this be achieved?

It is clear that the expansion of the total DMRG state $\Phi^{(N)}{}_{\mathrm{A,DMRG}}$ must be in terms of N-electron basis functions, which are themselves CI-type (or even FCI-type) basis functions $\Omega^{(N)}{}_K$,

$$\Psi^{(N)}_{A,el} = \Phi^{(N)}_{\mathrm{A,DMRG}} = \sum_K^{M'} B_K^{(A)}\, \Omega_K^{(N)}(\{\Phi_I^{(N)}\}) \tag{2}$$

in order to span the same space. However, the explicit constitution of DMRG basis functions in terms of configurations $\Phi^{(N)}{}_I$,

$$\Omega_K^{(N)} = \sum_L D_L^{(K)}\, \Phi_I^{(N)}(\psi_{I_1}, \psi_{I_2}, \ldots, \psi_{I_N}) \tag{3}$$

can only be known in those cases for which a FCI or CASSCF reference calculation is still possible [14].

Some expansion coefficients $D^{(K)}{}_L$ may be very close to zero, but they will still define a fixed ratio between all configurations considered. Thus, in order to have full flexibility of the individually contributing configurations $\Phi^{(N)}{}_I$, the required number of DMRG states M' must not be too small and depends, of course, on the electronic structure to be studied. The number of DMRG basis functions M' can only be equal to one in the case for which $\Omega^{(N)}{}_1$ *is* the FCI solution.

Now, assume that we are able to represent the electronic Hamiltonian in the basis of DMRG states $\Omega^{(N)}{}_K$. Then the $B^{(A)}{}_K$ are obtained simply by diagonalization of this Hamiltonian matrix (we do not discuss how the construction of the Hamiltonian is actually achieved, but may refer, for instance, to Ref. [14] for more details). The remaining (and most difficult) task to tackle is how to obtain the N-electron contraction coefficients $\mathbf{D}^{(K)} = \{D^{(K)}{}_L\}$ for the configurations in Eq. (3) *without ever knowing all configurations explicitly* (as otherwise the advantage of DMRG fades away). In order to avoid the exponentially growing number of N-electron basis functions this has to be handled stepwise. I.e., the N-electron basis has to be successively constructed and then contracted. The contraction of the determinants must be done in such a way that the total electronic state A can be represented best. Hence, we have to answer the question of how a *contracted* N-electron basis can represent the uncontracted FCI basis optimally. White showed in his early papers [3,4] that an optimum representation can be found by exploiting features of the reduced many-particle density matrix. The reduced many-particle density matrix in the DMRG context must not be confused with one- and two-particle density matrices often encountered in quantum chemistry. It is the density matrix of a pure state where basis states in Fock space spanned by some orbitals are integrated out. The eigenstates of the reduced density matrix with the largest eigenvalues minimize the quadratic norm between the uncontracted and contracted basis to obtain maximal overlap with the exact state.

In order to construct the DMRG basis functions $\Omega^{(N)}{}_K$, we proceed stepwise. First, we choose an active space of orbitals from which all configurations $\Phi^{(N)}{}_I$ could be constructed in principle. Their actual number, however, is too large to be manageable. Hence, we must construct the $\Omega^{(N)}{}_K$ functions by first considering only a few spin orbitals (the procedure can be easily generalized to spatial orbitals). Taking the first spin orbital from the active space, we can represent two states in this one-electron basis: occupied and unoccupied. Then we consider the next spin orbital from the active space which contributes two possible states (again: occupied and unoccupied) from which $2^2 = 4$ total states can be constructed. All these states defined on the first two spin orbitals can still be explicitly represented (also, note the different particle numbers of these states in Fock space). Considering the next four spin orbitals (yielding in total six spin orbitals), we have already $2^6 = 64$ total basis states with occupation numbers ranging from zero to six electrons. Now, this is already quite large (in the case of spatial orbitals we would have $4^6 = 4096$ many-electron states already) and we shall find a way to reduce the number of states defined on this subsystem of orbitals, which we shall call the *active subsystem* as it is constructed systematically.

As pointed out above, the eigenstates of the reduced density matrix are the means to accomplish the reduction (decimation). But then, we need to calculate the total electronic state first from the total Hamiltonian. Since we have considered only one part of the active space of orbitals, namely those orbitals that define the active subsystem, we are lacking any information about the remaining orbitals in the active space, which we call the *complementary subsystem* [14] and which functions as an environment to the active subsystem.

In general, we will not be able to construct the many-electron states on the large complementary subsystem of orbitals for the very same reason that we could not solve this task for the total active space of orbitals in the first place. There is thus only one option: we need to guess these states and hope that we can find a way to make sure that the algorithm is able to pick up all relevant configurations defined on the complementary subsystem, a feature that will crucially determine DMRG convergence as we shall discuss in later sections of this review.

So, we express the electronic Hamiltonian in terms of a basis constructed as a *tensor product* from states $\Omega^{(N')}{}_{AS,I}$ defined on the active subsystem and from (guessed) states $\Omega^{(N'')}{}_{CS,J}$ defined on the complementary subsystem

$$\Psi^{(N)}_{A,el} = \Phi^{(N)}_{A,\text{DMRG}} = \sum_{IJ} W^{(A)}_{IJ} \left[\Omega^{(N')}_{AS,I} \otimes \Omega^{(N'')}_{CS,J}\right] \tag{4}$$

where we have as a natural boundary conditions for the electrons represented by these states $N'+N'' = N$. This equation also explains why we have to take into account all states of different particle number defined on the two subsystems, $N',N'' \leq N$.

Diagonalizing the total Hamiltonian in the basis of all $\Omega^{(N)}{}_K \rightarrow \Omega^{(N')}{}_{AS,I} \otimes \Omega^{(N'')}{}_{CS,J}$ contracted functions yields the CI-type coefficients $W^{(A)}{}_{IJ}$, which are, of course, related to the $B^{(A)}{}_K$ coefficients of Eq. (2). Although these coefficients are obtained as a vector, they feature two indices, I and J, in order to denote that they are defined with respect to a product basis. From the construction of the total electronic state A in terms of a tensor product basis of two determinants defined for the two subsystems that build up the complete electronic system it can be shown that DMRG iteratively refines a matrix product state [15–29].

In a computer program, all $W^{(A)}{}_{IJ}$ form a single eigenvector for the electronic state A, but since it is clear from the construction of the total Hamiltonian matrix what N'-electron active-subsystem state I has been combined (multiplied) with what N''-electron complementary-subsystem state J, the composite index IJ of each entry in the eigenvector is known. From this eigenvector, we now compute the reduced density matrix by summation over all complementary-subsystem states for a given state I of the active subsystem. We therefore have for the active-subsystem reduced density operator

$$\begin{aligned} \hat{\rho}_{AS} &= \text{tr}_{CS} \left|\Phi^{(N)}_{A,\text{DMRG}}\right\rangle \left\langle \Phi^{(N)}_{A,\text{DMRG}}\right| \\ &= \sum_{IKJ} W_{IJ} W_{KJ} \left|\Omega^{(N')}_{AS,I}\right\rangle \left\langle \Omega^{(N')}_{AS,K}\right| . \end{aligned} \tag{5}$$

where the coefficients W_{IJ} and W_{KL} are chosen to be real. An element of the reduced density matrix $\{\rho^{(AS)}{}_{IK}\}$ is then given by

$$\rho^{(AS)}_{IK} = \left\langle \Omega^{(N')}_I \middle| \hat{\rho}_{AS} \middle| \Omega^{(N')}_K \right\rangle = \sum_J W_{IJ} W_{KJ}. \tag{6}$$

Next, we include into our systematic construction procedure also all one-electron states defined on the next orbital to be incorporated into our explicitly constructed (i.e., `active') subsystem. This enlarges the dimension of all many-electron states that can be represented on the enlarged active subsystem from $m$ to $2m$ in the case of spin orbitals (and to $4m$ in the case of spatial orbitals). We can then use the same procedure to again reduce (i.e., to `renormalize') the size of the basis back to only m contracted many-electron states by (1) setting up the total Hamiltonian, (2) diagonalizing it, (3) constructing the active-subsystem reduced density matrix, (4) selecting those m eigenvectors with largest eigenvalue to set up the renormalization matrix, and (5) to renormalize the many-electron states defined on the active subsystem with this matrix.

Hence, in each DMRG iteration step, we have a modified basis representation for the many-electron states composed from the states defined on AS and on CS, i.e.,

$$\Omega_{AS,I}^{(N',i)} = \sum_{L}^{m} R_{AS,L}^{(I,i)} \Phi_{L}^{(N')} \tag{7}$$

$$\Omega_{CS,J}^{(N'',i)} = \sum_{L}^{m} R_{CS,L}^{(J,i)} \Phi_{L}^{(N'')} \tag{8}$$

with weights $R_{AS,L}^{(i)}$ and $R_{CS,L}^{(i)}$ changing in each DMRG iteration step i. Note that M' in Eq. (2) refers to the number of DMRG states before decimation and is therefore given by $M' = (2m)\times(2m) = 4m^2$ (or $M' = (4m)\times(4m) = 16m^2$ in the case of spatial orbitals). The changes of the weights are, of course, determined by the eigenvectors of the reduced density matrix [14]. The procedure described terminates when the last orbital is picked from the set defined by the active space. Then, the active subsystem comprises all orbitals and no environment (complementary subsystem) is left over. Once this is achieved, a so-called *sweep* has been completed, which we occasionally also call a completed macroiteration step. Consequently, we have then obtained a first approximation to the electronic state by sequential renormalization steps. From the description of the algorithm given so far, it is clear that the quality of the renormalized basis $\{\Omega^{(N)}{}_K\}$ will depend on how good our guess was for the states defined on the complementary subsystem, which had been used to construct the total Hamiltonian matrix affecting thus its eigenstates and hence also the reduced density matrix.

So, the algorithm must be supplemented by a means to improve the complementary (environment) states. This can be done by a rather ingenious trick: The sequence of orbitals picked up step-by-step by the algorithm defines the *orbital ordering* (a purely technical term of the DMRG algorithm). After having completed the first sweep, the orbital ordering has been fixed. Now, we reverse this ordering and start with the orbital included in the last step to proceed with the algorithm. We take the second last orbital to explicitly construct many-electron states on this new active subsystem. For the construction of the total Hamiltonian we require renormalized states defined on the new complementary subsystem, i.e., on the active space without the two orbitals that define the new active subsystem. But these complementary states we can take from the second last step in our first sweep if stored in this first sweep. In such a way, the active subsystem constructed step wise in the previous sweep is now taken as the complementary subsystem. And if we iterate this process, i.e., sweeping with change of direction along the orbital ordering, we can sequentially improve the states defined on the complementary subsystem (the active-subsystem states are always constructed systematically as described above).

Once these macroiterations are converged, the final DMRG basis functions $\{\Omega^{(N)}{}_K\}$ have been found. Since their structure in terms of determinants cannot be known for active spaces larger than about 18 electrons in 18 active spatial orbitals, the DMRG algorithm produces renormalized operator representations in this basis instead (rather than the renormalized basis itself). Since all observables, and especially the energy, can be expressed in second quantization employing annihilation and creation operators defined for the orbitals of the active space, all what needs to be done is to renormalize the matrix representation of these creation and annihilation operators. All observables can then be constructed from these matrices. Having said this, it is obvious how to construct the matrix representation of the total Hamiltonian constructed from active and complementary subsystem states. Since creation and annihilation operators for the orbitals have been defined with respect to the active and complementary subsystems' bases, they can simply be used to set up the electronic Hamiltonian for the complete active space via direct products and matrix additions.

Now we know how to systematically construct a pre-contracted N-electron basis that can eventually pick up all relevant configurations, if the environment guess in the first step allows for this. The basis itself is (in general) never explicitly constructed and instead we have access only to operators expressed in second quantization in terms of this basis.

3. DMRG in Quantum Chemistry

The very first, pioneering papers on the quantum-chemical DMRG have already been mentioned in the Introduction. We shall now review the achievements that have been accomplished since those early days. Daul et al. [30] performed DMRG calculations for methane, which reached FCI accuracy in the chosen one-electron basis. In 2003, Legeza et al. calculated the ionic-neutral curve crossing of LiF [31]. It could be shown that the avoided crossing of the two singlet states is correctly described by the DMRG algorithm. Chan and Head-Gordon performed a DMRG calculation on the water molecule with 41 active orbitals [32]. In a follow-up article Chan described the parallelization of their DMRG code [33]. Chan also compared DMRG and coupled-cluster results for the potential energy curve of N_2 [34]. Already in 2001, Mitrushenkov et al. [35] have compared DMRG and FCI results for the equilibrium structure and dissociation energies of several diatomic molecules, such as Be_2, HF, and N_2.

In 2006, Hachmann et al. implemented a quadratic scaling DMRG algorithm for long, one-dimensional molecules [25]. These one-dimensional molecules resemble closely the one-dimensional lattices studied by the condensed-matter community, for which DMRG was originally developed. Quadratic scaling could be achieved by prescreening the two-electron integrals and neglecting those which lie below a certain threshold.

The first scalar-relativistic DMRG calculations were performed by Moritz et al. for the potential energy curve of the ground state and $^1\Sigma^+$ first excited state of cesium hydride [36]. For the cesium atom, relativistic effects cannot be neglected. The scalar relativistic effects were incorporated by means of the generalized Douglas-Kroll-Hess protocol [37] up to tenth order [38,39]. The decoupling procedure leads to a block diagonalization of the Dirac Hamiltonian which can be obtained by a sequence of suitably chosen unitary transformations [40]. It is clear that a fully relativistic, i.e., four-component DMRG calculation is quite straightforward to implement as only the four-component many-electron Hamiltonian in second quantization has to be employed [11]. The spectroscopic constants of the DMRG potential energy curve agree well with multi-reference CI and coupled-cluster reference data.

4. Peculiarities of DMRG

4.1 Orbitals and Convergence

Daul et al. [30] investigated the effect of different orbital bases on DMRG calculations. They employed canonical Hartree-Fock orbitals, localized Hartree-Fock orbitals, and Kohn-Sham orbitals. Quite contrary to the authors' expectation, the localization procedure did not significantly change or improve the convergence behavior or the converged energy as was originally anticipated from the experience of the condensed-matter community.

In a systematic study, the influence of various types of orbitals and different orbital orderings in a DMRG calculation was examined by Mitrushenkov et al. [41]. Canonical CASSCF orbitals delivered significantly better results than the localized orbitals. Nevertheless, the localized orbitals appear to be particularly effective in the weak interaction region. Mitrushenkov et al. also tried to improve the orbital ordering by selecting those orbitals with largest interactions as measured by the squared Coulomb two-electron integral divided by the difference in orbital energies and placing them near each other.

In the spirit of the orbital interaction scheme proposed by Mitrushenkov et al. [41], Legeza and Sólyom applied the von Neumann entropy to quantify the importance of a subsystem configuration [42]. In quantum information theory, the von Neumann entropy measures the entanglement of a bipartite system, i.e., between the active subsystem and the complementary subsystem. Legeza and Sólyom determined the entropies for subsystems of one single orbital in the Hubbard Hamiltonian and placed the orbitals with largest entropies in the middle of the orbital chain. The same ansatz was also proposed for the DMRG application in quantum chemistry where the molecular orbitals with the highest entropy correspond to the frontier orbitals. But in a follow-up article, the authors could demonstrate that their ansatz for the orbital ordering does not necessarily optimize the convergence behavior of the DMRG algorithm [43].

A similar approach based on the von Neumann entropy was also chosen by Rissler et al. to improve the orbital ordering [23]. The authors devised a scheme to compute the entanglement between any two orbitals from the von Neumann entropy. The idea is to place strongly interacting orbitals next to each other in order to reduce the entanglement of the entire system. Their optimized orbital sequences gave in all cases lower converged energies than the reference ordering where the orbitals are ordered by Hartree-Fock energies. The disadvantage of this methods is, of course, that preceding DMRG calculations have to be performed in order to devise the optimized orbital ordering.

It is expected that through the localization procedure, the complex electron correlation problem can be reduced to a good degree to nearest-neighbor-like interactions in elongated molecules. In molecules with lower entanglement between system and environment, the number of DMRG states can then be chosen rather small while still accurate results are obtained [25].

Mitrushenkov et al. [41] suggested to reformulate the DMRG algorithm in a non-orthonormal basis in order to facilitate and improve the orbital localization which should then lead to a better convergence. Chan and Van Voorhis adapted their DMRG code to support non-orthogonal molecular orbitals [44]. These authors implemented a biorthogonal basis where several modifications were necessary to deal with non-Hermitian operators.

Already in 2002, Chan and Head-Gordon provided the most comprehensive overview of the DMRG algorithm for quantum chemistry [45]. They examined the DMRG algorithm in terms of computational cost, memory usage, and disk storage. They also realized the slow convergence of DMRG when applied to quantum chemical systems. To solve the convergence issue, an optimized ordering was proposed by reordering the orbitals to block-diagonalize the one-electron integral matrix using the reverse Cuthill-McKee algorithm. Another important aspect of their work was the application of white noise to avoid local energy minima.

Local minima may occur when the initial guess of the basis of the complementary subsystem is insufficient. This means that the truncated basis on the CS might not contain important states to represent the total state appropriately. Due to the nature of the DMRG algorithm, such states will never be picked up in the wave function because they will have no contribution in the reduced density matrix and are therefore lost. The white noise allows to collect these basis states. Another strategy to prevent local minima on a more sound theoretical basis is the perturbative correction suggested by White [46]. The perturbative correction tries to construct the initial guess of the *CS* basis in such a way that all states on the *AS* have a counterpart in the *CS* basis. Both methods do not guarantee to converge to the global minimum.

We investigated convergence of the plain DMRG algorithm without any such precautions to guarantee the pick-up of relevant many-electron states in the environment. Then, convergence to local minima can be observed, but does not always occur. In particular, it depends on the ordering of orbitals, i.e., on the sequence of orbitals to be picked up sequentially by the increasing active subsystem. For this comparison we prepared optimized orbital orderings via a genetic optimization procedure [47]. While noise affects the environment states in a random manner, we studied [48] the feasibility of explicit environment guesses as well as of systematically increased environment states. A decomposition of the DMRG basis functions in terms of Slater determinants as described in section 2 has been carried out in Ref. [14], in order to study and eventually better understand energy convergence during DMRG iterations.

4.2 Orbital Optimization

DMRG-SCF implementations have been proposed by Zgid and Nooijen and by Ghosh et al. [49,50]. The DMRG-SCF algorithm optimizes the orbitals in the active space similar to the CASSCF method [51]. In the two implementations a similarity-transformed Hamiltonian was constructed like in coupled-cluster theory. The orbital-rotation amplitudes are obtained from a coupled-cluster type exponential parameterization using single excitations only in the exponential.

To evaluate the resulting energy expression, the one-particle and two-particle density matrices are required. Zgid and Nooijen as well as Ghosh et al. presented two only slightly different approaches on how to obtain the two-particle density matrices [52,50]. Note that the one-particle density matrices can be calculated at any position of the sweep without additional memory cost. The approach taken by Zgid and Nooijen is slightly more efficient than the one by Ghosh et al. since the computational scaling can be reduced by forming simple intermediates.

Zgid and Nooijen combined the two-site DMRG algorithm to converge the DMRG wave function to the variational optimum and switch then to the one-site algorithm where the two-particle density matrix can be easily constructed. Zgid and Nooijen presented preliminary result on the Cr_2Mn_2 metal cluster with their DMRG-SCF implementation.

Ghosh et al. showed state-averaged DMRG-SCF calculations for β-carotene using the full π-valence active space consisting of 22 projected atomic orbitals and 22 electrons. They obtained the correct state ordering as in experiment but the excitation energies were generally overestimated which might be most likely due to the lack of the σ– π dynamic correlation.

The DMRG-SCF implementation can also be cast into an acceleration algorithm to improve the convergence in large-scale CASSCF calculations as described in a recent work by Yanai et al. [53]. In this method, the DMRG algorithm replaces the CI calculation step in a two-step CASSCF calculation to reduce the number of iterations and therefore of expensive exact diagonalization steps.

4.3 Variational Nature of the DMRG Ansatz

A variational electronic structure theory is of great benefit because the energy is always an upper bound to the exact energy. The quality of the trial wave function can be therefore related to the energy and one can systematically improve the result. The nature of the DMRG algorithm is variational since the algorithm restricts the Hilbert space in which the Hamiltonian is solved at each iteration step [9,45]. However, one cannot guarantee that the energy is successively lowered since there is no simple relationship between the Hilbert spaces spanned by states of two subsequent iterations due to the basis truncation. As a result, one observes small deviations in the energy during a sweep which is caused by local fluctuations from an increase in the flexibility of the many-particle basis.

4.4 Excited States and DMRG Response Theory

In 2007, Dorando et al. investigated the state-averaged harmonic Davidson algorithm to describe excited states [54]. The harmonic Davidson approach is an iterative technique that works with shifted and inverted operators which allow one to specifically target excited states within the DMRG algorithm. The advantage over the traditional state-averaged Davidson approach is that it is not necessary to calculate and represent all states between the ground state and excited state of interest. The authors calculated several low-lying excited states of acenes ranging from naphthalene to pentacene. The ground states of the acenes were previously studied by the same group [55].

An analytic response theory was also implemented into the DMRG algorithm by Dorando et al. [56]. The basis was the reformulation of the DMRG method in terms of a Lagrangian formalism [57]. Static and frequency-dependent response properties were implemented.

4.5 Spin and Symmetry Adaptation of DMRG

Zgid and Nooijen implemented a fully spin-adapted DMRG algorithm which allows one to target spin and spatial symmetry states of interest [58]. Spin adaptation in the DMRG method was already performed before by McCulloch and Gulácsi for the Kondo lattice model [59–61]. Their implementation is based on the Clebsch-Gordon transformation and the elimination of quantum numbers by the use of the Wigner-Eckart theorem. Zgid and Nooijen decided to sacrifice computational efficiency in order to avoid the Clebsch-Gordon transformation and obtain a simpler implementation of the spin adaptation. They made sure that always complete sets of multiplets are included in the DMRG wave function expansion. A second important condition is the adaptation of the eigenvectors of the modified reduced density matrix to be an eigenfunction of the total spin operator of either the active or complementary subsystem.

Another scheme to target spin states was used in the DMRG study of Moritz et al. on the calculation of the potential energy curve of CsH where spin contamination became a problem at large nuclear distances [36]. The spin contamination was eliminated by modifying the Hamiltonian matrix in such a way that states with a wrong total spin are shifted up in energy.

5. Applications in Transition Metal Chemistry

In a pioneering study, we investigated DMRG for the prediction of relative energies of transition metal complexes and clusters of different spin and molecular structure [62]. Accurate energies for the spin splitting of the singlet and triplet state of cobalt hydride as well as for nickel carbonyl complexes were obtained. We demonstrated that the DMRG algorithm is capable to solve the spin state problem in quantum chemistry for mononuclear transition metal complexes [63–66].

Based on our numerical DMRG studies of the mononuclear complexes, we made the interesting observation that the relative energy appears to converge faster than the total absolute energy when increasing the number of DMRG states [62]. We note that no white noise or perturbative correction was applied. Nevertheless, a smooth convergence to the reference energy by increasing successively the DMRG basis states could be achieved.

After the successful application of the DMRG algorithm to the spin state problem, we tackled the question whether a reliable calculation of relative energies on a given potential energy surface of the same total spin is possible. We elaborated that DMRG is also well suited for complex electronic structures that require large active spaces already for a qualitatively correct description of the ground state wave function. As an example, we investigated oxygen-bridged dicopper transition metal clusters previously studied by Cramer et al. [67,68].

From a qualitative picture of the electronic structure of binuclear transition metal clusters, we concluded that the one-particle active space must be doubled compared to the mononuclear transition metal complexes in order to retain a qualitative correct description of the wave function. At present, the only multireference method able to deal with such large active spaces is the DMRG algorithm. In Table 1, we compare the DMRG energies from our original work in Ref. [62] with new results obtained with our new DMRG program [69] where the environment basis is built up as described in Ref. [48] (results with this new DMRG program had already been published in our work on the Slater-determinant decomposition in Ref. [14]). For these DMRG calculations, the same active space, basis set and effective core potential as in Ref. [62] was used. From Table 1 it is clear that the guess of the environment has a large effect on the DMRG energies. We now achieve lower DMRG energies at smaller m values even without the application of noise or a perturbative correction. If we apply noise for $m = 128$ DMRG states, however, the absolute DMRG energies are lower but the relative energy lies in the same region.

Table 1. Relative energies of μ-η^2:η^2 peroxo and bis(μ-oxo) $[Cu_2O_2]^{2+}$ isomers with D_{2h} symmetry in an active space of 44 orbitals with 26 electrons. The superscript * denotes DMRG calculations where white noise is applied as described in Ref. [45] and bs stands for a broken-symmetry solution. The DMRG results by Kurashige and Yanai [70] were obtained for a larger active space of 32 electrons in 62 orbitals and with White's perturbative correction. While total electronic energies are given in Hartree (E_h), the relative energies ΔE are given in kJ/mol.

method	E_{bisoxo}/E_h	E_{peroxo}/E_h	ΔE
Reference energies			
CASSCF(8,8)[62]	–541.431 345	–541.425 641	–15.0
CASPT2(8,8)[62]	–541.505 663	–541.510 178	11.9
CASSCF(16,14)[67]	–541.503 07	–541.503 45	1.0
CASPT2(16,14)[67]	–542.062 08	–542.064 35	6.0
bs-B3LYP[67]	–544.194 19	–544.278 44	221.2
RASPT2(24,28)[71]			119.7
DMRG energies from Ref. [62]			
DMRG(m = 64)	–541.456 375	–541.479 969	61.9
DMRG(m = 100)	–541.463 697	–541.494 473	80.8
DMRG(m = 200)	–541.466 781	–541.496 680	78.5
DMRG(m = 400)	–541.467 527	–541.497 171	77.8
DMRG(m = 600)	–541.467 721	–541.497 274	77.6
DMRG(m = 800)	–541.467 794	–541.497 314	77.5
DMRG energies (new code [69])			
DMRG(m = 32)	–541.440 272	–541.478 196	99.6
DMRG(m = 44)	–541.446 006	–541.483 405	98.2
DMRG(m = 64)	–541.458 021	–541.497 468	103.6
DMRG(m = 128)	–541.473 082	–541.514 702	109.3
DMRG*(m = 128)	–541.476 645	–541.515 621	102.3
DMRG(m = 2400)[70]	–541.968 391	–542.025 139	149.0

When this review was finished, a paper by Kurashige and Yanai was published [70] which provides the best reference data available so far for the binuclear copper cluster. Their best results are incorporated in Table 1. Note the larger active space and the one order of magnitude larger number of renormalized states m. Their expensive reference DMRG calculations clearly support our findings made previously in Ref. [62] that the DMRG algorithm is ideally suited to provide a qualitatively correct description of the relative energies in transition metal chemistry of open-shell complexes and clusters. However, it is also clear that a low number m of DMRG basis states is desirable for feasibility reasons, though m = 128 is apparently not yet sufficient for quantitative agreement. Nevertheless, our small-m DMRG results are already in good agreement with RASPT2 results and significantly better than the DFT and CASSCF results given in Table 1.

One niche of the DMRG algorithm in quantum chemistry will be transition metal chemistry of open-shell complexes and clusters, especially when supplemented by methods that are capable of capturing the missing dynamic electron correlation due to the virtual orbitals neglected from the active space.

6. Conclusion

The application of DMRG in quantum chemistry during the past decade has seen tremendous progress as reviewed here.

While previous presentations of the DMRG algorithm often focussed on the original derivation based on the tensor-product nature of N-electron states, the relation to standard quantum chemical approaches like configuration interaction was not that obvious, which is the reason why we presented a derivation of the DMRG algorithm in terms of interactions of Slater determinants in this work.

The algorithm is now very well elaborated and for specific classes of molecules DMRG has already been established as the best method available. Not only extended π-systems of organic chemistry, which feature a large amount of static electron correlation, belong to these classes, DMRG is also very suitable for rather compact molecules like transition metal complexes and clusters, in which the correlated motion of electrons is spatially very much confined.

Acknowledgement

We gratefully acknowledge financial support by the German Science Foundation DFG through projects 1703/1–1 and 1703/1–2 of the priority programme SPP1145 and through a TH-Grant (TH-26 07–3) from ETH Zurich.

References

1. K. G. Wilson, Rev. Mod. Phys. **47** (1975) 773–840.
2. K. G. Wilson, Rev. Mod. Phys. **55** (1983) 583–600.
3. S. R. White, Phys. Rev. Lett. **69** (1992) 2863–2866.
4. S. R. White, Phys. Rev. B **48** (1993) 10345–10356.
5. S. R. White, R. L. Martin, J. Chem. Phys. **110** (1999) 4127–4130.
6. Z. Shuai, J. Bredas, S. Pati, S. Ramasesha, *Quantum confinement effects on the ordering of the lowest-lying excited states in conjugated polymers.* In *Optical probes of conjugated polymers.* Vol. 3145, Z. V. Vardeny L. J. Rothberg, (Eds.), Conference on Optical Probes of Conjugated Polymers, SAN DIEGO, CA, JUL 28–30, (1997).
7. D. Yaron, E. Moore, Z. Shuai, J. Bredas, J. Chem. Phys. **108** (1998) 7451–7458.
8. G. Fano, F. Ortolani, L. Ziosi, J. Chem. Phys. **108** (1998) 9246–9252.
9. U. Schollwöck, Rev. Mod. Phys. **77** (2005) 259–315.
10. K. A. Hallberg, Adv. Phys. **55** (2006) 477–526.
11. M. Reiher, A. Wolf, *Relativistic Quantum Chemistry.* Wiley-VCH: Weinheim (2009).
12. J. Olsen, P. Jørgensen, J. Simons, Chem. Phys. Lett. **169** (1990) 463–472.
13. T. Helgaker, P. Jørgensen, J. Olsen, *Molecular Electronic-Structure Theory.* John Wiley & Sons Ltd.: Chichester (2000).
14. G. Moritz, M. Reiher, J. Chem. Phys. **126** (2007) 244109.

15. A. Klümper, A. Schadschneider, J. Zittartz, J. Phys. A: Math. Gen. **24** (1991) L955-L959.
16. A. Klümper, A. Schadschneider, J. Zittartz, Z. Phys. B **87** (1992) 281–287.
17. M. Fannes, B. Nachtergaele, R. F. Werner, Commun. Math. Phys. **144** (1992) 443–490.
18. B. Derrida, M. Evans, V. Hakim, V. Pasquier, J. Phys. A: Math. Gen. **26** (1993) 1493–1517.
19. S. Östlund, S. Rommer, Phys. Rev. Lett. **75** (1995) 3537–3540.
20. S. Rommer, S. Östlund, Phys. Rev. B **55** (1997) 2164–2181.
21. M. Martin-Delgado, J. Rodriguez-Laguna, G. Sierra, Phys. Rev. B **65** (2002) 155116.
22. D. Perez-Garcia, F. Verstraete, M. M. Wolf, J. I. Cirac, Quantum Inf. Comput. **7** (2007) 401–430.
23. J. Rissler, R. M. Noack, S. R. White, Chem. Phys. **323** (2006) 519–531.
24. F. Verstraete, M. M. Wolf, D. Perez-Garcia, J. I. Cirac, Phys. Rev. Lett. **96** (2006) 220601.
25. J. Hachmann, W. Cardoen, G. K.-L. Chan, J. Chem. Phys. **125** (2006) 141101.
26. G. K.-L. Chan, J. J. Dorando, D. Ghosh, J. Hachmann, E. Neuscamman, H. Wang, T. Yanai, *An Introduction to the Density Matrix Renormalization Group Ansatz in Quantum Chemistry*. In: *Frontiers in Quantum Systems in Chemistry and Physics*. Vol. 18, S. Wilson, P. J. Grout, J. Maruani, G. DelgadoBarrio, and P. Piecuch (Eds.), (2008), 12th European Workshop on Quantum Systems in Chemistry and Physics, London, ENGLAND, AUG 30-SEP 05 2007.
27. I. P. McCulloch, J. Stat. Mech: Theory Exp. (2007) 1742–5468.
28. H.-Q. Zhou, R. Orus, G. Vidal, Phys. Rev. Lett. **100** (2008) 080601.
29. F. Verstraete, V. Murg, J. I. Cirac, Adv. Phys. **57** (2008) 143–224.
30. S. Daul, I. Ciofini, C. Daul, S. White, Int. J. Quantum Chem. **79** (2000) 331–342.
31. Ö. Legeza, J. Röder, B. A. Hess, Mol. Phys. **101** (2003) 2019–2028.
32. G. K.-L. Chan, M. Head-Gordon, J. Chem. Phys. **118** (2003) 8551–8554.
33. G. K.-L. Chan, J. Chem. Phys. (2004) 3172–3178. **120** R-34:G. K.-L. Chan, M. Kallay, J. Gauss, J. Chem. Phys. **121** (2004) 6110–6116.
35. A. O. Mitrushenkov, G. Fano, F. Ortolani, R. Linguerri, P. Palmieri, J. Chem. Phys. **115** (2001) 6815–6821.
36. G. Moritz, A. Wolf, M. Reiher, J. Chem. Phys. **123** (2005) 184105.
37. A. Wolf, M. Reiher, B. A. Hess, J. Chem. Phys. **117** (2002) 9215–9226.
38. M. Reiher, A. Wolf, J. Chem. Phys. **121** (2004) 2037–2047.
39. M. Reiher, A. Wolf, J. Chem. Phys. **121** (2004) 10945–10956.
40. M. Reiher, Theor. Chem. Acc. **116** (2006) 241–252.
41. A. O. Mitrushenkov, R. Linguerri, P. Palmieri, G. Fano, J. Chem. Phys. **119** (2003) 4148–4158.
42. Ö. Legeza, J. Sólyom, Phys. Rev. B **68** (2003) 195116.
43. Ö. Legeza, J. Sólyom, Phys. Rev. B **70** (2004) 205118.
44. G. K.-L. Chan, T. Van Voorhis, J. Chem. Phys. **122** (2005) 204101.
45. G. K.-L. Chan, M. Head-Gordon, J. Chem. Phys. **116** (2002) 4462–4476.
46. S. White, Phys. Rev. B **72** (2005) 180403.
47. G. Moritz, B. A. Hess, M. Reiher, J. Chem. Phys. **122** (2005) 024107.
48. G. Moritz, M. Reiher, J. Chem. Phys. **124** (2006) 034103.
49. D. Zgid, M. Nooijen, J. Chem. Phys. **128** (2008) 144116.
50. D. Ghosh, J. Hachmann, T. Yanai, G. K.-L. Chan, J. Chem. Phys. **128** (2008) 144117.
51. B. O. Roos, Adv. Chem. Phys. **69** (1987) 399.
52. D. Zgid, M. Nooijen, J. Chem. Phys. **128** (2008) 144115.
53. T. Yanai, Y. Kurashige, D. Ghosh, G. K.-L. Chan, Int. J. Quantum Chem. **109** (2009) 2179–2183.

54. J. J. Dorando, J. Hachmann, G. K.-L. Chan, J. Chem. Phys. **127** (2007) 084109.
55. J. Hachmann, J. J. Dorando, M. Aviles, G. K.-L. Chan, J. Chem. Phys. **127** (2007) 134309.
56. J. J. Dorando, J. Hachmann, G. K.-L. Chan, J. Chem. Phys. **130** (2009) 184111.
57. G. K.-L. Chan, Phys. Chem. Chem. Phys. **10** (2008) 3454–3459.
58. D. Zgid, M. Nooijen, J. Chem. Phys. **128** (2008) 014107.
59. I. P. McCulloch, M. Gulácsi, Aust. J. Phys. **53** (2000) 597–612.
60. I. P. McCulloch, M. Gulácsi, Philos. Mag. Lett. **81** (2001) 447–453.
61. I. P. McCulloch, M. Gulácsi, Europhys. Let. **57** (2002) 852–858.
62. K. H. Marti, I. M. Ondik, G. Moritz, M. Reiher, J. Chem. Phys. **128** (2008) 014104.
63. M. Reiher, O. Salomon, B. A. Hess, Theor. Chem. Acc. **107** (2001) 48.
64. M. Reiher, Inorg. Chem. **41** (2002) 6928–6935.
65. J. N. Harvey, Struct. Bonding **112** (2004) 151–183.
66. M. Reiher, Farad. Discuss. **135** (2007) 97–124.
67. C. J. Cramer, M. Woch, P. Piecuch, C. Puzzarini, L. Gagliardi, J. Phys. Chem. A **110** (2006) 1991–2004.
68. C. J. Cramer, M. Woch, P. Piecuch, C. Puzzarini, L. Gagliardi, J. Phys. Chem. A **111** (2007) 4871.
69. G. Moritz, K. H. Marti, M. Reiher, Qc-Dmrg-ETH 2.0, A program for quantum chemical DMRG calculations, Copyright 2007–2009, ETH Zurich.
70. Y. Kurashige, T. Yanai, J. Chem. Phys. **130** (2009) 234114.
71. P. A. Malmqvist, K. Pierloot, A. R. M. Shahi, C. J. Cramer, L. Gagliardi, J. Chem. Phys. **128** (2008) 204109.

Local Time-Dependent Coupled Cluster Response for Properties of Excited States in Large Molecules

By Danylo Kats and Martin Schütz*

Institute of Physical and Theoretical Chemistry, University of Regensburg, Universitätsstraße 31, 93040 Regensburg, Germany

(Received July 20, 2009; accepted September 16, 2009)

Local Correlation / Laplace Transform / Response Theory

The new, Laplace transform based multi-state local CC2 response method is compared to the previous single-state approach. The new method employs adaptive state specific local approximations for the eigenvectors of the Jacobian. As a result, it is much less dependent on the initial un-truncated CCS wavefunction and able to locate the relevant eigenstates also in difficult cases where the initial CCS eigenvectors are qualitatively wrong and where the original single-state method failed. In this paper we compare excitation energies, transition strengths, and dipole moments of the individual local approaches and the canonical reference calculation, for a set of different molecules and excited states.

1. Introduction

There is presently an increasing demand for reliable theoretical methods for calculating properties of electronically excited states (excitation energies, oscillator strength, density matrices, energy gradients w.r. to nuclear displacements, ...) of extended molecular systems. Applications of such methods spread from life - to material science. Time-Dependent Density Functional Theory (TD-DFT), which, due to its efficiency, unfortunately is often the sole applicable method for large systems, can presently not be considered as a reliable method. It is well known and documented that TD-DFT often provides a qualitatively wrong picture of the photophysics of a molecular system, particularly so if excitations to charge transfer (CT) or Rydberg states, or excitations of larger π systems are involved [1–3] .

* Corresponding author. E-mail: martin.schuetz@chemie.uni-regensburg.de

Z. Phys. Chem. **224** (2010) 601–616
© by Oldenbourg Wissenschaftsverlag, München

A very beautiful and general wavefunction based approach to excited states is (Time-Dependent) Coupled Cluster (CC) Response theory (cf. Ref. [4] and references therein). A hierarchy of CC models (CCS, CC2, CCSD, CC3) was proposed in that context. The most economical one of these, which does include dynamical correlation effects, is the CC2 model [5]. It treats singles and doubles substitution wavefunction parameters (amplitudes) as zeroth- and first-order parameters, respectively, in terms of the fluctuation potential (the difference between Hamiltonian and Fock operator). For excited states dominated by singles substitutions CC2 provides quite accurate excitation energies [6] and generally can well be thought of as reliable without the intrinsic flaws of TD-DFT. An efficient implementation of CC2 response theory, based on Density Fitting (DF) [7–11] of the orbital products in the Electron Repulsion Integrals (ERIs) is available in the TURBOMOLE package [12,13]. Since the method is based on (delocalized) canonical molecular orbitals the relevant data objects in that basis (integrals, amplitudes, amplitude response) are not sparse. Its computational cost thus scales as $O(N^5)$ with molecular size N.

Over the past few years a *local* CC2 response method with reduced scaling behavior was developed in our group [14,15]. Local correlation approaches have been used before with great success in the context of *ab initio* theories for electronic ground states to circumvent the inherent scaling problem of these methods[16–26]. The essential idea of local correlation approaches is to exploit the short-range character of dynamic electron correlation effects and to work with the naturally sparse, rather than artificially dense (an artifact of the delocalized canonical MO basis) data objects. Hence, the de-localized canonical MOs are replaced by spatially *localized* orbitals. In our local CC2 response method we employ the ansatz proposed by Pulay [17] . Mutually orthogonal localized molecular orbitals (LMOs) and non-orthogonal projected atomic orbitals (PAOs) are used to span the occupied and virtual orbital spaces of the underlying Hartree-Fock reference. Strong orthogonality thus is conserved, yet among the PAOs there are redundancies which must be eliminated at some stage of the calculation. *A priori* restrictions then are imposed on the amplitudes of doubly substituted configuration state functions (single substitutions remain un-truncated): Restricted LMO pair lists and pair specific excitation subspaces of PAOs (domains) are specified and amplitudes outside these lists/domains are *a priori* set to zero. For the amplitudes (of the electronic ground state) such a truncation is based on spatial locality arguments, in analogy to our previous local correlation methods for ground states. Only LMO pairs up to a certain inter-orbital distance and only double substitutions of LMOs by nearby PAOs are considered. For the amplitude response (subject to the frequency dependent perturbation), on the other hand, *a priori* truncations are far more complicated and the specification of proper pair lists and domains actually is one of the key challenges in the formulation of a local method for excited states [36]. Exited states may have substantial non-local CT character, thus spatial locality is an improper criterion for setting up restricted pair lists and domains. In the first version of our *local* CC2 response method, presented in Refs. [14,15], the pair lists and domains for the amplitude response were determined by analysis of un-truncated wavefunctions obtained at a simpler level of theory, e.g., CCS wavefunctions. The method yet turned out to be frail for cases where the simpler level of theory provides qualitatively wrong wavefunctions for the individual excited states of interest. Relevant states then might be missed.

Recently [27], we presented a new local CC2 response method, which is utilizing the Laplace transform [10,28,29] to decompose the energy denominators occurring in the doubles-doubles block of the CC2 Jacobian. This allows for a partitioning of the CC2 eigenvalue problem exactly as in the case of the canonical theory [12], yet with integrals and amplitudes in a local, rather than a delocalized canonical orbital basis. As a result, an *effective* eigenvalue problem involving only the un-truncated singles vector has to be solved, while the contribution of the doubles amplitude response to the effective Jacobian - singles vector product can be computed on-the-fly and does not enter the Davidson diagonalization. On this basis a multi-state diagonalization procedure with much improved convergence behavior (compared to the single-state root-homing procedure of the previous approach) can be devised, which still maintains state-specific pair domains for the individual states. Furthermore, in combination with Density Fitting of the ERIs it is possible without much effort to utilize the un-truncated information of diagonal pair amplitudes for constructing suitable pair lists and domains for the pair amplitude response in an adaptive way in the course of the diagonalization. This new Laplace based method (termed LT-DF-LCC2 response) repairs the shortcomings of our initial DF-LCC2 response method. It faithfully finds the relevant excited states also for cases where the initial un-truncated wavefunction obtained at lower level of theory is qualitatively wrong.

In this paper the new LT-DF-LCC2 response method is compared to the previous DF-LCC2 response. For a particularly pathological case it is demonstrated that LT-DF-LCC2 response, contrary to DF-LCC2 response, finds all the relevant excited states. Some first results on first-order properties and oscillator strengths obtained via LT-DF-LCC2 response are also presented and compared to the corresponding results of DF-LCC2 response.

2. Theory

In this section we briefly outline the basic theory of DF-LCC2 and LT-DF-LCC2 response. For a detailed discussion and the working equations we refer to Refs. [14,15,27].

2.1 Excitation energies

The LCC2 model is specified by the amplitude equations

$$0 = \Omega_{\mu_1} = \langle \tilde{\mu}_1 | \hat{H} + [\hat{H}, T_2^{\mathrm{loc}}] | 0 \rangle$$
$$0 = \Omega_{\mu_2}^{\mathrm{loc}} = \langle \tilde{\mu}_2^{\mathrm{loc}} | \hat{H} + [F, T_2^{\mathrm{loc}}] | 0 \rangle, \quad (1)$$

where doubles substitution operators T_2^{loc} and the related contravariant projection manifold $\langle \tilde{\mu}_2^{\mathrm{loc}} |$ are *a priori* confined to local subspaces (specified by restricted pair lists and domains). F and $\hat{H} = \exp(-T_1) H \exp(T_1)$ are the Fock -, and the similarity transformed Hamilton operators, respectively. Here and in the following all objects decorated with a hat involve dressed integrals (produced by the similarity transform of the related operators with $\exp(T_1)$), i.e., the T_1 contractions are absorbed in the transformed integrals. The excitation energies $\omega_{\bar{m}}$ of state $\Psi^{\bar{m}}$ are obtained by solving either the left or the right eigenvalue problem

$$\mathbf{A}\boldsymbol{R}^{\bar{m}} = \omega_{\bar{m}} \mathbf{M} \boldsymbol{R}^{\bar{m}}, \quad \boldsymbol{L}^{\bar{m}} \mathbf{A} = \omega_{\bar{m}} \boldsymbol{L}^{\bar{m}} \mathbf{M}, \quad (2)$$

involving the non-symmetric LCC2 Jacobian

$$A_{\mu_i \nu_j} = \begin{pmatrix} \langle \tilde{\mu}_1 | [\hat{H}, \tau_{\nu_1}] \exp(T_2^{\mathrm{loc}}) | 0 \rangle & \langle \tilde{\mu}_1 | [\hat{H}, \tau_{\nu_2}] | 0 \rangle \\ \langle \tilde{\mu}_2 | [\hat{H}, \tau_{\nu_1}] | 0 \rangle & \langle \tilde{\mu}_2 | [F, \tau_{\nu_2}] | 0 \rangle \end{pmatrix}, \quad (3)$$

and the metric $\mathbf{M}$. The τ_{μ_i} are covariant one- or two-particle excitation operators. Eqs. (2) are solved by Davidson diagonalization [30,31]. In each iteration the left/right matrix-vector product of the Jacobian $\times$ a basis vector of the Davidson subspace, e.g., for the right eigenvalue equation,

$$\mathfrak{V}_{\mu_i}^m = A_{\mu_i \nu_j} U_{\nu_j}^m \quad (4)$$

is calculated (m enumerates the basis vectors of the Davidson subspace). Here and in the following the Einstein convention is employed, i.e., repeated indices are summed up. Explicit summations are put in the equations only where we find it useful for clarity. In eq. (4) the doubles parts of the basis vector, $U_{\mu_2}^m$ and of the corresponding matrix-vector product, $\mathfrak{V}_{\mu_2}^m$ are confined to the specific local subspace of state $\Psi^{\bar{m}}$, which is related to basis vector $\mathbf{U}^m$. These state specific local subspaces obviously differ from those of the ground state calculation (cf. eq. (1)). In the original DF-LCC2 response method these subspaces are determined *a priori* once by analyses of the related initial (un-truncated) CCS eigenstate. Working equations for the left/right matrix-vector products, based on the DF approximation for the ERIs, are given in Refs. [14,15].

From the matrix-vector products the Davidson Jacobian

$$A'_{mn} = \tilde{U}_{\mu_i}^m \mathfrak{V}_{\mu_i}^n \quad (5)$$

is constructed/extended ($\tilde{U}_{\mu_i}^m$ denotes the contravariant of $U_{\mu_i}^m$). Solving the eigenvalue equation

$$\mathbf{A}'\mathbf{c} = \mathbf{S}'\mathbf{c}\omega', \text{ with}$$
$$S'_{mn} = \tilde{u}_a^i(m) S_{ab} u_b^i(n) + \tilde{U}_{ab}^{ij}(m) S_{aa'} U_{a'b'}^{ij}(n) S_{bb'}, \quad (6)$$

yields the approximations

$$R_{\mu_i}^{\bar{m}} = U_{\mu_i}^{n} c_{n\bar{m}} \tag{7}$$

and $\omega'_{\overline{m}}$ to eigenstate and eigenvalue of $\mathbf{A}$ for state $\Psi^{\overline{m}}$ with residual

$$\begin{aligned} \mathfrak{r}_a^i(\bar{m}) &= \mathfrak{v}_a^i(n)c_{n\bar{m}} - \omega'_{\bar{m}} S_{aa'} r_{a'}^i(\bar{m}), \\ \mathfrak{R}_{ab}^{ij}(\bar{m}) &= \mathfrak{V}_{ab}^{ij}(n)c_{n\bar{m}} - \omega'_{\bar{m}} S_{aa'} R_{a'b'}^{ij}(\bar{m}) S_{bb'}. \end{aligned} \tag{8}$$

In eqs. (6–8) we used capital and small letters for doubles and singles parts of basis-, eigen-, and residual vectors, respectively, e.g., $R_{\mu_1}^{\overline{m}} = r_a^i(\overline{m})$, $R_{\mu_2}^{\overline{m}} = R_{ab}^{ij}(\overline{m})$, *etc.* Furthermore, we used the conventional notation for the orbital indices, i.e., *i,j,k,...* and *a,b,c,...* label occupied LMOs, and virtual PAOs, respectively. $\mathbf{S}$ represents the PAO metric. The residuals, calculated according to eq. (8), determine the new (additional) basis vectors via first-order perturbation theory.

In the LT-DF-LCC2 response method presented recently [27] the Laplace transform identity $1/x = \int_0^\infty \exp(-xt)dt$ is utilized to partition the eigenvalue problem in eq. (2) along the doubles-doubles block $A_{\mu_2\nu_2}$, which is diagonal in the canonical orbital basis. A new eigenvalue equation then is obtained, which involves only the singles parts of the vectors,

$$\begin{aligned} A_{\mu_1\nu_1}^{\text{eff}}(\omega_{\bar{m}})R_{\nu_1}^{\bar{m}} &= \omega_{\bar{m}} M_{\mu_1\nu_1} R_{\nu_1}^{\bar{m}} = \\ &= A_{\mu_1\nu_1}R_{\nu_1}^{\bar{m}} - A_{\mu_1\xi_2}\int_0^\infty dt e^{-\Delta\epsilon_{\xi_2}t} e^{\omega_{\bar{m}}t} A_{\xi_2\nu_1}R_{\nu_1}^{\bar{m}} \\ &\approx A_{\mu_1\nu_1}R_{\nu_1}^{\bar{m}} - A_{\mu_1\xi_2}\sum_{q=1}^{n_q} w_q e^{-\Delta\epsilon_{\xi_2}t_q} e^{\omega_{\bar{m}}t_q} A_{\xi_2\nu_1}R_{\nu_1}^{\bar{m}}. \end{aligned} \tag{9}$$

The *effective* Jacobian $\mathbf{A}^{\text{eff}}(\omega_{\overline{m}})$ now depends on the excitation energies of the individual states $\Psi^{\overline{m}}$. In eq. (9) $\Delta\varepsilon_{\xi_2}$ denotes the canonical orbital energy denominator, whereas the t_q and w_q are the points and corresponding weights of the numerical Laplace integration. Only a few points are required for a sufficiently accurate quadrature. The doubles part of the eigenvector $R_{\mu_2}^{\overline{m}}$, manifested in the Laplace quadrature, no longer enters the Davidson diagonalization, and can be computed on the fly. Eq. (9), as it stands, holds for canonical orbitals only, and little appears to have been achieved over the original formulation in terms of the doubles energy denominators. However, eq. (9) is the springboard for a formulation in terms of local orbitals. After some algebraic manipulations it can be cast in the computationally amenable form of

$$\begin{aligned} A_{\mu_1\nu_1}^{\text{eff}}(\omega_{\bar{m}})R_{\nu_1}^{\bar{m}} &= A_{\mu_1\nu_1}R_{\nu_1}^{\bar{m}} - A_{\mu_1 iajb}\sum_{q=1}^{n_q}\text{sgn}(w_q)e^{\omega_{\bar{m}}t_q} \\ &\times Y_{ac}^{\text{v}}(q)Y_{bd}^{\text{v}}(q)\left(A_{ckdl\nu_1}R_{\nu_1}^{\bar{m}}\right)X_{ki}^{\text{o}}(q)X_{lj}^{\text{o}}(q) \\ &= \omega_{\bar{m}} M_{\mu_1\nu_1}R_{\nu_1}^{\bar{m}}. \end{aligned} \tag{10}$$

Here, the doubles index ξ_2 of the singles-doubles and doubles-singles blocks of the Jacobi matrix is written explicitely in terms of LMO and PAO indices. The quadrature point dependent transformation matrices occurring in eq. (10) are defined as

$$X^{\mathrm{o}}_{ij}(q) = W^{\dagger}_{\bar{i}i} e^{(\epsilon_{\bar{i}}-\epsilon_F)t_q+\frac{1}{4}\ln|w_q|} W_{\bar{i}j},$$

$$X^{\mathrm{v}}_{ab}(q) = Q^{\dagger}_{a\bar{a}} e^{(-\epsilon_{\bar{a}}+\epsilon_F)t_q+\frac{1}{4}\ln|w_q|} Q_{\bar{a}b}, \tag{11}$$

$$Y^{\mathrm{v}}_{ab}(q) = V_{ac} X^{\mathrm{v}}_{cd}(q) V^{\dagger}_{db},$$

where the matrices **W**, **Q**, and **V** do appear. **W** is the usual unitary transformation matrix specified by the chosen localization criterion and transforms from canonical occupied orbitals to LMOs, while **Q** transforms from canonical virtuals to PAOs ($\bar{i}, \bar{j}, \bar{k}$ ... and $\bar{a}, \bar{b}, \bar{c}$ ... refer to canonical occupied - and virtual orbitals). The matrix **V** is a pseudo-inverse of $\mathbf{S}_{\mathrm{PAO}}$ (cf. eq. (11) in Ref. [32]). Since $R^{\bar{m}}_{\mu_2}$, calculated as the sum over the quadrature points, is confined to a pair specific local subspace (pair domains), **V** corresponds to a pseudo-inverse of *the corresponding block* of the PAO metric $\mathbf{S}^{ij}_{\mathrm{PAO}}$ and thus is (along with $\mathbf{Y}^{\mathrm{v}}$) *pair specific.* For simplicity this pair dependence is not explicitely written out in eqs. (10, 11). It has been shown before [32] that quadrature point dependent transformation matrices as defined in eq.(11) are sparse.

$\varepsilon_F = (\varepsilon_{\mathrm{HOMO}}+\varepsilon_{\mathrm{LUMO}})/2$, finally, appearing in the exponentials, cancels in eq. (10), but ensures that the individual exponential factors are always smaller than one (for positive t_q).

Eq. (10) is solved by Davidson diagonalization, yet only singles basis vectors $U^m_{\mu_1}$ now are used to span the Davidson subspace. Hence, since the singles remain un-truncated in the LT-DF-LCC2 response method as specified, any reference to local subspaces is absent. Multi-state calculations are now straight forwardly possible, which is mandatory to reliably find all the relevant low-lying states and also helps to improve the convergence behavior. In our program the whole Davidson procedure is carried out in the canonical basis, transforming basis vectors $U^m_{\mu_1}$ to local basis just for calculating the matrix-vector product $\mathfrak{V}^m_{\mu_1}$, and back-transforming the latter again to canonical basis for further processing in the Davidson procedure. For the working equations of the matrix-vector product, again based on the DF approximation for the ERIs, we refer to Ref. [27]. Of relevance here is the intermediate U_{μ_2}, defined in canonical basis as

$$U_{\mu_2} = \frac{A_{\mu_2\nu_1} U^m_{\nu_1}}{\omega_{\bar{m}} - \Delta\epsilon_{\mu_2}}. \tag{12}$$

It can be calculated directly in the local basis as

$$U_{ab}^{ij} = -V_{ac}^{ij}V_{bd}^{ij}\sum_{q}^{n_q}\text{sgn}(w_q)e^{\omega_{\bar{m}}t_q}$$
$$\times X_{ce'}^{\text{v}}(q)V_{e'e}X_{df'}^{\text{v}}(q)V_{f'f}\left(A_{ekfl\nu_1}U_{\nu_1}^{m}\right)X_{ki}^{\text{o}}(q)X_{lj}^{\text{o}}(q), \tag{13}$$
$$= V_{ac}^{ij}V_{bd}^{ij}\bar{U}_{cd}^{ij}$$

The U_{μ_2} is confined to a local subspace, i.e., restricted pair lists and pair domains are specified for it. The index ranges *a,b,c,d* in eq. (13) thus are restricted to individual pair domains [*ij*], and the pair index *ij* is confined to the restricted pair list. The dependence of the pseudo-inverse **V** on a specific LMO pair *ij* now is explicitely indicated, and it is understood that the summation does not run over the repeated indices *i,j*.

In multi-state calculations it is necessary to dynamically adapt the state specific local subspaces to changes of the approximate eigenvectors in the course of the Davidson diagonalization. For example, the individual eigenstates may come energetically close and mix, or switch the energetical order (root flip).Pair lists and domains for U_{μ_2} thus should dynamically be re-specified. In our program this is done *in each refresh* of the Davidson diagonalization, i.e., when the Davidson subspace is reset to comprise just as many basis vectors as there are eigenstates to be calculated. Pair lists and domains then can be determined by (i) analysis of the singles part of the present approximation to the eigenstate $R_{\mu_1}^{\bar{m}}$, or (ii) by analysis of the *unrestricted* quantity $\bar{U}_{cd}^{ii}$ (diagonal pairs only), implicitly defined in eq. (13), and evaluated with the actual $R_{\mu_1}^{\bar{m}}$. For this second variant we use the acronym *Laplace domains*. It has the advantage that pair lists and domains are constructed directly from the data object, which is to be approximated. Even without local restrictions $\bar{U}_{cd}^{ii}$ can be calculated without too much effort by virtue of the DF approximation. For all the details about the (more complicated) Davidson procedure, including pair list and domain construction in the context of the new LT-DF-LCC2 response method we refer to Ref. [27].

2.2 Transition strengths and first-order properties

The calculation of transition strengths and (orbital-unrelaxed) first-order properties in the framework of the original DF-LCC2 response approach is extensively discussed in Ref. [15]. Here we will just provide the most relevant equations to indicate the modifications related to the LT-DF-LCC2 response method. A detailed discussion of the implemented working equations for LT-DF-LCC2 transition strengths and first-order properties is postponed to an upcoming article.

In order to compute transition strength and first-order properties of state $\Psi^{\bar{m}}$ also the left eigenvector $\mathbf{L}^{\bar{m}}$ and the ground state multipliers $\mathbf{\Lambda}$ are needed. The former is obtained for the LT-DF-LCC2 case by solving the left eigenvalue equation analogue of (10), which is of course much easier to converge, since the eigenvalues now are already known. The $\mathbf{\Lambda}$ are determined by the linear equation system

$$0 = \Lambda_{\nu_j} A_{\nu_j \mu_i} + \eta_{\mu_i}, \quad \text{with} \quad \eta_{\mu_i} = \langle 0|[\hat{H}, \tau_{\mu_i}]|0\rangle. \tag{14}$$

Again, by virtue of the Laplace trick, it can be partitioned to yield a linear equation system just for the singles part only, i.e.,

$$-\eta_{\mu_1} = \Lambda_{\nu_1} A_{\nu_1 \mu_1} - \Big(\sum_{q=1}^{n_q} \text{sgn}(w_q) X^{\text{o}}_{ki}(q) X^{\text{o}}_{lj}(q) \\ \times \left(\eta_{iajb} + \Lambda_{\nu_1} A_{\nu_1 iajb}\right) Y^{\text{v}}_{ac}(q) Y^{\text{v}}_{bd}(q)\Big) A_{ckdl\mu_1}. \tag{15}$$

The Λ_{μ_2}, manifesting in the Laplace quadrature (as negative sum over q), is computed on the fly and kept after convergence for later use (cf. eq. (18)).

The transition strength of a one-photon excitation $\overline{m} \leftarrow 0$ is computed as

$$S^{0\bar{m}}_{XY} = \frac{1}{2}\left(\mathbf{M}^{0\bar{m}}_X \mathbf{M}^{\bar{m}0}_Y + (\mathbf{M}^{0\bar{m}}_Y \mathbf{M}^{\bar{m}0}_X)^*\right), \tag{16}$$

with the left and right transition moments defined as

$$\mathbf{M}^{0\bar{m}}_X = \eta^X \mathbf{R}^{\bar{m}} + \bar{\mathbf{M}}^{\bar{m}}(\omega_{\bar{m}})\xi^X, \quad \mathbf{M}^{\bar{m}0}_X = \mathbf{L}^{\bar{m}}\xi^X. \tag{17}$$

The quantities $\boldsymbol{\eta}^X$ and $\boldsymbol{\xi}^X$ appearing in the definitions of the transition moments are

$$\eta^X_{\mu_1} = (\langle 0| + \Lambda_{\nu_i}\langle\tilde{\nu}_i|)\,[\hat{X}, \tau_{\mu_1}]|0\rangle \\ + \Lambda_{\nu_2}\langle\tilde{\nu}_2|[[\hat{X}, \tau_{\mu_1}], T_2^{\text{loc}}]|0\rangle, \tag{18}$$

$$\eta^X_{\mu_2} = \Lambda_{\nu_i}\langle\tilde{\nu}_i|[\hat{X}, \tau_{\mu_2}]|0\rangle, \quad \text{and}$$

$$\xi^X_{\mu_i} = \langle\tilde{\mu}_i|\hat{X} + [\hat{X}, T_2^{\text{loc}}]|0\rangle, \tag{19}$$

with $\hat{X} = \exp(-T_1)X\exp(T_1)$ representing the similarity transformed position operator. The multipliers $\overline{\mathbf{M}}^{\overline{m}}(\omega_{\overline{m}})$ appearing in eq. (17) are the solutions of the linear equation system

$$0 = \bar{M}^{\bar{m}}_{\nu_j}(\omega_{\bar{m}})\left(A_{\nu_j\mu_i} + \omega_{\bar{m}} M_{\nu_j\mu_i}\right) + F_{\mu_i\nu_j} R^{\bar{m}}_{\nu_j}, \tag{20}$$

which again involves the left vector-matrix product with **A**, and additionally, as the *rhs*, the right matrix-vector product of the matrix **F** with the right eigenvector. The matrix **F** is defined as

$$F_{\mu_1\nu_1} = \left(\langle 0| + \Lambda_{\rho_i}\langle\tilde{\rho}_i|\right)[[\hat{H}, \tau_{\mu_1}], \tau_{\nu_1}]|0\rangle, \quad F_{\mu_2\nu_2} = 0, \\ F_{\mu_1\nu_2} = \Lambda_{\rho_1}\langle\tilde{\rho}_1|[[\hat{H}, \tau_{\mu_1}], \tau_{\nu_2}]|0\rangle, \quad F_{\mu_2\nu_1} = F_{\mu_1\nu_2}, \tag{21}$$

The partitioning of equation system (20) by application of the Laplace trick yields here

$$-F_{\mu_1\nu_j}R^{\bar{m}}_{\nu_j} = \bar{M}^{\bar{m}}_{\nu_1}(\omega_{\bar{m}})\left(A_{\nu_1\mu_1} + \omega_{\bar{m}}M_{\nu_1\mu_1}\right)$$
$$-\Big(\sum_{q=1}^{n_q}\mathrm{sgn}(w_q)e^{-\omega_{\bar{m}}t_q}X^{\mathrm{o}}_{ki}(q)X^{\mathrm{o}}_{lj}(q)\left((\boldsymbol{FR}^{\bar{m}})_{iajb}\right. \tag{22}$$
$$\left.+\,\bar{M}^{\bar{m}}_{\nu_1}(\omega_{\bar{m}})A_{\nu_1 iajb}\right)Y^{\mathrm{v}}_{ac}(q)Y^{\mathrm{v}}_{bd}(q)\Big)A_{ckdl\mu_1},$$

again reducing it to an equation just for the singles part of the multipliers $\overline{\mathbf{M}}^{\overline{m}}(\omega_{\overline{m}})$. The doubles part of $\overline{\mathbf{M}}^{\overline{m}}(\omega_{\overline{m}})$, calculated on the fly as the negative of the sum over q part in eq. (22), is kept at the end for the converged equation for use in eq. (17).

The (orbital unrelaxed)first-order properties of state $\Psi^{\overline{m}}$ related to perturbations Y (e.g. position operator for dipole moments) are calculated as

$$\langle Y\rangle_{\bar{m}} = \langle Y\rangle + \mathbf{L}^{\bar{m}}\mathbf{A}^{Y}\mathbf{R}^{\bar{m}} + \boldsymbol{\Lambda}^{\bar{m}}\xi^{Y}. \tag{23}$$

$\langle Y \rangle = \boldsymbol{\Lambda}\ \xi^{Y}$ is the corresponding ground-state property (e.g. ground-state dipole moment). The vector $\mathbf{L}^{\overline{m}}\ \mathbf{A}^{Y} = {}^{L}\boldsymbol{\eta}^{Y}$ is almost identical to $\boldsymbol{\eta}^{Y}$ defined in eq. (18), with the sole difference that the first term of the singles part (which involves the bra reference state) is dropped and the ground-state multipliers $\boldsymbol{\Lambda}$ are replaced by the left eigenvector $\mathbf{L}^{\overline{m}}$. The multipliers for the excited state $\boldsymbol{\Lambda}^{\overline{m}}$ occurring in eq. (23) are obtained by solving the linear equation system

$$\Lambda^{\bar{m}}_{\nu_j}A_{\nu_j\mu_i} = -L^{\bar{m}}_{\nu_j}B_{\nu_j\mu_i\rho_k}R^{\bar{m}}_{\rho_k}. \tag{24}$$

Again, the matrix $\mathbf{L}^{\overline{m}}\mathbf{B} = {}^{L}\mathbf{F}$ is almost identical to the matrix $\mathbf{F}$ defined in eq. (21) with the sole difference of dropping the first term with the bra reference state in the singles-singles block and substituting $\boldsymbol{\Lambda}$ by $\mathbf{L}^{\overline{m}}$. Hence, the *rhs* of eq. (24) can be calculated using the same machinery as for $\mathbf{FR}^{\overline{m}}$ in eq. (20). Application of the Laplace trick and partitioning of equation system (24) finally yields

$$-{}^{L}F_{\mu_1\nu_j}R^{\bar{m}}_{\nu_j} = \Lambda^{\bar{m}}_{\nu_1}A_{\nu_1\mu_1} - \Big(\sum_{q=1}^{n_q}\mathrm{sgn}(w_q)X^{\mathrm{o}}_{ki}(q)X^{\mathrm{o}}_{lj}(q) \tag{25}$$
$$\times\left({}^{L}F_{iajb\ \nu_1}R^{\bar{m}}_{\nu_1} + \Lambda^{\bar{m}}_{\nu_1}A_{\nu_1 iajb}\right)Y^{\mathrm{v}}_{ac}(q)Y^{\mathrm{v}}_{bd}(q)\Big)A_{ckdl\mu_1}.$$

3. Test Calculations

The formalism outlined in Sec.2 for calculating excitation energies, transition strengths, and orbital-unrelaxed first-order properties is implemented in the MOLPRO package [33]. For a detailed discussion we refer to Refs. [14] and [15] (DF-LCC2), and [27] (LT-DF-LCC2 excitation energies). The working equations and implementation details of LT-DF-LCC2 transition strengths and properties will be published elsewhere. In this section we present results performed for a test set of different molecules and different excited states, in order to compare LT-DF-LCC2 and DF-LCC2 response. As a particularly difficult

Table 1a. Excitation energies in [eV] for different molecules and excited states. For the local calculations the deviations from the related canonical reference values $\Delta\omega = \omega_{loc} - \omega_{ref}$ are given (except explicitely stated otherwise). The labels BP and BP-LT stand for DF-LCC2 and LT-DF-LCC2 results employing BP domains, LD-LT for LT-DF-LCC2 results employing Laplace domains.

Basis		cc-pVDZ					aug-cc-pVDZ				
		ω_{ref}		$\Delta\omega$			ω_{ref}		$\Delta\omega$		
Molecule	st	char.	can.	BP	BP-LT	LD-LT	char.	can.	BP	BP-LT	LD-LT
"β-dipeptide"	S_1	$n \rightarrow \pi^*$	4.861	-0.030	-0.042	-0.012	$n \rightarrow \pi^*$	4.715	-0.013	-0.023	-0.005
	S_2	$n \rightarrow \pi^*$	5.825	-0.012	-0.020	0.017	$\pi \rightarrow$ Ry	5.635	0.032	-0.003	-0.033
	S_3	$\pi \rightarrow$ Ry	6.908	0.033	0.013	-0.009	$n \rightarrow \pi^*$	5.665	0.010	-0.022	-0.030
"Dipeptide"	S_1	$n \rightarrow \pi^*$	5.871	-0.019	-0.026	-0.004	$n \rightarrow \pi^*$	5.743	0.002	-0.025	-0.001
	S_2	$n \rightarrow \pi^*$	6.106	0.000	-0.019	-0.005	$n \rightarrow \pi^*$	5.953	-0.005	-0.025	-0.016
DMABN	S_1	$\pi \rightarrow \pi^*$	4.525	0.038	0.033	-0.033	$\pi \rightarrow$ Ry	4.495	0.013	0.004	-0.014
	S_2	$\pi \rightarrow \pi^*$	4.891	-0.064	-0.061	-0.055	$\pi \rightarrow$ Ry	5.085	0.026	0.023	-0.022
HPA	S_1	$\pi \rightarrow \pi^*$	4.984	0.013	0.010	-0.035	$\pi \rightarrow \pi^*$	4.816	0.005	0.013	-0.013
	S_2	$n \rightarrow \pi^*$	6.148	0.011	0.002	0.005	$\pi \rightarrow$ Ry	5.216	-0.001	0.023	-0.024
	S_3	$\pi \rightarrow \pi^*$	6.285	-0.024	-0.053	-0.024	$\pi \rightarrow$ Ry	5.726		0.037	0.006
p-cresol	S_1	$\pi \rightarrow \pi^*$	4.981	0.011	0.002	-0.042	$\pi \rightarrow$ Ry	5.145	0.005	0.018	-0.029
N-acetyl-	S_1	$n \rightarrow \pi^*$	5.862	-0.013	-0.020	-0.006	$n \rightarrow \pi^*$	5.732	0.013	-0.020	-0.005
glycine	S_2	$n \rightarrow \pi^*$	6.252	0.013	-0.003	-0.006	$n \rightarrow$ Ry	5.989		-0.023	-0.031
	S_3	$\pi \rightarrow \pi^*$	7.373	0.005	0.010	-0.005	$n \rightarrow \pi^*$	6.275	0.025	-0.032	-0.035

case we consider the lowest five excited states of the phenothiazine-isoalloxazine dyad studied recently [3], for which the initial CCS wavefunctions are entirely wrong. The previous DF-LCC2 response method fails to find the important CT state, whereas LT-DF-LCC2 indeed finds the five lowest states.

The calculations were performed with the cc-pVDZ and aug-cc-pVDZ AO basis sets, along with the related fitting basis sets optimized for DF-MP2. The LMOs were constructed according to Pipek-Mezey (PM) localization (for the aug-cc-pVDZ basis the most diffuse functions were discarded in the PM localization). The Laplace quadratures were carried out over three points, which turned out to be sufficient for quite accurate excitation energies [27]. For the LCC2 ground state calculations the pair lists were truncated at an LMO interorbital distance of 10 bohr, and the orbital domains determined according to the Boughton-Pulay (BP) procedure with a criterion of 0.98. All response calculations were carried out for pair lists denoted in Ref. [14] as $\forall(ij)$, $(im) \leq 5$,$(mn) \leq 5$, i.e., pair lists which include *all* pairs of *important orbitals* and other pairs up to an LMO interorbital distance of 5 bohr. For the DF-LCC2 calculations and the LT-DF-LCC2 calculations involving BP domains the important LMOs $i,j,...$ were specified as described in Ref. [14] with a criterion of $\kappa_e = 0.995$, while for the BP domains again a criterion of 0.98 was used. For the LT-DF-LCC2 calculations involving Laplace domains, on the other hand, the important orbitals were determined according to Ref. [27] employing a criterion of $\kappa_e = 0.999$. For the

Table 1b. Excitation energies in [eV] for different molecules and excited states. For the local calculations the deviations from the related canonical reference values $\Delta\omega = \omega_{loc} - \omega_{ref}$ are given (except explicitely stated otherwise). The labels BP and BP-LT stand for DF-LCC2 and LT-DF-LCC2 results employing BP domains, LD-LT for LT-DF-LCC2 results employing Laplace domains.

Basis		cc-pVDZ						aug-cc-pVDZ			
			ω_{ref}	$\Delta\omega$				ω_{ref}	$\Delta\omega$		
Molecule	st	char.	can.	BP	BP-LT	LD-LT	char.	can.	BP	BP-LT	LD-LT
1-phenyl-	S_1	$\pi \rightarrow \pi^*$	5.072	0.012	0.008	-0.029	$\pi \rightarrow \pi^*$	4.921	0.023	0.019	-0.020
pyrrole	S_2	$\pi \rightarrow \pi^*$	5.555	-0.002	0.006	-0.030	$\pi \rightarrow \pi^*$	5.309	0.011	0.009	-0.037
	S_3	$\pi \rightarrow \pi^*$	5.771	0.014	0.000	-0.007	$\pi \rightarrow$ Ry	5.434	0.008	0.011	-0.027
	S_4	CT	6.091	0.135	0.205	-0.001	CT	5.489	-0.004	-0.005	-0.044
Propanamide	S_1	$n \rightarrow \pi^*$	5.926	0.001	-0.032	0.003	$n \rightarrow \pi^*$	5.657	0.012	0.013	0.001
	S_2	$\pi \rightarrow \pi^*$	7.776	0.022	0.048	0.026	$\pi \rightarrow$ Ry	6.267	0.043	0.031	-0.009
Tyrosine	S_1	$\pi \rightarrow \pi^*$	4.995	0.022	0.025	-0.013	$\pi \rightarrow \pi^*$	4.834	0.013	0.029	-0.008
	S_2	$n \rightarrow \pi^*$	5.824	-0.001	-0.014	0.032	$\pi \rightarrow$ Ry	5.292	-0.014	0.036	-0.020
	S_3	$\pi \rightarrow \pi^*$	6.205	-0.003	-0.033	-0.011	$\pi \rightarrow$ Ry	5.488		0.041	0.005
trans-	S_1	$n \rightarrow \pi^*$	4.987	0.010	-0.036	0.006	$n \rightarrow \pi^*$	4.863	-0.028	-0.023	-0.034
urocanic	S_2	$\pi \rightarrow \pi^*$	5.207	-0.024	-0.021	-0.031	$\pi \rightarrow \pi^*$	4.931	-0.008	-0.011	-0.021
acid	S_3	$\pi \rightarrow \pi^*$	6.233	-	-0.057	-0.037	$\pi \rightarrow$ Ry	5.285	0.059	0.054	-0.016
	S_4	$\pi \rightarrow \pi^*$	6.269	0.033	0.062	0.021	$\pi \rightarrow \pi^*$	5.953		0.038	-0.010
	S_5	$\pi \rightarrow$ Ry	6.877	0.027	0.006	-0.017	$\pi \rightarrow$ Ry	6.009	0.031	0.036	-0.005
dyad	S_1	$\pi \rightarrow \pi^*$	3.190	-0.024	-0.29	-0.040	$\pi \rightarrow \pi^*$			3.065[a]	3.029[a]
	S_2	$n \rightarrow \pi^*$	3.412	-0.032	-0.052	-0.051	CT			3.249[a]	3.198[a]
	S_3	CT	3.513	-	-0.036	-0.071	$n \rightarrow \pi^*$			3.362[a]	3.382[a]
	S_4	$\pi \rightarrow \pi^*$	-	3.557[a]	3.541[a]	3.534[a]	$\pi \rightarrow \pi^*$				3.421[a]
	S_5	$n \rightarrow \pi^*$	3.629	-	3.678[a]	3.662[a]	$n \rightarrow \pi^*$				3.593[a]

a) Excitation energy instead of its deviation from ω_{ref}.

domains a threshold of 0.98 for important centers was used. With these specifications the sizes of the truncated pair lists and the domains were quite similar for DF-LCC2 and LT-DF-LCC2 response. All canonical reference calculations were calculated with the TURBOMOLE implementation of CC2 response [12,13]. The assignment of the individual states is based on the analysis of the related singles vectors in canonical basis.

Tables 1a/b compare the excitation energies of canonical CC2 vs. the three different local variants investigated here, i.e., DF-LCC2 and LT-DF-LCC2 employing the local approximations for excited states (involving BP domains) presented earlier [14], the latter in the context of an adaptive multi-state scheme [27], and LT-DF-LCC2 employing the newly proposed Laplace domains [27]. The deviations are all considerably below 0.1 eV with the sole exception of the CT state of 1-phenylpyrrole in the cc-pVDZ basis. For that very case LT-DF-LCC2 with Laplace domains is the only local method which remains well within this error bar. Another special case is the S_3 state of *trans*-urocanic acid, which

Table 2a. Norms (in a.u.) of the transition strength vectors $|\mathbf{S}^{0\bar{m}}| = ((S_{XX}^{0\bar{m}})^2+(S_{YY}^{0\bar{m}})^2+(S_{ZZ}^{0\bar{m}})^2)^{1/2}$. The results for the different local calculations are given as the norm of the related difference vector $|\delta\, \mathbf{S}^{0\bar{m}}| = |\, \mathbf{S}^{0\bar{m}}(\text{loc.}) - \mathbf{S}^{0\bar{m}}(\text{can.})|$. The labels BP and BP-LT stand for DF-LCC2 and LT-DF-LCC2 results employing BP domains, LD-LT for LT-DF-LCC2 results employing Laplace domains.

Basis			cc-pVDZ				aug-cc-pVDZ				
			$\lvert\mathbf{S}^{0\bar{m}}\rvert$	$\lvert\delta\,\mathbf{S}^{0\bar{m}}\rvert$				$\lvert\mathbf{S}^{0\bar{m}}\rvert$	$\lvert\delta\,\mathbf{S}^{0\bar{m}}\rvert$		
Molecule	st	char.	can.	BP	BP-LT	LD-LT	char.	can.	BP	BP-LT	LD-LT
"β-dipeptide"	S_1	$n\to\pi^*$	0.0041	0.0002	0.0001	0.0001	$n\to\pi^*$	0.0001	0.0001	0.0001	0.0002
	S_2	$n\to\pi^*$	0.0114	0.0000	0.0000	0.0008	$\pi\to$ Ry	0.1046	0.0106	0.0273	0.0005
	S_3	$\pi\to$ Ry	0.0964	0.0089	0.0046	0.0096	$n\to\pi^*$	0.0180	0.0113	0.0291	0.0002
"Dipeptide"	S_1	$n\to\pi^*$	0.0104	0.0005	0.0005	0.0002	$n\to\pi^*$	0.0092	0.0006	0.0003	0.0005
	S_2	$n\to\pi^*$	0.0068	0.0002	0.0001	0.0003	$n\to\pi^*$	0.0015	0.0001	0.0001	0.0004
DMABN	S_1	$\pi\to\pi^*$	0.3043	0.0226	0.0197	0.0085	$\pi\to$ Ry	0.2130	0.0052	0.0074	0.0036
	S_2	$\pi\to\pi^*$	5.3951	0.2645	0.2208	0.2509	$\pi\to$ Ry	0.0000	0.0000	0.0000	0.0000
HPA	S_1	$\pi\to\pi^*$	0.1903	0.0052	0.0040	0.0025	$\pi\to\pi^*$	0.1842	0.0038	0.0015	0.0048
	S_2	$n\to\pi^*$	0.0023	0.0009	0.0008	0.0009	$\pi\to$ Ry	0.0015	0.0006	0.0009	0.0007
	S_3	$\pi\to\pi^*$	1.1457	0.1238	0.0405	0.0601	$\pi\to$ Ry	0.0990		0.0045	0.0055
p-cresol	S_1	$\pi\to\pi^*$	0.2036	0.0076	0.0053	0.0051	$\pi\to$ Ry	0.0001	0.0001	0.0001	0.0001
N-acetyl-	S_1	$n\to\pi^*$	0.0112	0.0003	0.0003	0.0005	$n\to\pi^*$	0.0084	0.0003	0.0003	0.0004
glycine	S_2	$n\to\pi^*$	0.0044	0.0002	0.0000	0.0001	$n\to$ Ry	0.1329		0.0041	0.0041
	S_3	$\pi\to\pi^*$	0.2627	0.1001	0.0654	0.0229	$n\to\pi^*$	0.0000	0.0001	0.0000	0.0000

was not located with the previous DF-LCC2 approach, but now, within the framework of the new multi-state approach no longer constitutes a problem. A particular nasty example is the dyad mentioned above, where the initial CCS eigenvectors have hardly any resemblance to the converged CC2 solution. Especially, the important CT state (shifting electron density from phenothiazine to isoalloxazine) could not be located with the previous DF-LCC2 single-state approach (what was assigned in Ref. [14] as the CT state emerged later as only a partial CT state shifting electron density from phenothiazine to the linking benzene ring, i.e., state S_4 in Table 1b). The LT-DF-LCC2 multistate method, on the other hand, is able to find the CT state for the cc-pVDZ and the aug-cc-pVDZ basis, and for both domain and pair list variants. The S_4 state of the dyad was not found by the canonical reference calculation. The S_5 state, according to the local methods, carries a considerable (though not dominant) amount of CT character, which is absent in the canonical reference calculation. This is also reflected in a much smaller transition strength vector for the local case. However, the canonical ADC(2) method [34], which usually yields results rather similar to CC2 response, also features two states, i.e., S_2 and S_4, with dominant CT character. It appears that the dyad constitutes a difficult case indeed, not only for the local methods. The dyad calculation in the aug-cc-pVDZ basis involving five states takes about 219 (250) minutes CPU (elapsed) time per Davidson iteration on six cores Intel(R) Xeon(R) E5462 @ 2.80 GHz. For more detailed timings we refer to Ref. [27].

Table 2b. Norms (in a.u.) of the transition strength vectors $|\mathbf{S}^{0\bar{m}}| = ((S_{XX}^{0\bar{m}})^2+(S_{YY}^{0\bar{m}})^2+(S_{ZZ}^{0\bar{m}})^2)^{1/2}$. The results for the different local calculations are given as the norm of the related difference vector $|\delta\ \mathbf{S}^{0m}| = |\ \mathbf{S}^{0m}(\text{loc.})-\mathbf{S}^{0m}(\text{can.})|$. The labels BP and BP-LT stand for DF-LCC2 and LT-DF-LCC2 results employing BP domains, LD-LT for LT-DF-LCC2 results employing Laplace domains.

Basis		cc-pVDZ					aug-cc-pVDZ				
			$\|\mathbf{S}^{0\bar{m}}\|$	$\|\delta\ \mathbf{S}^{0\bar{m}}\|$				$\|\mathbf{S}^{0\bar{m}}\|$	$\|\delta\ \mathbf{S}^{0\bar{m}}\|$		
Molecule	st	char.	can.	BP	BP-LT	LD-LT	char.	can.	BP	BP-LT	LD-LT
1-phenyl-	S_1	$\pi \to \pi^*$	0.0235	0.0007	0.0016	0.0018	$\pi \to \pi^*$	0.0359	0.0026	0.0042	0.0015
pyrrole	S_2	$\pi \to \pi^*$	2.2771	0.0524	0.0574	0.0709	$\pi \to \pi^*$	2.1535	0.0449	0.0340	0.0253
	S_3	$\pi \to \pi^*$	0.0848	0.0191	0.0051	0.0045	$\pi \to$ Ry	0.0003	0.0055	0.0003	0.0004
	S_4	CT	0.0103	0.0017	0.0013	0.0016	CT	0.1302	0.0276	0.0210	0.0053
Propanamide	S_1	$n \to \pi^*$	0.0066	0.0010	0.0000	0.0013	$n \to \pi^*$	0.0006	0.0001	0.0001	0.0000
	S_2	$\pi \to \pi^*$	0.0363	0.0007	0.0007	0.0001	$\pi \to$ Ry	0.1217	0.0037	0.0036	0.0013
Tyrosine	S_1	$\pi \to \pi^*$	0.1654	0.0033	0.0039	0.0070	$\pi \to \pi^*$	0.1540	0.0045	0.0027	0.0086
	S_2	$n \to \pi^*$	0.0513	0.0242	0.0252	0.0088	$\pi \to$ Ry	0.0111	0.0044	0.0092	0.0087
	S_3	$\pi \to \pi^*$	1.5584	0.1178	0.0559	0.0820	$\pi \to$ Ry	0.1516	0.1474	0.0071	0.0078
trans-	S_1	$n \to \pi^*$	0.0008	0.0001	0.0001	0.0000	$n \to \pi^*$	0.0000	0.0000	0.0000	0.0000
urocanic	S_2	$\pi \to \pi^*$	5.2365	0.0724	0.0555	0.0239	$\pi \to \pi^*$	5.5697	0.0462	0.0367	0.0019
acid	S_3	$\pi \to \pi^*$	0.0067	---	0.0000	0.0002	$\pi \to$ Ry	0.0029	0.0005	0.0005	0.0002
	S_4	$\pi \to \pi^*$	0.5199	0.0029	0.0372	0.0424	$\pi \to \pi^*$	0.3640		0.0572	0.0536
	S_5	$\pi \to$ Ry	0.7529	0.0712	0.0169	0.0230	$\pi \to$ Ry	0.0305	0.0052	0.0024	0.0041
dyad	S_1	$\pi \to \pi^*$	2.3948			0.1357					
	S_2	$n \to \pi^*$	0.0325			0.0149					
	S_3	CT	0.0968			0.0118					
	S_4	$\pi \to \pi^*$	-			2.0273[a]					
	S_5	$n \to \pi^*$	1.1422			0.0853[a]					

a) Norm of transition strength vector instead of the norm of its deviation from the canonical reference vector.

Tables 2a/b compare the transition strength, computed for the test set with the three local variants, to the corresponding canonical CC2 reference results. The norm of the deviation vector, $|\delta\mathbf{S}^{0\bar{m}}| = |\ \mathbf{S}^{0\bar{m}}(\text{loc.})-\mathbf{S}^{0\bar{m}}(\text{can.})|$, is about 10% or less of the norm of the canonical reference, except for cases with tiny transition strengths. For some cases like the S_3 ($\pi \to \pi^*$) state of N-acetyl-glycine (cc-pVDZ), or the S_3 ($n \to \pi^*$) state of "β-dipeptide" (aug-cc-pVDZ) there is a marked improvement between the previous DF-LCC2 and the new LT-DF-LCC2 approach with Laplace domains.

Tables 3a/b finally present an analogous comparison for the changes in the dipole moment vectors on going from the ground- to the individual excited states. The dipole moments turn out to be more sensitive w.r. to the specification of the domains. Note that no domain extensions as in Ref. [15] are used here. Deviations of up to 20 % (or even more in a few cases) are observed. Particularly affected are states with Rydberg character. Somewhat disappointingly the La-

Table 3a. Norms (in a.u.) of the dipole difference vectors $|\mu^{0\bar{m}}|$ (excited state relative to ground state). The results for the different local calculations are given as the ratio (in %) $|\delta\,\mu^{0\bar{m}}|/|\mu^{0\bar{m}}|$ of the norm of the related difference vector (canonical minus local vector) relative to the canonical reference value. The labels BP and BP-LT stand for DF-LCC2 and LT-DF-LCC2 results employing BP domains, LD-LT for LT-DF-LCC2 results employing Laplace domains.

Basis		cc-pVDZ					aug-cc-pVDZ				
		$\|\mu^{0\bar{m}}\|$		$\|\delta\,\mu^{0\bar{m}}\|/\|\mu^{0\bar{m}}\|$			$\|\mu^{0\bar{m}}\|$		$\|\delta\,\mu^{0\bar{m}}\|/\|\mu^{0\bar{m}}\|$		
Molecule	st	char.	can.	BP	BP-LT	LD-LT	char.	can.	BP	BP-LT	LD-LT
"β-dipeptide"	S_1	$n \to \pi^*$	0.3875	1.68	1.42	22.48	$n \to \pi^*$	0.3721	7.40	5.78	28.63
	S_2	$n \to \pi^*$	0.7600	6.25	5.63	3.11	$\pi \to$ Ry	2.9470	7.26	6.84	5.36
	S_3	$\pi \to$ Ry	1.6283	2.86	0.69	9.25	$n \to \pi^*$	1.0765	7.54	25.38	5.60
"Dipeptide"	S_1	$n \to \pi^*$	0.7353	5.44	4.63	2.61	$n \to \pi^*$	0.9629	1.94	0.91	6.77
	S_2	$n \to \pi^*$	0.7401	5.74	4.25	4.29	$n \to \pi^*$	1.0227	16.58	14.50	5.90
DMABN	S_1	$\pi \to \pi^*$	0.9348	0.34	0.17	5.88	$\pi \to$ Ry	2.8072	0.61	0.32	4.04
	S_2	$\pi \to \pi^*$	2.0718	4.49	3.47	3.91	$\pi \to$ Ry	1.9141	4.21	0.03	1.79
HPA	S_1	$\pi \to \pi^*$	0.2418	8.87	7.41	2.95	$\pi \to \pi^*$	0.2030	19.91	15.04	5.43
	S_2	$n \to \pi^*$	0.6240	3.65	0.86	9.45	$\pi \to$ Ry	4.5105	3.36	3.28	5.60
	S_3	$\pi \to \pi^*$	0.8985	3.70	0.96	8.05	$\pi \to$ Ry	1.5189		30.95	31.93
p-cresol	S_1	$\pi \to \pi^*$	0.2562	9.73	6.81	4.22	$\pi \to$ Ry	4.2811	1.78	4.20	6.73
N-acetyl-	S_1	$n \to \pi^*$	0.7408	3.74	3.12	3.55	$n \to \pi^*$	0.9480	0.90	0.84	6.40
glycine	S_2	$n \to \pi^*$	0.5942	3.50	4.18	0.83	$n \to$ Ry	2.2175		0.45	3.22
	S_3	$\pi \to \pi^*$	1.9856	27.11	20.17		$n \to \pi^*$	0.5878	13.22	3.48	2.58

place domains presented here do not show an improvement over the BP domains over all, but rather the contrary. On the other hand, for quite some cases, the Laplace domains are indeed better, e.g., for the HPA, 1-phenylpyrrole, and tyrosine S_1 states, or the β-dipeptide S_3 state (aug-cc-pVDZ basis). Perhaps a merging of the two domain approaches may indeed lead to general improvement. Results from first test calculations already look quite promising.

4. Conclusions

In this contribution we compare our new local CC2 response method, which is utilizing the Laplace transform approach to decompose the energy denominators in the doubles-doubles block of the Jacobian, to the previous local CC2 response method. The new approach, in contrast to the old, works with adaptive pair lists and domains for the eigenvectors of the Jacobian (and generally, the basis vectors of the Davidson subspace), which makes it less dependent on the initial untruncated CCS wavefunction, and therefore, more robust. Two different variants of adaptive pair lists and domains are compared. Furthermore, multi-state calculations with state-specific local truncations now are possible. With the present implementation, calculations involving more than hundred atoms are routinely feasible [27]. Here we present excitation energies, transition strengths, and or-

Table 3b. Norms (in a.u.) of the dipole difference vectors $|\mu^{0\overline{m}}|$ (excited state relative to ground state). The results for the different local calculations are given as the ratio (in %) $|\delta\mu^{0\overline{m}}| / |\mu^{0\overline{m}}|$ of the norm of the related difference vector (canonical minus local vector) relative to the canonical reference value. The labels BP and BP-LT stand for DF-LCC2 and LT-DF-LCC2 results employing BP domains, LD-LT for LT-DF-LCC2 results employing Laplace domains.

Basis		cc-pVDZ					aug-cc-pVDZ				
			$\|\mu^{0\overline{m}}\|$	$\|\delta\mu^{0\overline{m}}\| / \|\mu^{0\overline{m}}\|$				$\|\mu^{0\overline{m}}\|$	$\|\delta\mu^{0\overline{m}}\| / \|\mu^{0\overline{m}}\|$		
Molecule	st	char.	can.	BP	BP-LT	LD-LT	char.	can.	BP	BP-LT	LD-LT
1-phenyl-	S_1	$\pi \rightarrow \pi^*$	0.8822	4.36	7.49	3.48	$\pi \rightarrow \pi^*$	1.0998	8.71	10.22	1.29
pyrrole	S_2	$\pi \rightarrow \pi^*$	2.3809	3.74	5.47	1.81	$\pi \rightarrow \pi^*$	2.2354	4.91	5.84	2.26
	S_3	$\pi \rightarrow \pi^*$	4.4506	1.78	0.48	0.28	$\pi \rightarrow$ Ry	0.4478	71.74	36.42	37.80
	S_4	CT	5.7148	0.65	0.75	1.34	CT	3.8053	4.14	8.29	7.26
Propana-mide	S_1	$n \rightarrow \pi^*$	0.7889	5.99	0.84	6.80	$n \rightarrow \pi^*$	1.0260	11.10	11.52	10.58
	S_2	$\pi \rightarrow \pi^*$	2.0787	2.79	2.52	16.15	$\pi \rightarrow$ Ry	3.1085	4.08	4.32	4.92
Tyrosine	S_1	$\pi \rightarrow \pi^*$	0.2217	7.31	6.64	0.69	$\pi \rightarrow \pi^*$	0.1916	6.73	10.26	1.66
	S_2	$n \rightarrow \pi^*$	0.5702	7.51	4.99	19.29	$\pi \rightarrow$ Ry	4.1572	6.63	15.37	18.21
	S_3	$\pi \rightarrow \pi^*$	1.0688	9.84	2.94	12.30	$\pi \rightarrow$ Ry	4.4276		8.04	11.62
trans-	S_1	$n \rightarrow \pi^*$	2.3098	1.12	1.50	1.87	$n \rightarrow \pi^*$	2.2847	1.49	0.57	0.45
urocanic	S_2	$\pi \rightarrow \pi^*$	2.2610	2.68	2.53	1.08	$\pi \rightarrow \pi^*$	2.2021	2.85	2.65	1.81
acid	S_3	$\pi \rightarrow \pi^*$	2.9996	-	0.93	1.73	$\pi \rightarrow$ Ry	5.7190	2.74	2.78	1.12
	S_4	$\pi \rightarrow \pi^*$	0.4262	5.97	2.35	14.45	$\pi \rightarrow \pi^*$	0.8143		6.33	28.47
	S_5	$\pi \rightarrow$ Ry	0.1163	282.60	37.36	31.79	$\pi \rightarrow$ Ry	3.1825	8.48	3.36	3.27
dyad	S_1	$\pi \rightarrow \pi^*$	1.4090			8.59					
	S_2	$n \rightarrow \pi^*$	4.1830			17.27					
	S_3	CT	8.3233			9.80					
	S_4	$\pi \rightarrow \pi^*$	-			2.4015[a]					
	S_5	$n \rightarrow \pi^*$	2.9833			3.2401[a]					

a) Norm of dipole difference vector instead of the norm of its deviation from the canonical reference vector.

bital-unrelaxed dipole moments, obtained for a set of different molecules and excited states. While the results obtained for excitation energies and transition strengths agree reasonably well with the canonical reference values, the situation is not yet satisfying for the dipole moments. Presently, work is in progress for improvements along this direction. A generalization to orbital-relaxed properties (and finally analytical gradients w.r. to nuclear displacements) generally is possible for Laplace transformed methods [29,35]. A further discussion on orbital-relaxed properties in the context of LT-DF-LCC2 response is deferred to a forthcoming paper.

Acknowledgement

Financial support from the Deutsche Forschungsgemeinschaft DFG (priority program SPP1145) is gratefully acknowledged.

References

1. A. Dreuw and M. Head-Gordon, Chem. Rev. **105** (2005) 4009.
2. S. Grimme and M. Parac, ChemPhysChem **3** (2003) 292.
3. K. Sadeghian and M. Schütz, J. Am. Chem. Soc. **129** (2007) 4068.
4. O. Christiansen, P. Jørgensen, and C. Hättig, Int. J. Quantum Chem. **68** (1998) 1.
5. O. Christiansen, H. Koch, and P. Jørgensen, Chem. Phys. Lett. **243** (1995) 409.
6. M. Schreiber, M.R. Silva-Junior, S.P.A. Sauer, and W. Thiel, J. Chem. Phys. **128** (2008) 134110.
7. E. J. Baerends, D. E. Ellis, and P. Ros, Chem. Phys. **2** (1973) 41.
8. J. L. Whitten, J. Chem. Phys. **58** (1973) 4496.
9. B. I. Dunlap, J. W. D. Connolly, and J. R. Sabin, J. Chem. Phys. **71** (1979) 3396.
10. M. Häser and J. Almlöf, J. Chem. Phys. **96** (1992) 489.
11. M. Feyereisen, G. Fitzgerald, and A. Komornicki, Chem. Phys. Letters **208** (1993) 359.
12. C. Hättig and F. Weigend, J. Chem. Phys. **113** (2000) 5154.
13. C. Hättig and A. Köhn, J. Chem. Phys. **117** (2002) 6939.
14. D. Kats, T. Korona, and M. Schütz, J. Chem. Phys. **125** (2006) 104106.
15. D. Kats, T. Korona, and M. Schütz, J. Chem. Phys. **127** (2007) 064107.
16. S. Saebø and P. Pulay, Annu. Rev. Phys. Chem. **44** (1993) 213.
17. P. Pulay, Chem. Phys. Lett. **100** (1983) 151.
18. C. Hampel and H.-J. Werner, J. Chem. Phys. **104** (1996) 6286.
19. M. Schütz, G. Hetzer, and H.-J. Werner, J. Chem. Phys. **111** (1999) 5691.
20. M. Schütz and H.-J. Werner, J. Chem. Phys. **114** (2001) 661.
21. M. Schütz, J. Chem. Phys. **113** (2000) 9986.
22. M. Schütz, J. Chem. Phys. **116** (2002) 8772.
23. M. Schütz and F. R. Manby, Phys. Chem. Chem. Phys. **5** (2003) 3349.
24. M. Schütz, H.-J. Werner, R. Lindh, and F. R. Manby, J. Chem. Phys. **121** (2004) 737.
25. C. Pisani, L. Maschio, S. Cassassa, M. Halo, M. Schütz, and D. Usvyat, J. Comput. Chem. **29** (2008) 2113.
26. T.B. Adler, H.-J. Werner, and F. R. Manby, J. Chem. Phys. **130** (2009) 054106.
27. D. Kats and M. Schütz, J. Chem. Phys. **131** (2009) 124117.
28. J. Almlöf, Chem. Phys. Letters **181** (1991) 319.
29. M. Häser, Theor. Chim. Acta **87** (1993) 147.
30. E. R. Davidson, J. Comp. Phys. **17** (1975) 87.
31. K. Hirao and H. Nakatsuji, J. Comp. Phys. **45** (1982) 246.
32. D. Kats, D. Usvyat, and M. Schütz, Phys. Chem. Chem. Phys. **10** (2008) 3430.
33. H.-J. Werner, P. J. Knowles, R. Lindh, F. R. Manby, M. Schütz et al, *Molpro*, version 2008.3, a package of ab initio programs, (2009), see http://www.molpro.net.
34. A.B. Trofimov and J. Schirmer, J. Phys. B-At. Mol. Opt. **28** (1995) 2299–2324.
35. S. Schweizer, B. Doser, and C. Ochsenfeld, J. Chem. Phys. **128** (2008) 154101.
36. T. Korona, H.-J. Werner, J. Chem. Phys. **118** (2003) 3006.

Extended Systems in Electrostatic Fields

By Michael Springborg[1,*], Violina Tevekeliyska[1], Bernard Kirtman[2], Benoît Champagne[3], and Yi Dong[1]

[1] Physical and Theoretical Chemistry, University of Saarland, 66123 Saarbrücken, Germany
[2] Department of Chemistry and Biochemistry, University of California, 93106 Santa Barbara, California, U.S.A.
[3] Départment de Chimie, Facultés Universitaires Notre-Dame de la Paix, 5000 Namur, Belgium

(Received July 30, 2009; accepted September 7, 2009)

Electrostatic Fields / Periodic Systems / Chain Compounds / Polarization / Hartree-Fock / Kohn-Sham

The basic principles behind a theoretical method for treating infinite, periodic systems exposed to an external electrostatic field are outlined. The approach, based on the vector-potential description of the external field, leads to single-particle Hartree-Fock or Kohn-Sham equations that differ from the field-free counterparts in several aspects. In particular, solving them is only possible through a careful so-called smoothing procedure. In that case it is possible to derive a numerically stable and efficient approach. Results of model studies as well as of the first *ab initio* calculations are reported in order to illustrate the approach. Finally, extensions and open issues are briefly discussed.

1. Introduction

The fact that materials respond in a characteristic way to external electromagnetic fields forms the basis of a large variety of spectroscopic methods for characterizing the materials. Moreover, the responses can also be exploited in different technological applications. In both cases, a detailed understanding and description of the interactions between the fields and the materials are mandatory. It may therefore surprise that even the simple case of an electrostatic field interacting with an extended system poses a number of fundamental theoretical, conceptual, and methodological challenges that only partially have been met satisfactorily. These have also been the subject of a still on-going project whose results shall be reviewed here [1–7].

* Corresponding author. E-mail: m.springborg@mx.uni-saarland.de

Z. Phys. Chem. **224** (2010) 617–630
© by Oldenbourg Wissenschaftsverlag, München

Specifically, we shall consider systems that are extended and regular. As regular systems we will consider those that, except for near the surfaces, consist of a large number of identical units. Extended systems are those that contain a central region, far from any surface, where the presence of the surfaces is not felt by the electrons. In theoretical studies extended and regular systems are most conveniently modelled as being infinite and periodic. In most studies of the properties of such materials, the only consequence of this approximation is a very large simplification in the calculations, but exactly when studying the responses to electromagnetic fields, unexpected complications show up that have to be treated with care. These are the issue of this presentation. For the case of simplicity we shall consider solely systems that are extended in one dimension and finite in the other two, but the generalization to systems that are extended in more dimensions should be clear.

2. Fundamentals

The presence of an external electric field can be described in two different ways, i.e., either through a scalar potential or through a vector potential. Ultimately, we shall base our discussion on the vector-potential formulation, but at first we shall illustrate some of the conceptual problems by looking at the scalar-potential formulation.

For a long, but finite, regular chain in the presence of the electrostatic field the total energy is given as

$$\mathcal{E}_{\text{tot}} = \langle\Psi|\hat{H}_{\text{tot}}|\Psi\rangle = \langle\Psi|\hat{H}_{\text{tot},0}|\Psi\rangle + \vec{E}_{\text{DC}} \cdot \vec{\mu}_{\text{tot}}. \tag{1}$$

Here, $\hat{H}_{\text{tot}}$ and $\hat{H}_{\text{tot},0}$ is the total Hamilton operator of the electrons and the nuclei of our system with and without the external field, respectively. Moreover, $|\vec{E}_{\text{DC}}|$ is the strength of the field and

$$\vec{\mu}_{\text{tot}} = \int \vec{r}\rho_{\text{tot}}(\vec{r})d\vec{r} = \vec{\mu}_n - \vec{\mu}_e \tag{2}$$

is the dipole moment of the system that has been split into a nuclear and an electronic part. Also the total charge density,

$$\rho_{\text{tot}}(\vec{r}) = \rho_n(\vec{r}) - \rho_e(\vec{r}), \tag{3}$$

is split into a nuclear and an electronic part.

In the thermodynamic limit for the long, regular chains of our interest, the first term on the right-hand side of Eq. (1) will be determined from the properties of the central region. It is, however, not obvious that this also is the case for the second term. It can be shown [7] that the units of the central region are neutral, whereas those near the terminations may be charged (in the neutral case, the charges at the two terminations will be identical but of opposite sign). Due to these charges, the dipole moment contains a contribution from the terminations that grows with size of the system and, therefore, does not become increasingly negligible in the thermodynamic limit. This points to one difficulty when approximating the system as being infinite and periodic: in that case the surfaces are removed, so that the part of the dipole moment from the charges at the terminations has to be included in a different way.

In the present work we shall assume that the Born-Oppenheimer approximation is valid and, accordingly, split the total energy into an electronic and a nuclear part.

$$\mathcal{E}_{\text{tot}} = \mathcal{E}_e + \mathcal{E}_n. \tag{4}$$

The nuclear part contains the electrostatic interactions among the nuclei and the interactions between the nuclei and the external field. On the other hand, the electronic part will be calculated using a single-determinant (i.e., Hartree-Fock) wavefunction for the electrons, and contains the kinetic energy of the electrons, their mutual electrostatic interactions, the interactions between the nuclei and the electrons, and the interactions between the electrons and the external field,

$$\mathcal{E}_e = \langle\Psi_e|\hat{H}_{e,0}|\Psi_e\rangle - \vec{E}_{\text{DC}} \cdot \vec{\mu}_e = \langle\Psi_e|\hat{H}_{e,0} - \vec{E}_{\text{DC}} \cdot \sum_i \vec{r}_i|\Psi_e\rangle = \langle\Psi_e|\hat{H}_e|\Psi_e\rangle. \tag{5}$$

Here, $\hat{H}_{e,0}$ is the field-free part of the electronic Hamiltonian, $\vec{r}_i$ is the position of the ith electron, and we have set $|e| = 1$. We add that similar arguments can be carried through for approaches based on the density-functional theory in the Kohn-Sham formulation.

For the long, but finite, chains the single-particle orbitals that form the many-particle Slater determinant satisfy the single-particle equations

$$\left[\hat{F} - \vec{E}_{\text{DC}} \cdot \vec{r}\right]\psi_j(\vec{r}) = \epsilon_j\psi_j(\vec{r}), \tag{6}$$

where $\hat{F}$ is the field-free single-particle operator. This equation points to the second fundamental difficulty when approximating the system as being infinite and periodic: the potential felt by the particles is unbounded and destroys the translational symmetry.

It is, however, possible to deal with both difficulties. At first, it can be shown [6,8,9] that, in the thermodynamic limit, the dipole moment per unit is determined from properties of the central region, i.e., largely independent of the terminations. Second, as we shall see below, it has become possible to derive the equations for particles of an infinite, periodic system that experience an external electrostatic field, although it turns out that it is highly non-trivial to solve these equations.

For an infinite, periodic system it is most natural to study the various properties (e.g., total energy and dipole moment) per repeated unit. In that case, the dipole moment translates into the polarization, $\vec{P}$. For the sake of simplicity we shall assume that the chain is parallel to the z axis and that also the field is parallel to this axis. We shall therefore only discuss the z component of the polarization and, thus, omit the vector symbol.

In the absence of the external field, infinite and periodic systems are most conveniently treated by making explicit use of the translational symmetry, whereby the electronic orbitals can be written as

$$\psi_j(k,\vec{r}) = e^{ikz}u_j(k,\vec{r}), \tag{7}$$

with j being a band index and $u_j(k,\vec{r})$ a function that has the lattice periodicity. In a typical calculation a finite set of K equidistant k points in the interval $[-\pi/a;\pi/a]$ is employed. Here a is the lattice constant, and the k spacing is

$$\Delta k = \frac{2\pi}{Ka}. \tag{8}$$

It is common practice to approximate the orbitals through linear combinations of pre-chosen basis functions; here we shall use atom-centered basis functions. Then,

$$\begin{aligned} \psi_j(k,\vec{r}) &= \sum_p C_{pj}(k)\chi_p(k,\vec{r}), \\ \chi_p(k,\vec{r}) &= \frac{1}{\sqrt{K}}\sum_l e^{ikal}\chi_{lp}(\vec{r}) \end{aligned} \tag{9}$$

with χ_{lp} being the pth basis function of the lth unit.

About 15 years ago, Resta and Vanderbilt [10–12] proposed various forms for the electronic part of the polarization. One of those is

$$P_{\mathrm{R}} = \frac{a}{\pi}\mathrm{Im}\ \ln\ \det \underline{\underline{S}}^{+} = -\frac{a}{\pi}\mathrm{Im}\ \ln\ \det \underline{\underline{S}}^{-}, \tag{10}$$

where the $(NK)\times(NK)$ $S^{\pm}$ matrices contain elements of the type

$$S^{\pm}_{j_1 j_2}(k_1,k_2) = \langle\psi_{j_1}(k_1,\vec{r})|e^{\pm i\Delta kz}|\psi_{j_2}(k_2,\vec{r})\rangle. \tag{11}$$

$N = 2B$ is the number of electrons per repeated unit, with B being the number of doubly occupied bands (we assume that there is a gap between occupied and empty orbitals). An alternative form was also proposed,

$$P_{\mathrm{KSV}} = \frac{2i}{K}\sum_{k=1}^{K}\sum_{j=1}^{B}\langle u_j(k)|\frac{\partial}{\partial k}u_j(k)\rangle. \tag{12}$$

First about one decade later we were able to demonstrate how one may derive these expressions [1]. Since this derivation is non-trivial, it shall not be repeated here.

Using Eq. (9) we may write P_{KSV} as

$$\begin{aligned} P_{\mathrm{KSV}} &= \frac{2}{K}\sum_{j=1}^{B}\sum_{k=1}^{K}\sum_{l} e^{ikal}\sum_{pq}|C_{qj}(k)|^2\langle\chi_{0q}|z-la|\chi_{lp}\rangle \\ &\quad+\frac{2i}{K}\sum_{j=1}^{B}\sum_{k=1}^{K}\sum_{l} e^{ikal}\sum_{pq}C^{*}_{qj}(k)\langle\chi_{0q}|\chi_{lp}\rangle\frac{d}{dk}C_{pj}(k) \\ &\equiv P_{\rho}+P_{I}. \end{aligned} \tag{13}$$

This relation expresses the polarization as a sum of charge and current contributions. P_{ρ} is the expectation value of a potential that is piecewise linear and has the periodicity of the lattice, whereas P_I is not the expectation value of a potential since it involves the d/dk operator. P_I is related to the contribution from the terminations to the dipole moment of the finite chains and its existence is, accordingly, a consequence of approximating the system as being infinite and periodic.

From these expressions it is possible to calculate the electronic part of the static polarization in the absence of an external electrostatic field. It is seen that P_{R} has contributions from couplings between orbitals of neighbouring k points, whereas P_{KSV} involves a differentiation with respect to k, making its evaluation numerically demanding, but it is diagonal in k.

In the presence of the external field we may calculate the total energy per unit cell using the expression

$$\bar{\mathcal{E}}_e = \langle \Psi_e | \hat{H}_{e,0} | \Psi_e \rangle - E_{\mathrm{DC}} \cdot P_e \tag{14}$$

which is similar to that of Eq. (5), but with $\hat{H}_{e,0}$ being the Hamilton operator for the electrons of one unit cell and with P_e being one of the polarization expressions of Eqs. (10) or (12). Moreover, it equals the expression that has been used within the so called Modern Theory of Polarization [13–15].

In order to minimize $\bar{\mathcal{E}}_e$ we have derived an efficient approach [5] differing from the approaches of [14,15]. We minimize $\bar{\mathcal{E}}_e$ by varying the orbitals $\{\psi_j\}$, i.e., by varying the coefficients $C_{pj}(k)$ of Eq. (9). With P_{R} one arrives then at a matrix eigenvalue problem for which the part that originates from P_{R} depends in a highly non-linear way on the solutions to the equations. Moreover, the size of the eigenvalue problem is that of the Born-von-Kármán zone, i.e., that of K units. Instead, we have found that the equations that are obtained by using P_{KSV} are computationally easier to deal with [3,5]. These were found to be identical to the equations that were obtained earlier using the vector-potential approach [16,17] (which, accordingly, also could have been called the scalar-potential approach). In that case, one considers the time-dependent single-particle equations for the electrons moving in a time-dependent external electromagnetic field whose presence is described through a vector potential (in contrast to the scalar potential above). For a periodic system and considering, ultimately, the static limit one arrives at

$$\left\{\underline{\underline{F}}(k) - E_{\mathrm{DC}} \cdot \left[\underline{\underline{M}}(k) + i\underline{\underline{S}}(k)\frac{\partial}{\partial k}\right]\right\} \cdot \underline{C}_j(k) = \epsilon_j(k) \cdot \underline{\underline{S}}(k) \cdot \underline{C}_j(k) \tag{15}$$

where $C_j(k)$ is the (field-dependent) jth column of the matrix $C(k)$, and

$$\begin{aligned} S_{qp}(k) &= \sum_l e^{ikal} \langle \chi_{0q} | \chi_{lp} \rangle \\ M_{qp}(k) &= \sum_l e^{ikal} \langle \chi_{0q} | z - la | \chi_{lp} \rangle = \sum_l e^{-ikal} \langle \chi_{lq} | z | \chi_{0p} \rangle \\ F_{qp}(k) &= \sum_l e^{ikal} \langle \chi_{0q} | \hat{F} | \chi_{lp} \rangle \end{aligned} \tag{16}$$

are the overlap, unit cell dipole, and Fock or Kohn-Sham matrix elements, respectively.

An important point is that Eq. (15) possesses the lattice periodicity. However, because of the $\partial/\partial k$ term, Eq. (15) cannot directly be solved using standard approaches. This difficulty can be overcome by using

$$\underline{\underline{1}} = \underline{\underline{C}}^{\dagger}(k) \cdot \underline{\underline{S}}(k) \cdot \underline{\underline{C}}(k) \tag{17}$$

to obtain

$$\underline{\underline{S}}(k)\frac{\partial}{\partial k}\underline{\underline{C}}(k) = \left[\underline{\underline{S}}(k)\left(\frac{\partial}{\partial k}\underline{\underline{C}}(k)\right)\underline{\underline{C}}^{\dagger}(k)\underline{\underline{S}}(k)\right]\underline{\underline{C}}(k). \tag{18}$$

Then, just as the F matrix, the expression in the square brackets can be treated self-consistently. However, since the coefficients $C(k)$ are obtained numerically for each k point separately, they may possess phase factors that vary irregularly from one k point to the next, making the numerical evaluation of the expression in the square bracket in Eq. (18) difficult. To solve this problem we have proposed a smoothing procedure [3,5] that in numerous test calculations have proved to be numerically stable. We shall here not discuss the details of this approach but only give the main idea. For more details, the interested reader may consult refs. [3,5].

The smoothing procedure is based on adding extra phase factors to the orbital expansion coefficients,

$$C_{qj}(k) \to C_{qj}(k)e^{i\tilde{\phi}_j(k)}. \tag{19}$$

In a multi-step procedure, the phases are optimized by minimizing (for each band, j, separately)

$$\begin{aligned} Q_j \;=\; & \sum_k \sum_q \left| C_{qj}(k+\Delta k)e^{i\tilde{\phi}_j(k+\Delta k)} - C_{qj}(k)e^{i\tilde{\phi}_j(k)} \right|^2 \\ & + \lambda \sum_k \sum_q \left| C_{qj}(k+2\Delta k)e^{i\tilde{\phi}_j(k+2\Delta k)} - C_{qj}(k)e^{i\tilde{\phi}_j(k)} \right|^2 \end{aligned} \tag{20}$$

under the constraint

$$\tilde{\phi}_j(-\frac{\pi}{a}) = \tilde{\phi}_j(\frac{\pi}{a}) \tag{21}$$

and with $\tilde{\phi}_j(0)$ fixed. In numerical studies we found $\lambda = 0.1$ to be reasonable, meaning that the numerical differentiation of $C_{qj}(k)$ with respect to k is numerically stable. It is important to add that for $E_{\mathrm{DC}} \neq 0$ a parameter that quantifies the convergence in the iterative procedure of solving the Hartree-Fock or Kohn-Sham equations (defined, e.g., as the difference in input and output electron densities) cannot be made arbitrarily small. Instead, for increasing value of $|E_{\mathrm{DC}}|$ the smallest value of such a parameter increases. By modifying the smoothing procedure we found that the major effect was to change the range of values of E_{DC} for which the calculations could be brought to convergence. Converged calculations gave in all cases the same results, independent of the details of the smoothing procedure.

Subsequently, the derivatives of the orbital expansion coefficients with respect to k are calculated numerically using

$$\frac{\partial C_{qj}(k)}{\partial k} \simeq \frac{1}{\Delta k} \sum_{n=1}^{N_k} w_{n,N_k} \left[C_{qj}(k + n \cdot \Delta k) - C_{qj}(k - n \cdot \Delta k) \right] \tag{22}$$

with

$$C_{qj}(k + \frac{2\pi}{a}) = C_{qj}(k). \tag{23}$$

The coefficients $\{w_{n,Nk}\}$ are taken from Dvornikov [18].

The success of the smoothing procedure and the accuracy of the subsequent numerical evaluation of the derivatives are of paramount importance for the numerical stability of our computational approach. Therefore, the testing and formulation of these have been major challenges in the development of our method.

As mentioned above, the single-particle equations (15) can also be obtained from the scalar-potential description of the external field, whereby $P_{\rm KSV}$ is used as the expression for the polarization. The total energy can accordingly be written as in Eqs. (1) and (4). We separate $\hat{H}_{e,0}$ into one- and two-electron terms,

$$\hat{H}_{e,0} = \sum_i \hat{h}_0(\vec{r}_i) + \frac{1}{2} \sum_{i \neq j} \hat{v}(\vec{r}_i, \vec{r}_j). \tag{24}$$

Introducing the matrix elements

$$\begin{aligned} h_{0qp}(k) &= \sum_l e^{ikal} \langle \chi_{0q} | \hat{h}_0 | \chi_{lp} \rangle \\ v_{q_1q_2p_1p_2}(k_1, k_2) &= \frac{1}{K} \sum_{l_1,l_2,m_2} e^{ia(k_1l_1 + k_2l_2 - k_2m_2)} \\ &\quad \times \Big(\langle \chi_{0q_1} \chi_{m_2q_2} | \hat{v} | \chi_{l_1p_1} \chi_{l_2p_2} \rangle \\ &\quad - \langle \chi_{0q_1} \chi_{m_2q_2} | \hat{v} | \chi_{l_2p_2} \chi_{l_1p_1} \rangle \Big), \end{aligned} \tag{25}$$

$\mathcal{E}_e$ (for a given structure in the presence of an electrostatic field) can then be written as

$$\begin{aligned} \mathcal{E}_e &= \sum_k \sum_{pq} D_{qp}(k) h_{0qp}(k) \\ &\quad + \frac{1}{2} \sum_{k_1k_2} \sum_{p_1p_2q_1q_2} D_{q_1p_1}(k_1) D_{q_2p_2}(k_2) v_{q_1q_2p_1p_2}(k_1, k_2) \\ &\quad - E_{\rm DC} K P_{\rm KSV}, \end{aligned} \tag{26}$$

where $D_{qp}(k)$ is the (k-dependent) density matrix,

$$D_{qp}(k) = \sum_j C_{qj}(k) C^*_{pj}(k). \tag{27}$$

We have here not assumed spin degeneracy, so that the j summations are over all occupied spin-orbitals.

Of significant practical importance is that it has been possible [5,7] to obtain closed expressions for the derivatives of $\mathcal{E}_e$ with respect to, e.g., nuclear coordinates. Since this also is possible for $\mathcal{E}_n$, it becomes thereby possible to perform automatic structure optimizations. For the sake of completeness we shall here give the final expression for the derivatives of $\mathcal{E}_e$,

$$\begin{aligned}
\frac{\partial \mathcal{E}_e}{\partial x} &= \sum_k \sum_{pq} D_{qp}(k) \frac{\partial h_{0qp}(k)}{\partial x} \\
&+ \frac{1}{2} \sum_{k_1 k_2} \sum_{p_1 p_2 q_1 q_2} D_{q_1 p_1}(k_1) D_{q_2 p_2}(k_2) \frac{\partial v_{q_1 q_2 p_1 p_2}(k_1, k_2)}{\partial x} \\
&- \sum_k \sum_{pq} Q_{pq}(k) \frac{\partial S_{qp}(k)}{\partial x} \\
&- \frac{E_{\mathrm{DC}}}{2} i \sum_k \sum_{pq} [R_{qp}(k) - R^*_{pq}(k)] \frac{\partial S_{qp}(k)}{\partial x} \\
&- \frac{E_{\mathrm{DC}}}{2} \sum_k \sum_{pq} D_{qp}(k) \left(\frac{\partial M_{qp}(k)}{\partial x} + \frac{\partial M^*_{pq}(k)}{\partial x} \right)
\end{aligned} \tag{28}$$

in which

$$\begin{aligned}
R_{qp}(k) &= \sum_j C^*_{qj}(k) \frac{\partial C_{pj}(k)}{\partial k}, \\
Q_{qp}(k) &= \sum_j \epsilon_j(k) C^*_{qj}(k) C_{pj}(k)
\end{aligned} \tag{29}$$

have been introduced.

3. Model calculations

Ultimately, the goal is to be able to calculate the (electronic and structural) responses of extended materials to electrostatic fields using accurate, parameter-free, electronic-structure methods. However, in order to develop, check, and understand the theoretical methodologies outlined above, we have found it extremely useful to apply the approach to simple model systems. By constructing the model carefully it is possible to simulate a realistic electronic-structure calculation and perform many calculations both on large, finite systems and on infinite, periodic systems using the same model. Thereby, it could be verified that our approach for the infinite, periodic systems is able to reproduce the results for the large, finite systems. In this section we shall briefly present typical results of such model calculations. We consider a linear -A = B- chain, i.e., a chain with alternating atoms and alternating bond lengths. For the infinite, periodic system, two parameters describe the structure, i.e., the lattice constant, a, and a parameter u_0 that describes the bond-length alternation, so that alternating atoms

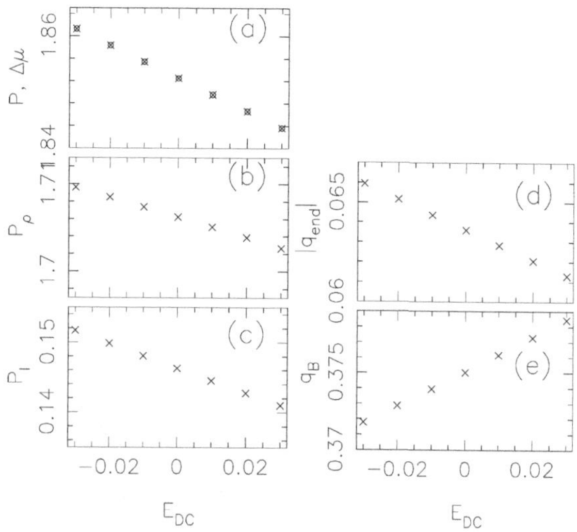

Fig. 1. Results of model calculations. The panels show (a) the polarization from finite (crosses) and periodic chain (circles) calculations, (b) P_ρ and (c) P_l from periodic chain calculations, (d) the charge of the terminations in the finite chain calculations (calculated by adding the Mulliken populations of all atoms of the left or right halfpart of the chain), and (e) the number of electrons on one of the two atoms in the central region. All quantities are shown as a function of E_{DC} for a fixed structure.

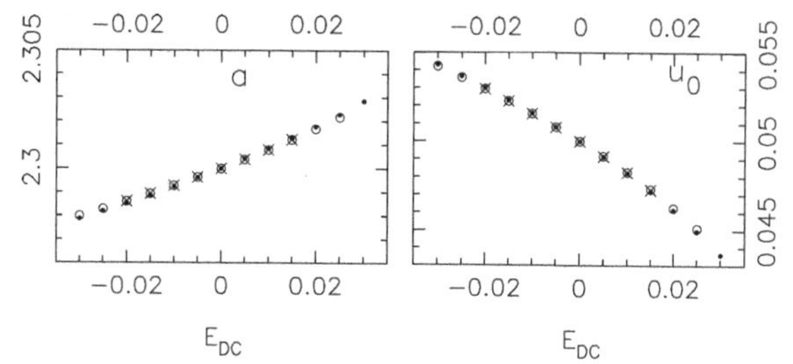

Fig. 2. The variation in the structure as a function of E_{DC}. The filled circles show results from periodic-chain calculations with $K = 51$ k points and the open circles and crosses show results from finite-chain calculations with (open circles) 30 or (crosses) 40 units. For the largest field strengths, Zener tunneling occurred for the finite chains.

are displaced by $+u_0$ and $-u_0$ away from the equidistant positions. Thus, the two bond lengths have the values $a/2-2u_0$ and $a/2+2u_0$.

In order to calculate the electronic properties we use a Hartree-Fock approximation. The system has 4 electrons per repeated unit, and the nuclear charges are $2|e|$. We use a basis set of orthonormal atom-centered functions with two functions per atom. The one-electron matrix element $\langle \chi_{mq}|\hat{h}_0|\chi_{lp}\rangle$ is assumed to be non-zero only for $(q,m) = (p,l)$ (in which case it is a constant) and for $m =$

$l\pm1$ (in which case it depends linearly on the interatomic distance). Of the two-electron matrix elements, only $\langle\chi_{lq}\chi_{lq}|\hat{v}|\chi_{lq}\chi_{lq}\rangle$ is assumed to be non-zero. Finally, $\langle\chi_{mq}|z|\chi_{lp}\rangle = \delta_{m,l}\delta_{q,p}z_{mq}$, where z_{mq} is the z coordinate of the atom where the function is centered.

In addition we include an elastic term in the nuclear energy which contains 2nd and 4th order terms in nearest- and next-nearest-neighbour bond lengths. The (four) constants entering this are determined by requiring that the structure of the infinite, periodic system in the absence of the electrostatic field has certain pre-chosen values for a and u_0.

In the first set of calculations we fixed the structure and calculated the electronic response to the external electrostatic field. We calculated the dipole moment for chains with 20 and 21 units and used the difference as an estimate for the polarization for the infinite, periodic system, which, in turn, was treated using $K = 51$ k points. The results are shown in Fig. 1. The agreement between the finite- and the infinite-chain values is perfect. Splitting the polarization into the charge and the current parts we see that, in the present case, they show a comparable variation as a function of E_{DC}. The parameter values of the model for the present calculations were chosen so that the system is strongly ionic. For more covalent systems we have found that the current term is the one showing the largest variation with E_{DC}. For the finite chains, we can study the total charge at the terminations and compare this with the charge distribution in the central part of the chain. The latter is, per definition, identical to the one obtained in the infinite-chain calculations. Both are shown in Fig. 1, too, where it is seen that these also show a variation with E_{DC} that is similar for the two. Notice, that the part of the dipole moment that is associated with the charge at the terminations for the finite chains is translated into the current term for the infinite, periodic chain.

In Fig. 2 we show that our approach also can be used in determining the structural responses to the electrostatic field. This figure shows the variation in the two structural parameters, a and u_0, as functions of E_{DC}. We have also included results from finite-chain calculations, whereby the structural parameters were extracted from middle bonds of a long chain with 30 or 40 AB units. Only for such fairly long chains are the calculations converged with respect to system size. Thus, calculations using the vector-potential approach constitute a highly attractive alternative. In the figure it is also interesting that the structural parameters depend non-linearly on the E_{DC}.

4. Ab initio calculations

As a first step towards a full *ab initio* treatment of periodic systems in external electrostatic fields, we have implemented our approach in the PLH code [19] that has been developed in Namur, Belgium. With this *ab initio* program pack-

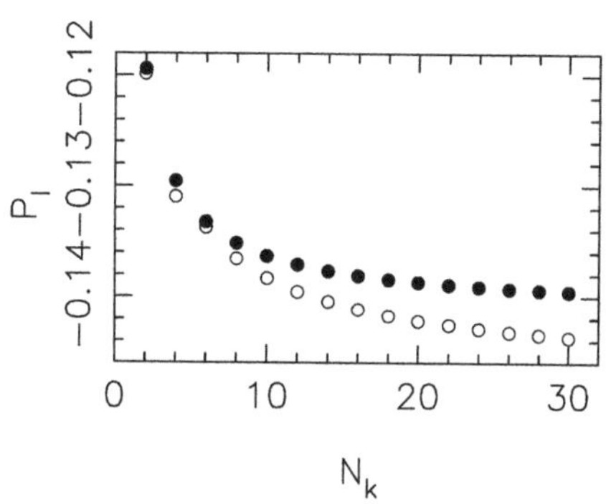

Fig. 3. P_I as a function of the number of k points used in the numerical differentiation, N_k without (filled circles) and with (open circles) the inclusion of an external electrostatic field (whose value is $E_{DC} = 0.0004$ a.u.).

age, Hartree-Fock calculations on quasi-one-dimensional systems can be carried through, using a basis set of Gaussians.

We studied a linear chain of H atoms with alternating bond lengths of 2 and 3 a.u. $K = 128$ equidistant k points were used in the k space sampling. As mentioned in Sec. 2, the critical step in our approach is the determination of current contribution P_I to the polarization and the corresponding terms in the single-particle Hartree-Fock equations. Both involve the smoothing and the numerical differentiation of the orbital expansion coefficients. In the absence of the external field, this affects `only' the final determination of P_I at the end of the iterative approach of solving the Hartree-Fock equations. On the other hand, for $E_{DC} \neq 0$ the Hartree-Fock equations themselves are affected by the smoothing and differentiation and, accordingly, P_I becomes much more sensitive to numerical noise related to the smoothing and differentiation.

In order to analyse this in more detail, we show in Fig. 3 the variation of P_I with the number N_k of k points that is used in the numerical differentiation [cf. Eq. (22)]. Of symmetry reasons, the total polarization vanishes for $E_{DC} = 0$, but this is not the case for the individual components, which in addition may depend on the origin of the coordinate system and of the choice of the unit cell. Therefore, when analysing the individual components like P_I, only their variation is of relevance.

From the figure we see that our approach leads to stable results that, due to the smoothing, are smooth functions of N_k and show a clear convergence behaviour. When comparing with our experience from the model calculations, the *ab initio* calculations seem to require a slightly larger number of k points both in the k space sampling (i.e., K) and in the numerical differentiation (i.e., N_k). We finally mention that the computational requirements for a calculation with a non-vanishing DC field were found to be of the order of 50% larger than those for a field-free calculation. Further results of the *ab initio* calculations on the linear hydrogen chains and on other systems will be presented elsewhere.

5. Concluding remarks

As a first step towards a general methodology for the theoretical treatment of extended systems exposed to electromagnetic fields we have here presented our approach for the case of electrostatic fields. Our approach can most conveniently be formulated within the vector-potential description of the external field, which provides a natural starting point when passing to time-dependent and/or magnetic fields. In the static limit we found that when the systems of our interest are approximated as being infinite and periodic, the operator for the polarization differs significantly from its analogue for finite systems. Therefore, the single-particle equations that describe the movement of the electrons in the external field are fundamentally different from those of the finite systems and it was necessary to develop special techniques to solve them. We have demonstrated in this short review that we have been able to develop such a method that, moreover, is numerically stable and efficient.

In order to demonstrate the applicability of our approach as well as to test and improve it we have found it extremely useful to focus on model studies of quasi-one-dimensional systems. We have presented here some results of such studies. However, ultimately the goal is to extend the approach to *ab initio* studies of real materials. Here, we have presented our first results in this direction. Although these results are primarily of academic interest, in the future we will apply our approach to more realistic systems. Moreover, extensions to higher dimensions will also be presented, although certain methodological problems related to the smoothing procedure in $\vec{k}$ space then have to be solved. Future developments include also the extensions to time-dependent and magnetic fields.

There are, however, still some fundamental issues related to the treatment of electrostatic fields that deserve some further attention and, therefore, briefly shall be mentioned here. At first, the phases of the wavefunctions, that usually are of minor or no importance, become important. The above-mentioned smoothing procedure is in fact a procedure for obtaining phases that lead to numerically stable calculations. However, by changing the orbital wavefunctions according to

$$\psi_j(k,\vec{r}) \rightarrow e^{i\phi_j(k)}\psi_j(k,\vec{r}), \tag{30}$$

the single-particle eigenvalues change according to

$$\epsilon_j(k) \rightarrow \epsilon_j(k) + E_{\text{DC}}\frac{d}{dk}\phi_j(k), \tag{31}$$

which means that the band structures are not unique and an interpretation based on the theorems of Koopmans and Janak may be difficult. Moreover, since

$$e^{i\phi_j(+\pi/a)} = e^{i\phi_j(-\pi/a)} \tag{32}$$

we have

$$\phi_j(+\pi/a) - \phi_j(-\pi/a) = \tilde{n}_j \cdot 2\pi \tag{33}$$

where $\tilde{n}_j$ is an unknown integer. Ultimately this means that the electronic polarization, i.e., P_R or P_{KSV} contains an unknown additive term, $\tilde{n} \cdot a$. The best way to determine the unknown integer $\tilde{n}$ is at the moment of writing the subject of intensive studies.

Due to the complexity of solving and interpreting the single-particle equations that result from our vector-potential approach, alternative strategies for extracting the responses of extended systems to electrostatic fields may be suggested. In a recent study [20] we have examined the case where the potential due to the external field is replaced by potentials that are piecewise linear. Our conclusion, however, is that such approaches are less general and computationally heavier than the treatment of long, but finite systems exposed to the true external potential. And ultimately the vector-potential approach presented here is by far the most efficient one.

In passing we add that, as an unexpected byproduct of our smoothing procedure, it was found that it could be useful in the development of a theoretical method for extracting band structures from calculations on a finite system which even did not have to be strictly regular [21].

Certain numerical findings may be related to the phase factors: it is well known that for molecular systems exposed to electrostatic fields Zener tunneling may take place, i.e., electrons may move from one end to the other of the molecule for sufficiently strong fields. The systems of our interest are clearly so large that Zener tunneling may occur. On the other hand, with P_{KSV} as the expression for the electronic part of the polarization the translational periodicity is not destroyed, meaning that a building-up of electron density in certain parts of the system is not possible. Numerical results suggest that instead the current part of the polarization, P_I, grows primarily due to an increase of the k dependence of the phase factors of the orbital expansion coefficients, $C_{pj}(k)$. A clear understanding of this point has, however, not yet been reached.

Finally, we mention briefly an issue that is related to, but not directly part of, the present work. It turns out that density-functional calculations with the currently commonly used approximate functionals tend to overestimate the linear and, particularly, non-linear responses of large systems to electrostatic fields. For a certain size these responses should approach the thermodynamic limit where they are proportional to the system size. Here, the density-functional calculations tend to predict a saturation at a too large size and a much too high value [22–24]. Whether the present approach, where we treat purely infinite systems instead of very large ones, can be used in exploring this so-called density-functional catastrophe, is at the moment of writing an open question.

Acknowledgement

This work is supported by the German Research Council (DFG) within the priority programme SPP 1145 through project Sp439/20. Moreover, one of the authors (MS) is very grateful to the International Center for Materials Research, University of California, Santa Barbara, for generous hospitality.

References

1. M. Springborg, B. Kirtman, and Y. Dong, Chem. Phys. Lett. **396** (2004) 404.
2. M. Springborg, B. Kirtman, Y. Dong, and V. Tevekeliyska, Lecture Series on Computer and Computational Sciences **7** (2006) 1234.
3. M. Springborg and B. Kirtman, J. Chem. Phys. **126** (2007) 104107.
4. M. Springborg and B. Kirtman, Am. Inst. Phys. Conf. Proc. **963** Vol. 2 (2007) 200.
5. M. Springborg and B. Kirtman, Phys. Rev. B **77** (2008) 045102. Eratum: Phys. Rev. B **77** (2008) 209101.
6. M. Springborg and B. Kirtman, Chem. Phys. Lett. **454** (2008) 105.
7. M. Springborg and B. Kirtman, Can. J. Chem. **87** (2009) 984.
8. D. Vanderbilt and R. D. King-Smith, Phys. Rev. B **48** (1993) 4442.
9. K. N. Kudin, R. Car, and R. Resta, J. Chem. Phys. **127** (2007) 194902.
10. R. D. King-Smith and D. Vanderbilt, Phys. Rev. B **47** (1993) 1651.
11. R. Resta, Rev. Mod. Phys. **66** (1994) 899.
12. R. Resta, Phys. Rev. Lett. **80** (1998) 1800.
13. R. W. Nunes and X. Gonze, Phys. Rev. B **63** (2001) 155107.
14. I. Souza, J. Íñiguez, and D. Vanderbilt, Phys. Rev. Lett. **89** (2002) 117602.
15. P. Umari and A. Pasquarello, Phys. Rev. Lett. **89** (2002) 157602.
16. B. Kirtman, F. L. Gu, and D. M. Bishop, J. Chem. Phys. **113** (2000) 1294.
17. D. M. Bishop, F. L. Gu, and B. Kirtman, J. Chem. Phys. **114** (2001) 7633.
18. M. Dvornikov, arXiv: math.NA/0306092.
19. J. M. Andre, D. H. Mosley, B. Champagne, J. Delhalle, J. G. Fripiat, J. L. Bredas, D. J. Vanderveken, and D. P. Vercauteren, in *Methods and Techniques in Computational Chemistry*. E. Clementi (Ed.) Stef, Cagliari (1993).
20. B. Kirtman, M. Ferrero, M. Rérat, and M. Springborg (to be published).
21. A. Pomogaeva, M. Springborg, B. Kirtman, F. L. Gu, and Y. Aoki, J. Chem. Phys. **130** (2009) 194106.
22. B. Champagne, E. A. Perpete, S. J. A. Gisbergen, E.-J. Baerends, J. G. Snijders, C. Soubra-Ghaoui, K. A. Robins, and B. Kirtman, J. Chem. Phys. **109** (1998) 10489.
23. S. J. A. van Gisbergen, P. R. T. Schipper, O. V. Gritsenko, E. J. Baerends, J. G. Snijders, B. Champagne, and B. Kirtman, Phys. Rev. Lett. **83** (1999) 694.
24. A. Ruszinszky, J. P. Perdew, and G. I. Sconka, Phys. Rev. A **78** (2008) 022513.

Exact Solutions for a Two-electron Quantum Dot Model in a Magnetic Field and Application to More Complex Sytems

By Manfred Taut* and Helmut Eschrig

IFW Dresden, Helmholtzstr. 20, 01069 Dresden, Germany

(Received July 30, 2009; accepted September 25, 2009)

Quantum Dots / Exact Solutions of Schrödinger Equation / Wigner Molecules / Current Density Functional Theory

We discussed exact solutions of the Schrödinger equation for a two-dimensional parabolic confinement potential in a homogeneous external magnetic field. It turns out that the two-electron system is exactly solvable in the sense, that the problem can be reduced to numerically solving one radial Schrödinger equation. For a denumerably infinite set of values of the effective oscillator frequency $\tilde{\omega} = \sqrt{\omega_0^2 + (\omega_c/2)^2}$ (where ω_0 is the frequency of the harmonic confinement potential and ω_c is the cyclotron frequency of the magnetic field) even analytical solutions can be given. Our solutions for three electrons are exact in the strong - and the weak correlation limit. For quantum dot lattices with Coulomb-correlations between the electrons exact solutions are given, provided the lattice constant is large compared with the dot diameters. We are investigating basic physical properties of these solutions like the formation and distortion of Wigner molecules, the dependence of the correlation strength from ω_0 and ω_c, and we show that in general there is no exact Kohn-Sham system for the semi-relativistic current density functional Theory.

1. Introduction

Exact (and somtimes analytical) solutions of the Schrödinger equation for realistic few-electron models of the quantum-dot-type provide a lot of unique opportunities. The physical essence of basic physical notions like the formation and distortion of Wigner molecules (WMs) and the consequences of inter-dot electron correlations in dot lattices can be understood more easily than with numerical brute-force approaches, which provide no formulas but only data. Moreover, these phenomena can be monitored over a wide range of external parameter

* Corresponding author. E-mail: m.taut@ifw-dresden.de

Z. Phys. Chem. **224** (2010) 631–649
© by Oldenbourg Wissenschaftsverlag, München

values, which allows us to tune the system continuously between the weakly and strongly correlated regime. Intermediate and strong correlations are of particular interest, because for weak correlations there are a lot of mean-field approaches available. In particular we can easily see what the basic difference between strongly correlated systems with low density (Wigner crystal) and high magnetic fields (fractional quantum Hall systems) is. Further, exact solutions allow us to check the precision of approximations like Hartree-Fock and Density Functional approaches and to reveal their weak and strong points. Unlike the comparison of approximate solutions for real systems with experiments, this approach has the advantage that all physical quantities (including those which are not experimentally accessible) can be considered, there are no experimental side effects, which obscure the comparison, and there are no discrepancies due to differences between the model and the real system. Last but not least, basic mathematical assumptions about the structure of solutions (like non-interacting v-representability in current density functional theory), which cannot be proven for general systems, can sometimes be rejected for special systems.

This paper is not a comprehensive review on quantum dots and quantum dot molecules and lattices. In particular it does not describe the approaches and the results from numerical diagonalizations in a complete set of basis functions, quasi-classical approaches for the Wigner limit, as well as quantum Monte Carlo-, density matrix renormalization group -, current density functional -, and Hartree-Fock approaches. Each of these methods warrants a separate review (see e.g. Ref.s [1–4] and references therein). Instead, this paper is focused on those systems which can be solved exactly or analytically, albeit the more complex sytems only in some limits for the external parameters.

A problem with some exactly solvable models is that they have to be sufficiently simple and the question is whether all their features are shared by real systems of greater complexity. Therefore, all approaches complement one another and they should be pursued in parallel. Apart from this aspect, our models for the two and three-electron quantum dots and lattices from two-electron dots are already interesting on its own.

2. Specification of the model and exact solutions

2.1 Model Hamiltonian

We consider a two-dimensional (2D) two-electron system (with Coulomb interaction) in a harmonic scalar potential $v^{ext}(r) = (1/2)\,\omega_0^2\,r^2$ and a magnetic field $\mathbf{B} = B\,\mathbf{e}_z$ represented by the vector potential (in symmetric gauge) $\mathbf{A}^{ext}(\mathbf{r}) = (1/2)\,\mathbf{B}\times\mathbf{r} = (1/2)\,B\,r\,\mathbf{e}_\alpha$. We introduced cylinder coordinates (r,α,z) with the cylinder axis perpendicular to the plane, to which the electron motion is confined. The Hamiltonian reads

$$
\begin{aligned}
H &= \sum_{i=1}^{2}\left\{\frac{1}{2}\left(\mathbf{p}_i + \frac{1}{c}\mathbf{A}^{ext}(\mathbf{r}_i)\right)^2 + \frac{1}{2}\,\omega_0^2\, r_i^2\right\} \\
&+ \frac{1}{|\mathbf{r}_2 - \mathbf{r}_1|} + H_{spin}\ ,
\end{aligned} \tag{1}
$$

where $H_{spin} = g^* \,\Sigma_{i=1}^{3}\, \mathbf{s}_i \,\cdot \mathbf{B}$, and atomic units $\hbar = m = e = 1$ are used throughout. This is a widely used effective Hamiltonian model for a two-electron quantum dot.

2.2 Exact solutions of the Schrödinger equation

The Schrödinger equation with the Hamiltonian (1) can be solved not only by reduction to the numerical solution of an (ordinary) radial Schrödinger equation [20], but even analytically for a discrete, but infinite set of effective frequencies $\tilde{\omega} = \sqrt{\omega_0^2 + (\omega_c/2)^2}$ [5], where we introduced the cyclotron frequency $\omega_c = B/c$.

If we introduce relative and center of mass (c.m.) coordinates

$$
\mathbf{r} = \mathbf{r}_2 - \mathbf{r}_1 \quad , \quad \mathbf{R} = \frac{1}{2}(\mathbf{r}_1 + \mathbf{r}_2) \tag{2}
$$

the Hamiltonian (1) decouples exactly.

$$
H = 2\,H_r + \frac{1}{2}\,H_R + H_{spin} \tag{3}
$$

The Hamiltonian for the c.m. motion agrees with the Hamiltonian of a non-interacting particle in a magnetic field

$$
H_R = \frac{1}{2}\left[\mathbf{P} + \frac{1}{c}\mathbf{A}_R\right]^2 + \frac{1}{2}\omega_R^2\, R^2 \tag{4}
$$

and only the relative Hamiltonian contains the electron-electron interaction

$$
H_r = \frac{1}{2}\left[\mathbf{p} + \frac{1}{c}\mathbf{A}_r\right]^2 + \frac{1}{2}\omega_r^2\, r^2 + \frac{1}{2r}\ , \tag{5}
$$

where we introduced rescaled parameters $\omega_R = 2\omega_0$, $\mathbf{A}_R = 2\mathbf{A}(\mathbf{R})$, $\omega_r = \frac{1}{2}\,\omega_0$, $\mathbf{A}_r = \frac{1}{2}\mathbf{A}(\mathbf{r})$ (the indices 'r' and 'R' refer to the relative and c.m. coordinate systems, respectively). The decoupling of H allows the ansatz

$$
\Phi = \xi(\mathbf{R})\,\varphi(\mathbf{r})\,\chi(s_1, s_2)\ , \tag{6}
$$

where $\chi(s_1,s_2)$ is the singlet or triplet spin eigen-function.

The eigen-functions of the c.m. Hamiltonian (4) have the form

$$
\xi = \frac{e^{iM\mathcal{A}}}{\sqrt{2\pi}}\,\frac{U_M(R)}{R^{1/2}} = \frac{e^{iM\mathcal{A}}}{\sqrt{2\pi}}\,R_M(R) \quad , \quad M = 0, \pm 1, \pm 2, \ldots\ , \tag{7}
$$

where the polar coordinates of the c.m. vector are denoted by $(R,\mathcal{A})$ and the radial functions $U_M(R)$ and $R_M(R)$ can be found in standard textbooks.

With the following ansatz for the relative motion

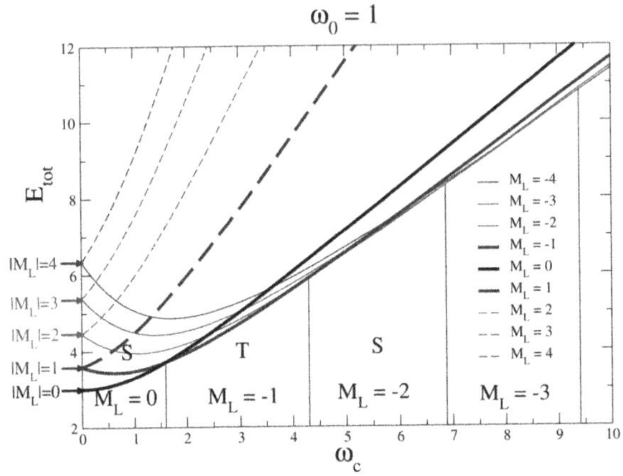

Fig. 1. Total energy for fixed confinement frequency $\omega_0 = 1$ versus cyclotron frequency ω_c (i.e. magnetic field). The c.m. system is always in the ground state with $M = 0$. The relative angular momentum m is varied. The vertical lines show where the total orbital angular momentum $M_L = M + m$ of the ground state changes. S and T indicates whether the ground state is singlet or triplet. Thick lines indicate states which can be NIVR.

$$\varphi = \frac{e^{im\alpha}}{\sqrt{2\pi}} \frac{u_m(r)}{r^{1/2}} \quad , \quad m = 0, \pm 1, \pm 2, \ldots \tag{8}$$

the Schrödinger equation $H_r\ \varphi(\mathbf{r}) = \varepsilon_r \varphi(\mathbf{r})$ gives rise to a radial Schrödinger equation for $u(r)$

$$\left\{-\frac{1}{2}\frac{d^2}{dr^2} + \frac{1}{2}\left(m^2 - \frac{1}{4}\right)\frac{1}{r^2} + \frac{1}{2}\,\widetilde{\omega}_r^2\, r^2 + \frac{1}{2r}\right\}u(r) = \widetilde{\epsilon}_r\, u(r) \ , \tag{9}$$

where the polar coordinates for the relative vector are denoted by (r,α), $\widetilde{\omega}_r = \frac{1}{2}\,\widetilde{\omega}$, $\widetilde{\varepsilon}_r = \varepsilon_r - \frac{1}{4}\,m\omega_c$, and $\omega_c = B/c$. The solutions are subject to the normalization condition $\int_o^\infty dr|u(r)|^2 = 1$. The Pauli principle demands that (because of the different particle exchange symmetry of the spin eigen-functions) the relative angular momentum m has to be even and odd in the singlet and triplet state, respectively. There is no constraint for the c.m. angular momentum M following from the Pauli principle. Because of the orthogonality of the coordinate transformation, the above described solutions are eigen-functions of the total orbital angular momentum with the eigenvalue $M_L = M{+}m$.

Fig. 1 shows that the modulus of the orbital angular momentum of the ground state (GS) grows stepwise with increasing magnetic field. This implies that the spin state oscillates between singlet and triplet [21]. The Zeeman term and quenching of the singlet state for higher magnetic fields is not included in Fig. 1. The c.m. excitations are not included as well, because they have no impact on the character of the ground state.

In Fig. 2 the magnetic field and the e-e-interaction are successively added to the levels in the confinement only. We observe that the magnetic field removes

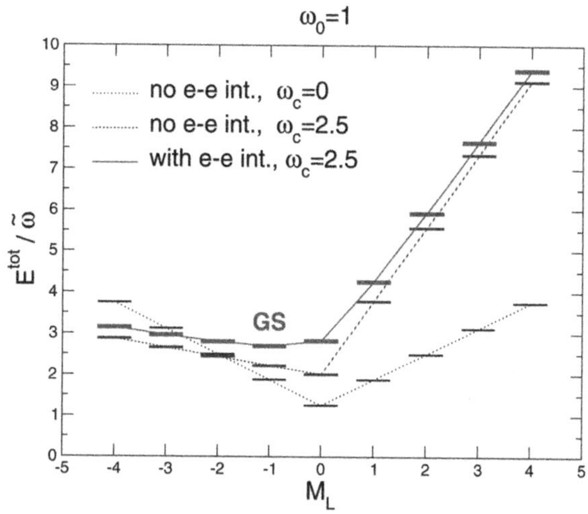

Fig. 2. Energy levels for fixed external fields versus total orbital angular momentum. The level of the GS is indicated. We started with noninteracting electrons in the confinement only, added the magnetic field and the e-e-interaction. (In the energy unit $\tilde{\omega}$ we used $\omega_c = 2.5$ for all three cases.).

the degeneracy with respect to the sign of M_L and breaks the symmetry with respect to up and downward directed fields. Angular momenta which are parallel to the field (positive) produce an magnetic moment which is anti-parallel to the field. They have a large interaction energy with the external field which shifts the levels upwards. In the opposite case the shift due to this contribution is downwards. Without e-e-interaction the GS has always $M_L = 0$. The shift due to the e-e-interaction is positive definite and it decreases with increasing $|M_L|$. This can be explained with the radial equation (9), which determines the contribution of the relative motion to the total energy. For large $M_L = m$ and small r the last (e-e-interaction) term is dwarfed by the second (centrifugal potential) term. For large r the third term (effective confinement) is dominating in all cases. In a classical picture this means that two electrons rotating with a high angular momentum are separated by the centrifugal force, so that the job of the e-e-interaction is already largely done and the addition of the e-e-interaction does not change much. It is this M_L-dependence of the e-e-interaction shift which moves the GS to smaller M_L with increasing ω_c.

2.3 Analytical solutions

In [5,6] it has been shown that the radial Schrödinger equation (9) has simple analytical solutions for a discrete, but infinite set of effective oscillator frequen-

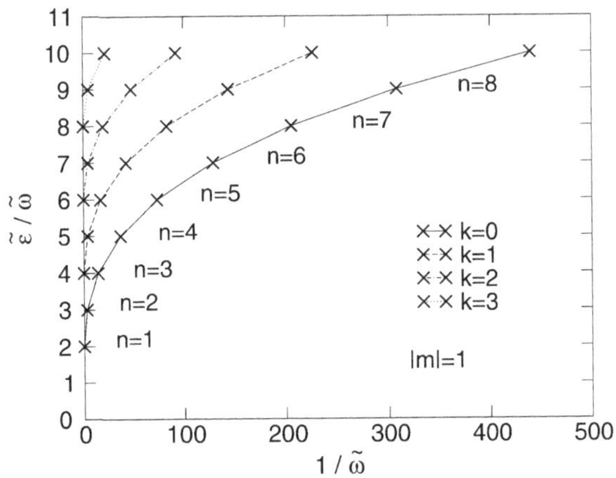

Fig. 3. Reduced energies (energy over effective oscillator frequency) versus inverse effective oscillator frequency for relative angular momentum $|m| = 1$. The crosses indicate solvable states. The lines are just a guide for the eye and they connect states with the same node number k. n is the same for all horizontal rows of crosses with the same ordinate.

cies $\tilde{\omega}$, the pattern of which for $|m| = 1$ can be seen in Fig. 3. The patterns for all $|m|$ look qualitatively similar. All solutions have the following form:

$$u(r) = r^{|m|+\frac{1}{2}}\, p(r)\, e^{-\frac{1}{4}\,\tilde{\omega}\, r^2} \quad , \tag{10}$$

where $p(r)$ is a finite polynomial of degree $(n-1)$. For 'non-soluble' systems, the polynomial has an infinite number of terms.

In this review we only provide the results for the simplest analytical solutions. For $n = 1$ there is only a solution for infinite $\tilde{\omega}$, which we call an asymptotic solution, because it is exact for $\tilde{\omega} \to \infty$

$$\frac{1}{\tilde{\omega}} = 0 \quad , \quad p(r) = 1 \tag{11}$$

This solution agrees not only with the Laughlin model wave function (WF), if the latter is applied to $N = 2$ and expressed in terms of the coordinates used here, but it is also the exact solution for non-interacting electrons (electrons without Coulomb interaction) in relative- and c.m. coordinates. The corresponding WF has no node and is a ground state.

For $n = 2$ there is one finite-field solution

$$\frac{1}{\tilde{\omega}} = (2\,|m| + 1) \quad , \quad p(r) = 1 + \frac{r}{(2\,|m| + 1)} \quad , \tag{12}$$

which is a ground state as well.

For $n = 3$ there is one asymptotic solution, which is a first excited state,

$$\frac{1}{\tilde{\omega}} = 0 \quad , \quad p(r) = 1 - \frac{r^2}{(|m| + 1)} \tag{13}$$

and one finite field solution, which is a ground state.

$$\frac{1}{\tilde{\omega}} = 2\,(4\,|m| + 3)\quad ,$$

$$p(r) = 1 + \frac{r}{(2\,|m| + 1)} + \frac{r^2}{2(2\,|m| + 1)(4\,|m| + 3)} \tag{14}$$

The exact energies ε of the relative coordinate systems corresponding to these eigen-functions can be obtained from one compact formula.

$$(\varepsilon - \frac{1}{2} m\,\omega_c) = (|m| + n)\,\tilde{\omega} \tag{15}$$

If we compare this result with the spectrum of a single electron in a quantum dot $(\varepsilon - \frac{1}{2}\,m\,\omega_c) = (|m| + 2k + 1)\,\tilde{\omega}$, where k is the node number or degree of excitation, then it becomes clear that in both cases $(\varepsilon - \frac{1}{2} m\,\omega_c)$ is a integer multiple of $\tilde{\omega}$. We can also say that an analytical solution exists for those values of $|m|$ and $\tilde{\omega}$, for which one state of the interacting system is degenerate with one state of the non-interacting system.

A great deal of approaches for the 2D electron gas in external fields is based on model wave functions (WF). For inspirations and checks in the limit N = 2 it would be desirable to have analytical solutions of the two-electron problem for a wide range of external field values. One idea is to use the special analytical solutions (10) with one of the exactly solvable polynomials given in (11–14) as discussed in Ref.[6]. If we use a special exact solution for a finite interval of external potential values we assume that the polynomial is independent of $\tilde{\omega}$ in this interval and depends only on $|m|$. Each of the choices (11–14) provides a different approximation to the exact solution. Now we check, what the precision of these choices over a wide range of external fields is.

Fig. 4 compares the precision of model WFs with different n. The negative poles in the curves for $log(\omega_0) = +2$ and n = 2 and 3 indicate the vicinity of those $\tilde{\omega}$, which provide exact solutions. It is seen that the solution for $n > 1$ is everywhere better than the solution for $n = 1$ (the latter is exact for infinite fields), and that all solutions become exact for $\tilde{\omega} \to \infty$. This means that the ansatz for the N-electron system proposed in [6] is definitely more precise than the Laughlin ansatz, if both are applied to the 2-electron system. We want to mention that the logarithm of the relative error in the energies calculated with the model WFs shows qualitatively the same behavior as the projection. It is only smaller in magnitude.

2.4 Exact densities

With (7) and (8), we obtain for the total density

$$n(\mathbf{r}) = 2 \int d\mathbf{r}'\ |\Phi(\mathbf{r}, \mathbf{r}')|^2 \tag{16}$$

the general expression

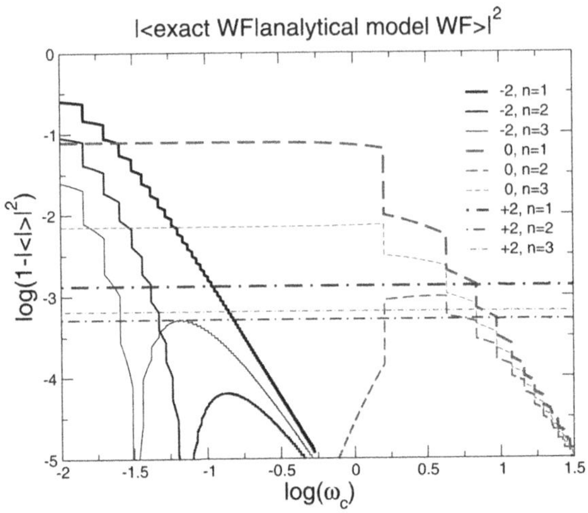

Fig. 4. Projection of analytical model wave functions with polynomials of degree n onto the exact WF. The logarithm of the deviation of the squared matrix element from unity is on the ordinate. Full, dashed and dash-dotted curves belong to $log(\omega_0)$ = -2, 0, and +2, respectively (see also first number in the legend). Thick, medium and thin lines belong to n = 1,2, and 3, respectively.

$$n(r) = \frac{1}{2\pi^2}\int_0^{2\pi} d\alpha \int_0^{\infty} dr' \left[R_M\left(\sqrt{r^2+\frac{1}{4}r'^2+rr'cos\alpha}\right)\right]^2 \left[u_m(r')\right]^2 \quad (17)$$

Because we are interested in the ground state only, we can safely use the c.m. state for $M = 0$: $R_0(R) = 2\sqrt{\tilde{\omega}}\ \exp(-\tilde{\omega}\,R^2)$ which allows to do one integration analytically leaving us with

$$n(r) = \frac{4\tilde{\omega}}{\pi}e^{-2\tilde{\omega}r^2}\int_0^{\infty} dr'\ e^{-(\tilde{\omega}/2)r'^2}\ I_0(2\tilde{\omega}rr')\left[u_m(r')\right]^2 \quad (18)$$

where $I_n(x)$ are the modified Bessel functions.

The general expression for the paramagnetic current density

$$\mathbf{j}^p(\mathbf{r}) = -i\int d\mathbf{r}' \left[\Phi^*(\mathbf{r},\mathbf{r}')\nabla\Phi(\mathbf{r},\mathbf{r}') - \Phi(\mathbf{r},\mathbf{r}')\nabla\Phi^*(\mathbf{r},\mathbf{r}')\right] \quad (19)$$

is somewhat complicated. Therefore, we give here only the formula for $M = 0$

$$\mathbf{j}^p(\mathbf{r}) = \mathbf{e}_\alpha\ m\ \frac{4\tilde{\omega}}{\pi}e^{-2\tilde{\omega}r^2}\int_0^{\infty} dr'\ e^{-(\tilde{\omega}/2)r'^2}\ \frac{I_1(2\tilde{\omega}rr')}{r'}\left[u_m(r')\right]^2 = \mathbf{e}_\alpha\ j^p(r) \quad (20)$$

As to be expected, the paramagnetic current density is proportional to the total angular momentum, points in azimuthal direction $\mathbf{e}_\alpha$, and the scalar $j^p(r)$ depends only on the distance r from the center and not from the azimuthal angle .

Although both formulas (18) and (20) rely on the functions $u_m(r)$, which are solutions of (9), the analytical behavior for $r \to 0$ can be expressed in terms of two positive definite integrals.

$$A_0 = \int_0^\infty dr\, e^{-(\widetilde{\omega}/2)r^2} \left[u_m(r)\right]^2 \tag{21}$$

$$A_2 = \int_0^\infty dr\, r^2\, e^{-(\widetilde{\omega}/2)r^2} \left[u_m(r)\right]^2 \tag{22}$$

After power series expansion of $I_n(x)$, we obtain

$$n(r) \rightarrow \frac{4\,\widetilde{\omega}}{\pi} e^{-2\widetilde{\omega}r^2}\left[A_0 + A_2\,\widetilde{\omega}^2\, r^2 + \cdots\right] , \tag{23}$$

$$j^p(r) \rightarrow m\frac{4\,\widetilde{\omega}^2}{\pi} e^{-2\widetilde{\omega}\,r^2}\; r\left[A_0 + \frac{1}{2}A_2\,\widetilde{\omega}^2\, r^2 + \cdots\right] . \tag{24}$$

For the origin this means that $n(0) = 4\widetilde{\omega}\; A_0/\pi$ is always finite and $j^p(0) = 0$ always vanishes. On the other hand, the derivative of the density at the origin $\frac{d\,n}{dr}(0) = 0$ vanishes, but the derivative of the paramagnetic current density $\frac{d\,j^p}{dr}(0) = m\;(4\widetilde{\omega}^2/\pi)\; A_0$ is finite, unless $m = 0$. Besides, there is a relation which does not involve the radial WFs explicitly.

$$\frac{d\,j^p}{dr}(0) = m\,\widetilde{\omega}\; n(0) \tag{25}$$

The exact vorticity, which has the form $\gamma(\mathbf{r}) = \mathbf{e}_z\;\gamma(r)$ reads in this limit

$$\gamma(r) \rightarrow m\,2\,\widetilde{\omega}\left(1 - \widetilde{\omega}^2\,\frac{A_2}{A_0}\; r^2 + \cdots\right) . \tag{26}$$

As will be seen in Sect.4, the limit $r \rightarrow 0$ is decisive for our proof of the violation of non-interacting v representability.

3. Formation of Wigner molecules and correlation strength

A illustrative classical picture for a WM in an environment with rotational symmetry is a rotating and vibrating electron molecule. For a two-electron system this is a dumbbell-like object. We are going to show that this configuration is a manifestation of strong e-e-correlations and it is formed in the limit of small ω_0 or large ω_c. The issue is: why is small ω_0 equivalent to large ω_c although the exact WF and consequently all distribution functions depend only on the effective confinement frequency $\widetilde{\omega} = \sqrt{\omega_0^2+(\omega_c/2)^2}$ where both ω_0 and ω_c have qualitatively the same influence. In particular, we will point out, how strong magnetic fields can cause strong correlations. This is not the same mechanism as for weak confinement (see also Ref.[8]).

For illustration we use the density (16), which provides the distribution of the electrons in space, and the pair correlation function

$$g(\mathbf{r}) = <\psi|\sum_{i<j}\delta(\mathbf{r}_i - \mathbf{r}_j - \mathbf{r})|\psi> \tag{27}$$

which determines the distribution of the distance between two electrons. Both quantities depend only from the modulus: $n(\mathbf{r}) = n(r)$ and $g(\mathbf{r}) = g(r)$. In our system the latter is given by the radial part of the relative WF alone.

$$g(r) = |\varphi(\mathbf{r})|^2 = \frac{[u(r)]^2}{2\pi r} \tag{28}$$

In Fig. 5 both quantities are shown for a few typical parameter sets. We have chosen scaled axes (tilted quantities) which allow us to show and compare different extreme cases in the same picture. On the abscissas the r-coordinate is scaled $\tilde{r} = r/\bar{r}$ with the average radius of the dot $\bar{r} = (1/2) \int d^2\mathbf{r}\ r\ n(r)$ and the ordinates are scaled in such a way that the norms are conserved: $\int d^2\tilde{\mathbf{r}}\ \ \tilde{n}(\tilde{r}) = 2$ and $\int d^2\tilde{\mathbf{r}}\ \ \tilde{g}(\tilde{r}) = 1$. For weak correlations $n(r)$ is peaked in the region of lowest potential energy (center of the dot) and $g(r)$ is spread out over the whole range of non-vanishing density, allowing in particular small distances between the electrons. This regime is realized in curves A and B applying to strong (ω_0 large) or medium confinement and small and medium ω_c. Strong correlations are connected with sharply peaked densities at non-zero r confining the electrons on a ring with radius $\bar{r}$. The pair correlation function is sharply peaked at a distance $2\ \bar{r}$ which is the diameter of the ring. This means that the electrons are localized at a ring and have virtually antipodal positions [5] what agrees with the above mentioned classical picture of a WM. This minimizes the e-e-interaction energy in the limits allowed by the potential confinement energy without enhancing the kinetic energy. In terms of external parameters this can be realized in two scenarios.

i) If ω_c is small (or zero), the confinement has to be weak (ω_0 small). This can be concluded from the comparison of curve B and D. For weak confinement the state is spread out widely and the density is low. Low density implies the dominance of the the e-e-interaction over the kinetic energy (last term in radial Schrödinger equation (9) versus the first term) and a state which minimizes the e-e-interaction. The condensed matter analog to this state is the Wigner crystal. All in all, strong correlations in systems with low densities are produced by the dominance of the the e-e-interaction over the kinetic energy.

ii) If ω_c is medium or large (and the confinement not too strong), then the angular momentum of the GS is strongly negative (see curve C and E). Large (modulus of) angular momentum means strong centrifugal potential (second term in radial Schrödinger equation (9)) which drives the electrons away from the center and produces the ring structure in the density. At the same time the remaining e-e-term maximizes the e-e-distance within the limits set by the density. This is most pronounced in curves C and E. The case of small ω_0 and large ω_c (which is not shown because of its numerical difficulties caused by of the extremely strong angular momentum) is even more strongly correlated. The condensed matter analog to this state is the fractional quantum Hall state. The state of curve C shows strong correlations as well, but does not have a large diameter ($\bar{r} = 0.6345$) and consequently a low density. This proves that low density is not

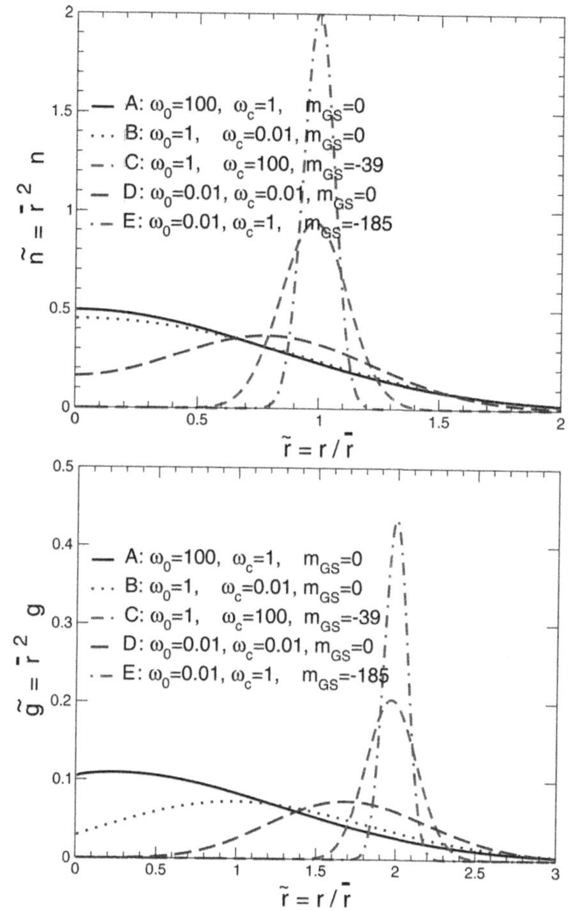

Fig. 5. Density (upper) and pair correlation function (lower) for several correlation strength. ω_0, ω_c, and the angular momentum of the ground state are given in the legend. The scaling parameters are: $\bar{r}$ = 0.0901(A), 1.0192(B), 0.6345(C), 14.5022(D), and 13.6472(E).

necessary for strong correlation, but a magnetic field can do the job alone. All in all, strong correlations in high magnetic fields are mainly produced by the high modulus of the angular momentum of the ground state.

From these considerations it follows that in our family of systems a suitable dimensionless quantitative definition of the correlation strength[1] can be set up by the mean square radius of the pair correlation function $\bar{r^2} = \int d^2\mathbf{r}\; r^2 g(r)$ and its half width $\Delta^2 r = \int d^2\mathbf{r}\; (r-\bar{r})^2 g(r)$, where $\bar{r} = \int d^2\mathbf{r}\; r g(r)$, according to

[1] Unlike in common quantum chemistry language, the *correlation strength* defined here comprises all effects beyond the Hartree approximation, in particular it includes the effect of exchange.

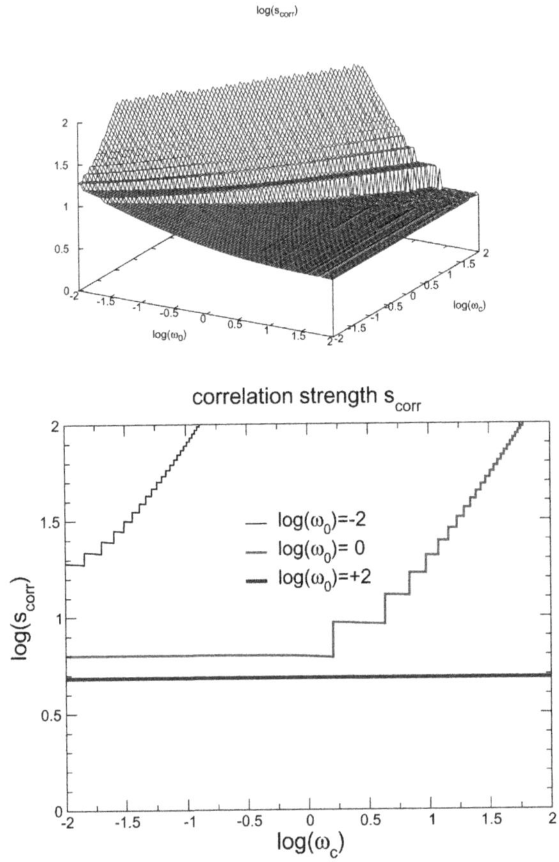

Fig. 6. Correlation strength versus external field parameters in log-scales for all axes. The steps, corresponding to parameters where the angular momentum of the GS changes (see Fig. 1), are fully resolved only in the lower panel, which shows curves for 3 discrete ω_0.

$$s_{corr} = \frac{\overline{r^2}}{\Delta^2 r} \quad . \tag{29}$$

This means, small half width and a peak at large r produce strong correlations. Fig. 6 shows this quantity as a function of the external field parameters. The steps are caused by a change in the angular momentum of the GS. It is obvious that for small ω_c a weakening of the confinement increases s_{corr}, but an increase of the magnetic field is much more effective, if it is connected to an increase of the modulus of the angular momentum of the GS, what happens in the region where the steps are found.

Matulis and Peeters [8] investigated the same issue using qualitative asymptodic expressions for the wave function instead of exact solutions. Nevertheless the resulting trends are the same, although their visual picture is different.

4. Violation of non-interacting v-representability of the exact solutions of the Schrödinger equation

The exact solutions of the special system considered in this review can be used to show that (unlike generally assumed) an exact Kohn-Sham (KS) system in the framework of semi-relativistic Current Density Functional Theory (CDFT) can exist only in special cases.

In Density Functional Theory (DFT) it can be shown that for the GS the external potential is a functional of the density [14] (see also textbooks [15,16] with more modern approaches)

$$v^{ext}(\mathbf{r}) \overset{C}{\leftarrow} \Psi \underset{\mathcal{D}^{-1}}{\overset{\mathcal{D}}{\rightleftharpoons}} n(\mathbf{r}) \tag{30}$$

which would imply non-interacting v-representability (NIRV) or the existence of an exact Kohn-Sham system for the GSs, if the interacting and the non-interacting systems would have a common set of ground state densities.

In the presence of a magnetic field and for (semi-relativistic) Current Density Functional Theory (CDFT), the generalization of $\mathcal{D}^{-1}$ for the ground state still exists, but Vignale and Rasolt [10,11] just presupposed the existence of the generalization of C [17] implying that NIVR and the existence of a KS scheme has not been proven. Capelle and Vignale [17], on the other hand, have shown that there can be several external potentials $\mathcal{V}^{ext}$ which provide the same WFs and densities

$$\begin{array}{lcl} \mathcal{V}_1^{ext}(\mathbf{r}) & \searrow & \\ \mathcal{V}_2^{ext}(\mathbf{r}) & \rightarrow & \Psi \rightleftharpoons \mathcal{N}(\mathbf{r}) \\ \cdots & \nearrow & \end{array} \tag{31}$$

where $\mathcal{V}^{ext}(\mathbf{r})$ and $\mathcal{N}(\mathbf{r})$ represent both external potentials ($v^{ext}(\mathbf{r})$ and $\mathbf{A}^{ext}(\mathbf{r})$) and both densities ($n(\mathbf{r})$ and $\mathbf{j}^p(\mathbf{r})$), respectively. Hence, C cannot exist anymore as an unique mapping and the question of NIVR cannot be answered in this way. However, the exact solutions of the special system considered in this review can be used to show that an exact Kohn-Sham system or NIVR can exist only in the following special cases.

All those states at non-zero B can be NIVR, which are continuously connected to the singlet and triplet ground states at $B = 0$ (see also Fig. 1). In more detail:

If the GS is a singlet (total orbital angular momentum M_L is even) both densities can be NIVR if the vorticity $\gamma(\mathbf{r}) = \nabla \times (\mathbf{j}^p(\mathbf{r}) \;/\; n(\mathbf{r}))$ of the exact

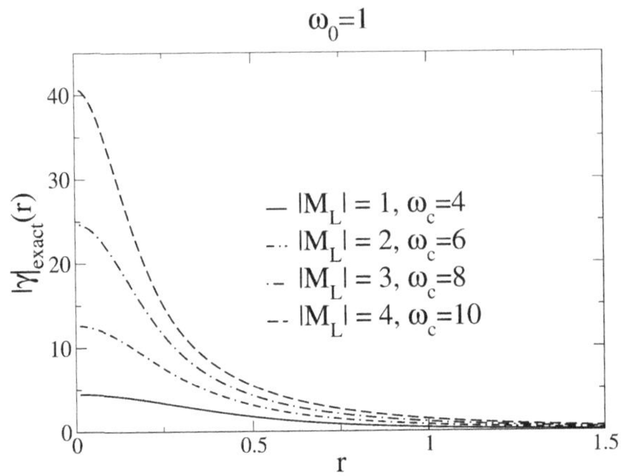

Fig. 7. The vorticities for $\omega_0 = 1$ and a few typical cyclotron frequencies ω_c where the state with negative M_L is the ground state. The sign of $\gamma_{exact}(r)$ agrees with the sign of M_L.

solution vanishes. For $M_L = 0$ this is trivially guaranteed because the paramagnetic current density vanishes. The vorticity based on the exact solutions for the higher M_L does not vanish, in particular for small r. In the limit $r \to 0$ this can even be shown analytically.

If the GS is a triplet (M_L is odd) and we assume circular symmetry for the KS system (the same symmetry as the real system) then only the exact states with $|M_L| = 1$ can be NIVR with KS states having angular momenta $m_1 = 0$ and $|m_2| = 1$.

Without specification of the symmetry of the KS system the condition for NIVR is that the small-r-exponents of the KS states are 0 and 1.

The proof of the statement for the singlet state is extremely simple and will be given here. The other proofs and more detailed information can be found in [9]. The question is if the (in this case doubly occupied) KS-WF $\varphi(\mathbf{r}) = R(\mathbf{r})\ e^{i\zeta(\mathbf{r})}$ can be chosen in such a way that the density and the gauge invariant vorticity of the non-interacting KS system and the exact solution agree.

$$n_{exact}(r) \overset{!}{=} n_{KS}(\mathbf{r}) = 2\ [R(\mathbf{r})]^2 \tag{32}$$

$$\gamma_{exact}(\mathbf{r}) \overset{!}{=} \gamma_{KS}(\mathbf{r}) = 0 \tag{33}$$

Eq. (33) defines the real part of the KS-WF. On the other hand, the vorticity of a two-electron singlet KS state vanishes exactly irrespective of the special form of $R(\mathbf{r})$ and $\zeta(\mathbf{r})$. Therefore, equation (34) can only be satisfied if the vorticity of the corresponding exact solution vanishes as well. Fig. 7 shows that this is not the case, in particular for small r the violation is massive. Eq. (27) provides $\gamma_{exact}(0) = 2\ M_L\ \tilde{\omega}$ which shows that the 'degree of violation' grows with growing $\tilde{\omega}$. The exact vorticity vanishes only for the state with zero angular momentum, which is the GS for small magnetic fields (see also Fig. 1).

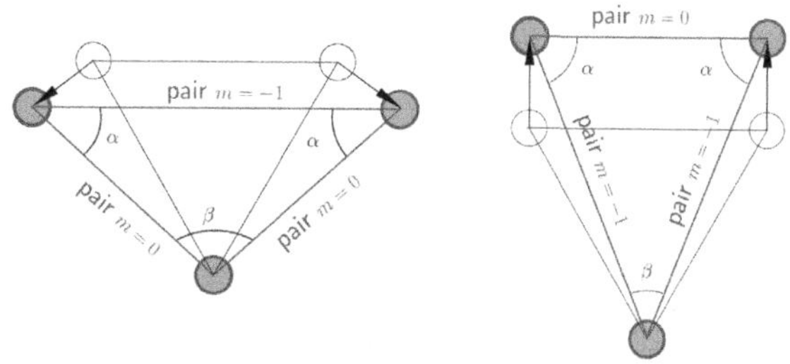

Fig. 8. Schematic picture of the distorted three-electron Wigner molecule with total orbital angular momentum $M_L = -1$ (left) and $M_L = -2$ (right). The thin lines depict the undistorted WM.

5. Distortion of the three-electron Wigner molecule

As shown above, the two-electron system is the simplest system which exhibits the phenomenon of the formation of WMs in finite systems. For three electrons there is another effect, namely a Jahn-Teller-like distortion of the WM [7] shown schematically in Fig. 8, which can be investigated using the solutions described above. The point is that in the strong correlation limit the three-electron system can be decoupled into three independent pairs, the Schrödinger equation for which agrees (apart from a renormalisation of the interaction parameters) with the Schrödinger equation for the relative coordinates in the two electron system [6,7].

The Hamiltonian of the three-electron system reads

$$H = \sum_{i=1}^{3}\left[\frac{1}{2}\left(\frac{1}{i}\nabla_i + \frac{1}{c}\mathbf{A}(\mathbf{r}_i)\right)^2 + \frac{1}{2}\omega_o^2 r_i^2\right] + \sum_{i<k}\frac{1}{|\mathbf{r}_i - \mathbf{r}_k|} \tag{34}$$

where the the Zeeman term is disregarded because it has no influence on the spacial distribution of the electrons, but shifts only the energies. We consider the unitary coordinate transformation from the original position vectors $\mathbf{r}_i$ to new ones $\mathbf{x}_i$

$$\begin{bmatrix} \mathbf{x}_1 \\ \mathbf{x}_2 \\ \mathbf{x}_3 \end{bmatrix} = \begin{bmatrix} 1/3 & a & b \\ b & 1/3 & a \\ a & b & 1/3 \end{bmatrix} \begin{bmatrix} \mathbf{r}_1 \\ \mathbf{r}_2 \\ \mathbf{r}_3 \end{bmatrix} \tag{35}$$

where $a = 1/3 - 1/\sqrt{3}$ and $b = 1/3 + 1/\sqrt{3}$. The corresponding inverse transformation provides for the difference coordinates in the e-e-interaction terms

$$\mathbf{r}_i - \mathbf{r}_j = \sqrt{3}\left(\mathbf{X} - \mathbf{x}_k\right) \tag{36}$$

where $(i,j,k) = (1,2,3)$ and cyclic permutations, and $\mathbf{X} \equiv \frac{1}{3}\,\Sigma_{i=1}^{3}\,\mathbf{x}_i$ is the center of mass (c.m.) in the new coordinates. It is a special feature of this transformation that the latter agrees with the c.m. **R** in the original coordinates. This transformation provides the equivalent Hamiltonian

$$H = \sum_{i=1}^{3} \left[\frac{1}{2} \left(\frac{1}{i} \nabla_i + \frac{1}{c} \mathbf{A}(\mathbf{x}_i) \right)^2 + \frac{1}{2}\, \omega_o^2\, x_i^2 + \frac{1}{\sqrt{3}}\, \frac{1}{|\mathbf{x}_i - \mathbf{X}|} \right] \tag{37}$$

Next we have to observe that in the strong correlation limit the uncertainty of the c.m. **X** is small compared with the expectation value of the new coordinates $\mathbf{x}_i$ (see the appendix of Ref.[7]). Therefore **X** can be considered as a small perturbation and in zero order in **X** the Hamiltonian (38) decouples into a sum of independent quasi-particle Hamiltonians

$$H^{(0)} = \sum_{i=1}^{3} h_i \tag{38}$$

The Schrödinger equation for the quasi-particles

$$h\, \varphi_q(\mathbf{x}) = \varepsilon_q\, \varphi_q(\mathbf{x}) \tag{39}$$

is similar to the Schrödinger equation for the Hamiltonian in the relative coordinates (5) and therefore it can be solved exactly. In terms of these solutions, the total energy is a sum of quasi-particle energies and the total orbital eigenfunction is a product of quasi-particle functions.

$$E_{q_1,q_2,q_3} \quad = \quad \varepsilon_{q_1} + \varepsilon_{q_2} + \varepsilon_{q_3} \tag{40}$$

$$\Phi_{q_1,q_2,q_3}(\mathbf{x}_1, \mathbf{x}_2, \mathbf{x}_3) \quad = \quad \varphi_{q_1}(\mathbf{x}_1) \cdot \varphi_{q_2}(\mathbf{x}_2) \cdot \varphi_{q_3}(\mathbf{x}_3) \tag{41}$$

where q_i comprises all quantum numbers. We can consider the quasi-particles as electron pairs, wherby their WFs $\varphi_q(\mathbf{x}_k)$ describe the distance $\mathbf{x}_k$ between two electrons (see Eq. (37)). The crucial point in explaining the distortion of the WM is the generalised Pauli exclusion principle for the quasi-particles states, i.e., the rules which determine the allowed combination of quantum numbers in Eqs. (41) and (42) in order to guarantee the anti-symmetry of the WF under electron transposition. These rules depend on the total spin (S) and orbital angular momentum (M_L) and they rule for some configurations the agreement of all three quantum numbers q_i out (see [6]). This means that the electron distances in all three electron pairs cannot agree giving rise to a distortion of the WM.

The final result is the following [7]:

In the ground state the electrons in an WM form an equilateral triangle (as might be expected from naive reasoning) only, if the state is a quartet ($S = 3/2$) and the orbital angular momentum is a magic quantum number ($M_L = 3\, m$; $m =$ integer). Otherwise the triangle in the ground state is isosceles. For $M_L = (3\, m{+}1)$ one of the sides is longer and for $M_L = (3\, m{-}1)$ one of the sides is shorter than the other two.

6. Coulomb correlations between Quantum dots

Other systems where the above described two-electron solutions play a crucial role are two-electron quantum dot molecules and quantum dot lattices, where the Coulomb correlation between electrons in differents dots is taken into account in the Van der Waals approximation [22]. This means that the diameter of the dots must be small compared with the distance between the dots and that the overlap between WFs of different dots should be negligible. Then the Coulomb interaction between the electrons at $\mathbf{r}_{nk} = \mathbf{R}_n^0 + \mathbf{u}_{nk}$ and $\mathbf{r}_{n'k'} = \mathbf{R}_{n'}{}^0 + \mathbf{u}_{n'k'}$ in different dots centered at $\mathbf{R}_n^0$ and $\mathbf{R}_{n'}{}^0$ with $(n \neq n')$ can be expanded in second order (dipole approximation) as

$$\frac{1}{|\mathbf{r}_{nk} - \mathbf{r}_{n'k'}|} = \frac{1}{|(\mathbf{R}_n^0 - \mathbf{R}_{n'}^0) + (\mathbf{u}_{nk} - \mathbf{u}_{n'k'})|} =$$

$$= \frac{1}{|\mathbf{R}_n^0 - \mathbf{R}_{n'}^0|} + \frac{1}{2}(\mathbf{u}_{nk} - \mathbf{u}_{n'k'}) \cdot \mathbf{T}(\mathbf{R}_n^0 - \mathbf{R}_{n'}^0) \cdot (\mathbf{u}_{nk} - \mathbf{u}_{n'k'}) + ...$$

where the dipole tensor $\mathbf{T}(\mathbf{R}) = (1/R^5)\ [\ 3\ \mathbf{R}\ {}^\circ\mathbf{R} - R^2\ \ \mathbf{I}\ \ ]$ has been introduced.

In the following the arrangement and number of dots is arbitrary, but for simpler notations we consider only dots with two electrons each. The bare confinement potential can vary from dot to dot. If we introduce for each dot a c.m. coordinate $\mathbf{R}_n = \mathbf{R}_n^0 + \mathbf{U}_n$ with $\mathbf{U}_n = (1/2)\ (\mathbf{u}_{n1} + \mathbf{u}_{n2})$ and a relative coordinate $\mathbf{r}_n = \mathbf{r}_{n2} - \mathbf{r}_{n1} = \mathbf{u}_{n2} - \mathbf{u}_{n1}$, then the total Hamiltonian decouples

$$H = H_{cm}(\{\mathbf{R}_n\}) + \sum_n H_{rel,n}(\mathbf{r}_n) \tag{42}$$

into a collective Hamiltonian

$$H_{cm} = \frac{1}{2}\left\{ \sum_n \frac{1}{2}\left[\mathbf{P}_n + \frac{2}{c}\mathbf{A}(\mathbf{U}_n)\right]^2 + 2\sum_{n,n'} \mathbf{U}_n \cdot \mathbf{C}_{n,n'} \cdot \mathbf{U}_{n'} \right\} \tag{43}$$

and a sum of individual intradot Hamiltonians

$$H_{rel,n} = 2\left\{ \frac{1}{2}\left[\mathbf{p} + \frac{1}{2c}\mathbf{A}(\mathbf{r})\right]^2 + \frac{1}{2}\ \mathbf{r} \cdot \mathbf{D}_n \cdot \mathbf{r} + \frac{1}{2\ r} \right\} \tag{44}$$

The force constant tensor $\mathbf{C}_{n,n'}$ of the collective Hamiltonian and the effective confinement tensor $\mathbf{D}_n$ of the decoupled intradot Hamiltonians contain both the bare confinement potential and a contribution from the dipole tensor from the interdot interaction [22].

The spectrum of the intradot excitations from (45) can be obtained with the methods for single quantum dots and its general features have been described in the previous sections. The Hamiltionian (44) describes magneto-phonon excitations. If the strength of the interdot interaction reaches a critical value, a magneto-phonon mode can become soft indicating a lattice instability. Such a case is shown in Fig. 9 for a rectangular periodic lattice. Solutions for a selection of dot dimers and periodic lattices are given and discussed in some detail in Ref.[22].

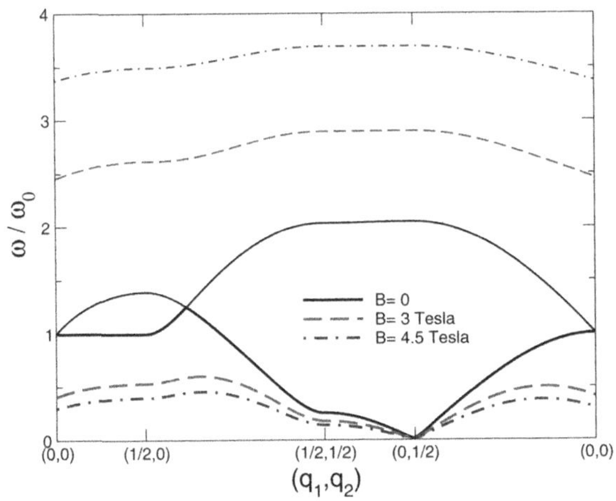

Fig. 9. Magneto-phonon dispersion in a rectangular quantum dot lattice with lattice constants $a_1/a_2 = 2$ for the critical interaction strength and for the three magnetic fields given in the legend. The abszissa gives the magneto-phonon frequency ω in units of the bare confinement frequency ω_0 of the dots. The wave vector (q_1,q_2) shown on the ordinate varies along the edge of the irreducible Brillouin zone.

Acknowledgement

This work was supported by the German Research Foundation (DFG) in the Priority Program SPP 1145.

References

1. L. Jacak, P. Hawrylak, and A. Wojs, Quantum Dots. Springer (1998).
2. P. A. Maksym et al., J. Phys.: Condens. Matter **12** (2000) R299.
3. S. M. Reimann, M. Manninen, Rev. Mod. Phys. **74** (2002) 1283.
4. C. Yannouleas, U. Landman, Rep. Prog. Phys. **70** (2007) 2067.
5. M. Taut, J. Phys. A **27** (1994) 1045 and J. Phys. A **27** (1994) 4723. (Misprints: in formula (8) the factor $r^{|M|}$ has to be replaced by $R^{|M|}$, in formula (10) in the term containing $\partial/\partial\alpha$ a factor $\frac{1}{2}$ is missing, and on the r.h.s. of (19a) and (20a) $\tilde{\omega}$ must be replaced by $\tilde{\omega}_r$.) For the method of exact solutions see also: M. Taut, Phys. Rev. A **48** (1993) 3561.
6. M. Taut; J. Phys.: Condens. Matter **12** (2000) 3689. (Misprints: in formula (21) the counter of the second term reads $(\beta r)^2$, in formula (73), $43^{1/3}$ has to be read as $4 \cdot 3^{1/3}$, and 4 lines before, in the definition of r_0, $\tilde{\omega}$ has to be replaced by $\tilde{\omega}^2$.) See also: M. Taut; Proceedings of the EP2DS Meeting, Ottawa (1999), published in Physica E **6** (2000) 479.
7. M. Taut, J. Phys.: Condens. Matter **21** (2009) 075302. See also: M. Taut; Proceedings of the EP2DS Meeting, Genova 2007, published in Physica E **40** (2008) 1062.
8. A. Matulis, F. M. Peeters, Solid State Commun. **117** (2001) 655.
9. M. Taut, P. Machon, and H. Eschrig, Phys. Rev. A **80** (2009) 022517.
10. G. Vignale, M. Rasolt, Phys. Rev. B **37** (1988) 10685.
11. G. Vignale, M. Rasolt, Advances in Quantum Chemistry **21** (1990) 235.

12. H. Eschrig, W. E. Pickett, Solid State Commun. **118** (2001) 123.
13. H. Eschrig, W. E. Pickett, J. Phys.: Condens. Matter **19** (2007) 315203.
14. P. Hohenberg, W. Kohn, Phys. Rev. **136B** (1964) 864.
15. H. Eschrig, The Fundamentals of Density Functional Theory. B. G. Teubner Verlagsgesellschaft (1996).
16. R. M. Dreizler, E. K. U. Gross, Density Functional Theory. Springer Verlag (1990).
17. K. Capelle, G. Vignale, Phys. Rev. B **65** (2002) 113106.
18. K. Capelle, G. Vignale, Phys. Rev. Lett. **86** (2001) 5546.
19. A. Wensauer, U. Rössler, Phys. Rev. B **69** (2004) 155301 and 155302.
20. U. Merkt, J. Huser and M. Wagner, Phys. Rev. B **43** (1991) 7320.
21. M. Wagner, U. Merkt and A. V. Chaplik, Phys. Rev. B **45** (1992) 1951.
22. M. Taut, Phys. Rev. B **62** (2000) 8126; and Phys. Rev. B **63** (2001) 1153

Adaptive Methods in Quantum Chemistry

By Heinz-Jürgen Flad*, Thorsten Rohwedder, and Reinhold Schneider

TU Berlin, Straße des 17. Juni 136, 10623 Berlin, Germany

(Received July 30, 2009; accepted September 9, 2009)

Adaptive Algorithms / Convergence Analysis / Best N-term Approximation / Coupled Cluster Theory / Error Estimators

In this article, we summarize the results of our numerical analysis for various adaptive schemes in electronic structure calculations. Starting from a general form of iterative schemes used in quantum chemistry, we discuss miscellaneous features which are of importance when passing to an adaptive version of an algorithm in order to provide rigorous estimates. The key features are maintainance of convergence under small perturbations while keeping control of the computational complexity.

As a prototype for correlation methods, we consider the configuration interaction (CI) method and discuss a quasi-optimal algorithm for computation of the ground state energy. Our analysis of the coupled cluster method shows that various properties of the CI method can be transferred hereunto as well. First numerical examples demonstrate the potential of the coupled cluster residual, measured in a norm on the dual space, as an error estimator. These studies are supplemented by our result for the best N-term approximation of electron-pair correlations by wavelets.

For Hartree-Fock and Kohn-Sham methods, we discuss our results of the convergence analysis for a direct minimization scheme, the best N-term approximation of orbitals and s^*-compressibility of the Fock operator which altogether may serve as a basis for an adaptive version of these methods.

1. Introduction

Concepts of adaptivity have found widespread use in quantum chemistry, ranging from the construction of *Gaussian-type orbital* (GTO) basis sets, see e.g. the development of correlation consistent bases [1–3], to linear scaling methods in *density functional theory* (DFT) [4–9], selective *configuration interaction* (CI) methods [10–18] and local correlation methods based on many-body perturbation or *coupled cluster* (CC) theory [19–22]. In most of these methods, the adaptive procedure is based on physical insights and empirical evidence from numerical

* Corresponding author. E-mail: flad@math.tu-berlin.de

Z. Phys. Chem. **224** (2010) 651–669
© by Oldenbourg Wissenschaftsverlag, München

simulations; a rigorous mathematical justification, however, is usually missing. Often, familiar concepts lose a lot of their original power if one tries to put them in a rigorous mathematical framework. As a typical example, let us mention GTO bases, for which highly sophisticated optimization schemes have been devised in quantum chemistry. However, even systematic GTO basis sets like the correlation consistent bases mentioned before do not meet the requirements of a Riesz basis frequently used in numerical analysis. Riesz bases are not only complete, but remain stable even in the limit of an infinitely dimensional space. Putting GTO bases in a rigorous setting, which can be done e.g. within *approximate approximation theory* [23], deprives them to a certain extent of their flexibility and adaptivity. Unfortunately, similar drawbacks can be brought forward to most of the adaptive methods mentioned before. Therefore, we will not shoulder the monumental and perhaps questionable task to provide a rigorous mathematical analysis for the adaptive approaches used in nowadays quantum chemistry. Instead it seems more favourable to us to perform the mathematical analysis of methods used in electronic structure calculations within settings familiar to numerical analysis. Of course, this might lead to adaptive methods which are not fully competitive from a practical point of view; for example, working with a systematic Riesz basis instead of GTO bases requires from the onset larger basis sets and the benefit of systematic improvement might be a distant prospect. However, we have the more realistic prospect that our rigorous analysis provides new and hopefully enlightening perspectives on standard adaptive methods, which we reckon cannot be obtained in another way.

The paper is organized as follows: In Section 2, we present a general outline of iterative adaptive algorithms and discuss some of the basic mathematical concepts involved in their numerical analysis. Turning to concrete examples in Section 3, we discuss adaptive CI and CC algorithms. Section 4 deals with the numerical analysis of *Hartree-Fock* (HF) and *Kohn-Sham* (KS) equations which paves the way for adaptive algorithms with optimal computational complexity. For the convenience of the reader we have summarized essential background material from functional analysis in Appendix A.

2. Design of adaptive algorithms

The following discussion of adaptive algorithms is based on the general assumption that the underlying problem can be treated in terms of an iterative solution scheme of the abstract form:

Iterative Algorithm (I)

Require: *Initial iterate* $\boldsymbol{x}^{(0)} \in V$;
for $n = 0, 1, \ldots$ *until convergence do*
(1) *From* $\boldsymbol{x}^{(n)}$ *and given data, compute update* $\Delta\boldsymbol{x}^{(n)}$.

(2) *Update* $\boldsymbol{x}^{(n+1)} = \boldsymbol{x}^{(n)} + \Delta\boldsymbol{x}^{(n)}$.
endfor

In most cases of interest, step (1) consists in matrix-vector multiplications or function evaluations to compute a residual-like quantity; for example, in the case of CI methods, this role is played by $\mathbf{r}^{(n)} = \mathbf{H}\,\mathbf{x}^{(n)} - E^{(n)}\mathbf{x}^{(n)}$, in CC methods it is the evaluation of the CC function $\mathbf{r}^{(n)} = \mathbf{g}(\mathbf{x}^{(n)})$, cf. Section 3 below. To compute the update, this residual is usually multiplied by a preconditioning matrix. In electronic structure calculations this role is often played by the inverse of the shifted Fock matrix, i.e. $\Delta\mathbf{x}^{(n)} = \mathbf{F}^{-1}\mathbf{r}^{(n)}$ Note that after step (2), some algorithms, e.g. those for HF or KS equations, require a supplementary step like normalization or orthogonalization of the iterates in order to fulfil additional constraint conditions.

The objective of an adaptive algorithm is the usage of sparse quantities in every step of algorithm (I) and thereby a rigorous control of the computational complexity. In a pseudo-algorithmic form, an adaptive algorithm may be formulated in the following way:

Adaptive Algorithm (II)

Require: *Sparse initial iterate* $\boldsymbol{x}^{(0)} \in V$;
for $n = 0, 1, \ldots$ *until convergence do*
(1) *APPLY-step: From* $\boldsymbol{x}^{(n)}$ *and compressed data, compute with optimal computational complexity an approximate update* $\Delta_\varepsilon\, \boldsymbol{x}^{(n)}$.
(2) *Update* $\boldsymbol{x}^{(n+1)} = \boldsymbol{x}^{(n)} + \Delta_\varepsilon\, \boldsymbol{x}^{(n)}$.
(3) *COARSE-step: If* $supp(\boldsymbol{x}^{(n+1)})$ *gets too big, sparsify* $\boldsymbol{x}^{(n+1)}$ *by thresholding.*
endfor

Such an adaptive algorithm features two approximation steps, i.e., the computation of an approximate residual instead of the exact one, and an intermediate coarsening step to keep sparsity of the iterates. The general goal is (i) to construct an adaptive algorithm in the fashion of (II), which converges to the limit of the exact algorithm (I) up to a controllable error tolerance ε while (ii) it is based on the usage of sparse quantities to control the computational complexity of the algorithm. Let us investigate these two points in more detail:

(i) Maintaining convergence: Let us assume for illustration that for the basic iteration scheme (I), we have linear convergence, i.e., the error reduces by a fixed factor $c < 1$ in each step,

$$\mathrm{err}(\mathbf{x}^{(n+1)}) = \mathrm{err}(\mathbf{x}^{(n)} + \Delta\mathbf{x}^{(n)}) \leq c\, \mathrm{err}(\mathbf{x}^{(n)}). \tag{2.1}$$

For a subadditive error measure, we may estimate the error of an approximate iterate $\mathbf{x}_\varepsilon^{(n+1)} = \mathbf{x}^{(n)} + \Delta_\varepsilon\mathbf{x}^{(n)}$ by

$$\mathrm{err}(\mathbf{x}_\epsilon^{(n+1)}) \;\leq\; \mathrm{err}(\mathbf{x}^{(n)} + \Delta\mathbf{x}^{(n)}) + \mathrm{err}(\Delta_\epsilon\mathbf{x}^{(n)} - \Delta\mathbf{x}^{(n)}), \tag{2.2}$$

so that convergence can be maintained (with a different constant $\tilde{c}$ still smaller than 1) if we are for instance able to control the approximation error by

$$\mathrm{err}(\Delta_\epsilon \mathbf{x}^{(n)} - \Delta \mathbf{x}^{(n)}) \;<\; \epsilon\, \mathrm{err}(\mathbf{x}^{(n)}) \;\text{ with } \tilde{c} = c + \epsilon < 1.$$

In many cases, it can be shown that the current error may be estimated from above and below by the residual (measured in an appropriate norm), so that the residual may be utilized to estimate the accuracy of the adaptive approximation needed in order to maintain convergence.

From a mathematical point of view, this procedure raises the following questions: To lay a basis for error estimates like (2.2), it has first of all to be shown that convergence results like (2.1) hold at least in a neighbourhood of the solution. Of course, this is also a question of its own interest. This significance of convergence analysis is contrasted by the scarcity of rigorous convergence analysis for methods utilized in quantum chemistry. Next, the relation between residuals and the actual errors mentioned before has to be established. Finally, we have to investigate the mutual interplay between errors in the iterates and the error in the quantity of interest, i.e., in our case the eigenvalues of the operators under consideration.

(ii) Control of complexity: Once convergence has been established for the full, still infinite dimensional problem, the principal idea is to restrict each iteration step in an approximate manner to a finite dimensional subspace in order to control the computational complexity of the whole scheme. This requires that all components of the iteration scheme allow for sparse approximations which are preserved during the operations of the iteration step.

(a) Sparse representability of iterates: In order to keep the number of nonzero coefficients $x_i^{(n)} \neq 0$ of an iterate $\mathbf{x}^{(n)}$ as small as possible, we need an adaptive approximation theory for the final solution and the intermediate iterates. For a certain class of stable Riesz bases, e.g. wavelets, this can be accomplished in the framework of *best-N-term approximation* theory. In this paragraph we give a short outline of this theory and discuss its application for adaptive algorithms. For a detailed exposition of this subject we refer to [24].

Given a function f which can be exactly represented in a Riesz basis

$$B := \{\zeta_i : i \in \mathcal{I}\}, \tag{2.3}$$

we denote by $\mathbf{f}$ the coefficient vector with respect to B. The smallest error

$$\sigma_N(\mathbf{f}) := \inf_{\mathbf{f}_N \in \Sigma_N} \|\mathbf{f} - \mathbf{f}_N\|_{\ell_2} \tag{2.4}$$

that can be achieved for the approximation of $\mathbf{f}$ by a linear combination of at most N basis functions, i.e. by functions taken from the nonlinear subset

$$\Sigma_N := \Big\{ \sum_{j \in \mathcal{J}} c_j \zeta_j \mid \mathcal{J} \subset \mathcal{I}, |\mathcal{J}| \leq N \Big\}, \tag{2.5}$$

is the error of the best N-term approximation of $\mathbf{f}$ in $\ell_2(\mathcal{I})$.

The approximation error $\sigma_N(\mathbf{f})$ for the coefficient vector is given with respect to the Euclidean norm on $\ell_2(\mathcal{I})$. For a Riesz basis B this error is equivalent to the approximation error of the function f in the Hilbert space L_2 or the Sobolev space H^1 (see Appendix A) depending on the weights assigned to the basis functions. In order to characterize all functions for which the error of their best N-term approximation decays at least as N^{-s} for some $s > 0$ we introduce the (quasi-) seminorm

$$|\mathbf{f}|_{\mathcal{A}^s} := \sup_{N \in \mathbb{N}} N^s \sigma_N(\mathbf{f}) \tag{2.6}$$

This seminorm can be used to define the best N-term approximation spaces

$$\mathcal{A}^s := \{\mathbf{f} \in \ell_2(\mathcal{I}) : \|\mathbf{f}\|_{\mathcal{A}^s} := \|\mathbf{f}\| + |\mathbf{f}|_{\mathcal{A}^s} < \infty\},$$

which are (quasi-)Banach spaces for all $s > 0$. Another way to view at this is the following: In order to approximate a given function f with coefficient vector $\mathbf{f} \in \mathcal{A}^s$ with an accuracy ε it requires at most $N_\varepsilon = \varepsilon^{-1/s} |\mathbf{f}|_{\mathcal{A}^s}^{1/s}$ nonzero entries. This fundamental relation

$$\text{accuracy } \varepsilon \quad \longleftrightarrow \quad (|\mathbf{f}|_{\mathcal{A}^s}/\varepsilon)^{1/s} \text{degrees of freedom} \tag{2.7}$$

reflects the desired sparsity of the iterates in $\ell_2(\mathcal{I})$ and serves as a guideline for the construction of adaptive algorithms.

Due to the norm equivalence of the function spaces L_2 or H^1 with the discrete $\ell_2(\mathcal{I})$ norm of the coefficient vector for an appropriate Riesz basis B, a best N-term approximation of the coefficient vector, which can be achieved by simply selecting the N largest coefficients, gives a quasi-best approximation of the corresponding function in Σ_N. In order to predict quasi-optimal convergence rates of adaptive algorithms it is therefore necessary to specify the best N-term approximation spaces $\mathcal{A}^s$ to which the final solution and all intermediate iterates belong. The seminorm (2.6) is often not convenient for this purpose. As an alternative it is possible to choose a subspace $\mathcal{A}_q^s \subset \mathcal{A}^s$ with $\frac{1}{q} = s + \frac{1}{2}$ which is only slightly smaller, i.e. $\mathcal{A}^{s'} \subset \mathcal{A}_q^s \subset \mathcal{A}^s$ for all $0 < s' < s$, but easier to handle. The subspaces $\mathcal{A}_q^s$ can be identified with certain Besov spaces [24] which provide an alternative to Sobolev spaces to characterize the smoothness of a function. In contrast to Sobolev spaces of higher order of regularity, the less stringent Besov spaces of comparable order allow e.g. for a finite number of discontinuities in the function or its derivatives provided that the function is sufficiently regular in between. This is of special significance for the point-like singularities at the electron-nuclear cusps.

Among the most popular realizations of Riesz bases are wavelets. The corresponding multiresolution analysis provides a convenient setting for the development of adaptive algorithms. We refer to the monographs [25,26] for a detailed exposition of this subject.

(b) Complexity of matrix-vector operations: Computation of the residual from an iterate $\mathbf{x}^{(n)}$ involves (possibly a series of) matrix-vector operations with an $M \times M$ matrix. For the sparse case, these operations may be executed at a favourable expense of $m \cdot \text{supp}(\mathbf{x}^{(n)})$ with $m \ll M$, while for the non-sparse case, $M \cdot \text{supp}(\mathbf{x}^{(n)})$ products have to be computed. If the matrix is *quasi-sparse*

or s^*-*compressible*, i.e., the coefficients decay in a certain manner, a more elaborate scheme selecting the important contributions to the product may be utilized to approximate matrix-vector multiplications at an expense no higher than $\mathcal{O}(\text{supp}(\mathbf{x}^{(n)}))$[27]. To rely on these schemes, an analysis of the properties of the involved operators, i.e. the Hamiltonian H resp. Fock operator, in the mathematical context of s^*-*compressibility* is in order. To explain this terminology, we assume a fixed basis set (2.3), which provides us with a Hamiltonian matrix $\mathbf{H}$ in the Galerkin discretization (which is infinite dimensional if we use a complete basis). A Hamiltonian matrix $\mathbf{H}$ is called s^*-compressible if for each $q \in \mathbb{N}$ an approximate matrix $\mathbf{H}_q$ can be constructed, that has in each row and column at most $\mathcal{O}(2^q)$ non vanishing entries and satisfies $\| \mathbf{H} - \mathbf{H}_q \| \lesssim 2^{-tq}$ in the spectral norm for any $0 < t < s^*$. Obviously, the s^*-compressibility of the Galerkin discretization of a certain operator also depends on the specific properties of the basis set. A similar consideration holds for the approximation spaces $\mathcal{A}^s$: The approximation properties of a given basis set can be characterized in terms of a constant $\bar{s}$, and if $s \geq \bar{s}$ the spaces $\mathcal{A}^s$ are essentially empty. Therefore, $\bar{s} \leq s^*$ is a necessary condition in order to fully benefit from the best N-term approximation rate.

(c) Maintaining sparsity: Even if (quasi-)sparse matrices (b) are used in the algorithm, the support of sparse iterates (a) may get bigger by each matrix-vector multiplication, in turn increasing the number of operations needed for the next iteration step. To keep iterates sparse, a $\mathcal{O}(\text{supp}(\mathbf{x}^{(n)}))$ thresholding procedure based on binary binning [28] may be utilized, corresponding to step (3) in algorithm (II).

It is the purpose of our work to investigate the various aspects mentioned in (i) and (ii) for some standard algorithms used in quantum chemistry. Considered together, it enables the design of quasi-optimal algorithms, i.e. algorithms which up to a fixed constant use the minimal number N of non-null coefficients needed for a prescribed target accuracy ε (determined via (2.7) by the best-N-term approximation), while their computational complexity scales linearly with N.

3. Adaptive correlation methods

Adaptivity in quantum chemical correlation methods, i.e., the selection of important excitation amplitudes, is either based on screening procedures using perturbative arguments or uses locality criteria together with local basis functions. Typical examples for the first approach are canonical orbital based methods like CIPSI [10–12] or [16,17] and selective MR-CI methods [13–15]. The second approach is represented by the local correlation methods of Werner and coworkers [19,22] or by the work of Ayala and Scuseria [20,21]. A combination of both approaches is the selective and local CC method of Auer and Nooijen [18]. In the first part of this section, we give for the conceptually simple case of CI a concrete prototype example for a mathematically rigorous adaptive algorithm which uses a residual-based selection criterion and is quasi-optimal in the sense discussed before. Turning to the CC equations, we summarize our recent results about their convergence behaviour and present some preliminary numerical results for an adaptive CC algorithm. Finally, we discuss best-N-term approximation theory for pair-correlation methods.

3.1 A prototype quasi-optimal CI algorithm

For the CI method, corresponding to a Galerkin approach for the Hamiltonian H, we devised in [29] an adaptive algorithm for the computation of the ground state wave function. The basic algorithm (I) is in this case a Richardson/preconditioned gradient eigensolver in which

$$\Delta\mathbf{x}^{(n)} = -\alpha\mathbf{F}^{-1}(\mathbf{H}\mathbf{x}^{(n)} - E^{(n)}\mathbf{x}^{(n)})$$

with a fixed parameter α. In the discrete case, convergence results are well known, cf. Ref. [31]. We considered the case of an infinite dimensional function space with iterates $\Phi^{(n)}$. For this we proved that linear convergence (2.1) is guaranteed in a neighbourhood of the solution. Furthermore, we showed that this result is retained in the sense of (2.2) with the use of approximate applications $APPLY(H,\Phi^{(n)},\varepsilon)$ of the Hamiltonian, cf. step (2) of Algorithm II. The accuracy ε of the operator application is herein in each step proportional to the size of (an approximation of) the current residual. In [30], we have also extended the following sharp result [31] about the convergence of eigenvalues to our adaptive algorithm:

Theorem 1 *Let $\Phi \in H^1(\mathbb{R}^{3N})$, $\|\Phi\|_{L_2} = 1$, such that the associated Rayleigh quotient $E = \langle H\,\Phi, \Phi\rangle$ lies between the k-th and (k+1)-th eigenvalue of H, $E_k \leq E < E_{k+1}$ below the essential spectrum of H. Denote by*

$$\|\Phi\|_{\mathfrak{H}} := \langle (H+\mu I)\Phi, \Phi\rangle^{\frac{1}{2}}, \quad \|\phi\|_{\mathfrak{H}^{-1}} := \langle (H+\mu I)^{-1}\phi, \phi\rangle^{\frac{1}{2}}$$

the norms induced by the shifted Hamiltonian for $\Phi \in H^1$ and $\phi \in H^{-1}$, respectively. Let the shifted Fock matrix be scaled such that

$$\|I - \alpha F^{-1}H\|_{\mathfrak{H}} \leq \gamma_P < 1. \tag{3.1}$$

If we bound the error in the operator application by

$$\|H\Phi - APPLY(H,\Phi,\varepsilon)\|_{\mathfrak{H}^{-1}} \quad \leq \quad \gamma_\xi \|H\Phi - E\Phi\|_{\mathfrak{H}^{-1}},$$

where $\gamma = \gamma_P + \gamma_\xi < 1$, and obtain the next (inexact) iterate by normalizing

$$\Phi' = \Phi - \alpha F^{-1}(APPLY(H,\Phi,\varepsilon)), \tag{3.2}$$

there holds for the associated energy $E' = \langle H\,\Phi', \Phi'\rangle / \langle \Phi', \Phi'\rangle$ either $E' < E_k$ or $E_k \leq E' < E_{k+1}$. In the latter case,

$$\frac{E' - E_k}{E_{k+1} - E'} \leq q^2(\gamma, E_k, E_{k+1})\frac{E - E_k}{E_{k+1} - E}.$$

Herein, the reduction factor $q < 1$ depends only on the eigenvalue gap E_k/E_{k+1} and the quality (3.1) of the shifted Fock matrix as preconditioner.

A reliable and accurate estimator for the (in principle infinite dimensional) residual $H\Phi - E\,\Phi$ has been designed in Ref. [30]. This estimator thus provides lower and upper bounds for the error of the iterates. The main achievement of our work [29] is a complexity analysis for a concrete realization *MINIEIG*(**H**,**F**,ε) of the adaptive algorithm (II) that shows asymptotically optimal complexity in the following sense:

Theorem 2 *Assume that for a chosen basis B, the Galerkin matrices* $\boldsymbol{H}$ *and* $\boldsymbol{F}$ *are* s^**-compressible for some* $s^* > 0$ *and that (3.1) holds. Denote the exact ground state solution by* (Ψ, E_0)*, the coordinate vector of* Ψ *by* $\boldsymbol{u}$ *and by* $P_{\boldsymbol{u}}$ *the projector on span*$\{\boldsymbol{u}\}$*. Then for any target accuracy* $\varepsilon > 0$*, the scheme MINIEIG*(**H**,**F**,ε)*, consisting of the update step (3.2) and a thresholding procedure to coarse the iterates (see [29]), produces after finitely many steps an approximate coefficient vector/energy pair* $(\boldsymbol{x}(\varepsilon), E(\varepsilon))$ *satisfying*

$$\|(I - P_{\mathbf{u}})\mathbf{x}(\varepsilon))\|_{\ell_2} \leq \varepsilon, \quad |E_0 - E(\varepsilon)| \leq \varepsilon, \tag{3.3}$$

where $I - P_{\boldsymbol{u}}$ *is the projector on the orthogonal complement of* $\boldsymbol{u}$*. Moreover, if* $\boldsymbol{u} \in \mathcal{A}^s$ *for some* $s < s^*$*, then one has for the overall computational and storage requirements*

$$\#\text{flops}\,\mathbf{x}(\varepsilon),\ \#\,\text{supp}\ \mathbf{x}(\varepsilon) \ \leq\ C \cdot (\|\mathbf{u}\|_{\mathcal{A}^s}/\varepsilon)^{1/s} \tag{3.4}$$

and $\|\boldsymbol{x}(\varepsilon)\|_{\mathcal{A}^s} \leq C \cdot \|\boldsymbol{u}\|_{\mathcal{A}^s}$*, where the constants are independent of* ε *and* $\boldsymbol{u}$ *but depend only on s when s approaches* s^**, so that MINIEIG is an asymptotically quasi-optimal algorithm is in the discussed sense.*

The s^*-compressibility of the Fock operator F has been studied in Ref. [32]. Our main result is briefly discussed in Section 4 together with the best N-term approximation of orbitals. For the Hamiltonian H we can establish s^*-compressibility, however, this result is presumably not optimal and a more refined analysis is required.

3.2 Convergence of the CC equations

We consider the linked CC formulation in canonical molecular obitals with the HF reference determinant Ψ_0 and excitation operators of the form $T = \Sigma_{\nu \in \mathcal{J}} t_\nu X_\nu$, where e.g. the twofold excitations, $X_\mu = a_r^\dagger a_s^\dagger a_l a_k$, are indexed by $\mu \cong \binom{l,k}{r,s} \in \mathcal{J}$. They are given by

$$f_\mu(\mathbf{t}) = \langle \Psi_\mu, e^{-T} H e^{T} \Psi_0 \rangle = \epsilon_\mu t_\mu + \langle \Psi_\mu, e^{-\sum t_\nu X_\nu} U e^{\sum t_\nu X_\nu} \Psi_0 \rangle \ = \ 0, \tag{3.5}$$

where ε_μ stems from the shifted Fock operator

$$\mathbf{F} = (\langle \Psi_\nu, [\mathcal{F}, X_\mu] \Psi_0 \rangle)_{\nu,\mu \in \mathcal{J}} = \text{diag}(\epsilon_\mu)$$

and

$$\mathbf{g}(\mathbf{t}) = \left(\langle \Psi_\mu, e^{-\sum t_\nu X_\nu} U e^{\sum t_\nu X_\nu} \Psi_0 \rangle \right)_{\nu,\mu \in \mathcal{J}}$$

is the corresponding fluctuation potential.

It is shown in [33] that the projected CC equation admits a solution provided that the reference determinant is sufficiently closed to an exact non-degenerate wave function Ψ, and that it allows for a quasi-optimal error estimate. Herein, the error is not estimated with respect to the L_2 norm, but rather in an non-equivalent energy norm. In terms of the unknown amplitudes, this norm can be expressed by

$$\|\mathbf{t}\|_V^2 := \sum_{\mu,\nu\in\mathcal{J}} t_\mu t_\nu \langle \Psi_\nu, [\mathcal{F}, X_\mu]\Psi_0\rangle = \sum_\mu \epsilon_\mu |t_\mu|^2 \,. \tag{3.6}$$

This result is established by proving *strong monotonicity* of the unlinked coupled cluster formulation, which under reasonable conditons provides the same solution (see [33]). Strong monotonicity for $\mathbf{f}$ means that there is a constant $\gamma > 0$ for which

$$\langle \mathbf{f}(\mathbf{v}) - \mathbf{f}(\mathbf{w}), \mathbf{v} - \mathbf{w}\rangle \geq \gamma \|\mathbf{v} - \mathbf{w}\|_V^2 \quad \text{for all } \mathbf{v}, \mathbf{w} \in V \tag{3.7}$$

For a Lipschitz continuous strongly monotone function $\mathbf{f}$, it is known that there exists a damping factor $0 < \alpha \leq 1$ such that the damped quasi-Newton method

$$\mathbf{t}^{n+1} := \mathbf{t}^n - \alpha \mathbf{F}^{-1}\mathbf{g}(\mathbf{t}^n), \tag{3.8}$$

converges linearly. Usually, this iteration is used with $\alpha = 1$, obtaining a single Picard fixed point iteration. For this scheme, linear convergence is guaranteed under the more restrictive contraction assumption $\|\mathbf{g}(\mathbf{t})\|_V \leq q \|\mathbf{t}\|_V$ with a contraction rate $0 < q < 1$. If the convergence of this scheme fails the damped variant is applied instead. We would like to highlight that convergence can be improved by subspace acceleration techniques like DIIS [34].

Improved error bounds for the energy can be provided by consideration of the variational CC Lagrangian

$$L(\mathbf{t}, \mathbf{a}) = E(\mathbf{t}) + \langle \mathbf{f}(\mathbf{t}), \mathbf{a}\rangle \,,\; E(\mathbf{t}) = \langle \Psi_0, H(T_2 + \frac{1}{2}T_1^2)\Psi_0\rangle$$

with a Lagrange-multiplier $\mathbf{a} \in V$, see [33,34] for further details. So far, we have not exploited the potential of the duality approach, which from the perspective of accuracy can be seen as an improvement of perturbation techniques. Essentially, the iteration scheme for the CC equations is quite similar to the one for the CI approach.

3.3 Adaptive CC algorithms

We have modified the fixed point algorithm (3.8), which is commonly used in CC algorithms, into an adaptive algorithm in the style of Algorithm II:

Adaptive CC Algorithm (ACC)

Initial iterate: $t^{(0)} = \mathbf{0}$, *initial accuracies:* $\varepsilon_0, \bar{\varepsilon}_0$.
for $n = 0, 1, \ldots$ *until convergence do*
(1) $\Delta_{\varepsilon n}\, t^{(n)} = APPLY\,(t^{(n)}, F^{-1}g, \varepsilon_n)$
(2) *Update* $\bar{t}^{(n+1)} = t^{(n)} + \Delta_{\varepsilon n}\, t^{(n)}$.
(3) $t^{(n+1)} = COARSE\, \bar{t}^{(n+1)}, \bar{\varepsilon}_n)$.

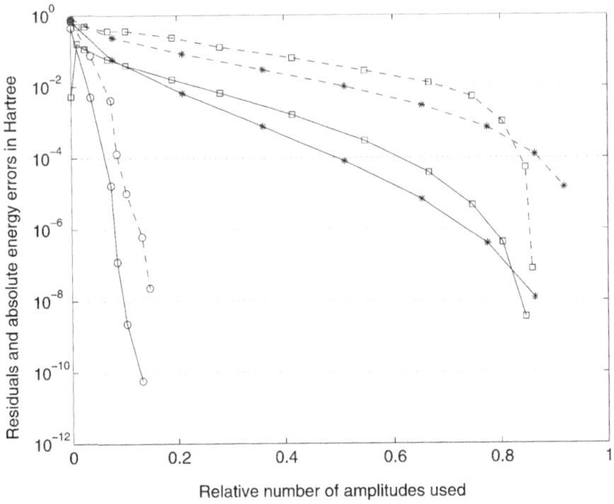

Fig. 1. Absolute error in energy (solid line) and residuals (dashed line) during the iteration of the ACC algorithm plotted against the ratio $N_{coa}/(N_{occ}^2 N_{virt}^2)$ of selected T_2 amplitudes. Results are shown for H_2O (aug-cc-pVTZ, □), N_2 (cc-pVQZ, ○) and glycine (Dunning-TZ, *). Each symbol represents an outer iteration step of the ACC algorithm.

(4) *Choose* ε_{n+1}, $\bar{\varepsilon}_{n+1}$ *proportional to current residual;*
continue until residual is of desired accuracy.
endfor

Herein, the evaluation of the fixed point residual $\Delta \mathbf{t}^{(n)} = \mathbf{r}^{(n)} = \mathbf{F}^{-1}\mathbf{g}\ (\mathbf{t}^{(n)})$ is replaced by the routine $APPLY(\mathbf{t}^{(n)}, \varepsilon_n)$ which determines $\mathbf{r}^{(n)}$ up to accuracy ε_n. Afterwards, a thresholding step $COARSE(\mathbf{t}^{(n)}, \bar{\varepsilon}_n)$ outputs an $\bar{\varepsilon}$-approximation $\mathbf{t}_{n+1}$ of $\mathbf{t}$ having a quasi-minimal number of non-null entries to keep the amplitude vector as sparse as possible. On the selected amplitudes produced in this way, we iterated the CC equations 3–5 times before calculating the next full residual. For an iterate $\mathbf{t}$ with support of size N_{coa} instead of the full set of $N_{occ}^2\ N_{virt}^2$ amplitudes, the complexity of the matrix-vector multiplications involved in the evaluation of the full CC function is thus reduced by a factor $N_{coa}/(N_{occ}^2\ N_{virt}^2)$; for example, the computationally most demanding term in CCSD scales as $N_{coa}\ N_{virt}^2$ instead of $N_{occ}^2\ N_{virt}^4$.

We implemented an adaptive CCSD algorithm based on the NWChem package [35,36], in which the residual is first calculated exactly and coarsened subsequently, corresponding to a selection of most significant excitation amplitudes in a large basis set. Motivated by considerations common in finite element methods, we investigated if the dual norm $\rho = \|\mathbf{f}(\mathbf{t})\|_{\mathbf{F}^{-1}}$ of the residual $\mathbf{f}(\mathbf{t})$ (see Appendix A) may be utilized as an error estimator. In Figure 1, the convergence of the adaptive algorithm and the relation between the amount of single and double amplitudes $\mathbf{t}_\varepsilon$ and the accuracy of the energy is exemplified for some molecules using canonical orbital bases. The current coarsening threshold ε_n was chosen proportional to ρ in each iteration step. Note that although only a part of the amplitudes, predicted by the residual, is used, we have in principle a smooth convergence to the full limit. In the canonical orbital basis, the computational work to be done is reduced only moderately to about 60%, a saving that is of course offset by the extra effort needed for the coarsening step (although in our calculations, further savings result for N_2 from additional symmetries not exploited in NWChem). The more promising case of localized orbitals will be studied in the near future in cooperation with A. Auer. Although our calculations are so far limited to the CCSD approach, it is in principle possible and desirable to include contributions of higher excitations.

3.4 Best N-term approximation for pair correlations

We have also studied the best N-term approximation of pair-correlation functions which appear in the continuous formulation of perturbation theory [37] and in the Jastrow factors [38,39] used in *quantum Monte Carlo* (QMC) calculations. In order to apply best N-term approximation theory we have to assume an asymptotic smoothness property in a neighbourhood of the electron-electron cusp.

Assumption 1 *The two-particle correlation function $\tau^{(2)}$ belongs to $C^\infty(\mathbb{R}^3 \times \mathbb{R}^3 \setminus D)$, with $D := \{(x,y) \in \mathbb{R}^3 \times \mathbb{R}^3 : x = y\}$. Furthermore it satisfies the asymptotic smoothness property*

$$|\partial_{\mathbf{x}}^{\alpha} \partial_{\mathbf{y}}^{\beta} \tau^{(2)}(\mathbf{x}, \mathbf{y})| \leq c_{\alpha,\beta} |\mathbf{x} - \mathbf{y}|^{1-|\alpha|-|\beta|}, \tag{3.9}$$

for $x \neq y$ and $|\alpha| + |\beta| \geq 1$, in any bounded neighbourhood $\Omega \times \Omega \subset \mathbb{R}^3 \times \mathbb{R}^3$.

Here we have used the standard short-hand notation for mixed partial derivatives

$$\partial^{\beta} := \frac{\partial^{\beta_1}}{\partial x_1^{\beta_1}} \frac{\partial^{\beta_2}}{\partial x_2^{\beta_2}} \frac{\partial^{\beta_3}}{\partial x_3^{\beta_3}}, \tag{3.10}$$

with absolute value of the multi-index $|\beta| := \beta_1 + \beta_2 + \beta_3$. A rigorous justification of our assumption is presently under investigation, however, it is obviously satisfied by the standard Jastrow factors employed in QMC calculations [38,39] and by the r_{12} terms used in explicitly correlated methods [40]. With our regularity assumption we obtained the following lemma, cf., Ref. [41].

Lemma 1 *Suppose $\tau^{(2)}$ satisfies Assumption 1. Then $\tau^{(2)}$ belongs to $A_q^{\alpha}(H^1(\Omega \times \Omega))$ for $\alpha < \frac{1}{2}$ and $\frac{1}{q} = \alpha + \frac{1}{2}$.*

It has been actually shown that this lemma remains valid under a more realistic assumption which contains contributions from electron-nuclear and electron-electron-nuclear cusps which appear in the Fock expansion and in Hylleraas CI calculations [42].

Because Lemma 1 does not hold for $\alpha \geq \frac{1}{2}$ for any physically reasonable $\tau^{(2)}$, cf. [41], it is only possible to achieve a convergence rate $\sigma_N(\tau^{(2)}) \sim$

$N^{-1/2+\varepsilon}$, for any $\varepsilon > 0$, with respect to the norm of the Sobolev space $H^1(\Omega \times \Omega)$. This means that the energy converges with $O(N^{-1+\varepsilon})$ which is twice the convergence rate that can be achieved with standard pair-correlation methods, like CISD or CCSD, in combination with VXZ-GTO basis sets [43,44].

4. Adaptive HF and DFT algorithms

In HF and KS theory, we are faced with nonlinear eigenvalue equations

$$F_{\Psi}^{HF}\psi_i = \lambda_i\psi_i, \text{ resp. } F_{\Psi}^{KS}\psi_i = \lambda_i\psi_i, \quad \lambda_i \in \mathbb{R}, \quad i = 1, \ldots, N, \tag{4.1}$$

for an orthonormal set $\Psi = (\psi_1, \ldots, \psi_N)$ of single particle wave functions $\psi_i(\mathbf{x}) \in V$, where $V = H^1(\Omega)$ is the space of one time weakly differentiable functions on $\Omega = \mathbb{R}^3$ for closed shell models and $\Omega = \mathbb{R}^3 \times \{\pm\frac{1}{2}\}$ otherwise, see Appendix A.

4.1 Convergence of the direct minimization algorithm

The Fock operator F_{Ψ}^{HF} and the Kohn-Sham operator F_{ψ}^{KS} in (4.1) are derived from the first order conditions for a minimum of the associated energy functionals $\mathcal{J}_{HF}(\Phi)$, $\mathcal{J}_{KS}(\Phi)$, see e.g. [34]. Therefore, instead of solving the eigenvalue equations (4.1), it is also possible to consider HF and KS methods from the minimization point of view, cf. Ref. [45]. In [46] we treated these methods as a classical optimization task of minimizing a functional $\mathcal{J}: V^N \to \mathbb{R}$ under orthogonality constraints, which may be tackled by direct minimization algorithms like the following one:

Projected Gradient Descent Algorithm (PGD)

Require: *Initial iterate* $\Phi^{(0)} \in V$;
evaluation of $F_{\Phi^{(n)}}\Phi^{(n)}$ *and of preconditioner(s)* B_n^{-1}
(see comments below)
Iteration:
for $n = 0, 1, \ldots$ *do*
(1) *Compute* $F_{\Phi^{(n)}}\,\varphi_i^{(n)}$, $i = 1,\ldots,N$.
(2) *Update* $(\lambda_{i,j})_{i,j=1}^N = \langle F_{\Phi^{(n)}}\,\varphi_i^{(n)}, \varphi_j^{(n)}\rangle \in \mathbb{R}^{N\times N}$.
(3) *Let* $\hat{\varphi}_i^{(n+1)} := \varphi_i^{(n)} - \alpha B_n^{-1}(F_{\Phi^{(n)}}\,\varphi_i^{(n)} - \Sigma_{j=1}^N \lambda_{i,j}\,\varphi_j^{(n)})$.
(4) *Compute orthonormal set* $\Phi^{(n+1)} = (\varphi^{(n+1)})_{i=1}^N$ *for which*
span $\Phi^{(n+1)} = span\{\hat{\varphi}_i^{(n+1)} | i = 1,\ldots,N\}$ *(see comments below).*
endfor

The algorithm PGD is implemented in the recent wavelet-basis-based density functional code bigDFT [47], which itself is a part of the open source ABINIT package [48–51]. Computations show that the convergence of the algorithm is quite robust, also if the HOMO-LUMO gap is relatively small [51]. For implementation of the PGD algorithm, the choice of preconditioners B_n: $V \to V'$ is crucial; see [46] for various possibilities. This is reflected in the theoretical requirements that B_n has to be an elliptic symmetric operator B_n: $V \to V'$, which induces a norm on H^1 that is equivalent to that induced by F^{HF} or F^{KS}; for example, one can use approximations of the shifted Laplacian, i.e., $B_n \approx \alpha(-\frac{1}{2}\Delta + C)$, the choice also taken in the BigDFT project. As regards the orthogonalization of $\hat{\varphi}_i$, $i = 1,\ldots,N$, required in step (4), there are various favorable possibilities [46]; nevertheless, it provides the computational bottleneck in nowadays direct minimization HF/DFT computations [51]. For this PGD algorithm, we investigated the questions raised in point (i) of Section 2 and proved the following statements under a certain ellipticity assumption imposed on the Hessian of the respective functional $\mathcal{J}_{HF}(\Phi)$ or $\mathcal{J}_{KS}(\Phi)$, see below.

Linear reduction of the error: We denote by $\mathcal{D}_\Psi$ the L_2-projector on the sought subspace spanned by $\Psi = (\psi_1,\ldots,\psi_N)$. A measure for the energy error of an iterate $\Phi^{(n)}$ is given by

$$\|(\mathcal{I} - \mathcal{D}_\Psi)\Phi^{(n)}\|_{V^N}, \tag{4.2}$$

where the norm $\|.\|_V^N$ is equivalent but, depending on the preconditioner, not necessarily identical to the $(H^1)^N$-norm, see Appendix A. For the first iterate $\Phi^{(0)}$ sufficiently close to Ψ, there is a constant $\chi < 1$ such that

$$\|(\mathcal{I} - \mathcal{D}_\Psi)\Phi^{(n+1)}\|_{V^N} \;\leq\; \chi \cdot \|(\mathcal{I} - \mathcal{D}_\Psi)\Phi^{(n)}\|_{V^N}. \tag{4.3}$$

Note that in contrast to this, the *self-consistent-field* (SCF) iteration scheme, another prominent choice in HF and DFT, is faced with convergence problems [52], for the remedy of which advanced techniques have to be invoked to guarantee convergence [53].

Residuals as error estimators: The vector of "subspace residuals"

$$R^{(n)} = (r_1^{(n)},\ldots,r_N^{(n)}),\quad r_i^{(n)} = F_{\Phi^{(n)}}\phi_i^{(n)} - \sum_{j=1}^{N} \lambda_{i,j},\phi_j^{(n)},$$

computed in step (3) of the algorithm, also provides an efficient and reliable error estimator in the sense that for the error $\|(\mathcal{I}-\mathcal{D}_\Psi)\Phi^{(n)}\|_{V^N}$ sufficiently small, there are constants c, $C > 0$ such that

$$c\|R^{(n)}\|_{(V^N)'} \;\leq\; \|(\mathcal{I} - \mathcal{D}_\Psi)\Phi^{(n)}\|_{V^N} \;\leq\; C\|R^{(n)}\|_{(V^N)'}. \tag{4.4}$$

Herein, $\|.\|_{(V^N)'}$ is the dual norm (see Appendix A) that is equivalent to the norm on V^N only in the discrete case.

Quadratic convergence of the energy: If the orthogonality constraints are satisfied and the functional is two times continuously differentiable, the error in the energy depends quadratically on the approximation error of the minimizer Ψ, i.e.,

$$\mathcal{J}(\Phi) - \mathcal{J}(\Psi) \;\lesssim\; \|(I - \mathcal{D}_\Psi)\Phi^{(n)}\|_{V^N}^2 . \tag{4.5}$$

The differentiability condition is verified for HF since it depends polynomially on Φ. For DFT, the exchange correlation potential is not known exactly, so the question remains open for the general case.

The previous results apply to the continuous space $V = H^1(\mathbb{R}^3)$ as well as to V standing for a corresponding finite dimensional ansatz subspace of H^1, independent of discretization parameters like the size of the basis set. The mentioned ellipticity assumption for the Hessian of $\mathcal{J}$ under which our results hold is related with fundamental questions like uniqueness of a solution up to unitary transformations; for a simplified linearized HF model, it may be guaranteed if a HOMO-LUMO gap exists, see [46]. Although this assumption in our work cannot be verified rigorously, the proposed convergence behavior is observed by all benchmark computations. Note also that since the PGD algorithm is gradient directed, a line search based on the Armijo rule will guarantee convergence in principle, even without a coercivity condition [54,55].

4.2 Best N-term approximation for HF orbitals

The HF equation is of outstanding significance in electronic structure calculations. It is therefore remarkable that almost no rigorous results were known about the approximation properties of the solutions, the so-called HF orbitals. It is only known for the H atom [56,57] that GTO bases provide almost exponential convergence and there exists merely numerical evidence [58] that this result extends to general atoms and molecules. We have studied best N-term approximation spaces for HF orbitals and the corresponding one-electron reduced density matrix. More precisely, our results uses the level-shifting algorithm. According to the analysis of Cancès and Le Bris [53], this SCF iteration scheme converges and therefore satisfies our requirement (i) for an adaptive algorithm. The corresponding iterates and the final SCF solutions are smooth except at the nuclei. In order to apply best N-term approximation theory we require all intermediate and final solutions to be *asymptotically smooth* in a neighbourhood of a nucleus, i.e.,

$$|\partial^\beta_{\mathbf{x}}\phi_n(\mathbf{x})| \leq C_\beta |\mathbf{x} - \mathbf{A}_k|^{1-|\beta|}, \text{for } \mathbf{x} \neq \mathbf{A}_k, \text{ and } |\beta| \geq 1. \tag{4.6}$$

The asymptotic smoothness property essentially means that we get control on the partial derivatives of the orbitals near the nuclei. It is an immediate consequence of the following theorem [59]:

Theorem 3 *All intermediate iterative solutions* $\{\phi_n^{(i)}\}_{n=1,\dots,N/2}$ *as well as the final SCF solutions* $\{\phi_n\}_{n=1,\dots,N/2}$ *of the closed-shell HF equations, obtained via the level-shifting algorithm, exhibit Taylor asymptotics (polar coordinates) in a neighbourhood* Ω_k *of any of the* $k = 1,\dots K$ *nuclei, i.e.,*

$$\phi = \omega(r) \sum_{j=0}^{l-1} c_j(\varphi, \theta) r^j + \Phi_l,$$

with $\Phi_l \in C^m_B(\Omega_k)$ *for* $l > m$, *provided that the initial guess* $\{\phi_n^{(0)}\}_{n=1,\dots,N/2}$ *possesses this property. Here* $\omega \in C_0^\infty(\Omega_k)$ *denotes a cut-off function which is equal to* 1 *in a neighbourhood of the nucleus. Furthermore we have* $\phi \in \mathcal{S}(\mathbb{R}^3 \setminus \cup_{k=1}^{K} \Omega_k)$.

This theorem together with Lemma 1, from Ref. [60] enables general conclusions about the best N-term approximation of HF orbitals and the corresponding one-electron reduced density matrix.

Theorem 4 *The SCF solutions* ϕ_i *from Theorem 3 and the corresponding one-electron reduced density matrix belongs to* $\mathcal{A}_q^s(H^1)$ *for each* $s > 0$ *and* $\frac{1}{q} = s + \frac{1}{2}$.

Theorem 4 shows that for a sufficiently regular univariate wavelet basis with $p > s+1$ vanishing moments, the orbitals ϕ_i and one-electron reduced density matrix can be approximated in the corresponding *anisotropic* 3d or 6d tensor product wavelet basis with optimal convergence rate $\sigma_N(\phi) \sim N^{-s}$ with respect to the norm of the Sobolev space H^1. For the corresponding *isotropic* 3d tensor product wavelet basis, the optimal convergence rate reduces to $\sigma_N(\phi) \sim N^{-s/3}$ for the orbitals.

We are now able to compare the best N-term approximation rates with our result for the s^*-compressibility of the discrete Fock operator in an isotropic wavelet basis which satisfies certain Bernstein and Jackson estimates in the Sobolev space $W_\infty^t(\mathbb{R}^3)$, cf. Ref. [32] for further details. It has been shown that the infinite Galerkin matrix of the Fock operator is s^*-compressible with $s^* = t/3$. For example, in the case of spline wavelets of order $d = t+1$ and with $p \geq t+1$ vanishing moments we can achieve N-term approximation rates $s < t/3$, i.e., $\bar{s} = t/3$ cf. Section 2 (ii), which nicely fits together with $s^* = t/3$. The factor $1/3$ enters because of the usage of isotropic tensor product wavelet bases. These considerations are necessary but not sufficient to demonstrate that an adaptive wavelet algorithm with optimal computational complexity exists in the sense of Ref. [27].

5. Conclusions

We have presented our program for the numerical analysis of adaptive iterative methods in electronic structure calculations. Key features include the convergence analysis of iterative algorithms, sparse approximations of the final solution and of intermediate iterates, and sparse approximate discretizations of the Hamiltonian. Partial results of this extensive program include adaptive CI and CC methods and a fairly comprehensive study of the Hartree-Fock model. However much remains to be done in order to use our results as a basis for coherent adaptive methods which allow for controllable accuracies and complexity the one hand while they are competitive to already existing adaptive algorithms, based on physical and chemical insights and practical experience, on the other hand.

Appendix A: Mathematical concepts and notions

This section is meant to explain some of the mathematical notions fundamentally related with the treatment of the electronic Schrödinger equations from the mathematician's point of view. We refer e.g. to the monograph [61] for a detailed exposition of the subject.

Hilbert space, stable bases: The basic space for electronic structure calculation is some Hilbert space $\mathcal{H} = L_2$ of square integrable functions (in the Lebesgue sense). On this space, a basis B_0 may be chosen, and we may then express every $\Psi \in \mathcal{H}$ by basis functions Φ_i from B_0, $\Psi = \Sigma c_i \Phi_i$. If the ℓ_2-norm of the coefficient vector $\mathbf{c} = (c_i)$ of the solution then reflects (up to constants) the L_2-norm of the corresponding wave function, i.e.,

$$c \cdot \|\mathbf{c}\|_{\ell_2} \leq \|\Psi\|_{\mathcal{H}} \leq C \cdot \|\mathbf{c}\|_{\ell_2}, \tag{A.1}$$

with c,C independent of Ψ, B_0 is a so-called *stable* or *Riesz basis*. Note that (A.1) does not need to hold although B_0 is linearly independent; for an orthonormal basis, it is fulfilled as equality, the so-called *Parseval identity*. In general, a relation like in (A.1) between two norms is termed *equivalence of norms*.

Unboundedness, Sobolev space H^1, weak formulation: Operators containing differential operators, as for instance the Hamiltonian or the Fock/KS operator, are unbounded and not defined everywhere on $\mathcal{H}$; this poses some severe formal problems on the one hand, reflecting on the other hand in the fact that the discretized equations get more and more ill-conditioned the bigger the basis set is chosen. The common procedure for a Schrödinger-like equation, say $H\Psi = E\Psi$, is to switch to the *bilinear form* $\langle H\Psi, \Phi\rangle$ associated with H, which is maximally defined on the *Sobolev space* H^1 of functions from $\mathcal{H}$ for which a first derivative exists (in the weak sense) and is in L_2, that is, $\Psi \in H^1$ iff the norm

$$\|\Psi\|_{H^1} = \left(\int |\Psi(x)|^2 dx + \int |\nabla\Psi(x)|^2 dx\right)^{\frac{1}{2}}$$

(which is not equivalent to the L_2 norm) stays finite. The equation then is treated in the weak formulation

$$\langle H\Psi, \Phi\rangle = E\langle\Psi, \Phi\rangle \text{ for all } \Phi \in B_1, \tag{A.2}$$

where B_1 is a basis in the restricted ansatz space H^1. Choosing a basis B_1 which is stable with respect to H^1, i.e. (A.1) holds with the H^1-norm instead of the L_2-norm, enables an estimate of the H^1-norm of the iterates from the ℓ_2-norm of their coefficient vector.

An interesting fact is that in many cases, the norm associated to the shifted energy $\|\Psi\|_{\mathfrak{H}} = \langle (H + \mu I)\, \Psi, \Psi\rangle^{1/2}$ is actually equivalent to the H^1-norm. We therefore often have a relation between the error of the iterates, measured in the H^1-norm, and the error of the energy (which often is quadratic, see e.g. (4.5) and analogous results from [29] for CI). Note that in contrast, no such equivalence holds for the L_2/ℓ_2 norms from (A.1), rendering the coefficients of Ψ in an L_2-basis useless for error estimation, especially when the basis sets get large.

Dual norms, Sobolev space H^{-1}, residual estimators: In the above setting, the operator of interest, say H, can be viewed as an operator assigning to an element $\Psi \in H^1$ the linear functional $\psi: f \to \langle H\Psi, f\rangle$. H is therefore mapping to the *dual space* H^{-1} of H^1, i.e., the space of continuous linear functionals acting on H^1. An important consequence of this is that the residual, say $r = H\Psi - E\Psi$, is an element of H^{-1} and has to be measured in a corresponding norm. This may be done by taking the norm $\|r\|_{A^{-1}} := \langle A^{-1}r, r\rangle^{1/2}$ where A is a bounded, strictly positive operator mapping $H^1 \to H^{-1}$. A typical choice is for example the shifted Laplacian Δ or the shifted Fock operator F. In this norm it is often possible to obtain an equivalence to the H^1-error of the iterates which often provides in turn a reliable and efficient estimator for the error in energy.

Acknowledgement

The authors would like to thank Prof. A. A. Auer (Chemnitz) for providing his version of the NWChem code and computational resources as well as for discussions. Furthermore, we would like to thank Prof. W. Hackbusch (Leipzig) and Dipl. Phys. A. Zeiser (Berlin) for useful discussions.

References

1. T. H. Dunning, Jr., J. Chem. Phys. **90** (1989) 1007.
2. R. A. Kendall, T. H. Dunning, Jr. and R. J. Harrison, J. Chem. Phys. **96** (1992) 6796.
3. D. E. Woon and T. H. Dunning, J. Chem. Phys. **98** (1993) 1358; ibid. **100** (1994) 2975; **103** (1995) 4572.
4. X.-P. Li, R. W. Nunes and D. Vanderbilt, Phys. Rev. B **47** (1993) 10891.
5. M. S. Daw, Phys. Rev. B **47** (1993) 10895.
6. S. Goedecker and L. Colombo, Phys. Rev. Lett. **73** (1994) 122.
7. W. Kohn, Phys. Rev. Lett. **76** (1996) 3168.
8. J. M. Millam and G. E. Scuseria, J. Chem. Phys. **106** (1997) 5569.
9. C. Ochsenfeld and M. Head-Gordon, Chem. Phys. Lett. **270** (1997) 399.
10. B. Huron, J. P. Malrieu and P. Rancurel, J. Chem. Phys. **58** (1973) 5745.
11. S. Evangelisti, J. P. Daudey and J. P. Malrieu, Chem. Phys. **75** (1983) 91.
12. R. J. Harrison, J. Chem. Phys. **94** (1991) 5021.
13. R. J. Buenker and S. D. Peyerimhoff, Theoret. Chim. Acta **35** (1974) 33.
14. B. Engels, J. Chem. Phys. **100** (1994) 1380.
15. M. Hanrath and B. Engels, Chem. Phys. **225** (1997) 197.
16. C. F. Bunge, J. Chem. Phys. **124** (2006) 014107.
17. C. F. Bunge and R. Carbó-Dorca, J. Chem. Phys. **125** (2006) 014108.
18. A. A. Auer, M. Nooijen, J. Chem. Phys. **125** (2006) 24104.
19. M. Schütz, G. Hetzer and H.-J. Werner, J. Chem. Phys. **111** (1999) 5691.
20. G. E. Scuseria and P. Y. Ayala, J. Chem. Phys. **111** (1999) 8330.
21. P. Y. Ayala and G. E. Scuseria, J. Comput. Chem. **21** (2000) 1524.
22. M. Schütz and H.-J. Werner, J. Chem. Phys. **114** (2001) 661.
23. V. Maz'ya and G. Schmidt, *Approximate Approximations, Mathematical Surveys and Monographs*, Vol. 141, American Mathematical Society (2007).
24. R. A. DeVore, Acta Numerica **7** (1998) 51.

25. I. Daubechies, *Ten Lectures on Wavelets*, CBMS-NSF Regional Conference Series in Applied Mathematics **61** (1992).
26. S. Mallat, *A Wavelet Tour of Signal Processing* Academic Press, San Diego (1998).
27. A. Cohen, W. Dahmen, R. DeVore, Math. Comp. **70** (2001) 27.
28. A. Barinka, Doctoral Dissertation, RWTH Aachen 2004.
29. W. Dahmen, T. Rohwedder, R. Schneider, A. Zeiser, Numer. Math. **110** (2008) 277.
30. T. Rohwedder, R. Schneider, A. Zeiser, *Perturbed preconditioned inverse iteration for operator eigenvalue problems with applications to adaptive wavelet discretization*, Adv. Comput. Math. to appear 2009, available online: DOI 10.1007/s10444-009-9141-8.
31. A. Knyazev and K. Neymeyr, SIAM J. Matrix Anal. Appl. **31** (2009) 621.
32. H.-J. Flad and R. Schneider, *s^*-compressibility of discrete Hartree-Fock equations*, preprint TU Berlin 2009.
33. R. Schneider, Num. Math. **113** (2009) 433.
34. T. Helgaker, P. Jørgensen and J. Olsen, *Molecular Electronic-Structure Theory*, Wiley, New York (1999).
35. E. J. Bylaska, W. A. de Jong, N. Govind, K. Kowalski, T. P. Straatsma, M. Valiev, D. Wang, E. Apra, T. L. Windus, J. Hammond, P. Nichols, S. Hirata, M. T. Hackler, Y. Zhao, P.-D. Fan, R. J. Harrison, M. Dupuis, D. M. A. Smith, J. Nieplocha, V. Tipparaju, M. Krishnan, Q. Wu, T. Van Voorhis, A. A. Auer, M. Nooijen, E. Brown, G. Cisneros, G. I. Fann, H. Fruchtl, J. Garza, K. Hirao, R. Kendall, J. A. Nichols, K. Tsemekhman, K. Wolinski, J. Anchell, D. Bernholdt, P. Borowski, T. Clark, D. Clerc, H. Dachsel, M. Deegan, K. Dyall, D. Elwood, E. Glendening, M. Gutowski, A. Hess, J. Jaffe, B. Johnson, J. Ju, R. Kobayashi, R. Kutteh, Z. Lin, R. Littlefield, X. Long, B. Meng, T. Nakajima, S. Niu, L. Pollack, M. Rosing, G. Sandrone, M. Stave, H. Taylor, G. Thomas, J. van Lenthe, A. Wong, and Z. Zhang, "NWChem, A Computational Chemistry Package for Parallel Computers, Version 5.1" (2007), Pacific Northwest National Laboratory, Richland, Washington 99352–0999, USA. A modified version.
36. R. A. Kendall, E. Aprà, D. E. Bernholdt, E. J. Bylaska, M. Dupuis, G. I. Fann R. J. Harrison, J. Ju, J. A. Nichols, J. Nieplocha, T. P. Straatsma, T. L. Windus, A. T. Wong, Comput. Phys. Commun. **128** (2000) 260.
37. K. Szalewicz, B. Jeziorski, H. J. Monkhorst and J. G. Zabolitzky, J. Chem. Phys. **78** (1983) 1420.
38. C. J. Umrigar, K. G. Wilson, and J. W. Wilkins, Phys. Rev. Lett. **60** (1988) 1719.
39. K. E. Schmidt, and J. W. Moskowitz, J. Chem. Phys. **93** (1990) 4172.
40. D. P. Tew and W. Klopper, J. Chem. Phys. **123** (2005) 074101.
41. H.-J. Flad, W. Hackbusch and R. Schneider, ESAIM: M2AN **41** (2007) 261.
42. D. E. Freund, B. D. Huxtable and J. D. Morgan III, Phys. Rev. A **29** (1984) 980.
43. T. Helgaker, W. Klopper, H. Koch and J. Noga, J. Chem. Phys. **106** (1997) 9639.
44. A. Halkier, T. Helgaker, P. Jørgensen, W. Klopper, H. Koch, J. Olsen and A. K. Wilson, Chem. Phys. Lett. **286** (1998) 243.
45. D. C. Allen, T. A. Arias, J. D. Joannopoulos, M. C. Payne, M. P. Teter, Rev. Mod. Phys. **64** (1992) 1045.
46. R. Schneider, T. Rohwedder, J. Blauert and A. Neelov, J. Comp. Math. **27** (2009) 360.
47. http://www-drfmc.cea.fr/sp2m/L_Sim/BigDFT/index.en.html
48. ABINIT is a common project of the Université Catholique de Louvain, Corning Incorporated, and other contributors, for further details see http://www.abinit.org
49. X. Gonze, J.-M. Beuken, R. Caracas, F. Detraux, M. Fuchs, G.-M. Rignanese, L. Sindic, M. Verstraete, G. Zerah, F. Jollet, M. Torrent, A. Roy, M. Mikami, Ph. Ghosez, J.-Y. Raty, D. C. Allan, Comput. Mater. Sci. **25** (2002) 478.
50. X. Gonze, G.-M. Rignanese, M. Verstraete, J.-M. Beuken, Y. Pouillon, R. Caracas, F. Jollet, M. Torrent, G. Zerah, M. Mikami, Ph. Ghosez, M. Veithen, J.-Y. Raty, V.

Olevano, F. Bruneval, L. Reining, R. Godby, G. Onida, D. R. Hamann, D. C. Allan, Zeit. Kristallogr. *220* (2005) 558.
51. L. Genovese, A. Neelov, S. Goedecker, T. Deutsch, S. A. Ghasemi, A. Willand, D. Caliste, O. Zilberberg, M. Rayson, A. Bergman, R. Schneider, J. Chem. Phys. **129** (2008) 014109.
52. H. B. Schlegel and J. J. W. McDouall, in *Computational Advances in Organic Chemistry*, Kluwer Academic (1991) p. 167.
53. E. Cancès and C. Le Bris, M2AN **34** (2000) 749.
54. L. Armijo, Pacific J. Math **16** (1966) 1.
55. C. Geiger and C. Kanzow, *Theorie und Numerik restringierter Optimierungsaufgaben*, Springer, Berlin (2002).
56. W. Kutzelnigg, Int. J. Quantum Chem. **51** (1994) 447.
57. D. Braess, J. Approx. Theory **83** (1995) 93.
58. A. Halkier, T. Helgaker, P. Jørgensen, W. Klopper and J. Olsen, Chem. Phys. Lett. **302** (1999) 437.
59. H.-J. Flad, R. Schneider and B.-W. Schulze, Math. Methods Appl. Sci. **31** (2008) 2172.
60. H.-J. Flad, W. Hackbusch and R. Schneider, ESAIM: M2AN **40** (2006) 49.
61. W. Hackbusch, *Elliptic Differential Equations: Theory and Numerical Treatment*, Springer, Berlin (1992).

A Relativistic Four- and Two-component Generalized-active-space Coupled Cluster Method

By Lasse K. Sørensen[1], Timo Fleig[2,*], and Jeppe Olsen[3]

[1] Institut für Theoretische Chemie, Heinrich-Heine-Universität Düsseldorf, Universitätsstr. 1, 40225 Düsseldorf, Germany
[2] Laboratoire de Chimie et Physique Quantiques, I.R.S.A.M.C., Université Paul Sabatier Toulouse III, Route de Narbonne 118, 31062 Toulouse, France
[3] Theoretical Chemistry, University of Aarhus, Langelandsgade 140, 8000 Aarhus, Denmark

(Received August 17, 2009; accepted September 24, 2009)

Relativistic Coupled Cluster / Multi-reference / General Excitation Level / 4- and 2-component / Heavy Elements / Relativistic Electronic Structure

We present an overview of our recent developments of relativistic generalized-active-space coupled cluster and its initial applications to heavy-element systems. The new genuinely string-based method may treat coupled cluster expansions of general order and simulate multi-reference expansions by including higher excitations in properly constructed active spinor spaces. The four-component Dirac-Coulomb Hamiltonian or any approximation to it may be employed, and the underlying spinors are assumed and required to be time-reversal partners. The capability of the new approach is demonstrated and discussed in sample applications to HBr and BiH.

1. Introduction

Relativistic coupled cluster (CC) calculations for molecules containing heavy elements can presently not attain the same accuracy as non-relativistic CC calculations for molecules containing light elements. The limiting factors are the number of electrons which need to be correlated, the size of the required one-particle basis sets, the more complicated electronic structure of many heavy-element compounds, and the less-developed methodology for relativistic CC calculations. Also and especially in the relativistic domain, general-order and multi-reference approaches are highly desirable. This pertains to the occurrence of unpaired electrons, not only when dissociation processes are to be described but very often in

* Corresponding author. E-mail: timo.fleig@irsamc.ups-tlse.fr

Z. Phys. Chem. **224** (2010) 671–680
© by Oldenbourg Wissenschaftsverlag, München

the ground- and excited-state electronic structures close to equilibrium geometries. General-order and multi-reference implementations are therefore related, and advances in relativistic CC methodology are likely to touch upon both these issues.

There is still no general consensus on how a multi-reference coupled cluster theory should be formulated. Modern developments including iterative excitation levels higher than CC doubles, e.g. [1–5] and various multi-reference (MR) approaches [2,6–11] have been presented in the non-relativistic framework. Generalizations of molecular CC approaches to a relativistic formalism, which is here to be understood as also including spin-orbit interaction terms, remain rare.

For the study of atomic structure relativistic CC methods have a longer tradition. Recently, various relativistic CC approaches have been applied, e.g. to hyperfine structure and excitation energies [12], spacetime variation of the fine-structure constant [13], or atomic parity nonconservation [14,15]. Concerning molecules, Hirata et al. [16] describe higher-order electron correlation methods including coupled cluster approaches where spin-orbit interaction is introduced by means of relativistic effective core potentials with spin-orbit effective potentials. The 4-component implementations by Visscher et al., Kramers-restricted [17] and unrestricted [18], respectively, are not generally applicable to open-shell/multi-reference states. The only genuinely relativistic multi-reference approaches reported to the date are the Fock-Space (FS) CC implementations by Landau et al. [19] and Visscher et al. [20]. These methods, as the Fock-Space approach in general, use a common orbital basis for all of the occurring ionized systems. Relativistic FSCC approaches have been applied to systems with few unpaired electrons and in this domain are likely to be the most accurate relativistic molecular many-body methods currently available.

Alternatively, the multi-reference problem may be approached by introducing active orbital spaces with variable occupation constraints (e.g. exploiting the concept of Generalized Active Spaces (GAS)) in the construction of CC model spaces [2,21], leading to variants of so-called state-specific (SS) MRCC. These may be regarded as simulations of genuine MRCC. The GASCC formulation, however, retains the advantages of single-reference approaches such as the commutativity of cluster operators and allows for a flexible definition and robust treatment of open-shell and multi-reference problems [22].

2. Generalized-Active-Space (GAS) coupled cluster method

The original idea for the type of multi-reference CC approach presented here is ascribed to Oliphant and Adamowicz [10,23]. The present approach is based on the generalized implementation of these ideas by Olsen [2,22] where the projection manifold is extended to simulate excitations from additional reference functions beside those from the Fermi vacuum state. Figure 1 gives an illustration of how the CC model spaces are constructed. We resort to a simple case with

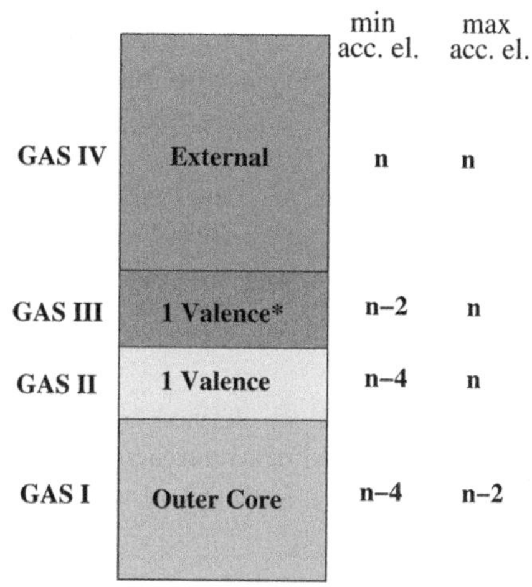

Fig. 1. Example of a Generalized Active Space model for a CC wave function. *n* is the total number of electrons, min./max. acc. el. specifies the minimum/maximum number of accumulated electrons after consideration of this subspace.

undefined number of occupied Kramers pairs[1] in GAS I, one valence (e.g. bonding) occupied Kramers pair in GAS II, another unoccupied Kramers pair, e.g. of antibonding character, in GAS III, and an undefined number of unoccupied Kramers pairs in the external space GAS IV. The arising model space is best understood by explicitly considering the CC excitation manifold $\langle\mu|$ obtained by these definitions. Since the number of holes in GAS I is restricted to one or two, the excitation manifold can be written as

$$\begin{aligned} \langle\mu_{\text{GASCC}}| \;=\; & \left\langle\mu^{S(\text{III}^1)}\right| + \left\langle\mu^{S(\text{IV}^1)}\right| + \left\langle\mu^{D(\text{III}^2)}\right| + \left\langle\mu^{D(\text{IV}^2)}\right| + \left\langle\mu^{D(\text{III}^1+\text{IV}^1)}\right| \\ & + \left\langle\boldsymbol{\mu}^{\mathbf{T(III^1+IV^2)}}\right| + \left\langle\boldsymbol{\mu}^{\mathbf{T(III^2+IV^1)}}\right| + \left\langle\boldsymbol{\mu}^{\mathbf{Q(III^2+IV^2)}}\right|. \end{aligned} \tag{1}$$

The term $\langle\mu^{S(\text{III}^1)}|$ indicates that relative to the (in this case) closed-shell reference state, where GAS I and GAS II are fully occupied, Single (S) replacements with one electron occupying GAS III have been included and the created hole may either be in GAS I or in GAS II (in the above example. This of course may be refined by adjusting the occupation constraints of the GAS). Likewise, the other terms of the extended excitation manifold are interpreted. Particular attention should be paid to the three terms given in boldface symbols: $\langle\boldsymbol{\mu}^{\mathbf{T(III^1+IV^2)}}|$ denotes triple excitations with respect to the reference state where one particle resides in the internal space (GAS III) and two particles reside in the external space (GAS IV). Compared with the term $\langle\mu^{D(\text{IV}^2)}|$ which "correlates" the reference state it is

[1] A Kramers pair is comprised by two spinors related by time-reversal symmetry, see e.g. reference [24].

seen that in the Triples term a single excitation is "correlated", as Double excitations into the external space are combined with these Single excitations. In an analogous manner, the term $\langle\mu^{Q(III^2+IV^2)}|$ describes the correlation of the one doubly-excited state within spaces II and III. The term $\langle\mu^{T(III^2+IV^1)}|$ which is also not present in a single-reference CCSD model space describes the relaxation of the one internal double excitation.

We may therefore express the excitation manifold in a more compact form as

$$\langle\mu_{\text{GASCC}}| = \left\langle\Psi^{\text{Ref}}\right| \hat{\tau}^{\dagger}_{\mu_{\text{GAS}}} \tag{2}$$

where $\hat{\tau}_{\mu\text{GAS}}$ denotes all the excitations defined by the GAS constraints and $\left|\Psi^{\text{Ref}}\right\rangle$ is a restricted (Dirac-) Hartree-Fock reference wave function. The cluster operators are constructed accordingly, leading to a wave function ansatz

$$\left|\Psi^{\text{GASCC}}\right\rangle = exp(\sum_{\mu} t_{\mu}\hat{\tau}_{\mu_{\text{GAS}}}) \left|\Psi^{\text{Ref}}\right\rangle. \tag{3}$$

Since all occuring amplitudes are optimized here, this ansatz is therefore a generalization of the SSMRCC by Oliphant and Adamowicz [10,23] and an extension due to the introduction of GAS.

The drawback of this formalism - in contrast to FSCC or the state-universal ansatz [8,25] or a new multi-reference exponential ansatz by Hanrath [7] - is its variance with respect to the choice of Fermi vacuum since the various conceivable reference functions are not strictly treated on the same footing. The straightforward remedy, a separate cluster expansion for every reference function, leads to the state-universal Hilbert-space CC theory [25] which is more problematic in practice due to the intruder state problem [26]. The loss of Fermi vacuum invariance, however, does not appear to be a crucial penalty in most applications and problems can usually be avoided by proper choices of active spaces, as demonstrated in references [21,27].

3. Relativistic implementation

Our implementation of the described GAS Coupled Cluster is interfaced to a local version of the DIRAC quantum chemistry program package [28]. We may employ the four-component Dirac-Coulomb Hamiltonian or - without modification of the CC code - any approximation to it, for instance the infinite-order two-component (IOTC) Hamiltonian [29] which avoids the explicit treatment of small-component degrees of freedom in the spinor optimization.

The cluster operators $\hat{T} = \sum_{\mu} t_{\mu}\ \hat{\tau}_{\mu\text{GAS}} = \sum_{m}\ \hat{T}_m$ are now generalized to the relativistic framework, which entails the possibility of flipping the Kramers projection[2] along with the excitation [31]:

[2] The Kramers projection is an auxiliary quantity taking the value $(1/2)$ for an unbarred and $-(1/2)$ for a barred spinor, see e.g. reference [30].

$$
\begin{aligned}
\hat{T}_1 &= \sum_{ia}\left\{t_i^a\hat{\tau}_i^a + t_{\bar{i}}^a\hat{\tau}_{\bar{i}}^a + t_i^{\bar{a}}\hat{\tau}_i^{\bar{a}} + t_{\bar{i}}^{\bar{a}}\hat{\tau}_{\bar{i}}^{\bar{a}}\right\} \\
\hat{T}_2 &= \sum_{ij,ab}\left\{\frac{1}{4}t_{ij}^{ab}\hat{\tau}_{ij}^{ab} + \frac{1}{2}t_{\bar{i}j}^{ab}\hat{\tau}_{\bar{i}j}^{ab} + \frac{1}{2}t_{i\bar{j}}^{ab}\hat{\tau}_{i\bar{j}}^{ab} + \frac{1}{2}t_{ij}^{\bar{a}b}\hat{\tau}_{ij}^{\bar{a}b} + \frac{1}{2}t_{ij}^{a\bar{b}}\hat{\tau}_{ij}^{a\bar{b}}\right. \\
&\quad + \frac{1}{4}t_{\overline{ij}}^{ab}\hat{\tau}_{\overline{ij}}^{ab} + t_{i\bar{j}}^{\bar{a}b}\hat{\tau}_{i\bar{j}}^{\bar{a}b} + t_{i\bar{j}}^{a\bar{b}}\hat{\tau}_{i\bar{j}}^{a\bar{b}} + t_{\bar{i}j}^{\bar{a}b}\hat{\tau}_{\bar{i}j}^{\bar{a}b} + t_{\bar{i}j}^{a\bar{b}}\hat{\tau}_{\bar{i}j}^{a\bar{b}} + \frac{1}{4}t_{ij}^{\overline{ab}}\hat{\tau}_{ij}^{\overline{ab}} \qquad (4) \\
&\quad \left. + \frac{1}{2}t_{\overline{ij}}^{\bar{a}b}\hat{\tau}_{\overline{ij}}^{\bar{a}b} + \frac{1}{2}t_{\overline{ij}}^{a\bar{b}}\hat{\tau}_{\overline{ij}}^{a\bar{b}} + \frac{1}{2}t_{\bar{i}j}^{\overline{ab}}\hat{\tau}_{\bar{i}j}^{\overline{ab}} + \frac{1}{2}t_{i\bar{j}}^{\overline{ab}}\hat{\tau}_{i\bar{j}}^{\overline{ab}} + \frac{1}{4}t_{\overline{ij}}^{\overline{ab}}\hat{\tau}_{\overline{ij}}^{\overline{ab}}\right\}
\end{aligned}
$$

etc.

where the summations run over Kramers pairs. We define as an important quantity for our algorithm the excitation class, which is a unique set of excitations given by the number of Kramers flips in the term, the number of particles involved, and the combination of active spaces of the occurring indices. In the case of spin orbitals, the Kramers flipping would correspond to the introduction of terms accounting for spin-orbit interaction. In the present, this special case is generalized to be applicable with any kind of one-particle space of Kramers-paired spinors. The approach is therefore Kramers restricted but does not exploit time-reversal symmetry relations for the CC amplitudes, only for the required one- and two-particle integrals. A fully Kramers adapted formalism has been shown to be of vastly increased complexity [32] despite of the reduction of the number of free parameters. Double point group symmetry has been implemented for the real-valued [33,34] matrix groups D_{2h}^*, D_2^* and C_{2v}^*. This ensures a completely real-valued formalism, also when spin-orbit interaction is included. Higher symmetry, e.g. linear molecular symmetry, has not yet been considered. Linear molecules or atoms are currently treated in one of the above real-valued subgroups.

The central quantity to be evaluated in the course of a CC optimization is the CC vector function Ω_μ for a given element μ of the excitation manifold:

$$
\Omega_\mu = \left\langle \mu \left| e^{-\hat{T}}\hat{H}e^{\hat{T}} \right| \mathrm{Ref} \right\rangle \qquad (5)
$$

The evaluation of eq. (5) will be discussed in greater detail in the following subsections.

3.1 CC based on configuration interaction expansions

The evaluation of the CC vector function proceeds in an analogous fashion to the one described in reference [2]. In the present, however, it is obtained in corresponding steps which are based on relativistic CI expansions [35] and string manipulations. The relativistic implementation [31] can be summarized as follows:

Step 1: $|a\rangle = e^{\hat{T}}|\mathrm{Ref}\rangle$, expansion of the reference vector where $\hat{T}$ is the relativistic operator from eq. (4).

Step 2: $|b\rangle = \hat{H}\,|a\rangle$, linear transformation of the expanded reference vector where $\hat{H}$ is the Dirac-Coulomb Hamiltonian or an approximation to it.

Step 3: $|c\rangle = e^{-\hat{T}}\,|b\rangle$, expansion of the transformed reference vector.

Step 4: $\Omega_\mu = \langle\mu|c\rangle$, evaluation of transition density matrix elements where μ designates the excitation manifold in the above relativistic framework.

Although this CI-expansion based formalism is not very efficient, it is generally applicable and exploits the advantages of string-based relativistic CI.

Some care has to be taken regarding the term "string-based" in this context. Strictly speaking, the afore-presented approach is string-based because the CI expansions are performed in a string-based manner, as expounded in reference [35]. However, genuinely string-based CC is usually understood as determining connections and performing contractions based directy on the manipulation of (spinor) strings, not intermediate CI expansions. An approach of the latter type has recently been developed [36] and will be described in the following subsection.

3.2 Genuine string-driven coupled cluster

As expounded in references [2,22,31] CI-based CC exhibits an inefficient computational scaling as $O^{n+2}\ V^{n+2}$ where O is the number of occupied spinors, V the number of virtual spinors and n is the highest CC excitation level in the calculation. We have therefore implemented an evaluation algorithm for the CC vector function which displays the scaling of conventional CC approaches, as $O^n\ V^{n+2}$. It is imperative to gain on efficiency, since in particular in the treatment of heavy-element electronic structure many electrons (often subvalence and/or outer core, e.g. in [27]) per atom have to be correlated explicitly for results of high accuracy.

The new algorithm presented in reference [36] for evaluating eq. (5) relies on a general way of determining the contraction of some non-relativistic or relativistic Hamiltonian operator with a given set of non-relativistic or relativistic cluster operators. All operators are represented by strings of basic second quantized operators and the general commutation properties of these are exploited in the required manipulations.

Baker-Campbell-Hausdorff expansion according to $e^{-\hat{T}}\,\hat{H}\,e^{\hat{T}} = \hat{H} + [\hat{H},\hat{T}] + (1/2)\,[[\hat{H},\hat{T}],\hat{T}] + (1/6)\,[[[\hat{H},\hat{T}],\hat{T}],\hat{T}] + \ldots$ yields the starting point for the loops of the algorithm. The outermost loop determines the excitation class [37] i of the Hamiltonian $\hat{H}(i)$, which is given by a specification of the Kramers projection of the individual operator strings and the orbital subspaces of the corresponding second-quantized operators. Depending on the number of deexcitations in $\hat{H}(i)$ the maximum number of commutators for a given excitation class i is fixed. Loops over the excitation classes $j,k,\ldots$ of the cluster operators $\hat{T}$ occurring in the nested commutator, e.g. $[[\hat{H}(i),\hat{T}(j)],\hat{T}(k)]$, are performed and the possible contractions (their number amounts to $n!$, where n is the nesting order of the given commutator) for this commutator are carried out. The non-vanishing connections (which are comprised by all fully connected terms) between the excitation manifold and a given nested commutator acting on the reference state are thus determined by the involved excitation classes of both the excitation manifold and the cluster operators occurring in the commutator.

The relativistic generalization of this algorithm [36] includes the full implementation of double point group symmetry and an interface to transformed integrals over Kramers-paired spinors. From a computational point of view, this generalization leads to a significantly increased demand. Beside the obvious increase in the number of independent amplitudes, there are two crucial points in the relativistic algorithm which deserve special attention: (1) Due to the possibility of flipping the Kramers projection in the relativistic case (see eqn. 4) the number of excitation classes may be significantly increased [36]. (2) As a consequence the number of non-vanishing connections with the excitation manifold $\langle\mu_{\mathrm{GASCC}}|$ may also be greatly increased leading to computational bottlenecks in the relativistic case. Nevertheless, the recent implementation has made relativistic calculations with higher excitations (triples and greater) and larger numbers of correlated electrons possible for atoms and small molecules containing heavy elements. The commutator-based CC implementation is size extensive (valence extensive [38,39], referring to the active space, i.e. all Kramers pairs which reside in GAS which are not required to be doubly occupied) for electronic ground states due to the termwise size-extensivity of CC theory. This is also true for the CI-based implementation since this is merely an (inefficient) alternative to implementing the same CC vector function.

4. Sample applications

In order to demonstrate the capabilities of the GASCC implementation in its current form we present two applications to heavy-element hydrides, HBr and BiH.

4.1 HBr molecule

The results in table 1 have been obtained with an uncontracted basis set of triple zeta quality and using closed-shell Dirac-Coulomb Hartree-Fock 4-component spinors [31].

Correlating the $3d$ electrons on bromine leads to a significant contraction of the bond which has been anticipated by Visscher et al. [41]. Upon correlating 18 electrons the difference between the Configuration Interaction and the CC model becomes sizeable, especially for the harmonic frequency which already at the CCSD level is corrected by more than 50 cm^{-1}. Correcting for spin-orbit interaction in an additive fashion (Δ_{SO} has been obtained from the difference between MRCCSD(8) calculations including and neglecting spin-orbit coupling) is a viable procedure in the case of HBr. However, with the new and improved commu-

Table 1. Spectral constants of HBr using various CC models and the corresponding CI model with (SO) and without (SOF) spin-orbit interaction and correlating (n) electrons.

Method	R_e [Å]	ω_e [cm^{-1}]	D_e [eV]
SO CCSDT (8)	1.4176	2647	
SOF CCSDT (18)	1.4137	2665	4.01
SO CISD (18)	1.4064	2736	
SO CCSD (18)	1.4119	2683	
SO CCSD(T) (18)	1.4139	2664	
SOF CCSDT (18) +Δ_{SO}	1.4138	2659	3.87
SO CCSDT (18)	1.4140	2663	3.87
Exp.[1]	1.4144	2649.0	3.92

[1] Huber and Herzberg [40]

Table 2. Spectral constants of BiH using various CC models and compared to experiment.

Method	corr. el.	Amplitudes	R_e[Å]	ω_e [cm^{-1}]
MRCCSD(6in5)	6	620 677	1.8259	1676.12
CCSDT	16	64 191 680	1.7932	1708.51
Exp.[1]				1699.52
Exp.[2]			1.80867	1697.62

[1] Urban et al., Infrared spectroscopy, reference [43]
[2] Bopegedera et al., Diode-laser spectrometer, reference [44]

tator-based implementation of the CC vector function the full calculation SO CCSDT (18) has become possible. The results at this computational level are - not surprisingly - very similar to the results obtained with the SO CCSD(T) (18) model. CC models including full iterative triple excitations have, however, been proven to be of high value for obtaining balanced complete potential energy curves in high-accuracy calculations, e.g. as demonstrated in a study of the LiCs molecule [27]. We assign the largest fraction of the remaining deviations from the experimental values to basis set incompleteness and the basis set superposition error.

4.2 BiH molecule

The spectroscopic properties of the BiH molecule are affected strongly by both electron correlation and special relativity [42]. This is mainly due to the contraction of the $6s_{1/2}$ and $6p_{1/2}$ spinors on the bismuth atom. We have therefore chosen this heavier molecule as a more demanding test system for our recent implementation [36]. Basis sets of triple zeta quality and closed-shell Dirac-Coulomb Hartree-Fock 4-component spinors have been used.

Two specific CC models are presented in table 2 which we consider as important for future applications. MRCCSD(6in5) comprises an active space of 5 Kramers pairs in which higher excitations than the external Singles and Doubles are allowed. This model yields a good cost/accuracy ratio. For highest accuracy, however, the 5*d* electrons on Bi have to be included in the correlation treatment which has been done in the case of CCSDT. The deviation of the harmonic vibrational frequency is only around 10 cm^{-1} here, but the number of amplitudes is already quite large and an intermediate model of the type MRCCSD(6in5) where 16 electrons are correlated appears to be desirable. The application of such models requires the use of a larger number of active orbital spaces leading to long computation times, as discussed above. Improvements of the current implementation along these lines are under way.

5. Conclusions

We have accomplished the implementation of a rigorous and yet practical relativistic general-order coupled cluster method based on string representations. It may employ one-particle functions (spinors) which include spin-orbit interaction *a priori* and therefore treats special relativity in electronic-structure theory in a rigorous manner. The recently implemented commutator-based evaluation algorithm for the coupled cluster vector function exhibits the computational scaling of conventional CC methods. Therefore, fully relativistic calculations with currently on the order of 20 to 30 correlated electrons have become possible, where large one-particle basis sets and internal or external excitation levels higher than double excitations may be employed. Current work involves both algorithmic improvements as well as explorations of the applicability to more complicated electronic structure, e.g. to electronic states with several unpaired electrons, and the treatment of electronically excited states.

References

1. M. Kállay and P. Surján, J. Chem. Phys. **115** (2001) 2945.
2. J. Olsen, J. Chem. Phys. **113** (2000) 7140.
3. P.-D. Fan and S. Hirata, J. Chem. Phys. **124** (2006) 104108.
4. S. Hirata, J. Chem. Phys. **121** (2004) 51.
5. M. Kállay and J. Gauss, J. Chem. Phys. **123** (2005) 214105.
6. M. Kállay, P. Szalay, and P. Surján, J. Chem. Phys. **117** (2002) 980.
7. M. Hanrath, J. Chem. Phys. **123** (2005) 084102.
8. K. Kowalski and P. Piecuch, Mol. Phys. **102** (2004) 2425.
9. M. L. Abrams and C. D. Sherrill, Chem. Phys. Lett. **404** (2005) 284.
10. N. Oliphant and L. Adamowicz, J. Chem. Phys. **96** (1991) 3739.
11. I. Hubac, J. Pittner, and P. Cársky, J. Chem. Phys. **112** (2000) 8779.
12. C. Sur, R. K. Chaudhuri, B. K. Sahoo, B. P. Das, and D. Mukherjee, J. Phys. B, **41** (2008) 065001.
13. A. Borschevsky, E. Eliav, Y. Ishikawa, and U. Kaldor, Phys. Rev. A, **74** (2006) 062505.
14. B. K. Sahoo, R. Chaudhuri, B. P. Das, and D. Mukherjee, Phys. Rev. Lett. **96** (2006) 163003.

15. G. Gopakumar, B. P. Das, R. K. Chaudhuri, D. Mukherjee, and K. Hirao, J. Chem. Phys. **126** (2007) 014301.
16. S. Hirata, T. Yanai, R. J. Harrison, M. Kamija, and P.-D. Fang, J. Chem. Phys. **126** (2007) 024104.
17. L. Visscher, K. G. Dyall, and T. J. Lee, Int. J. Quantum Chem.: Quantum Chem. Symp. **29** (1995) 411.
18. L. Visscher, T. J. Lee, and K. G. Dyall, J. Chem. Phys. **105** (1996) 8769.
19. A. Landau, E. Eliav, and U. Kaldor, Chem. Phys. Lett. **313** (1999) 399.
20. L. Visscher, E. Eliav, and U. Kaldor, J. Chem. Phys. **115** (2001) 9720.
21. A. Köhn and J. Olsen, J. Chem. Phys. **125** (2006) 174110.
22. J. W. Krogh and J. Olsen, Chem. Phys. Lett. **344** (2001) 578.
23. N. Oliphant and L. Adamowicz, J. Chem. Phys. **94** (1991) 1229.
24. H. J. Aa. Jensen, K. G. Dyall, T. Saue, and K. Fægri, J. Chem. Phys. **104** (1996) 4083.
25. B. Jeziorski and H. J. Monkhorst, Phys. Rev. A, **24** (1981) 1668.
26. R. J. Bartlett and M. Musial, Rev. Mod. Phys. **79** (2007) 291.
27. L. K. Sørensen, T. Fleig, and J. Olsen, J. Phys. B, **42** (2009) 165102.
28. DIRAC, a relativistic ab initio electronic structure program, Release DIRAC08 (2008), written by L. Visscher, H. J. Aa. Jensen, and T. Saue, with new contributions from R. Bast, S. Dubillard, K. G. Dyall, U. Ekström, E. Eliav, T. Fleig, A. S. P. Gomes, T. U. Helgaker, J. Henriksson, M. Ilias, Ch. R. Jacob, S. Knecht, P. Norman, J. Olsen, M. Pernpointner, K. Ruud, P. Saek, and J. Sikkema (see http://dirac.chem.sdu.dk).
29. M. Ilias and T. Saue, J. Chem. Phys. **126** (2007) 064102.
30. J. Thyssen, T. Fleig, and H. J. Aa. Jensen, J. Chem. Phys. **129** (2008) 034109.
31. T. Fleig, L. K. Sørensen, and J. Olsen, Theoret. Chem. Acc. **118** (2007) 347, erratum in: Theor. Chem. Acc. **118** (2007) 979.
32. T. Fleig, Phys. Rev. A, **77** (2008)062503.
33. T. Saue and H. J. Aa. Jensen, J. Chem. Phys. **111** (1999) 6211.
34. L. Visscher and T. Saue, J. Chem. Phys. **113** (2000) 3996.
35. T. Fleig, J. Olsen, and L. Visscher, J. Chem. Phys. **119** (2003) 2963.
36. L. K. Sørensen, J. Olsen, and T. Fleig, *Kramers-restricted generalized active space coupled cluster. Implementation and initial application on BiH.* In preparation.
37. T. Fleig, *Relativistic string-based electron correlation methods.* In *Relativistic Methods for Chemists.* M. Barysz and Y. Ishikawa (Eds.), Springer Verlag, Berlin (2010) in press.
38. M. Hanrath, Chem. Phys. **356** (2008) 31.
39. D. Mukherjee and S. Pal, Adv. Quantum Chem. **20** (1989) 291.
40. K. P. Huber and G. Herzberg, *Molecular Spectra and Molecular Strucure, IV. Constants of Diatomic Molecules.* Van Nostrand, New York (1979).
41. L. Visscher, J. Styszyński, and W. C. Nieuwpoort, J. Chem. Phys. **105** (1996) 1987.
42. K. Balasubramanian, Chem. Phys. Lett. **114** (1985) 201.
43. R. D. Urban, P. Polomsky, and H. Jones, Chem. Phys. Lett. **181** (1991) 485.
44. A. M. R. P. Bopegedera, C. R. Brazier, and P. F. Bernath, Chem. Phys. Lett. **162** (1989) 301.

Canonical Tensor Products as a Generalization of Gaussian-type Orbitals

By Sambasiva Rao Chinnamsetty[1], Mike Espig[1], Heinz-Jürgen Flad[2,*], and Wolfgang Hackbusch[1]

[1] Max-Planck-Institut für Mathematik in den Naturwissenschaften, Inselstraße 22, 04103 Leipzig, Germany
[2] Institut für Mathematik, Technische Universität Berlin, Straße des 17. Juni 137, 10623 Berlin, Germany

(Received July 30, 2009; accepted September 24, 2009)

Tensor Product Approximation / Density-fitting Method / Two-electron Integrals

We propose a possible generalization of Gaussian-type orbital (GTO) bases by means of canonical tensor products. The present work focus on the application of tensor products as an alternative to conventional GTO based density fitting schemes. Tensor product approximation leads to highly nonlinear optimization problems which require sophisticated algorithms. We give a brief description of the optimization problem and algorithm. The present work extends our previous paper [S. R. Chinnamsetty, M. Espig, B. N. Khoromskij, W. Hackbusch and H.-J. Flad, J. Chem. Phys. **127** (2007), 084110], where we discussed tensor product approximations of the electron density and the Hartree potential, to orbital products which are required for the exchange part of Hartree-Fock and in post Hartree-Fock methods. We provide a detailed error analysis for the Coulomb and exchange terms in Hartree-Fock calculations. Furthermore, a comparison is given between all-electron and pseudopotential calculations.

1. Introduction

In quantum chemistry, single-electron wavefunctions, the so-called orbitals, appear as solutions of various mean-field Schrödinger equations, e.g., Kohn-Sham or Hartree-Fock equations. Furthermore, these orbitals provide a natural starting point for the construction of Slater determinants which serve as a basis to represent many-electron wavefunctions which incorporate electron correlations. Since the early days of quantum chemistry, the optimal choice of basis functions for the approximation of orbitals has been a controversial issue which was finally

* Corresponding author. E-mail: flad@math.tu-berlin.de

Z. Phys. Chem. **224** (2010) 681–694
© by Oldenbourg Wissenschaftsverlag, München

decided by the majority of quantum chemists in favour of atomic centered Gaussian-type orbital (GTO) bases. Originally suggested by Boys [1], GTOs were first applied by Boys et al. [2] and Preuss [3] as basis sets for electronic structure calculations. From a physical point of view, GTOs seem to be only the second-best choice because they do not satisfy the required asymptotic properties close to a nucleus and at infinity. Compared with Slater-type orbital (STO) basis functions, which have the right asymptotic behaviour, GTOs, however, have the invaluable advantage of computational simplicity. This means that all the required integrals can be calculated analytically, cf., the monograph [4]. Furthermore it has been proven by Kutzelnigg and Braess [5,6] that the ground state of the hydrogen atom can be approximated with almost exponential, i.e., $O(e^{-c\sqrt{n}})$, convergence rate, by n GTO basis functions. Whether this result remains valid in the case of Hartree-Fock calculations for general molecules seems to be an open issue, although there exists some numerical evidence in favour of it, cf., Ref. [7].

Several well-known mathematical objections can be raised against GTO basis sets. These basis sets suffer from linear dependencies which means that they represent essentially over-complete bases so-called frames. As a consequence they are not stable in the sense of a Riesz basis [8] in our envisaged Hilbert spaces L^2 and H^1. These spaces consist of square integrable functions which in addition, in the case of the Sobolev space H^1, have square integrable first derivatives. However, everybody who tries to compete with GTOs using even the most sophisticated systematic basis functions, like wavelets[1], get puzzled from the almost unbelievable efficiency of GTOs at least for moderate accuracies. Although GTO bases are not a subject of standard approximation theory, their properties have been studied within approximate approximation theory [10]. Therein one has to give up the goal to turn GTOs into a systematic basis which can in principle achieve arbitrary accuracies. Instead it is advisable to content oneself with a certain residual error that enters in a controlled way into the construction of the basis set. This enables e.g., the construction of approximate multiscale bases similar to wavelets which provide a systematic error convergence rate up to a certain predetermined accuracy. Any attempt to turn GTOs into "systematic bases" is at the expense of loosing some flexibility which is actually a prominent feature of GTO bases. Therefore, the present work goes into a different direction. Instead of restricting flexibility, we seek to generalize GTOs by means of tensor product approximations of greatest possible generality.

GTOs are not only used for the expansion of orbitals but also for products of orbitals and electron densities. This requires the so-called auxiliary GTO basis sets which attracted considerable attention within density fitting schemes, also known as resolution of the identity, in order to reduce the computational complexity for Hartree-Fock and Kohn-Sham methods in particular for the Hartree potential [11–16] and also for two-electron integrals [17–22] in general. These techniques have been further applied to various post Hartree-Fock methods [23–

[1] The approximation of orbitals in wavelet bases has been studied in Ref. [9].

28]. Related approaches are the Cholesky decomposition of two-electron integrals [29–31] and the pseudo-spectral method [32,33].

In the present work we want to consider possible generalizations of (auxiliary) GTO basis sets via tensor product approximations with "optimal" separation rank. Our approach does not restrict to orbitals, orbital products, and electron densities but can be also applied to their convolutions with the Coulomb potential, like for the Hartree potential. This means that we are looking for the best possible approximation of some function f in $\mathbb{R}^3$ in the canonical format

$$f(\mathbf{x}) \approx \sum_{k=1}^{\kappa} h_k^{(1)}(x_1)\, h_k^{(2)}(x_2)\, h_k^{(3)}(x_3), \tag{1.1}$$

with $\mathbf{x} := (x_1,x_2,x_3)$, for fixed separation rank κ where the univariate functions $h_k^{(i)}(x_i)$, $k = 1,\ldots,\kappa$ for $i = 1,2,3$, so-called Kronecker factors, have to be chosen in an "optimal" manner. In particular, there are no orthogonality constraints imposed. Within the mathematical literature, such an approximation by a tensor product of a certain rank became known as canonical decomposition (CANDECOMP) or parallel factors (PARAFAC) model, cf., Ref. [34]. In the following we refer to κ as the Kronecker rank, a notion that has been borrowed from linear algebra [35] where it denotes the least possible decomposition rank of a tensor. We apply the more general notion of a separation rank to tensor product decompositions (1) for which the rank is supposed to be suboptimal, at least in some approximate manner.

Any function represented by GTOs can be considered as a tensor product in the format (1) where the separation rank simply corresponds to the number of primitve Cartesian GTO of the basis set. In this sense, we consider tensor products as a generalization of GTO bases. It is the topic of the present work to study "optimal" tensor product approximations of such a function with fixed Kronecker rank κ which is much smaller than the size of the corresponding GTO basis set. Together with an appropriate tensor product approximation for the Coulomb potential this also enables fast convolutions in the tensor format which are required e.g., for the Hartree and exchange potentials. The tensor product format does not only provide separable representations for convolutions, moreover it offers the possibility to consider their further approximation with "optimal" separation rank. In this sense our approach goes beyond conventional GTO based density-fitting schemes.

The present work provides a general introduction into the subject where we present new results extending our previous work [36], hereafter denoted Paper I, which mainly focused on the Hartree potential. This paper is organized as follows: in Section 1.1 we give a concise overview on tensor product formats and related optimization problems. For illustrative purposes we first discuss some simple single-electron systems in Section 1.2 which is followed in Section 2 by tensor product approximations for individual orbitals. Next, we proceed in Section 2 to products of orbitals and the electron density. The latter, already discussed in Paper I, is positive everywhere and therefore easier to approximate than products between different orbitals which are required e.g., by the exchange potential and might have a rather complicated oscillatory behaviour near nuclei. In Section 2.1 we discuss our results using tensor product approximations to compute Hartree and exchange potentials for some small molecules. Although our focus is on the Hartree-Fock model, we want to mention that this approach can be easily extended to post Hartree-Fock methods similar to conventional GTO density-fitting schemes where already many successful applications have been reported in the literature. Finally we provide in Section 3 a comparison between all-electron and pseudo densities from calculations with pseudopotentials. This provides some insight how sensible tensor product approximations respond to electron-nuclear cusps.

1.1 Tensor product approximation

In the field of linear algebra, tensor products attracted considerable interest within the last few years [34,37]. A problem of general interest is to obtain an analogue for the singular value decomposition (SVD) of matrices for tensors of order > 2, cf., Ref. [38]. Here and in the following order denotes the number of Kronecker factors in each term of the tensor product. Within the present work we restrict to order three tensors. According to the Eckart-Young theorem, a SVD provides the best rank κ approximation of a matrix by restriction to the largest κ singular values. There exits no canonical definition of a SVD analogue for tensors with order > 2 and various formats have been discussed in the literature [39–41]. The ultimate goal is to obtain best possible separable approximations of general tensors for a given separation rank. Furthermore, this task must be achieved with moderate computational effort and controlled accuracy. Most of the work has been done in a discrete setting which is equivalent to choosing a certain discretization for the Kronecker factors. In the following we give all of our expressions for functions. The reader should, however, keep in mind that for the actual computations these functions are represented on discrete grids, cf., Paper I for further details.

A popular alternative to the canonical format (1) is the Tucker format

$$f(\mathbf{x}) \approx \sum_{k_1,k_2,k_3=1}^{\varrho} b_{k_1,k_2,k_3} u_{k_1}^{(1)}(x_1)\, u_{k_2}^{(2)}(x_2)\, u_{k_3}^{(3)}(x_3), \tag{1.2}$$

where we can assume, without loss of generality, orthogonal Kronecker factors in each direction, i.e., $\langle u^{(i)}{}_{ki}, u^{(i)}{}_{li} \rangle = \delta_{ki,li}$, $k_i,l_i = 1,\dots,\rho$ for $i = 1,2,3$. The Tucker rank ρ specifies the size of the core tensor $\mathbf{b} := \{b_{k1,k2,k3}\}_{k1,k2,k3\,=\,1,\dots,\rho} \in \mathbb{R}^{\rho\times\rho\times\rho}$ with ρ^3 entries. In the representation (2), the separation rank of the Tucker tensor product is ρ^3, however, by a simple reordering we can obtain a canonical tensor product with separation rank ρ^2. For the approximation of orbitals, orbital products and electron densities, the canonical and Tucker formats seem to be most appropriate [36,42,43]. Within the present work we restrict ourselves to the canonical format and refer to Ref. [43] for applications of the Tucker format to electronic structure calculations. Recently several new tensor formats [44–47] have been proposed in the literature which are especially of interest for higher order tensors. These formats might be of interest for post Hartree-Fock methods and density matrix theory. The usefulness of a tensor product format depends not only on its approximation properties but also on the possibility to apply differential or integral operators in this format in an efficient manner, cf., Refs. [48–51]. Furthermore, it is crucial to provide fast algorithms which compute the actual approximations (1) and (2) in the canonical format [52,53] and Tucker format [54–57], respectively.

The error of a best separable rank κ approximation is defined via the least-squares problem

$$\sigma_\kappa(f) := \inf_{h_k^{(i)} \in L^2(\mathbb{R})} \left\| f - \sum_{k=1}^{\kappa} h_k^{(1)} \otimes h_k^{(2)} \otimes h_k^{(3)} \right\|_{L^2(\mathbb{R}^3)}, \tag{1.3}$$

where the error is measured in the L^2 norm, i.e., $\| f \|_{L^2}^2 := \int | f(\mathbf{x}) |^2 \, d^3x$. The correct mathematical formulation of (3) requires the introduction of the symbol $\otimes$ for an abstract tensor product between functions without explicit reference to the variables. Our present algorithm for the least-squares problem (3) is based on Newton's method [52,53] and requires that the function f is already given in a tensor product format, i.e.,

$$f(\mathbf{x}) = \sum_{k=1}^{K} g_k^{(1)}(x_1)\, g_k^{(2)}(x_2)\, g_k^{(3)}(x_3), \tag{1.4}$$

like a linear combination of GTOs with initial separation rank K. It can be shown, cf., [41,52], that the optimal Kronecker factors $h_k^{(i)}$, $k = 1,\ldots, \kappa$, are contained in the subspace $U_i := \operatorname{span}\, \{g_k^{(i)}\}_{k=1,\ldots,K}$ for $i = 1,2,3$ which reduces the optimization problem to finite dimension. The computational complexity of the various steps in the Newton algorithm is

$$O\left(\kappa(\kappa + K) + \kappa^3 + (\kappa + K) \sum_{i=1}^{3} \dim U_i\right), \tag{1.5}$$

where $\kappa \ll K$ has been assumed, cf., Ref. [52,53]. In the worst case we have $\dim U_i = K$ which results in $O(K^2)$ complexity for an orthogonalization step prior to the actual Newton optimization. However in our applications it turned out that often $\dim U_i \ll K$ after almost linear dependent functions have been removed from the set $\{g_k^{(i)}\}_{k=1,\ldots,K}$.

The variational problem (3) is typically ill-posed, cf., Ref. [41], which means that a minimizer does not always exist. Nevertheless, it is possible to specify a set of functions $h_k^{(i)}(x_i)$ for which the error is arbitrarily close to σ_κ. However, it might happen that this set is ill-conditioned in a sense discussed in Ref. [53] and requires some additional precautions, cf., [41,52], for further details. Despite this shortcoming, one should bear in mind that for our envisaged applications it is only of relevance to achieve a certain accuracy with a reasonably small Kronecker rank κ and it is rather irrelevant whether this representation actually corresponds to an exact minimum. Another problem arises from a multitude of local minima into which a numerical algorithm for the solution of the variational problem (3) might get stucked. Newton's method, e.g., converges quadratically in an appropriate local neighbourhood of a minimum irrespective whether it is local or global. Therefore the overall performance of this method depends critically on the initial guess which has to be constructed in a suitable way. In order to obtain the final best rank κ approximation (3) for the function f we construct successively a sequence of best rank $k = 1,\ldots,\kappa-1$ approximations f_k. From f_k we can easily obtain a rank one approximation g_k of the remainder $f - f_k$ via the alternating least squares (ALS) method, starting from a cross approximation, cf. Ref. [53] for further details. There exist several reasonable cross approximations. Actually, we consider a set of cross approximations which after the ALS step lead to possibly different optimal rank one approximations of the remainder. From this set we take the best one to form the initial guess $f_{k+1}^{(0)} := f_k + g_k$ for the best rank k+1 approximation. This relatively cheap procedure helps us to avoid being trapped into poor local minima.

The ultimate goal of the present work is to develop efficient schemes for the computation of two-electron integrals arising from localized occupied and virtual orbitals. This comprises applications to local correlation methods [58,59] and linear-scaling Kohn-Sham or Hartree-Fock methods [60] based on localized orbitals. In order to benefit from tensor formats, separable approximations of the Coulomb potential are required. These can be obtained from the familiar representation of the Coulomb potential via a Gaussian transform, we refer to Ref. [61] for further details. In Paper I, we have discussed a wavelet based approach for the canonical format which furthermore provides sparse approximations of the Kronecker factors. An alternative approach based on the Tucker format has been studied in Ref. [43]. Concerning our envisaged application, the L^2 norm underlying the least-squares problem (3) is not optimal. It is well known from various studies of density fitting schemes [16,18] that the Coulomb norm gives considerably better results. We have argued in Paper I that the Coulomb norm is similar to the norm of the Sobolev space H^{-1}. It is work in progress to modify the Newton algorithm in order to allow for H^{-1} and various other norms.

1.2 Canonical tensor products versus GTOs

In order to judge the applicability of best separable rank κ approximations in quantum chemistry, we have studied orbital compression rates for a few single-electron systems. First we solved the Schrödinger equation using large uncontracted GTO basis sets. In the second step we have generated best separable rank κ approximations of the wavefunctions by minimizing the least-squares functional (3) and calculated the corresponding variational energy. The simplest single-electron system is the H atom. Errors in energy for best separable rank κ approximations obtained with the Newton algorithm are shown in Fig. 1 a). These have been compared with best radial Gaussian approximations of the same

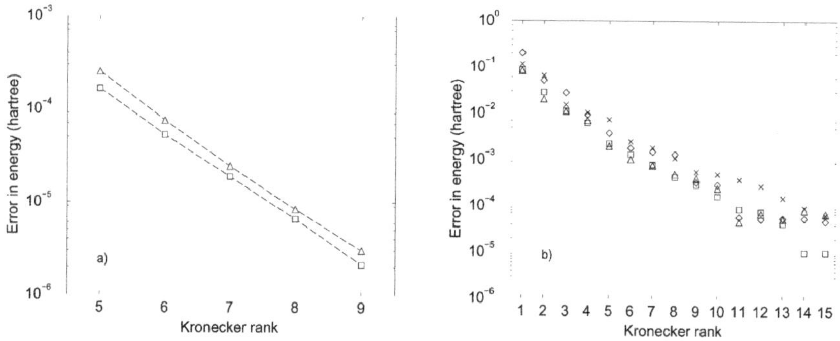

Fig. 1. Error in total energy (hartree) versus Kronecker rank κ of best separable approximations for one-electron wavefunctions. a) Best separable $L^2(\mathbb{R}^3)$ (□) and best radial Gaussian $L^\infty(\mathbb{R}_+)$ (Δ) approximations for the H atom. b) Best separable rank κ approximations for various positively charged dimers (□) H_2^+ ($K = 58$), (◊) HHe^{2+} ($K = 58$), and trimers (Δ) H_3^{2+} ($K = 126$, triangle), (×) H_3^{2+} ($K = 75$, linear).

rank. The latter were optimized with respect to a weighted $L^\infty(\mathbb{R}_+)$ norm. It can be seen that both approaches lead to similar approximation errors, where the best separable rank κ approximation performs only slightly better. Our results indicate that GTOs already provide almost optimal separable approximations for the H atom. However, this does not seem to be the case for systems with several nuclei as can be seen from Fig. 1 b) where significant compression rates have been achieved with respect to standard GTO basis sets. The exponents of the uncontracted (8*s*4*p*3*d*) GTO basis were taken from V5Z basis sets [62]. Once the Schrödinger equation has been solved in this basis set, we compressed the wavefunctions for different Kronecker ranks $1 \leq \kappa \leq 15$. Thereafter, the set of rank-1 tensors $\{h_k^{(1)}(x_1)\, h_k^{(2)}(x_2)\, h_k^{(3)}(x_3)\}$ from the tensor product expansion (1) was taken as a new basis in which the Schrödinger equation was solved again. It should be mentioned that the approximation error of best separable rank κ approximations might depend on the orientation of the molecule with respect to the coordinate axes. The dimers H_2^+, HHe^{2+}, and a linear trimer H_3^{2+} were therefore oriented along the diagonal in order to avoid such kind of rank reductions due to symmetry. Furthermore, we have considered a triangular structure for H_3^{2+} with each nucleus located on an axis. The resulting errors in energy are shown in Fig. 1 b). It can be seen that the overall convergence for these systems is rather similar and an error in energy $\leq 10^{-4}$ hartree, which roughly corresponds to the basis set error of the GTO bases, can be achieved at Kronecker rank $\kappa \leq 15$. For comparison, the initial separation ranks of the GTO bases, ranging from $K = 58$ for H_2^+ up to $K = 126$ for H_3^{2+} (triangular structure), are considerably larger.

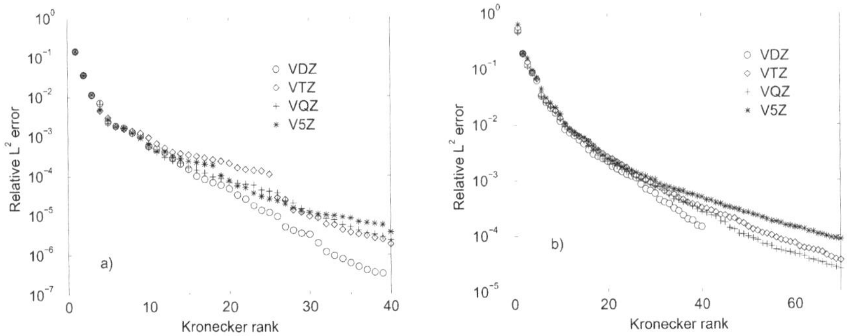

Fig. 2. Relative L^2-error, i.e., $\sigma_\kappa(f)/\|f\|_{L^2}$ of tensor product approximations for orbitals represented by various GTO basis sets, ranging from VDZ up to V5Z. Results are shown for the core orbital (a) and the highest occupied orbital (b) of CH_4.

2. Orbitals and their products

The simple examples in the previous section illustrate that canonical tensor product approximations might be beneficial for molecular orbitals. An interesting question is how does the error for a certain Kronecker rank depend on the size of the GTO basis set. We have considered tensor product approximations for the orbitals of CH_4 represented in VDZ, VTZ, VQZ and V5Z basis sets. It can be seen from Fig. 2 that the approximation error for core and valence orbitals is not sensitive to the quality of the GTO basis set. However there is a marked difference between core and valence orbitals. The canonical tensor product approximation actually converges much faster for the 1*s* core orbital than for valence orbitals. It seems that the nuclear cusps do not restrain, at least in leading order, the convergence of the approximation. This is not surprising in view of the almost exponential convergence of GTO bases [5,6] mentioned before. Obviously, the spatially more extended valence orbitals with their richer structure require considerably higher Kronecker ranks in order to achieve a certain accuracy.

2.1 Density fitting via canonical tensor products

For our envisaged applications, canonical tensor product approximations of orbitals are of minor interest. Instead we are mainly interested in orbital products and the total electron density which appear in general two-electron integrals and the Hartree potential, respectively. Therefore our approach is closely related to conventional density fitting schemes based on auxiliary GTO basis sets. The pivotal question we have to answer is whether we can achieve a substantial reduction of the initial separation rank which might be worth the additional effort caused by the compression step. Presently, it seems to be premature to outline a detailed algorithm for a density fitting scheme based on canonical tensor products. However we can already highlight certain specific features of such a scheme.

(i) We expect our approach to be applicable to large molecules using localized orbitals. This is in agreement with the prevalent view adapted in local correlation methods [58,59] which are based on localized occupied and virtual orbitals. More recently, linear-scaling Kohn-Sham and Hartree-Fock methods based on localized orbitals [60] have been studied.

(ii) Each tensor product approximation of a certain orbital product can be used many times in order to compute all the corresponding two electron integrals.

(iii) Further significant reductions of the separation rank can be achieved at the expense of the second compression step after the convolution of the orbital product with the Coulomb potential. This might be especially interesting for those orbital products which give contributions to a large number of two-electron integrals. We refer to Paper I where the compression of the Hartree potential has been studied in detail.

In Fig. 3, we have shown the relative L^2 errors of orbital products for the molecules CH_4, H_2O, C_2H_2, C_2H_6 and CH_3OH. The maximum errors for all molecules are of similar magnitude except for C_2H_2 where the maximal errors are considerably larger. A closer look reveals that the largest relative errors are from products between σ and π orbitals. It can be easily visualized that these products have the most complicated structures, however, their absolute values are comparatively small. Another significant difference can be observed for products where core orbitals contribute which have much smaller relative errors than the rest. This can be understood from our discussion of the compression of individual orbitals where core orbitals were shown to have much smaller errors for a given Kronecker rank than valence orbitals.

The canonical tensor product approximation for products of orbitals can be directly applied to compute the corresponding Coulomb and exchange contributions. Some results concerning the error of the Hartree-Fock energy for different Kronecker ranks are listed in Table 1. It can be seen that the absolute errors for $\kappa = 45$ are well below 1 mhartree, except for CH_3OH where a slightly larger Kronecker rank, i.e., $\kappa = 75$, is required. A more detailed analysis of the error has been shown in Fig. 4 for CH_3OH, where we have considered the Coulomb and exchange contributions of each orbital separately, i.e.,

$$\mathcal{J}_a := \sum_{b=1}^{N/2} (aa|bb), \quad \mathcal{K}_a := \sum_{b=1}^{N/2} (ab|ab). \tag{2.1}$$

Increasing the Kronecker rank from $\kappa = 45$ to $\kappa = 75$ reduces the error for most of the orbitals by an order of magnitude. It can be seen that the errors are roughly the same for all orbitals where the errors for the exchange part are smaller than for the Coulomb contribution.

3. Pseudopotentials

For standard GTO basis sets, a relatively small number of contracted primitive GTOs are sufficient for the representation of core orbitals. In our tensor terminol-

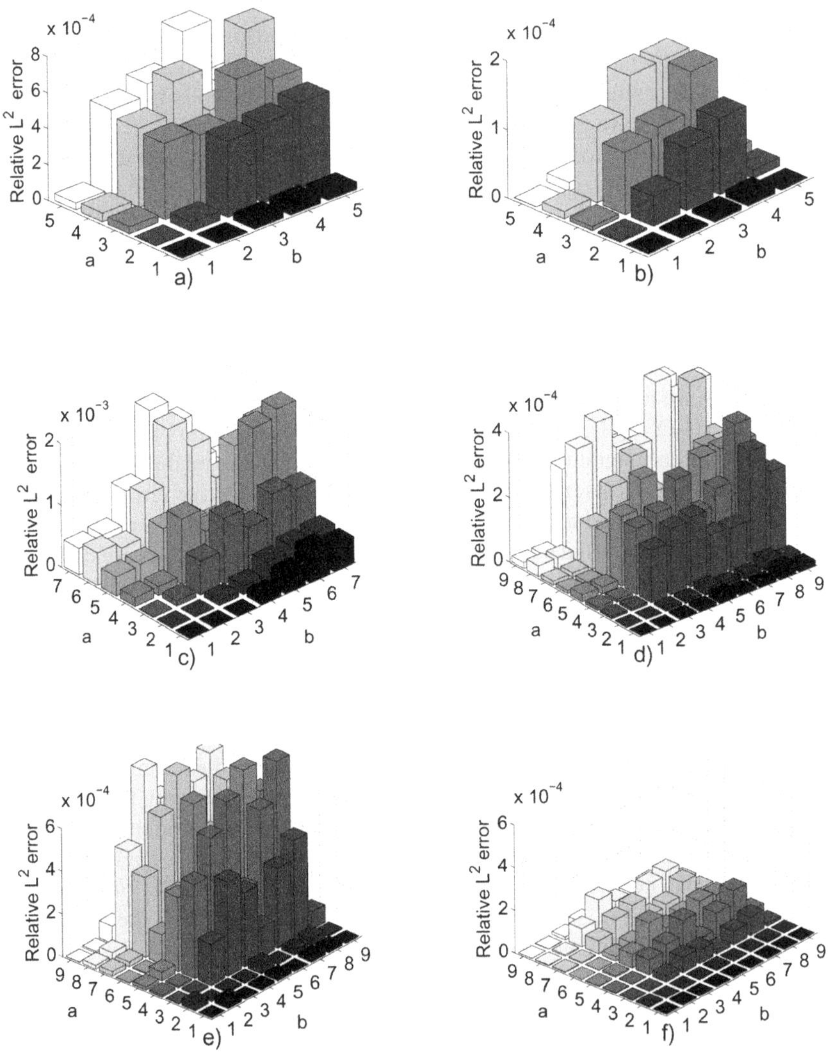

Fig. 3. Relative L^2 error, i.e., $\sigma_\kappa(f)/\|f\|_{L^2}$ of canonical tensor product approximations of orbital products for the molecules CH_4 (a), H_2O (b), C_2H_2 (c), C_2H_6 (d) and CH_3OH (e,f). Results are shown for Kronecker rank $\kappa = 45$ (a-e) as well as $\kappa = 75$ (f) in the case of CH_3OH. All orbitals were originally represented in VDZ basis sets.

ogy this means that they can be efficiently approximated by low rank tensor products. An alternative approach is to replace the core electrons by pseudopotentials. This has the further advantage that it also allows for smaller valence basis sets because for pseudo valence orbitals the orthogonality constraint with

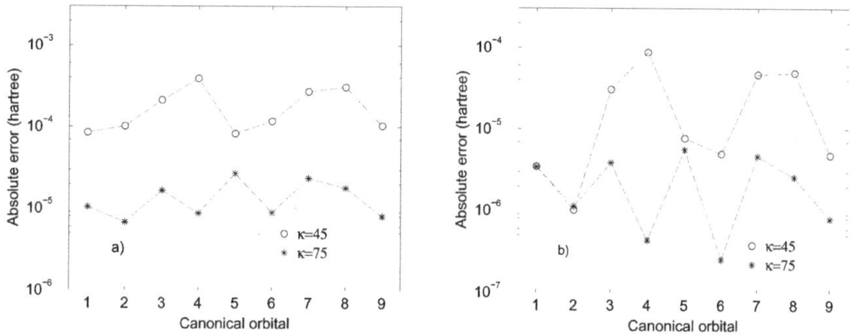

Fig. 4. Individual contributions of the canonical orbitals of CH_3OH to the error of Coulomb (a) and exchange (b) energy, cf. (1), in the canonical tensor product density fitting scheme. Results are shown for Kronecker ranks $\kappa = 45$ and $\kappa = 75$.

Table 1. Error in the Hartree-Fock energy $E_{HF}^{GTO} - E_{HF}^{\kappa}$ (hartree) for different Kronecker ranks κ of the canonical tensor product density fitting scheme.

κ	H_2O	CH_4	C_2H_2	C_2H_6	CH_3OH
45	2.92×10^{-4}	7.26×10^{-4}	1.20×10^{-3}	1.05×10^{-3}	3.13×10^{-3}
55	1.64×10^{-4}	5.34×10^{-4}	6.45×10^{-4}	2.54×10^{-4}	1.18×10^{-3}
75	-	-	-	-	2.16×10^{-4}

respect to the core orbitals must not be retained. We have studied the effects of pseudopotentials on the tensor product approximation of the electron density for the simple molecules CH_4 and SiH_4. Semilocal energy-consistent pseudopotentials [63] have been used for C and Si. These pseudopotentials still contain a singular Coulomb potential with respect to the effective charge of the atomic core. For Si, we have also tested a new type of smooth pseudopotential which has been developed especially for applications in quantum Monte Carlo calculations [64]. This pseudopotential has no singular term anymore and results in a smoother behaviour of the pseudo orbitals near nuclei.

In order to study the effect of pseudopotentials on canonical tensor product approximations, we considered only the total (pseudo) electron density instead of individual orbital products. The tensor product approximation was then used to calculate the Hartree potential according to the procedure described in Paper I. We have listed in Table 2 the error in the Hartree-Fock energy due to the density fitted Hartree potential for different Kronecker ranks. It can be seen that for CH_4 the errors for the pseudopotential and the all-electron calculation are almost the same. Going from CH_4 to SiH_4 increases the size of the core region however absolute errors up to $\kappa = 20$ are similar. At higher ranks the errors for SiH_4 are on the average by a factor of 6 smaller. For comparison, the smooth pseudopotential has larger errors, varying between a factor of 2 to 7, than the standard pseudopotential. Our results indicate that tensor product ap-

Table 2. Error in the Hartree-Fock energy $E_{HF}^{GTO} - E_{HF}^{\kappa}$ (hartree) for different Kronecker ranks κ of the tensor product approximation of the (pseudo) electron density. Results are shown for all-electron (AE) and pseudopotential (PP) calculations.

	CH_4	SiH_4		
κ	PP[a]	AE	PP[a]	PP[b]
15	1.92×10^{-3}	3.25×10^{-3}	-1.30×10^{-3}	-
20	6.64×10^{-4}	5.67×10^{-4}	5.17×10^{-4}	-1.23×10^{-3}
25	6.05×10^{-4}	5.41×10^{-4}	-9.23×10^{-5}	1.75×10^{-4}
30	1.38×10^{-4}	5.15×10^{-4}	1.98×10^{-5}	1.33×10^{-4}
35	1.00×10^{-4}	9.40×10^{-5}	2.37×10^{-5}	5.77×10^{-5}
40	8.70×10^{-5}	8.44×10^{-5}	1.55×10^{-5}	5.14×10^{-5}
45	3.89×10^{-5}	5.50×10^{-5}	1.39×10^{-5}	4.35×10^{-5}
50	4.62×10^{-5}	5.31×10^{-5}	5.91×10^{-6}	4.34×10^{-5}

[a] Energy consistent pseudopotentials from Ref. [63].
[b] Pseudopotential without singular term from Ref. [64].

proximations are not very sensitive with respect to nuclear cusps of the electron density.

4. Conclusions

We have studied an alternative to conventional GTO based density fitting schemes using canonical tensor product approximations for orbitals, products of orbitals and the electron density. It has been demonstrated that a considerable reduction of the initial separation rank with respect to GTO bases has been achieved, despite the fact that our optimization scheme uses the L^2 norm which is considered not to be optimal. The present work represents the first step into the largely unexplored landscape of tensor product approximations in quantum chemistry. Nevertheless, the highly successful GTO based density fitting schemes set high standards for all future work in this area. Taking into account the substantial computational effort to determine canonical tensor product approximations, it will require considerable efforts to develop a tensor product based density fitting scheme which becomes competitive to GTO auxiliary bases. It is our appraisement, however, that research in tensor product approximation is still at the very beginning and significant progress can be expected in the near future.

Acknowledgement

The authors gratefully acknowledge PD Dr. B. Khoromskij, V. Khoromskaia, Dr. L. Grasedyck, and S. Schwinger (Leipzig) as well as Prof. R. Schneider (Berlin) for useful discussions. This work was supported by the Deutsche Forschungsgemeinschaft (SPP 1145).

References

1. S. F. Boys, Proc. R. Soc. London Ser. A **200** (1950) 542.
2. S. F. Boys, G. B. Cook, C. M. Reeves and I. Shavitt, Nature **178** (1956) 1207.
3. H. Preuss, Z. Naturforsch. A **11** (1956) 323.
4. T. Helgaker, P. Jørgensen and J. Olsen, Molecular Electronic-Structure Theory, Wiley, New York (1999).
5. W. Kutzelnigg, Int. J. Quantum Chem. **51** (1994) 447.
6. D. Braess, J. Approx. Theory **83** (1995) 93.
7. A. Halkier, T. Helgaker, P. Jørgensen, W. Klopper and J. Olsen, Chem. Phys. Lett. **302** (1999) 437.
8. I. Daubechies, Ten Lectures on Wavelets, CBMS-NSF Regional Conference Series in Applied Mathematics **61** (1992).
9. H.-J. Flad, W. Hackbusch, and R. Schneider, ESAIM: M2AN **40** (2006) 49.
10. V. Maz'ya and G. Schmidt, Approximate Approximations, Mathematical Surveys and Monographs, Vol. 141, American Mathematical Society (2007).
11. E. J. Baerends, D. E. Ellis and P. Ros, Chem. Phys. **2** (1973) 41.
12. H. Sambe and R. H. Felton, J. Chem. Phys. **62** (1975) 1122.
13. B. I. Dunlap, J. W. D. Connolly and J. R. Sabin, J. Chem. Phys. **71** (1979) 3396.
14. K. Eichkorn, O. Treutler, H. Öhm, M. Häser and R. Ahlrichs, Chem. Phys. Lett. **240** (1995) 283.
15. F. R. Manby and P. J. Knowles, Phys. Rev. Lett. **87** (2001) 163001.
16. F. R. Manby, P. J. Knowles and A. W. Lloyd, J. Chem. Phys. **115** (2001) 9144.
17. C. Van Alsenoy, J. Comput. Chem. **9** (1988) 620.
18. O. Vahtras, J. Almlöf and M. W. Feyereisen, Chem. Phys. Lett. **213** (1993) 514.
19. S. Ten-no and S. Iwata, Chem. Phys. Lett. **240** (1995) 578.
20. R. A. Kendall and A. Früchtl, Theor. Chem. Acc. **97** (1997) 158.
21. F. Weigend, Phys. Chem. Chem. Phys. **4** (2002) 4285.
22. R. Polly, H.-J. Werner, F. R. Manby and P. J. Knowles, Mol. Phys. **102** (2004) 2311.
23. M. Feyereisen, G. Fitzgerald and A. Komornicki, Chem. Phys. Lett. **208** (1993) 359.
24. D. E. Bernholdt and R. J. Harrison, Chem. Phys. Lett. **250** (1996) 477.
25. F. Weigend and M. Häser, Theor. Chem. Acc. **97** (1997) 331.
26. F. Weigend, M. Häser, H. Patzelt and R. Ahlrichs, Chem. Phys. Lett. **294** (1998) 143.
27. H.-J. Werner, F. R. Manby and P. J. Knowles, J. Chem. Phys. **118** (2003) 8149.
28. M. Schütz and F. R. Manby, Phys. Chem. Chem. Phys. **5** (2003) 3349.
29. N. H. F. Beebe and J. Linderberg, Int. J. Quantum Chem. **12** (1977) 683.
30. F. Aquilante, R. Lindh and T. B. Pedersen, J. Chem. Phys. **127** (2007) 114107.
31. L. Boman, H. Koch, A. S. de Merás, J. Chem. Phys. **129** (2008) 134107.
32. B. H. Greeley, T. V. Russo, D. T. Mainz, R. A. Friesner, J.-M. Langlois, W. A. Goddard III, R. E. Donnelly, Jr., M. N. Ringnalda, J. Chem. Phys. **101** (1994) 4028.
33. T. J. Martinez and E. A. Carter, in Modern Electronic Structure Theory, Vol. II, D.R. Yarkony, Ed., World Scientific, Singapore, (1995) p. 1132.
34. T. G. Kolda and B. W. Bader, SIAM Review **51**, to appear September 2009.
35. A. Smilde, R. Bro and P. Geladi, Multi-way Analysis with Applications in the Chemical Sciences, John Wiley & Sons, Chichester, (2004).
36. S. R. Chinnamsetty, M. Espig, B. N. Khoromskij, W. Hackbusch and H.-J. Flad, J. Chem. Phys. **127** (2007) 084110.
37. C. F. Van Loan, J. Comput. Appl. Math. **123** (2000) 85.
38. V. De Silva and L.-H. Lim, SIAM J. Matrix Anal. Appl. **30** (2008) 1084.
39. L. De Lathauwer, B. De Moor and J. Vandewalle, SIAM J. Matrix Anal. Appl. **21** (2000) 1253.
40. L. De Lathauwer, B. De Moor and J. Vandewalle, SIAM J. Matrix Anal. Appl. **26** (2004) 295.

41. M. Espig and W. Hackbusch, The efficient best approximation with tensors in the canonical polyadic tensor representation. Preprint MPI-MIS 2009.
42. S. R. Chinnamsetty, Ph. D. thesis, Universität Leipzig 2008.
43. B. N. Khoromskij, V. Khoromskaia, S. R. Chinnamsetty, H.-J. Flad J. Comput. Phys. **228** (2009) 5749.
44. W. Hackbusch and S. Kühn, A new scheme for the tensor representation. Preprint No. **2**, MPI MIS 2009.
45. I. V. Oseledets, Compact matrix form of the d-dimensional tensor decomposition. Preprint No. **1**, INM RAS 2009.
46. I. V. Oseledets E. E. Tyrtyshnikov Breaking the curse of dimensionality, or how to use SVD in many dimensions. Preprint No. **2**, INM RAS 2009.
47. I. V. Oseledets, Tensors inside of matrices give logarithmic complexity. Preprint No. **4**, INM RAS 2009.
48. W. Hackbusch and B. N. Khoromskij, J. Complexity, **23** (2007) 697.
49. B. N. Khoromskij and V. Khoromskaia, Cent. Eur. J. Math. **5** (2007) 523.
50. W. Hackbusch and B. N. Khoromskij, SIAM J. Matrix Anal. Appl. **30** (2008) 1233.
51. B. N. Khoromskij, Comput. Vis. Sci. **11** (2008) 259.
52. M. Espig, Ph. D. thesis, Universität Leipzig 2008.
53. M. Espig and W. Hackbusch, A new modified Newton's method for the efficient low tensor rank approximation with sums of elementary tensors in high dimensions. Preprint MPI-MIS 2009.
54. I. V. Oseledets and D. V. Savostianov, Comput. Math. Math. Phys. **46** (2006) 1641.
55. I. V. Oseledets, D. V. Savostianov and E. E. Tyrtyshnikov, SIAM J. Matrix Anal. Appl. **30** (2008) 939.
56. H.-J. Flad, B. N. Khoromskij, D. V. Savostyanov and E. E. Tyrtyshnikov. Russian J. Numer. Anal. Math. Modelling, **23** (2008) 329.
57. B. N. Khoromskij and V. Khoromskaia, SIAM J. Sci. Comp. **31** (2009) 3002.
58. M. Schütz, G. Hetzer and H.-J. Werner, J. Chem. Phys. **111** (1999) 5691.
59. G.E. Scuseria and P.Y. Ayala, J. Chem. Phys. **111** (1999) 8330.
60. S. Goedecker, in Handbook of Numerical Analysis, Vol. X, C. Le Bris and P. G. Ciarlet, eds., Elsevier, Amsterdam, (2003).
61. H.-J. Flad, W. Hackbusch, B.N. Khoromskij, and R. Schneider, in: Matrix Methods: Theory, Algorithms and Applications, V. Olshevsky and E. Tyrtyshnikov eds., (World Scientific Publishing, Singapore, 2010) pp. 313.
62. T.H. Dunning, Jr., J. Chem. Phys. **90** (1989) 1007.
63. A. Bergner, M. Dolg, W. Küchle, H. Stoll and H. Preuss, Mol. Phys. **80** (1993) 1431.
64. M. Burkatzki, C. Filippi and M. Dolg, J. Chem. Phys. **126** (2007) 234105.

Analytic Calculation of First-order Molecular Properties at the Explicitly-correlated Second-order Møller-Plesset Level

By Sebastian Höfener[1], Christof Hättig[2], and Wim Klopper[1,*]

[1] Karlsruhe Institute of Technology (KIT), Theoretical Chemistry Group, Kaiserstraße 12, 76131 Karlsruhe, Germany
[2] Ruhr-Universität Bochum, Lehrstuhl für Theoretische Chemie, Universitätsstraße 150, 44801 Bochum, Germany

(Received July 29, 2009; accepted October 12, 2009)

First-order Molecular Properties / Analytic Gradients / Møller-Plesset Perturbation Theory / Explicitly-correlated Wave Functions

Formulae are derived and implemented for the analytic calculation of first-order molecular properties at the level of explicitly-correlated second-order Møller-Plesset perturbation theory. In this theory, which is denoted as MP2-F12 theory, Slater-type geminals are used to expand the first-order wave function. A second-order perturbation theory correction for single excitations into a complementary auxliary basis set is also included. At the MP2-F12 level, it seems sufficient to restrict the analytic calculation of energy derivatives to the level of standard approximation A of MP2-F12 theory and to assume the extended Brillouin condition to hold. Smooth and rapid convergence towards the basis-set limit is observed for the dipole moments of a selection of small closed- and open-shell molecules when calculated at the RI-MP2-F12/ 2A*[T+V] + CABS singles level in augmented correlation-consistent polarized valence double-, triple-, and quadruple-zeta basis sets that have been optimized especially for use in MP2-F12 theory.

1. Introduction

Over the past 50 years, the numerical methods of quantum chemistry have developed into what can be said to be a "quantitative quantum chemistry", that is, a tool for providing quantitative data about chemistry, sufficiently accurate to reliably predict, confirm, or reject experimental observations and measurements [1].

Important components of this quantitative tool are the coupled-cluster theory since the early 1980s and the correlation-consistent cc-pVXZ basis sets devel-

* Corresponding author. E-mail: klopper@kit.edu

© by Oldenbourg Wissenschaftsverlag, München

oped by Dunning and others since the late 1980s [2]. However, although very successful, coupled-cluster calculations in cc-pVXZ basis sets suffer from a notoriously slow basis-set convergence, that is, slow convergence with the cardinal number X = 2, 3, 4, ... towards the basis-set limit obtained for X = ∞.

This slow convergence can be traced back to the inability of Slater determinants build from one-electron functions to appropriately describe the Coulomb hole around the electron-electron coalescence point. The basis-set convergence can, however, be accelerated significantly by using explicitly-correlated coupled-cluster wave functions that account for the electron-electron cusp and describe the Coulomb hole in an accurate manner.

It is the purpose of the present article to briefly review the recent advances in explicitly-correlated theory (R12 and F12 approaches, *vide infra*) and to summarize the achievents made in this field with respect to the analytic calculation of molecular properties including nuclear gradients. We begin by discussing the early R12 theory of Kutzelnigg and proceed by reviewing the more recent F12 theory and the calculation of analytic derivatives.

1.1 Early R12 theory

In 1985, Kutzelnigg [3] suggested to supplement conventional configuration-interaction wave functions with two-electron basis functions of the type $r_{12}\phi_i(1)\phi_j(2)$, where r_{12} is the distance between the electrons 1 and 2, and where ϕ_i and ϕ_j are occupied orbitals of the Hartree-Fock reference determinant. For molecular many-electron systems, this approach was first implemented at the level of second-order Møller-Plesset perturbation theory and was denoted as MP2-R12 theory [4]. Similar implementations at the configuration-interaction, coupled-electron-pair approximation [5], and coupled-cluster levels (CC-R12) followed shortly thereafter [6,7].

Using linear r_{12} terms in the many-electron wave function dramatically improved the convergence towards the limit of a complete one-electron basis. In Dunning's [8] correlation-consistent cc-pVXZ basis, for example, the basis-set error had been reduced from $\propto X^{-3}$ for conventional expansions [9] to $\propto X^{-7}$ for R12-type wave functions [10].

Until 2002, the R12 approaches had employed linear r_{12} terms and a single set of atomic basis functions, that is, Gaussians. This basis set was used both for expanding the Hartree-Fock orbitals and for the approximate resolution of the identity (RI) that was inserted into the three- and four-electron integrals to factorize these into sums of products of two-electron integrals. This required the use of very large basis sets. Nevertheless, highly accurate correlation energies of benchmark quality were obtained for small systems, that is, for few-electron atoms and molecules [11].

The R12 methods were used mainly to obtain MP2 and coupled-cluster energies close to the limit of a complete basis set. On a few occasions, static electric properties such as dipole moments, polarizabilities, and first and second hyperpolarizabilities were computed by numerical differentiation, for example for the four-electron systems Be [12], LiH [13], and He_2 [14]. Other examples include the numerical calculation of the dipole moments of BH and HF [15] and the relativistic two-electron Darwin energies of Ne and HF [16]. Molecular geometries were usually kept fixed, but Refs. [17] and [18] provide examples of partial geometry optimizations of the (protonated) water dimer by means of point-wise energy calculations. More recently, but still using R12 wave functions, such a point-wise numerical approach was applied to obtain the basis-set-limit equilibrium geometries [19] of a series of closed- and open-shell molecules as well as their harmonic vibrational frequencies [20].

1.2 Modern F12 theory

The introduction of an auxiliary basis set (ABS) for the intrinsic RI approximation in 2002 [10] revolutionized the R12 approach. This new ABS approach was soon further improved by switching to a complementary auxiliary basis set (CABS) [21]. Density-fitting (DF) techniques were developed for the two-electron integrals that occur in R12 theory [22], the linear r_{12} terms were replaced by much more efficient Slater-type geminals (STGs) of the form $\exp(-\zeta r_{12})$ [23], an approximation was introduced to avoid two-electron integrals over the commutator between the operators of kinetic energy and r_{12} [24], and it was proposed to keep the amplitudes of the R12 geminals fixed at values that follow from Kato's cusp condition [25].

When using STGs or other functions of r_{12}, Møller-Plesset perturbation theory is referred to as MP2-F12 theory (in place of MP2-R12 theory) to highlight the fact that the function $f(r_{12})$ is used in place of r_{12} itself. Today, MP2-F12 theory is applied using the aforementioned techniques and approximations that have become available since about 2003. These include STGs, CABS, DF, the commutator approximation, and fixed amplitudes [26–32]. The MP2-R12 and MP2-F12 theories have recently been reviewed in Ref. [33].

Although the focus of the present article is on MP2-F12 theory, we notice that explicitly-correlated coupled-cluster theory, for example CCSD-F12, has developed in a similar manner as MP2-F12 theory since about 2004. With respect to coupled-cluster theory, we refer the reader to Refs. [34–48] and the references therein.

Not only at the R12 level but also at the F12 level, equilibrium geometries and harmonic vibrational frequencies have been obtained by means of point-wise energy calculations [49–52].

1.3 Analytic gradients

Recently, Kordel *et al.* implemented the analytic calculation of the density matrix for relaxed first-order molecular properties at the level of MP2-R12 theory [53]. Also the analytic calculation of nuclear gradients was accomplished at the MP2-

R12 level [54], however without using an auxiliary basis set (neither ABS nor CABS).

In these previous works by Kordel and co-workers, [53,54] only Ansatz 1 of MP2-R12 theory was considered. Nuclear gradients were implemented using standard approximation A, that is, at the MP2-R12/1A level. Furthermore, the same basis of atomic orbitals used to expand the Hartree-Fock orbitals was used for the RI approximation to simplify the theory. And finally, whereas Kordel *et al.* conducted their implementation in the Dalton program [55] at the MP2-R12 level, the present article addresses the implementation of the analytic calculation of nuclear gradients of density-fitted RI-MP2-F12 theory in the Turbomole program [56].

2. Theory

2.1 The MP2-F12 Hylleraas functional

In terms of spin orbitals from an unrestricted Hartree-Fock (UHF) reference determinant, the Ansatz for the first-order wave function is

$$\begin{aligned} |\Psi^{(1)}\rangle &= \left(\hat{\mathcal{T}}_2^{(1)} + \hat{\mathcal{T}}_{2'}^{(1)}\right)|\mathrm{HF}\rangle \\ &= \sum_{i>j}\sum_{a>b} t_{ij}^{ab}|_{ij}^{ab}\rangle + \sum_{i>j}\sum_{\mathbf{k}>\mathbf{l}} c_{ij}^{\mathbf{kl}} \sum_{\alpha\beta} \bar{F}_{\mathbf{kl}}^{\alpha\beta}|_{ij}^{\alpha\beta}\rangle, \end{aligned} \tag{1}$$

with $\bar{F}_{\mathbf{kl}}^{\alpha\beta} = \langle\mathbf{kl}|f_{12}\hat{Q}_{12}|\alpha\beta\rangle$, $f_{12} = \exp(-\zeta r_{12})$, and $\hat{Q}_{12}$ is the projection operator that ensures strong orthogonality. In the following, $|xy\rangle$ denotes an antisymmetrized product of spin orbitals. The lower-case indices $i,j,k,...$ denote active occupied orbitals. Bold letters refer to the F12 excitations, usually into a geminal constructed from a pair of occupied orbitals

$$\hat{Q}_{12}f_{12}|\mathbf{kl}\rangle. \tag{2}$$

Capital letters indicate all occupied orbitals (frozen + active). $c_{ij}^{\mathbf{kl}}$ are the amplitudes of the F12 excitations, which can be determined either variationally or held fixed according to Ten-no's rational generator approach [25].

Throughout this article, we use Ansatz 2 for the projection operator,

$$\hat{Q}_{12} = (1 - \hat{O}_1)(1 - \hat{O}_2) - \hat{V}_1\hat{V}_2, \tag{3}$$

where $\hat{O} = \Sigma_I |I\rangle\langle I|$ is the projector onto all occupied orbitals and $\hat{V} = \Sigma_a |a\rangle\langle a|$ the projector onto all virtual orbitals. Inserting the Ansatz for $\Psi^{(1)}$ into the Hylleraas functional yields

$$\mathcal{H}_{\text{MP2-F12}} = \sum_{i>j}\sum_{a>b} t_{ij}^{ab}\left[2\bar{g}_{ab}^{ij} + (\epsilon_a + \epsilon_b - \epsilon_i - \epsilon_j)t_{ij}^{ab}\right] + \sum_{i>j}\sum_{\mathbf{k}>\mathbf{l}} c_{ij}^{\mathbf{kl}}\left[2\bar{V}_{\mathbf{kl}}^{ij} + \sum_{\mathbf{m}>\mathbf{n}} \bar{B}_{\mathbf{kl},\mathbf{mn}}^{(ij)} c_{ij}^{\mathbf{mn}}\right] + 2\sum_{i>j}\sum_{a>b} t_{ij}^{ab} \sum_{\mathbf{k}>\mathbf{l}} \bar{C}_{\mathbf{kl}}^{ab} c_{ij}^{\mathbf{kl}}, \quad (4)$$

with the antisymmetrized (indicated by a bar) integrals

$$\bar{g}_{ab}^{ij} = \langle ab|r_{12}^{-1}|ij\rangle, \quad (5)$$

$$\bar{C}_{\mathbf{kl}}^{ab} = \langle \mathbf{kl}|f_{12}\hat{Q}_{12}(\hat{F}_1 + \hat{F}_2)|ab\rangle \overset{\text{EBC}}{\approx} 0, \quad (6)$$

$$\bar{V}_{\mathbf{kl}}^{ij} = \langle \mathbf{kl}|f_{12}\hat{Q}_{12}r_{12}^{-1}|ij\rangle, \quad (7)$$

$$\bar{B}_{\mathbf{kl},\mathbf{mn}}^{(ij)} = \langle \mathbf{kl}|f_{12}\hat{Q}_{12}(\hat{F}_1 + \hat{F}_2 - \epsilon_i - \epsilon_j)\hat{Q}_{12}f_{12}|\mathbf{mn}\rangle \overset{A}{\approx} \frac{1}{2}\langle \mathbf{kl}|\left(f_{12}\hat{Q}_{12}[\hat{T}_{12}, f_{12}] + [f_{12}, \hat{T}_{12}]\hat{Q}_{12}f_{12}\right)|\mathbf{mn}\rangle. \quad (8)$$

$\hat{F}_1$ and $\hat{F}_2$ are the Fock operators for electrons 1 and 2, and $\hat{T}_{12}$ is the sum of the operators of kinetic energy for electrons 1 and 2. The MP2-F12 energy is obtained by minimizing $\mathcal{H}_{\text{MP2-F12}}$ with respect to the amplitudes t_{ij}^{ab} and $c_{ij}^{\mathbf{kl}}$.

The off-diagonal block Eq. (6) of the Fock matrix vanishes ($C_{\mathbf{kl}}^{ab} = 0$) if we assume that the extended Brillouin condition (EBC) [33] holds. This is equivalent with assuming that the canonical virtual orbitals are exact eigenfunctions of the Fock operator. We attach an asterisk to the method's acronym if the EBC is assumed (here: 2^*).

In approximation A [33], terms with the exchange operator involved are neglected, including the orbital energies. The remaining contributions are shown in Eq. (8). Considerable amounts of computation time are saved by computing the integrals over the operator $[\hat{T}_{12}, f_{12}]$ in Eq. (8) using the approach proposed by Kedžuch [24]. In this approach, the exact operator is approximated with for example the matrix representation of $\hat{T} + \hat{V}$ (the bare nuclear Hamiltonian), which we use by default in our MP2-F12 calculations with Turbomole.

In addition to improving the description of elecron correlation, we also include a correction to the Hartree-Fock energy obtained from treating single excitations into the CABS by perturbation theory,

$$E_{\text{CABS}} = \sum_{I,p'} c_I^{p'} F_{Ip'} = \sum_{I,p'} \frac{F_{Ip'}^2}{\epsilon_I - \epsilon_{p'}}, \quad (9)$$

where the CABS is denoted as $\{p'\}$. It is understood that the Hartree-Fock basis functions as well as the CABS have been rotated among themselves in such a manner that the Hartree-Fock...Hartree-Fock and CABS...CABS blocks of the Fock matrix **F** are diagonal, and the diagonal elements are given by the orbital energies ε_I and $\varepsilon_{p'}$. This provides a balanced description (in terms of accuracy) of the Hartree-Fock reference and the correlation part. The $F_{ap'}$ off-diagonal elements, which give rise to certain coupling terms [35,46,57], have so far been neglected in our work.

2.2 Level of approximation

With respect to the implementation of analytic nuclear gradients in Turbomole, we aim at the MP2-F12/2*A [T+V] level. Despite the many approximations that are involved to allow for efficient calculations, this level captures all main effects for both properties and geometries, as explained in the following.

The most important improvement is obtained by switching from Ansatz 1 to 2 and from R12 to F12 theory. In terms of basis-set quality, it has been shown that quintuple-zeta quality results are obtained using Ansatz 2 of F12 theory in just a triple-zeta basis [34].

Table 1 demonstrates that the differences between the equilibrium geometries obtained with the Ansätze 2 and 2* are very small, for both approximations A and B. Note that the effects of changing the basis set from cc-pVTZ-F12 to cc-pVQZ-F12 are larger than the differences between the four MP2-F12 Ansätze. The results furthermore show that in basis sets of cc-pVTZ-F12 quality and better, the differences between approximations A and B are insignificant, such that it seems sufficient to only consider standard approximation A.

Table 2 indicates that the CABS singles term should be included when considering the calculation of nuclear gradients, especially in comparably small (*e.g.*, cc-pVDZ-F12) basis sets. The CABS singles effects are an order of magnitude larger than the effects of using one of the Ansätze 2A, 2A*, 2B, or 2B*. When including the CABS singles correction, the first derivative of the MP2-F12 energy with respect to a perturbation α to the one-electron Hamiltonian can be written as

$$\left.\frac{\mathrm{d}E_{\text{MP2-F12}}}{\mathrm{d}\alpha}\right|_{\alpha=0} = \sum_{\mu\nu}\left(z_{\mu\nu} + d_{\mu\nu}^{\text{MP2}} + d_{\mu\nu}^{\text{F12}} + d_{\mu\nu}^{\text{CABS}}\right)\left.\frac{\partial h_{\mu\nu}}{\partial\alpha}\right|_{\alpha=0}. \tag{10}$$

$\mathbf{d}^{\text{MP2}}$ is the unrelaxed density matrix of standard MP2 theory. Because we restrict our implementation to standard approximation A, the F12-part of the unrelaxed density matrix vanishes, $\mathbf{d}^{\text{F12}} = \mathbf{0}$ [53]. The CABS singles part of the unrelaxed density matrix, however, is nonzero,

$$d_{IJ}^{\text{CABS}} = -\sum_{p'} c_I^{p'} c_J^{p'},\; d_{p'q'}^{\text{CABS}} = \sum_I c_I^{p'} c_I^{q'},\; d_{Ip'}^{\text{CABS}} = d_{p'I}^{\text{CABS}} = c_I^{p'}. \tag{11}$$

For simplicity, the density matrix is given in the MO basis. Note that the basis set $\{\mu\}$ in Eq. (10) is the union of the orbital basis and CABS.

To calculate relaxed first-order properties, the response of the Hartree-Fock orbitals has to be taken into account. This is achieved via the Lagrangian multipliers $\mathbf{z}$, which are obtained when solving the linear z-vector equation [53]

Table 1. Effect of the Ansatz (2A, 2A*, 2B, or 2B*) of MP2-F12 theory on the equilibrium geometry of a few selected molecules. The CABS singles contribution is not included.

Basis	Molecule	Parameter	2A	2A*	2B	2B*
cc-pVDZ-F12	CH_2 ($^1A_1$)	R_{CH}/pm	110.38	110.36	110.35	110.33
		$\angle_{HCH}$/°	102.03	102.02	102.02	102.02
	CO	R_{CO}/pm	113.79	113.82	113.80	113.81
	H_2O	R_{OH}/pm	96.11	96.12	96.10	96.11
		$\angle_{HOH}$/°	104.16	104.16	104.18	104.18
	HF	R_{FH}/pm	92.14	92.15	92.14	92.14
	NH_3	R_{NH}/pm	101.25	101.26	101.22	101.22
		$\angle_{HNH}$/°	106.39	106.34	106.47	106.43
	OH	R_{OH}/pm	96.92	96.92	96.91	96.91
cc-pVTZ-F12	CH_2 ($^1A_1$)	R_{CH}/pm	110.25	110.25	110.26	110.26
		$\angle_{HCH}$/°	102.24	102.24	102.22	102.22
	CO	R_{CO}/pm	113.41	113.40	113.42	113.41
	H_2O	R_{OH}/pm	95.90	95.90	95.90	95.90
		$\angle_{HOH}$/°	104.36	104.35	104.35	104.34
	HF	R_{FH}/pm	91.94	91.95	91.95	91.95
	NH_3	R_{NH}/pm	101.00	100.99	101.00	101.00
		$\angle_{HNH}$/°	106.97	106.96	106.97	106.96
	OH	R_{OH}/pm	96.71	96.71	96.72	96.71
cc-pVQZ-F12	CH_2 ($^1A_1$)	R_{CH}/pm	110.22	110.21	110.22	110.22
		$\angle_{HCH}$/°	102.29	102.28	102.28	102.28
	CO	R_{CO}/pm	113.38	113.37	113.39	113.38
	H_2O	R_{OH}/pm	95.85	95.85	95.85	95.85
		$\angle_{HOH}$/°	104.39	104.39	104.39	104.38
	HF	R_{FH}/pm	91.86	91.86	91.86	91.86
	NH_3	R_{NH}/pm	100.96	100.97	100.97	100.97
		$\angle_{HNH}$/°	107.00	107.00	106.99	106.99
	OH	R_{OH}/pm	96.67	96.66	96.67	96.67

$$\sum_{b,J} z_{bJ}\left(2\delta_{ab}\delta_{IJ}(\epsilon_b - \epsilon_J) + A_{bJaI}\right) = A_{aI}, \tag{12}$$

where A_{bJaI} is the coupled-perturbed Hartree-Fock (CPHF) matrix. We proceed by discussing the contributions to the right-hand side A_{aI}.

2.3 Orbital-relaxed first-order properties

In Ref. [53], only Ansatz 1 of MP2-F12 theory was considered, because the projection operator in this Ansatz is not effected by changes in the Hartree-Fock orbitals. In Ansatz 2, however, the projection operator of Eq. (3) is approximated as

$$\hat{Q}_{12} \approx \hat{Q}_{12}^{\mathrm{RI}} = 1 - \hat{P}_1\hat{P}_2 - \hat{O}_1\hat{P}_2' - \hat{P}_1'\hat{O}_2. \tag{13}$$

Table 2. Equilibrium geometries of selected molecules optimized at the MP2-F12/2A and MP2-F12/2B levels including the CABS singles contribution.

Basis	Molecule	Parameter	2A	2B
cc-pVDZ-F12	CH_2 ($^1A_1$)	R_{CH}/pm	110.31	110.28
		$\angle_{HCH}/°$	102.16	102.15
	CO	R_{CO}/pm	113.48	113.49
	H_2O	R_{OH}/pm	95.89	95.88
		$\angle_{HOH}/°$	104.29	104.31
	HF	R_{FH}/pm	91.89	91.89
	NH_3	R_{NH}/pm	101.04	101.01
		$\angle_{HNH}/°$	106.84	106.92
	OH	R_{OH}/pm	96.70	96.69

Whereas the projector $\hat{P}'$ onto the CABS is invariant under orbital rotations, the projector $\hat{O}$ onto the space spanned by the occupied Hartree-Fock orbitals is not.

In the following, we shall distinguish beween inner and outer derivatives, where the outer derivatives are those with respect to the orbital indices of the matrices **B** and **V** and the inner derivatives those with respect to the projection operator $\hat{O}$. For simplicity, we only consider (active) occupied orbitals for the F12 excitations. Bold letters are thus omitted in the following.

In approximation A, we can write the F12 part of the Hylleraas functional of Eq. (4) as

$$\begin{aligned} \mathcal{H}_{F12} &= \sum_{k>l}\sum_{m>n} D_{kl,mn}\bar{B}_{kl,mn} + 2\sum_{k>l}\sum_{m>n} c_{kl}^{mn}\bar{V}_{mn}^{kl}, \\ D_{kl,mn} &= \sum_{i>j} c_{ij}^{kl}c_{ij}^{mn}. \end{aligned} \tag{14}$$

The outer derivatives lead to the contribution [53]

$$A_{ai}^{\text{outer}} = 2\sum_{l}\sum_{m>n}\left(D_{il,mn}\bar{B}_{al,mn} + c_{il}^{mn}\bar{V}_{mn}^{al} + c_{mn}^{il}\bar{V}_{al}^{mn}\right). \tag{15}$$

Inner derivatives arise from the contribution

$$-\langle mn|f_{12}\left(\hat{O}_1\hat{P}_2' + \hat{P}_1'\hat{O}_2\right)r_{12}^{-1}|kl\rangle \tag{16}$$

to the matrix element $\bar{V}_{mn}^{kl}$ from which we obtain

$$A_{aI}^{\text{inner,V}} = -2\sum_{k>l}\sum_{m>n} c_{kl}^{mn}\sum_{p'}\left(\bar{f}_{mn}^{Ip'}\bar{g}_{ap'}^{kl} + \bar{f}_{mn}^{ap'}\bar{g}_{Ip'}^{kl}\right), \tag{17}$$

with $\bar{f}_{pq}^{rs} = \langle pq|f_{12}|rs\rangle$. The corresponding contribution from the projection operators $\hat{O}_1$ and $\hat{O}_2$ in the matrix element $\bar{B}_{kl,mn}$ is

$$A_{aI}^{\text{inner,B}} = -\sum_{k>l}\sum_{m>n} D_{kl,mn}\sum_{p'}\left(\bar{f}_{mn}^{Ip'}\bar{t}_{ap'}^{kl} + \bar{f}_{mn}^{ap'}\bar{t}_{Ip'}^{kl}\right), \tag{18}$$

with $\bar{t}_{pq}^{rs} = \langle pq|[\hat{T}_{12}, f_{12}]|rs\rangle$. Note that in our implementation, the integrals $\bar{t}_{pq}^{rs}$ are evaluated using the matrix representation of the bare nuclear Hamiltonian (*cf.* Section 2.2).

Table 3. Basis-set combinations used in the present work. MP2-fitting basis sets (cbas) are labeled with 'C-'.

Notation	X	Basis-set name	CABS	cbas
aVXZ	D, T, Q, 5	aug-cc-pVXZ	C-aug-cc-pVQZ	C-aug-cc-pV5Z
aV6Z		aug-cc-pV6Z	C-aug-cc-pVQZ	C-aug-cc-pV6Z
aVXZ-F12[a]	D, T, Q	aug-cc-pVXZ-F12	cc-pVXZ-F12	C-aug-cc-pwCV5Z

[a] The aVXZ-F12 basis was obtained by adding the diffuse functions of the aug-cc-pVXZ basis to the cc-pVXZ-F12 basis.

The final contribution to A_{al} comes from the CABS singles correction,

$$\begin{aligned} A_{al}^{\text{CABS}} &= 2\sum_{p',q'} d_{p'q'}\,\bar{g}_{Ip'}^{aq'} - 2\sum_{K,L} d_{KL}\,\bar{g}_{IK}^{aL} + 2\sum_{p'} d_{Ip'}F_{p'a} \\ &+ 2\sum_{K,p'} d_{Kp'}\left(\bar{g}_{Ip'}^{aK} + \bar{g}_{ap'}^{IK}\right). \end{aligned} \quad (19)$$

Inactive occupied-active occupied orbital rotations must be taken into account in the frozen-core approximation. The corresponding equation is analogous to Eq. (15) and is obtained by replacing the virtual index a by a frozen-occupied index (*cf.* Ref. [53]).

3. Example calculations

3.1 Computational details

The analytic calculation of relaxed first-order properties has been implemented in the Ricc2 module [58,59] of the Turbomole package [56]. The geometries of CH_2 ($^1A_1$), H_2O, NH_3, HF, and CO were taken from Ref. [60] while R_{ClO} = 155.60 pm and R_{OH} = 97.00 pm were used for ClO and OH, respectively. Dunning's aug-cc-pVXZ basis sets [8,61] were used with X = D, T, Q, 5, 6. As CABS, we used the MP2-fitting C-aug-cc-pVQZ [62] basis. For the density-fitting approximation, we used the MP2-fitting C-aug-cc-pVXZ (X = 5, 6) and C-aug-cc-pwCV5Z basis sets [63] (cbas). We have also used Peterson's cc-pVXZ-F12[a] basis set with X = D, T, Q [64], to which we added additional diffuse functions (aug-cc-pVXZ-F12) taken from Dunning's correlation consistent sets [61] (since the cc-pVXZ-F12 sets are already rather diffuse, the additional diffuse functions were sometimes strongly overlapping with existing functions, but this did not lead to numerical problems). The MP2-F12 calculations in the aug-cc-pVXZ-F12 basis sets were carried out with the corresponding optimized cc-pVXZ-F12 CABS [65]. Table 3 summarizes the basis-set combinations used.

Depending on the Ansatz and approximations, Fock matrix elements of the type $F_{pq'}$ are needed in MP2-F12 theory. Such matrix elements are computed invoking the RI-*JK* approximation using an exchange-fitting basis [66] (note that this approximation is not used in the underlying Hartree-Fock calculation, in which only elements F_{pq} are needed). No exchange-fitting basis (jkbas) is needed for standard approximation A of MP2-F12 theory, but on a few occasions, we have computed first-order properties at the MP2-F12/2B level by numerical differentiation. In these calculations, we used the corresponding exchange-fitting basis set of Weigend [66] to represent the exchange operator for the extended Fock matrix.

A linear combination of six Gaussian geminals was used to represent the STG, with coefficients and exponents taken from Ref. [67]. The STG exponent was 1.4 a_0^{-1} except for the calculations in Peterson's (aug-)cc-pVXZ-F12 F12 basis sets, where we used the exponents recommended by Peterson [64] (0.9, 1.0, and 1.1 a_0^{-1} for X = D, T, and Q, respectively).

With respect to numerical differentiation, results were obtained from calculations in finite electric fields of strength $F = \pm 0.0005\ E_h/(ea_0)$. The dipole moment was in this case obtained from a central-differences formula. Furthermore, the convergency criteria for the Hartree-Fock calculation were tightened to $10^{-10}\ E_h$ change in energy (scfconv). The convergence criterion for the density matrix norm was also set to 10^{-10} (denconv).

Open-shell molecules were treated in a fully unrestricted formalism for both the Hartree-Fock (UHF) and correlation part (UMP2-F12).

Although the inclusion of spin-flipped geminals is necessary to fulfill the cusp conditions for *s* and *p* waves [68,69], we have used the diag option as described in Ref. [70] at the UMP2-F12 level for fixed amplitudes, since this was the only method available at the time of writing this article. Spin-flipped geminals are currently being implement in Turbomole for fixed as well as optimized amplitudes, using contracted geminals for the latter [71]. All calculations were carried out in the frozen-core approximation.

3.2 Results and discussion

From the orbital-relaxed one-electron density matrix, we have computed the dipole moments for a selection of closed- and open-shell molecules (Table 4) in the aVXZ-F12 basis sets of Peterson [64]. In the present work, we are interested in the performance of the MP2-F12 method with respect to the convergence of the calculated dipole moment to the basis set limit of valence-shell MP2 theory. We therefore compare our computed dipole moments with the basis set limit of MP2 theory, but not with experimental values. For such a comparison between theory and experiment, high-order correlation and vibrational effects should be taken into account [72]. To estimate the basis set limits, we have performed various calculations in large basis sets for closed- and open-shell molecules (Table 5). Applying the X^{-3} extrapolation of Helgaker *et al.* [9] to the aV5Z and aV6Z correlation contributions to the dipole moment yields values that are probably within $0.5 \times 10^{-3}\ ea_0$ of the basis-set limit. Our best estimates are obtained at the MP2-F12/2A* [T+V] + CABS singles level. Based on the excellent agreement with the MP2-F12/2B values, we assume that these estimates are accurate

Table 4. Dipole moments (in ea_0) obtained at the valence-shell MP2-F12/2A* level for a few selected closed-shell and open-shell molecules.

Basis	CABS singles	Amp.[a]	H_2O	NH_3	OH	ClO
aVDZ-F12	No	Opt.	0.74572	0.61631	0.66716	0.49248
	No	Fix.	0.74538	0.61620	0.66611	0.49291
	Yes	Opt.	0.74144	0.61543	0.66400	0.48890
	Yes	Fix.	0.74110	0.61532	0.66294	0.48933
aVTZ-F12	No	Opt.	0.73765	0.61167	0.66175	0.49018
	No	Fix.	0.73756	0.61159	0.66153	0.49045
	Yes	Opt.	0.73674	0.61129	0.66113	0.48952
	Yes	Fix.	0.73665	0.61122	0.66091	0.48981
aVQZ-F12	No	Opt.	0.73575	0.61047	0.66048	0.48819
	No	Fix.	0.73573	0.61044	0.66042	0.48827
	Yes	Opt.	0.73570	0.61044	0.66046	0.48824
	Yes	Fix.	0.73567	0.61041	0.66040	0.48832
Limit			0.7353	0.6102	0.6603	0.4884

[a] The F12 amplitudes are either optimized (Opt.) or kept fixed (Fix.).

Table 5. Estimates of the valence-shell MP2 basis-set limits for dipole moments (in ea_0) of a few selected closed-shell and open-shell molecules.

Method	Basis	H_2O	NH_3	OH	ClO
MP2	aV(56)Z[a]	0.73504	0.61000	0.65993	0.48845
MP2-F12/2B[b]	aV5Z	0.73534	0.61020	0.66031	0.48877
MP2-F12/2A*	aV6Z	0.73536	0.61024	0.66028	0.48829
MP2-F12/2A*+CABS[c]	aV6Z	0.73534	0.61023	0.66026	0.48845

[a] The Hartree-Fock contribution was calculated in the aV6Z basis while the correlation contribution was X^{-3} extrapolated from the aV5Z and aV6Z results as described in Ref. [15].
[b] From numerical differentiation.
[c] CABS singles included.

to within 1×10^{-4} ea_0. In the large aV6Z Basis, the CABS singles contribute only little to the dipole moments, from which we conclude that the underlying Hartree-Fock limit has been approached closely ($\approx 5\times10^{-5}$ ea_0).

Table 4 shows that with the aVXZ-F12 (X = D, T, Q) basis sets, the results converge smoothly and rapidly to the basis-set limit. Including CABS singles, the MP2-F12/2A*/aVDZ-F12 level yields results to within ca. 0.004 ea_0 of the basis-set limit. In this basis, the CABS singles contribution is significant. It is of the same order of magnitude as the remaining basis-set error at this level.

In the aVTZ-F12 basis, the CABS singles contributions are an order of magnitude smaller than in the aVDZ-F12 basis. Also the differences between the results obtained with optimized and fixed F12 amplitudes are an order of magnitude smaller than in the aVDZ-F12 basis.

Table 6. Dipole moment (in ea_0) of HBr obtained at the valence-shell MP2 level[a,b].

Method	MP2	MP2-F12	MP2-F12 + CABS singles
aVDZ	0.35349	0.36836	0.34976
aVTZ	0.33133	0.34539	0.34308
aVQZ	0.33647	0.34363	0.34256
aV5Z	0.33892	0.34305	0.34232

[a] At internuclear distance R_{HBr} = 140.75 pm.
[b] MP2-F12 refers to the MP2-F12/2A* [T+V] level.

The aVQZ-F12 basis yields converged results, that is, with remaining basis-set errors of $O(10^{-4})$ ea_0. Neither the CABS singles nor the optimized–fixed differences are significant at this level.

Finally, Table 6 shows results for the HBr molecule in the standard aVXZ basis sets with X = D, T, Q and 5. Obviously, the basis-set convergence of the MP2 dipole moments is much accellerated by the F12 terms. Whereas the basis-set error at the MP2/aVTZ level is of the order of 0.01 ea_0, this error is reduced to about 0.001 ea_0 at the MP2-F12/2A* level. In the aVTZ basis, the MP2-F12 approach benefits from the CABS singles contribution, which becomes less important in the aVQZ and aV5Z basis sets. Futhermore, whereas the MP2-F12+CABS approach yields very satifactory results in the aVTZ, aVQZ, and aV5Z basis sets, the aVDZ basis appears to be inadequate (note that at the MP2/aVDZ level, the calculated dipole moment is larger than the anticipated basis-set limit of ca. 0.342 ea_0).

Concerning computation times, we note that the analytic calculation of first-order properties is m times more involved than a single-point energy calculation, where m depends on the one-particle basis set. For a molecule such as methanol, the factor m is roughly 9, 7, and 5 for the basis sets aVDZ-F12, aVTZ-F12, and aVQZ-F12, respectively.

4. Conclusions

Numerical geometry optimizations at various levels of MP2-F12 theory have shown that it seems legitimate to assume the extended Brillouin condition to simplify the analytic calculation of energy derivatives while it is worthwhile to account for single excitations into the complementary auxiliary basis. Furthermore, it seems sufficient to restrict this analytic calculation to standard approximation A of MP2-F12 theory. Smooth and rapid convergence towards the basis-set limit is observed for the dipole moments of a selection of small molecules when calculated at the MP2-F12/2A* level in augmented cc-pVXZ-F12 basis sets (aVXZ-F12).

Acknowledgement

The development and implementation of the MP2-F12 method, including the calculation of analytic energy derivatives, have continuously been supported by the Deutsche Forschungsgemeinschaft's Priority Programme SPP 1145 through grant No. KL 721/2 since 2003. S. H. gratefully acknowledges support by the Deutsche Telekom Stiftung. We thank Florian A. Bischoff, Elena Kordel, David P. Tew, and Cristian Villani for valuable discussions.

References

1. T. Helgaker, W. Klopper, D. P. Tew, Mol. Phys. **106** (2008) 2107.
2. T. H. Dunning Jr, K. A. Peterson, D. E. Woon, in *Encyclopedia of Computational Chemistry*, edited by P.v.R. Schleyer, *et al.* (Wiley, New York, 1998), Vol. 1, p. 88.
3. W. Kutzelnigg, Theor. Chim. Acta **68** (1985) 445.
4. W. Klopper, W. Kutzelnigg, Chem. Phys. Lett. **134** (1987) 17.
5. W. Klopper, R. Röhse, W. Kutzelnigg, Chem. Phys. Lett. **178** (1991) 455.
6. J. Noga, W. Kutzelnigg, W. Klopper, Chem. Phys. Lett. **199** (1992) 497.
7. J. Noga, W. Kutzelnigg, J. Chem. Phys. **101** (1994) 7738.
8. T. H. Dunning Jr., J. Chem. Phys. **90** (1989) 1007.
9. T. Helgaker, W. Klopper, H. Koch, J. Noga, J. Chem. Phys. **106** (1997) 9639.
10. W. Klopper, C. C. M. Samson, J. Chem. Phys. **116** (2002) 6397.
11. J. Noga, P. Valiron, W. Klopper, J. Chem. Phys. **115** (2001) 2022; *ibid.* **115** (2001) 5690 (Erratum); *ibid.* **117** (2002) 2989 (Erratum).
12. D. Tunega, J. Noga, W. Klopper, Chem. Phys. Lett. **269** (1997) 435.
13. D. Tunega, J. Noga, Theor. Chem. Acc. **100** (1998) 78.
14. M. Jaszuński, W. Klopper, J. Noga, J. Chem. Phys. **113** (2000) 71.
15. A. Halkier, W. Klopper, T. Helgaker, P. Jørgensen, J. Chem. Phys. **111** (1999) 4424.
16. A. Halkier, T. Helgaker, W. Klopper, J. Olsen, Chem. Phys. Lett. **319** (2000) 287.
17. A. A. Auer, T. Helgaker, W. Klopper, Phys. Chem. Chem. Phys. **2** (2000) 2235.
18. W. Klopper, J. G. C. M. van Duijneveldt-van de Rijdt, F. B. van Duijneveldt, Phys. Chem. Chem. Phys. **2** (2000) 2227.
19. M. Heckert, M. Kállay, D. P. Tew, W. Klopper, J. Gauss, J. Chem. Phys. **125** (2006) 044108.
20. D. P. Tew, W. Klopper, M. Heckert, J. Gauss, J. Phys. Chem. A **111** (2007) 11242.
21. E. F. Valeev, Chem. Phys. Lett. **395** (2004) 190.
22. F. R. Manby, J. Chem. Phys. **119** (2003) 4607.
23. S. Ten-no, Chem. Phys. Lett. **398** (2004) 56.
24. S. Kedžuch, M. Milko, J. Noga, Int. J. Quantum Chem. **105** (2005) 929.
25. S. Ten-no, J. Chem. Phys. **121** (2004) 117.
26. H.-J. Werner, T. B. Adler, F. R. Manby, J. Chem. Phys. **126** (2007) 164102.
27. H.-J. Werner, F. R. Manby, J. Chem. Phys. **124** (2006) 054114.
28. F. R. Manby, H.-J. Werner, T. B. Adler, A. J. May, J. Chem. Phys. **124** (2006) 094103.
29. S. Höfener, F. A. Bischoff, A. Glöß, W. Klopper, Phys. Chem. Chem. Phys. **10** (2008) 3390.
30. F. A. Bischoff, S. Höfener, A. Glöß, W. Klopper, Theor. Chem. Acc. **121** (2008) 11.
31. T. B. Adler, H.-J. Werner, F. R. Manby, J. Chem. Phys. **130** (2009) 054106.
32. F. A. Bischoff, S. Wolfsegger, D. P. Tew, W. Klopper, Mol. Phys. **107** (2009) 963.
33. W. Klopper, F. R. Manby, S. Ten-no, E. F. Valeev, Int. Rev. Phys. Chem. **25** (2006) 427.

34. D. P. Tew, W. Klopper, C. Neiss, C. Hättig, Phys. Chem. Chem. Phys. **9** (2007) 1921; *ibid.* **10** (2008) 6325 (Erratum).
35. T. B. Adler, G. Knizia, H.-J. Werner, J. Chem. Phys. **127** (2007) 221106.
36. D. P. Tew, W. Klopper, C. Hättig, Chem. Phys. Lett. **452** (2008) 326.
37. A. Köhn, G. W. Richings, D. P. Tew, J. Chem. Phys. **129** (2008) 201103.
38. E. F. Valeev, Phys. Chem. Chem. Phys. **10** (2008) 106.
39. M. Torheyden and E. F. Valeev, Phys. Chem. Chem. Phys. **10** (2008) 3410.
40. J. Noga, S. Kedžuch, J. Šimunek, S. Ten-no, J. Chem. Phys. **128** (2008) 174103; *ibid.* **130** (2009) 029901 (Erratum).
41. E. F. Valeev and T. D. Crawford, J. Chem. Phys. **128** (2008) 244113.
42. T. Shiozaki, M. Kamiya, S. Hirata, E. F. Valeev, Phys. Chem. Chem. Phys. **10** (2008) 3358.
43. T. Shiozaki, M. Kamiya, S. Hirata, E. F. Valeev, J. Chem. Phys. **129** (2008) 071101.
44. D. Bokhan, S. Bernadotte, S. Ten-no, Chem. Phys. Lett. **469** (2009) 214.
45. T. Shiozaki, M. Kamiya, S. Hirata, E. F. Valeev, J. Chem. Phys. **130** (2009) 054101.
46. G. Knizia, T. B. Adler, and H.-J. Werner, J. Chem. Phys. **130** (2009) 054104.
47. A. Köhn, J. Chem. Phys. **130** (2009) 131101.
48. T. B. Adler, H.-J. Werner, J. Chem. Phys. **130** (2009) 241101.
49. J. J. Lee, S. Höfener, W. Klopper, T. N. Wassermann, M. A. Suhm, J. Phys. Chem. C **113** (2009) 10929.
50. G. Rauhut, G. Knizia, H.-J. Werner, J. Chem. Phys. **120** (2009) 054105.
51. P. Botschwina, R. Oswald, G. Knizia, H.-J. Werner, Z. Phys. Chem. **223** (2009) 447.
52. D. P. Tew, W. Klopper, R. A. Bachorz, C. Hättig, in *Handbook of High-Resolution Spectroscopies*. F. Merkt, M. Quack (Eds.), Wiley, Chichester (in press).
53. E. Kordel, C. Villani, W. Klopper, J. Chem. Phys. **122** (2005) 214306.
54. E. Kordel, C. Villani, W. Klopper, Mol. Phys. **105** (2007) 2565.
55. Dalton, a molecular electronic structure program, release 2.0, 2005; available from: http://www.kjemi.uio.no/software/dalton/dalton.html.
56. TURBOMOLE V6.0 2009, a development of Universität Karlsruhe (TH) and Forschungszentrum Karlsruhe GmbH, 1989–2007, Turbomole GmbH, since 2007; available from http://www.turbomole.com.
57. G. Knizia, H.-J. Werner, J. Chem. Phys. **128** (2008) 154103.
58. C. Hättig, F. Weigend, J. Chem. Phys. **113** (2000) 5154.
59. C. Hättig, J. Chem. Phys. **118** (2003) 7751.
60. W. Kopper, J. Chem. Phys. **120** (2004) 10890.
61. R. A. Kendall, T. H. Dunning Jr., R. J. Harrison, J. Chem. Phys. **96** (1992) 6796.
62. F. Weigend, A. Köhn, C. Hättig, J. Chem. Phys. **116** (2002) 3175.
63. C. Hättig, Phys. Chem. Chem. Phys. **7** (2005) 59.
64. K. A. Peterson, T. B. Adler, H.-J. Werner, J. Chem. Phys. **128** (2008) 084102.
65. K. E. Yousaf, K. A. Peterson, J. Chem. Phys. **129** (2008) 184108.
66. F. Weigend, J. Comput. Chem. **29** (2007) 167.
67. D. P. Tew, W. Klopper, J. Chem. Phys. **123** (2005) 074101.
68. D. Bokhan, S. Ten-no, J. Noga, Phys. Chem. Chem. Phys. **10** (2008) 3320.
69. D. Bokhan, S. Bernadotte, S. Ten-no, J. Chem. Phys. **131** (2009) 084105.
70. TURBOMOLE V6.0 2009, User's Manuel, Turbomole GmbH, Karlsruhe, 2009, Section 9.5.
71. D. P. Tew, W. Klopper, Mol. Phys. **108** (2010) 315.
72. K. L. Bak, J. Gauss, T. Helgaker, P. Jørgensen, J. Olsen, Chem. Phys. Lett. **319** (2000) 563.

Author Index

Keywords